Smart Spaces

Intelligent Data-Centric Systems

Smart Spaces

Edited by

Zhihan Lyu

Department of Game Design, Uppsala University, Campus Gotland, Visby, Sweden

Series Editor: Fatos Xhafa

Universitat Politècnica Catalunya, Barcelona, Spain

Associate Series Editor: Houbing Song

University of Maryland, Baltimore County (UMBC), Baltimore, MD, United States

Academic Press is an imprint of Elsevier
125 London Wall, London EC2Y 5AS, United Kingdom
525 B Street, Suite 1650, San Diego, CA 92101, United States
50 Hampshire Street, 5th Floor, Cambridge, MA 02139, United States

ISBN 978-0-443-13462-3

For information on all Academic Press publications
visit our website at https://www.elsevier.com/books-and-journals

Publisher: Mara Conner
Editorial Project Manager: Emily Thomson
Production Project Manager: Prasanna Kalyanaraman
Cover Designer: Miles Hitchen

Typeset by STRAIVE, India

Contents

Contributors

Dmitriy Alexeev
Department of Information Security, Kostroma State University, Kostroma, Russia

Álvaro Alonso
Departamento de Ingeniería de Sistemas Telemáticos, Escuela Técnica Superior de Ingenieros de Telecomunicación, Universidad Politécnica de Madrid, Madrid, Spain

Andreas Andreou
Department of Computer Science, University of Nicosia and University of Nicosia Research Foundation, Nicosia, Cyprus

Francisco Arcas
Catholic University of Murcia (UCAM), Polytechnic School, Murcia, Spain

Christopher Arukwe
Greyspot Consult Research Unit, Pretoria, South Africa

Marianna Charitonidou
Faculty of Art Theory and History, Athens School of Fine Arts, Athens, Greece

Zheyi Chen
Department of Computer and Information Sciences, Towson University, Towson, MD, United States

Javier Conde
Departamento de Ingeniería de Sistemas Telemáticos, Escuela Técnica Superior de Ingenieros de Telecomunicación, Universidad Politécnica de Madrid, Madrid, Spain

Andreea Danielescu
Accenture Labs, San Francisco, CA, United States

T.T. Dhivyaprabha
Centre for Machine Learning and Intelligence, Avinashilingam Institute for Home Science and Higher Education for Women, Coimbatore, India

Leighton Evans
Swansea University, Swansea, Wales, United Kingdom

Tomáš Gajdošík
Department of Tourism, Faculty of Economics, Matej Bel University, Banská Bystrica, Slovakia

Zuzana Gajdošíková
Department of Tourism, Faculty of Economics, Matej Bel University, Banská Bystrica, Slovakia

Adamu Hussaini
Department of Computer and Information Sciences, Towson University, Towson, MD, United States

M. Krishnaveni
Department of Computer Science, Centre for Machine Learning and Intelligence, Avinashilingam Institute for Home Science and Higher Education for Women, Coimbatore, India

P.V. Hareesh Kumar
Naval Physical & Oceanographic Laboratory, Defence Research and Development Organization, Kochi, India

S. Lakshminarayana
Department of CSSE, College of Engineering, Andhra University, Visakhapatnam, Andhra Pradesh, India

S. Lekshmi
Naval Physical & Oceanographic Laboratory, Defence Research and Development Organization, Kochi, India

Hengshuo Liang
Department of Computer and Information Sciences, Towson University, Towson, MD, United States

Weixian Liao
Department of Computer and Information Sciences, Towson University, Towson, MD, United States

Sonsoles López-Pernas
School of Computing, Faculty of Science, Forestry and Technology, University of Eastern Finland, Joensuu, Finland

Chao Lu
Department of Computer and Information Sciences, Towson University, Towson, MD, United States

Zhihan Lyu
Department of Game Design, Faculty of Arts, Uppsala University, Visby, Sweden

Aditi Maheshwari
Accenture Labs, San Francisco, CA, United States

Matúš Marciš
Department of Tourism, Faculty of Economics, Matej Bel University, Banská Bystrica, Slovakia

Evangelos K. Markakis
Department of Electrical and Computer Engineering, Hellenic Mediterranean University, Heraklion, Crete, Greece

Constandinos X. Mavromoustakis
Department of Computer Science, University of Nicosia and University of Nicosia Research Foundation, Nicosia, Cyprus

H. Patricia McKenna
AmbientEase and the UrbanitiesLab, Victoria, BC, Canada

Andres Muñoz
University of Cadiz, Higher Polytechnic School, Cádiz, Spain

Andres Munoz-Arcentales
Departamento de Ingeniería de Sistemas Telemáticos, Escuela Técnica Superior de Ingenieros de Telecomunicación, Universidad Politécnica de Madrid, Madrid, Spain

Emeka Ndaguba
Department of Building & Human Settlements Development, Faculty of Engineering, the Built Environment and Technology, Nelson Mandela University, Port Elizabeth, South Africa

Małgorzata Pańkowska
Faculty of Informatics and Communication, Department of Informatics, University of Economics in Katowice, Katowice, Poland

Cheng Qian
Department of Computer and Information Sciences, Towson University, Towson, MD, United States

Mian Qian
Department of Computer and Information Sciences, Towson University, Towson, MD, United States

Michal Rzeszewski
Adam Mickiewicz University in Poznań, Poland

Joaquín Salvachúa
Departamento de Ingeniería de Sistemas Telemáticos, Escuela Técnica Superior de Ingenieros de Telecomunicación, Universidad Politécnica de Madrid, Madrid, Spain

Oleg Shchekochikhin
Department of Analytics at PJSC "Softline", Moscow, Russia

Valeria Shvedenko
T-INNOVATIK, Limited Liability Company, St. Petersburg, Russia

Vladimir N. Shvedenko
FSBUN All-Russian Institute of Scientific and Technical Information of the Russian Academy of Sciences (VINITI RAS), Moscow, Russia

Navjot Sidhu
Catholic University of Murcia (UCAM), Polytechnic School, Murcia, Spain

G. Sree Lakshmi
EEE Department, CVR College of Engineering, Hyderabad, Telangana, India

P. Subashini
Department of Computer Science, Centre for Machine Learning and Intelligence, Avinashilingam Institute for Home Science and Higher Education for Women, Coimbatore, India

Fernando Terroso-Saenz
Catholic University of Murcia (UCAM), Polytechnic School, Murcia, Spain

Pu Tian
Department of Computer and Information Sciences, Towson University, Towson, MD, United States

Washington Velasquez
Escuela Superior Politécnica del Litoral, ESPOL, FIEC, Guayaquil, Ecuador

Mikhail Vilenskii
Saint-Petersburg State University of Architecture and Civil Engineering, St. Petersburg, Russia

Guobin Xu
Department of Computer Science, Morgan State University, Baltimore, MD, United States

Wei Yu
Department of Computer and Information Sciences, Towson University, Towson; Department of Computer Science, Morgan State University, Baltimore, MD, United States

Mariusz Żytniewski
Faculty of Informatics and Communication, Department of Informatics, University of Economics in Katowice, Katowice, Poland

CHAPTER

Smart spaces: A review

1

Zhihan Lyu

Department of Game Design, Faculty of Arts, Uppsala University, Visby, Sweden

1 Introduction

A smart space is a work or living space embedded with computing and information equipment and multimodal sensing devices. This type of environment allows its inhabitants to easily access the services offered by computers through an intuitive and responsive interface [1]. Smart spaces are equipped with sensors, user interfaces, and other applications that allow them to identify the user and their context to provide a personalized experience. Therefore smart space solutions are gradually entering different application fields, each with corresponding specific characteristics [2–4]. The potential for energy savings in the space management industry is significant with the help of technological advancements. A growing number of commercial real estate owners are grappling with the challenge of energy mismatch. Much time and many resources are being invested in constructing eco-friendly structures and transforming entire cities to accommodate urban living. The Internet of Things (IoT) is useful in this context.

Both digital and physical devices are widely used in smart environments. Formerly separate information space and physical space will blend together, as ubiquitous computing makes it possible for computing and information services to exist almost anywhere in a manner suitable for human use. Given the significance of smart spaces in the study of ubiquitous computing, many current studies on the topic are underway worldwide [5]. The basic tenet of ubiquitous computing is the provision of computing services to users at all times and in any location, as well as the seamless integration of computing into people's daily lives. The term "smart space" refers to a spatial realization of the ideals behind ubiquitous computing. It brings computing into people's everyday lives and actively provides computing services to them, based on their location and other contextual factors.

The smart space market is segmented into solutions and services. The services business segment is projected to expand at a high compound annual growth rate during the forecast period. Successful customer connections are built through professional and managed services [6,7]. In addition, these services assist enterprises in

Smart Spaces. https://doi.org/10.1016/B978-0-443-13462-3.00009-1

terms of resource use, enhanced project execution, and streamlined corporate operations [8]. Consulting services, for example, will be in high demand as the Internet of Things (IoT), artificial intelligence (AI), and deep learning spread around the world. The IoT, AI, cloud computing, and sensor technology are all briefly discussed, along with their respective applications, in this review. Further, the communication technology backbone of the smart space and the need for it to play its part reliably and effectively are investigated.

2 Computer-related technologies in smart spaces

2.1 IoT technology

The IoT is closely related to visions such as AI, the Internet of Everything, and Digital Twins. Understanding the fundamentals of the IoT is a crucial first step for urban planners interested in learning more about the future of urban intelligence, spatial aspects, and new technological application possibilities. Besides the local network, Qiu et al. [9] noted that the IoT environment has access to the whole web and all of its many services and resources. There are now numerous mobile actors in the IoT, each of which may make its own decisions without interference from other devices.

In the smart environment, the terminal devices, networks, platforms, and application scenarios make up the core components of the IoT architecture (Fig. 1). In most cases, these four parts are simplified into the perception, transport, and platform layers to build an engineering architecture. Strictly speaking, IoT only refers to the

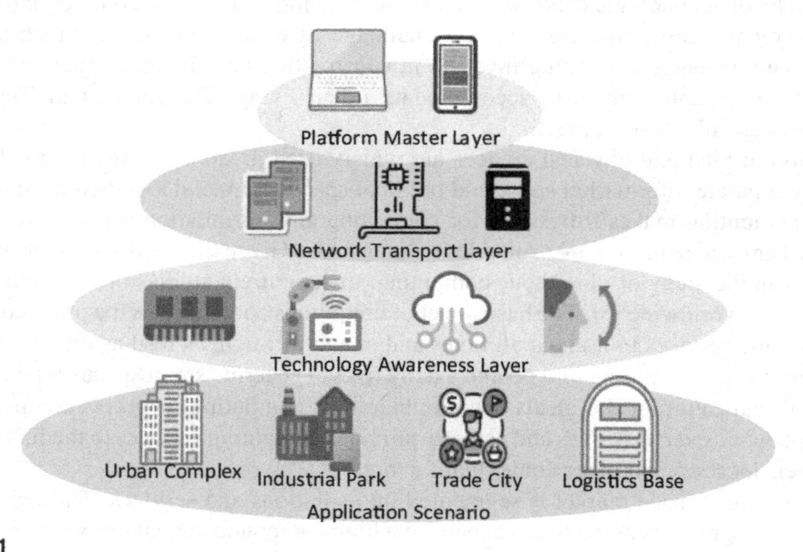

FIG. 1

Infrastructure of the IoT in smart spaces.

network component, which includes the tasks performed by the transport layer. According to Calderón-Gómez et al. [10], an IoT design based on the concept of microservices, in which each device provides one or more services in the form of encapsulated components, is required when the focus is on services. There is always one communication endpoint per service using a unique protocol for gaining access to the service. In a microservices-based smart space architecture, devices perform their duties by interacting with other devices and services. The optimal utilization of resources occurs when users activate and disable certain services, and they should impose policies on services to guarantee their safety and permit self-management of services.

In the smart city, the most common type of smart space, IoT devices improve the dependability of public transit. Deng et al. [11] argue that environmental protection, services, industry, and commerce smart cities use information and communication technologies to perceive, analyze, and organize various information about urban operations to intelligently respond to issues like people's livelihoods. Sensor data allows bus firms to accurately forecast their routes in advance and present passengers with up-to-date information on their vehicles.

Ashraf [12] conducted a hierarchical analysis of the influence of services like smart transportation, smart energy, smart infrastructure, smart health, smart agriculture, and smart entertainment on smart cities and old traditional cities in light of the growing prevalence of IoT across industries. They found that about 98% of users were satisfied with living in a smart city and with applying these smart services. Cyber-physical system security was investigated by Tyagi and Sreenath [13], who offered an overview of the difficulties and potential solutions across a number of different domains, such as the energy sector, transportation, the environment, and healthcare. Affia and Aamer [14] proposed that the IoT can greatly assist smart warehouses. It is a common misconception that the warehouse has nothing to do with IoT technology. However, for a long time, the warehouse's role has been to contain items and act as a hub of intelligence to boost the supply chain's efficiency.

The IoT is also becoming increasingly critical to the development of cybermedicine. For instance, the smart mobile medical electronic service system in the smart space is also inseparable from the assistance of the IoT [15,16]. For example, the data sensed by the devices of body area networks and personal area networks are depicted in Fig. 2 as part of the overarching concept of the IoT-supported patient personal mobile medical system in the smart space. Wearable and implanted medical technology, including electrocardiogram monitors, insulin pumps, accelerometers, and radio frequency identification tags, are all part of the body area network.

2.2 Sensor and multisensor fusion technology

With the aid of sensor-provided data, smart spaces provide end customers with remote service as one of their primary roles. According to Pivoto et al. [17], the IoT was born from the meeting point of new technologies and developing markets. The sensors are the heart of the perception layer and the foundation of the IoT.

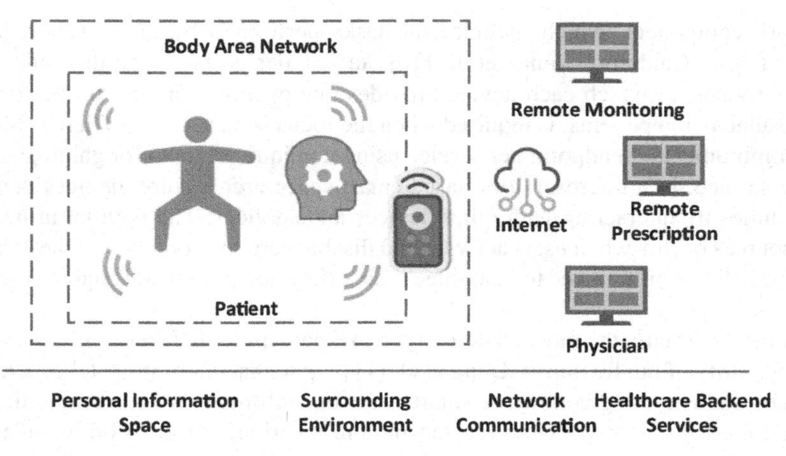

FIG. 2

IoT-enabled mobile medical system in a smart space.

Connectivity is the key to the successful growth of true intelligence in a smart home. Sensors serve as "bridges" between conventional household equipment such as doors, windows, lights, etc. Smart homes can be made a reality with the help of various sensor devices, and their owners can experience the full range of benefits that come with living in the digital age.

In tandem with the progress in science and technology, the evolution of sensors has led to greater levels of intelligence and compatibility. The IoT can offer scenario-based solutions for the entire home, as noted by Oliveira et al. [18]. The solution encompasses a wide range of frequently occurring activities within a household, including but not limited to lighting, door and window control, security, appliance management, media playback, setting the mood, detecting health issues, monitoring the environment, and more. AI relies on the clever application of sensors to achieve natural interaction between household equipment and humans [19,20], granting equipment "human-like" intelligence and enhancing the "smartness" of smart homes. The ability to send out digital signals is the most useful aspect of smart sensors, since it facilitates further processing and computation. Fig. 3 depicts the signal output mode of the smart sensor.

Data is used to determine the state of affairs in smart spaces, according to Bouchabou et al. [21]. As a result, the sensor must supply reliable operating data to prevent data failures from leading to system problems. There should be adequate ways to deal with unknown, ambiguous, and erroneous data due to the high quantity of erroneous noise that sensors can acquire when collecting data. It is possible that some smart space configurations may not make use of certain design or operational aspects. The sensors' placement should also guarantee a balanced capture of the phenomenon of interest. For example, a temperature sensor to measure ambient temperature must be kept away from sources of heat. Equipment worn by users in order to collect data from the crowd is subject to the same rules.

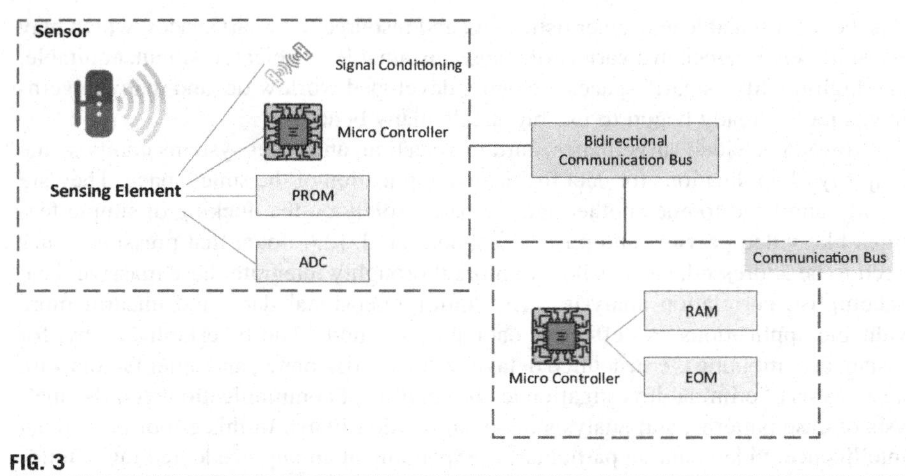

FIG. 3

Signal output patterns of sensors in smart spaces.

Smart spaces require the consolidation of data from several sources. When data from several sensors of the same type or different types are automatically analyzed, optimized, and synthesized, accurate, trustworthy, and full fusion information and conclusions can be drawn from a single source. As pointed out by Zhang et al. [22], multisensor fusion technology overcomes the restrictions of information expression in a single sensor and sidesteps the information blind spot in a single sensor. It facilitates the development of sound evaluations and choices by increasing the precision of multisource information processing outcomes. By synthesizing and processing data from several sources, sensor information fusion technology can reveal the underlying relationships and patterns among seemingly unrelated pieces of data, allowing for a complete picture of the sensed item and better decision-making. Redundancy, complementarity, speed of acquisition, and low costs are all features of the combined data from various sensors.

2.3 Cloud computing and big data technology

Using cloud computing and big data technology aids in acquiring and processing a vast amount of data, which is essential in developing smart places for users. According to Bibri [23], the use of big data computing in "smart spaces" represents a "new paradigm" that is dramatically altering the way contemporary cities are run, managed, planned, and developed. This technology shapes and drives several urban domains, particularly those concerned with maximizing resource usage, minimizing environmental risks, reacting to socioeconomic requirements, and enhancing residents' quality of life and well-being in an increasingly urbanizing world. The widespread adoption of fundamental tools for analyzing large amounts of data demonstrates that this model has spread throughout all sectors of smart cities. Sundarakani et al. [24] revealed that big data is the next frontier of urban research, since it is the most scalable and synergistic asset and resource for smart spaces, which may boost its performance in a variety of ways. Sustainable, efficient, resilient, equitable, and high-quality "smart" spaces are being developed worldwide, and many governments have already begun to use big data's many benefits.

Nowadays, video surveillance, card intersection, and other systems compose the majority of applications for data fusion and application of the smart space. They are mostly unrelated to one another and are based solely on the docking of simple features like video preview and push. Borsboom et al. [25] noted that pressing issues need to be addressed, such as how to more thoroughly integrate these massive data, accomplish correlation analysis across multidimensional data, and unearth more valuable applications. An effective data analysis model can be established by, for instance, combining the structured data on vehicles, personnel, and other factors with an analysis of criminal investigation leads, analysis of communication records, analysis of case patterns, and analysis of social opinion items. In this era of perceptual intelligence, video data, in particular, is expanding at an unprecedented rate. Building a flexible and efficient processing platform needs to address two issues. The first is the lightning-fast amassing, maintenance, and retrieval of vast amounts of data; the second is the lightning-fast processing and mining analysis of massive amounts of data, including data structured conversion, intelligent identification, search, and mining.

Based on their research, dos Reis et al. [26] concluded that more powerful computers or systems were necessary to carry out the localization task by evaluating the accuracy and computational cost of computer vision techniques used to localize mobile robots. Phan and Kim [27] claim that the home gateway is the fundamental component of the conventional smart house. The home gateway is the central hub of communication between all the connected devices in the house. As the existing house gateway is replaced by cloud computing as the fundamental system, a home gateway with as few moving parts as possible and minimal power consumption can collect data from a wide range of sensors. Then, it uploads those data points to a cloud server and awaits commands from there to take charge of the smart home. The primary benefit is that it simplifies and clarifies the home gateway's responsibilities, making the home gateway more adaptable and standardized. A user's home system can send data in real time, which the cloud servers can use to make broad-scale adjustments. In addition, they can store massive amounts of data for data mining in the future, helping to improve the system as a whole and encourage the growth of connected disciplines [28–30].

In the smart home system connected by the cloud, the system's operator provides smart home services; sellers of said equipment and consumers at large can be counted among the system's user base. Manufacturers of this equipment need merely produce the necessary hardware following the system's needs. Data is sent to the cloud data center through the home gateway, and the device hardware acts on commands received from the cloud data center. Individuals in their homes can access the processed data from the cloud data center via a number of smart terminals and

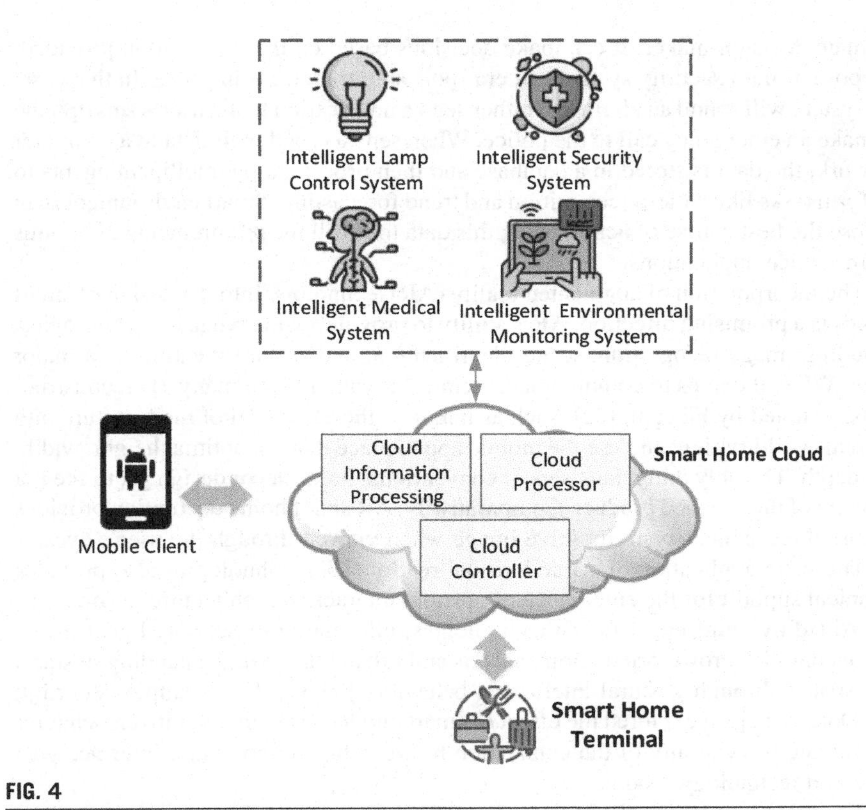

FIG. 4

Home furnishing system in a smart space based on cloud computing.

propose their own needs or control schemes. Fig. 4 depicts the system architecture of home equipment in a smart space based on cloud computing.

2.4 AI and VR technology

Smart spaces use AI to detect when people are present. They monitor sensor data for signs of human involvement and notify the user of any suspicious behavior. It was noted by Li et al. [31] that AI relies on data analysis technologies to glean insights from a wide variety of datasets and establish connections among them. AI relies on voice-driven technology for speech recognition, enabling users to carry on simple conversations with it—for example, by inquiring about the weather, placing an online order, or calling a cab. AI is utilized to accomplish tasks such as facial identification, emotion recognition, biometric recognition, and scene understanding in picture recognition. It is a scientific method for gauging and analyzing human behavior and the anatomy and physiology of the human body. AI takes the reins as the ultimate decision-maker. It can make decisions based on the information provided. Suppose a smart security system camera spots an intruder, for instance. In that case, the system will sound an alarm and either send a notification to the user's smartphone or make an emergency call to the police. When sensors send their data to a computer network, the data is stored in a database and then processed by intelligent agents to perform tasks like pattern recognition and trend forecasting. Smart environments can choose the best course of action using this data to fulfill the requirements of various smart device applications.

The incorporation of augmented reality (AR) technology into the design of smart spaces is a promising direction. AR's ability to provide a wide range of technologies, including image recognition, scene identification, and feature tracking, is a major draw. When it comes to commonplace items, AR can produce many 3D reconstructions, as noted by Fu et al. [32], such as integrate the 3D model of the furniture into the real-world setting, and use the mobile app to place it at the optimal height, width, and depth. The only thing that users of conventional home décor design get to see is a drawing of the finished product. Comparatively, AR smart home decoration provides a more three-dimensional, intuitive image when viewed through the user's eyes.

The wide application of AI and virtual reality (VR) technologies also provides technical support for the emergence of gestural interaction architecture in smart cities. Aided by multisensor fusion technologies, information-based intelligent facilities accomplish crowd orientation analysis and urban human understanding in smart city spaces through gestural interaction behaviors [33,34]. For example, Voordijk and Dorrestijn [35] explored the effect of smart city technologies on citizen behavior to evaluate the viability of the connection between human-computer interface gestures and technology models.

The magnificent visual effects and multidimensional spatiotemporal data of the virtual map are made visible by the application of virtual reality-supported gestural interaction behavior to the virtual smart space of smart cities in the interactive scenario. These enhancements are beneficial for improving users' efficient understanding of map data and their enthusiasm for discovering new areas on the map [36–38]. Natural gesture interaction allows for a more fluid, intuitive, and inventive workflow, and expands the user's potential operational space. Furthermore, the user is freed from the need to rely on hardware devices like a mouse and keyboard, allowing them to instead concentrate on the semantic and specific interaction content expressed by natural gestures. As a result, the task architecture should be modified to better accommodate the gesture interface by combining the characteristics of natural gesture interaction [39]. Park et al. [40] applied deep learning for human-computer interaction in mixed-reality environments, enabling coarse-to-fine localization of user activities in virtual maps of smart cities through the use of multimodal gestures, including gaze and head gestures. Newbury et al. [41] developed a VR system for embedded maps and embedded gesture interfaces. They discovered that the interaction method could be used with a variety of immersive maps, which boded well for encouraging the development of new, niche forms of interaction for VR environments.

3 Communication mechanism in IoT-supported smart space development

3.1 Satellite communication technology

Satellite communications allow for an astronomical view of the situation, while the ground-based infrastructure is also under constant surveillance. With the help of satellite communication technology, the smart light poles that China has begun installing this year will eventually evolve into miniature satellite base stations and provide even more innovative services to city dwellers.

In addition, the satellite network can greatly enhance signal coverage to provide a network experience similar to that of inside WiFi, even in outdoor settings. When placed on the ground, smart light poles provide direct satellite connectivity for mobile phones, watches, and other devices. People on the city's outskirts may still use the Internet in an emergency, proving that there are no black holes in the network's extensive coverage.

On modern urban roads, the road conditions are complex and changeable. Because it is difficult to synchronize data on traditional maps in real time, drivers are sometimes forced to deal with delays, detours, and yaw. Initiating a space network for vehicle-road coordination is possible using 5G and IoT technologies in conjunction with microbase stations like smart light poles. From that point, a car's navigation system isn't only a positioning tool; it can also read road signs, connect with other vehicles via roadside infrastructure, and more. A car collision or other emergency that necessitates a sudden stop can be avoided to the maximum extent possible if the owner is made aware of it in real time. Moreover, the satellite network has extensive coverage and is highly reliable; it can allow simultaneous large-scale vehicle-road coordination and uploading real-time road surface information to the space network [42,43]. Such a geographical network can offer vehicle-machine systems with trustworthy reference data.

3.2 6G communication perception integration

6G communication perception integration (CPI) technology achieves reciprocal benefits through the efficient coordination of communication and perception, as pointed out by Wang et al. [44]. This is a key strategic direction for developing the information and communication sector. This technology enhances the efficiency of communication and perception, facilitates the pooling of scarce resources, and fortifies the links between the real and virtual worlds through enhanced perceptual and informational interactions in communication networks based on picture diffraction. The ultimate goal of 6G is to improve capabilities in worldwide coverage, synesthesia integration, smart interaction, and smart universal energy and to serve as the technical driving force for smart space and smart transportation. Improved communication capabilities are one of 6G CPI's sensor-based design goals. Fig. 5 displays the latest architecture of 6G CPI technology.

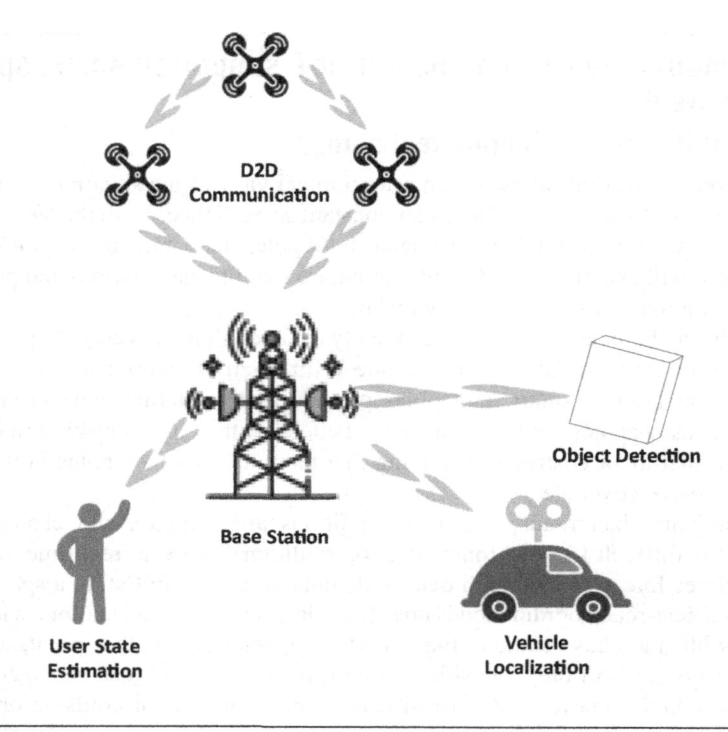

FIG. 5

6G CPI technology.

The 6G CPI system uses communication and perception fusion technology to enable accurate positioning services. The reference signal in the connection is used to determine the device's position. In addition, the reflected wireless signal's time delay, angle, and Doppler information are used to calculate the device's speed and direction. More reliable positioning data can be obtained through the extensive combination of wireless communication and perception technology, leading to enhanced positioning precision. High-precision positioning has many applications beyond just the communication system, as noted by Lv and Kumar [45]. Some examples include the operation and automatic driving of drones, the control of robot motion, the use of augmented reality in the workplace, the management of a smart factory or logistics center, and the transportation of smart goods.

According to Kim et al. [46], synesthesia fusion technology in smart spaces can perceive and assess user attributes and status, network performance and status, terminal performance and status, and ambient attributes and status. Smart spaces are currently able to increase network resource utilization, decrease energy consumption, and manage and intelligently schedule wireless spectrum resources, network computing resources, slices, etc. in a more flexible and efficient manner. For instance, technologies such as positioning, image, and gesture/action detection, in conjunction with sophisticated algorithms and cloud/edge computing capabilities, are utilized to determine user characteristics and status.

In ad hoc wireless communication scenarios, such as those involving unmanned aerial vehicles and the Internet of Vehicles, Ksouri et al. [47] argued that flexible networking should be implemented promptly in response to attributes and status changes in the environment. For instance, collisions can be reduced, and neighbor-finding convergence can be accelerated by using sensing in communication-aware integrated systems to gather a priori information on the distribution of surrounding nodes. This allows for the rapid formation of a network, the implementation of flexible and efficient multiple access, switching, routing, etc., and the enhancement of network performance.

4 Security issues in smart spaces

Building and maintaining secure networks is crucial to the wisdom space network security center. These procedures lay the groundwork for a futuristic smart city, ensure the success of the digital society, and provide the means for its citizens to experience the benefits of modern urban living fully. Awareness of the importance of keeping one's network secure must be fostered more actively. Thus a security risk assessment system, a network security organization system, and a network security protection system are all set up as part of a centralized security infrastructure for the smart city. Moreover, the city's diverse sectors need to work together to improve network security by sharing and coordinating information and enhancing their capabilities in the areas of organization building, security protection, and risk assessment.

The vital plaintext data of smart city building enterprise systems are vulnerable to fraudulent modification or theft during transmission. In particular, information sent through a wireless network can easily be intercepted and deciphered by malicious actors. Attackers can either insert harmful code into the intercepted data or outright spoof requests in order to fool government terminals. The integrity, veracity, and security of sensitive government information are all at risk from these assaults. According to Liu et al. [48], ensuring the privacy and integrity of sensitive business communications calls for using encryption technologies like Secure Sockets Layer or Internet Protocol Security Virtual Private Network, both of which adhere to the highest standards in the industry. Huang et al. [49] explored the consensus mechanism of software-defined blockchain in IoT and proposed a supervisory consensus scheme based on improved Delegated Proof of Stake-Practical Byzantine Fault Tolerance. They further enabled the supervision of IoT systems by ranking the credit rating of consensus nodes in blockchain networks through a dynamic grouping algorithm of trustworthiness. The scheme has been proven to give a practical reference for the study of blockchain consensus mechanisms in IoT by guaranteeing the highest level of data transmission consistency for all nodes in the network.

The wireless transmission of data is widely used in smart space. Serious implications, such as the paralysis of wireless communication networks and the theft and forgery of personal user information, might result from the exposed wireless signal being interfered with and taken by attackers. Both possible security concerns and wireless detection happen primarily at the perception layer of the IoT. On the one hand, the data from the sensor nodes are being taken. A sensor node often consists of little more than a memory with some computational power. Unauthorized parties have easy access to the data stored on sensor nodes. As a result of the attacker stealing, copying, or otherwise tampering with the label information of the sensor node, the node's identity is fraudulently counterfeited, and the node's credibility and effectiveness are compromised. Network attackers utilize network detection tools to execute network scans in order to acquire information about the network, including hosts, subnets, ports, and protocols. To the same extent, there are currently available tools that can search for IoT devices. Currently, available IoT devices rely heavily on wireless communication standards, including ZigBee, Z-Wave, Bluetooth-LE, and Wi-Fi802.11. Those vulnerabilities will likely be used in wireless reconnaissance and detection attacks.

Several networks, such as those based on the Internet Protocol (IP) and those based on user identities, are typically involved in any given routing operation. Ultimately, the terminal identity is mapped to an IP address in order to accomplish IP-based unified routing, as noted by Ren et al. [50]. Security risks to smart space hardware from unauthorized hardware infiltration can be mitigated, and unauthorized users of the system can be prevented through the implementation of perfect routing control technology. Network security access technology is crucial to the reliable operation of sensors at the perception layer of the IoT. When a threat is detected, the routing program at the perception layer will automatically block the node or initiate self-destruction, preventing the intruder from obtaining data from the routing node. Technologies, including lightweight cryptographic algorithms, cryptographic protocols, and security encryption, are currently employed to perform routing security control.

5 Conclusion

In the current iteration, smart spaces provide a programming paradigm that may be applied to the development of a wide variety of pervasive computing environments via software engineering techniques. Today, there is a growing convergence between smart spaces and the IoT. More precisely, players in a given IoT environment can collaborate on developing sophisticated digital services thanks to the information exchange made possible by smart spaces. Service identification (detection of user demands), service construction (automated preparation of vast volumes of data), and service perception are the focal points of this digital offering (providing derived information for user decision-making).

This chapter concisely overviews the most critical smart space technologies, such as the IoT, sensors, multisensor fusion, cloud computing, and big data. In addition, the communication process for creating smart spaces based on the IoT is detailed. In particular, the forthcoming 6G CPI technology can leverage sensing to improve communication performance in smart spaces. This review discusses the development of technologies connected to smart spaces to shed light on the significance and future research directions of smart spaces in the age of intelligent perception. Further study can investigate how best to analyze smart spaces' lightweight data structures, all while advocating for the spaces' ongoing optimization.

References

[1] C.A. Graves, T.P. Negron, M. Chestnut II, et al., Studying smart spaces using an "Embiquitous" computing analogy, IEEE Pervasive Comput. 14 (2) (2015) 64–68.

[2] D.W. Jackson, Y. Cheng, Q. Meng, et al., "Smart" greenhouses and pluridisciplinary spaces: supporting adolescents' engagement and self-efficacy in computation across disciplines, Discip. Interdscip. Sci. Educ. Res. 4 (1) (2022) 6.

[3] M. Kashef, A. Visvizi, O. Troisi, Smart city as a smart service system: human-computer interaction and smart city surveillance systems, Comput. Hum. Behav. 124 (2021) 106923.

[4] J. Chin, V. Callaghan, S.B. Allouch, The Internet-of-Things: reflections on the past, present and future from a user-centered and smart environment perspective, J. Ambient Intell. Smart Environ. 11 (1) (2019) 45–69.

[5] F. Al-Turjman, 5G-enabled devices and smart-spaces in social-IoT: an overview, Futur. Gener. Comput. Syst. 92 (2019) 732–744.

[6] G. Lentaris, K. Maragos, I. Stratakos, et al., High-performance embedded computing in space: evaluation of platforms for vision-based navigation, J. Aerosp. Inf. Syst. 15 (4) (2018) 178–192.

[7] N.H. Tien, R.J.S. Jose, B. RafalKuc, et al., Customer care and customer relationship maintenance at Gamuda Land Celadon City real estate project in Vietnam, Turk. J. Comput. Math. Educ. 12 (14) (2021) 4905–4915.

[8] P. Lecomte, iSpace: principles for a phenomenology of space user in smart real estate, J. Prop. Invest. Financ. 38 (4) (2020) 271–290.

[9] J. Qiu, Z. Tian, C. Du, et al., A survey on access control in the age of internet of things, IEEE Internet Things J. 7 (6) (2020) 4682–4696.

[10] H. Calderón-Gómez, L. Mendoza-Pittí, M. Vargas-Lombardo, et al., Evaluating service-oriented and microservice architecture patterns to deploy eHealth applications in cloud computing environment, Appl. Sci. 11 (10) (2021) 4350.

[11] T. Deng, K. Zhang, Z.J.M. Shen, A systematic review of a digital twin city: a new pattern of urban governance toward smart cities, J. Manag. Sci. Eng. 6 (2) (2021) 125–134.

[12] S. Ashraf, A proactive role of IoT devices in building smart cities, Internet Things Cyber-Phys. Syst. 1 (2021) 8–13.

[13] A.K. Tyagi, N. Sreenath, Cyber physical systems: analyses, challenges and possible solutions, Internet Things Cyber-Phys. Syst. 1 (2021) 22–33.

[14] I. Affia, A. Aamer, An internet of things-based smart warehouse infrastructure: design and application, J. Sci. Technol. Policy Manag. 13 (1) (2021) 90–109.

[15] E.F. Orumwense, K. Abo-Al-Ez, Internet of Things for smart energy systems: a review on its applications, challenges and future trends, AIMS Electron. Electr. Eng. 7 (1) (2023) 50–74.

[16] Z. Lv, Security of internet of things edge devices, Softw. Pract. Exp. 51 (12) (2021) 2446–2456.
[17] D.G.S. Pivoto, L.F.F. de Almeida, R. da Rosa Righi, et al., Cyber-physical systems architectures for industrial internet of things applications in Industry 4.0: a literature review, J. Manuf. Syst. 58 (2021) 176–192.
[18] L. Oliveira, J.J.P.C. Rodrigues, S.A. Kozlov, et al., MAC layer protocols for Internet of Things: a survey, Future Internet 11 (1) (2019) 16.
[19] A. Haroun, X. Le, S. Gao, et al., Progress in micro/nano sensors and nanoenergy for future AIoT-based smart home applications, Nano Express 2 (2) (2021) 022005.
[20] B. Dong, Q. Shi, Y. Yang, et al., Technology evolution from self-powered sensors to AIoT enabled smart homes, Nano Energy 79 (2021) 105414.
[21] D. Bouchabou, S.M. Nguyen, C. Lohr, et al., A survey of human activity recognition in smart homes based on IoT sensors algorithms: taxonomies, challenges, and opportunities with deep learning, Sensors 21 (18) (2021) 6037.
[22] J.J. Zhang, Z.Y. Ye, K.F. Li, Multi-sensor information fusion detection system for fire robot through back propagation neural network, PLoS One 15 (7) (2020) e0236482.
[23] S.E. Bibri, The sciences underlying smart sustainable urbanism: unprecedented paradigmatic and scholarly shifts in light of big data science and analytics, Smart Cities 2 (2) (2019) 179–213.
[24] B. Sundarakani, A. Ajaykumar, A. Gunasekaran, Big data driven supply chain design and applications for blockchain: an action research using case study approach, Omega 102 (2021) 102452.
[25] D. Borsboom, M.K. Deserno, M. Rhemtulla, et al., Network analysis of multivariate data in psychological science, Nat. Rev. Methods Primers 1 (1) (2021) 58.
[26] M.C. dos Reis, A. Sávio, A.R. Alexandria, et al., New trends on computer vision applied to mobile robot localization, Internet Things Cyber-Phys. Syst. 2 (2022) 63–69.
[27] L.A. Phan, T. Kim, Breaking down the compatibility problem in smart homes: a dynamically updatable gateway platform, Sensors 20 (10) (2020) 2783.
[28] H.B. Hassen, N. Ayari, B. Hamdi, A home hospitalization system based on the Internet of things, Fog computing and cloud computing, Inform. Med. Unlocked 20 (2020) 100368.
[29] S. Ravikumar, D. Kavitha, IoT based home monitoring system with secure data storage by Keccak–Chaotic sequence in cloud server, J. Ambient Intell. Humaniz. Comput. 12 (2021) 7475–7487.
[30] B.N. Alhasnawi, B.H. Jasim, P. Siano, et al., A novel real-time electricity scheduling for home energy management system using the internet of energy, Energies 14 (11) (2021) 3191.
[31] L. Li, S. Rong, R. Wang, et al., Recent advances in artificial intelligence and machine learning for nonlinear relationship analysis and process control in drinking water treatment: a review, Chem. Eng. J. 405 (2021) 126673.
[32] K. Fu, J. Peng, Q. He, et al., Single image 3D object reconstruction based on deep learning: a review, Multimed. Tools Appl. 80 (2021) 463–498.
[33] X. Lou, X.A. Li, P. Hansen, et al., Hand-adaptive user interface: improved gestural interaction in virtual reality, Virtual Reality 25 (2021) 367–382.
[34] S.H.I. Yuanyuan, L.I. Yunan, F.U. Xiaolong, et al., Review of dynamic gesture recognition, Virtual Real. Intell. Hardw. 3 (3) (2021) 183–206.
[35] H. Voordijk, S. Dorrestijn, Smart city technologies and figures of technical mediation, Urban Res. Pract. 14 (1) (2021) 1–26.

[36] S.C. Yeh, E.H.K. Wu, Y.R. Lee, et al., User experience of virtual-reality interactive interfaces: a comparison between hand gesture recognition and joystick control for XRSPACE MANOVA, Appl. Sci. 12 (23) (2022) 12230.

[37] M. Colley, P. Jansen, E. Rukzio, et al., Swivr-car-seat: exploring vehicle motion effects on interaction quality in virtual reality automated driving using a motorized swivel seat, Proc. ACM Interact. Mobile Wearable Ubiquitous Technol. 5 (4) (2021) 1–26.

[38] M.I. Gul, I.A. Khan, S. Shah, et al., Can nonliterates interact as easily as literates with a virtual reality system? A usability evaluation of VR interaction modalities, Systems 11 (2) (2023) 101.

[39] Y.J. Huang, K.Y. Liu, S.S. Lee, et al., Evaluation of a hybrid of hand gesture and controller inputs in virtual reality, Int. J. Hum. Comput. Interact. 37 (2) (2021) 169–180.

[40] K.B. Park, S.H. Choi, J.Y. Lee, et al., Hands-free human–robot interaction using multimodal gestures and deep learning in wearable mixed reality, IEEE Access 9 (2021) 55448–55464.

[41] R. Newbury, K.A. Satriadi, J. Bolton, et al., Embodied gesture interaction for immersive maps, Cartogr. Geogr. Inf. Sci. 48 (5) (2021) 417–431.

[42] N.U.L. Hassan, C. Huang, C. Yuen, et al., Dense small satellite networks for modern terrestrial communication systems: benefits, infrastructure, and technologies, IEEE Wirel. Commun. 27 (5) (2020) 96–103.

[43] K.R. Ozyilmaz, A. Yurdakul, Designing a Blockchain-based IoT with Ethereum, swarm, and LoRa: the software solution to create high availability with minimal security risks, IEEE Consum. Electron. Mag. 8 (2) (2019) 28–34.

[44] Z. Wang, Y. Du, K. Wei, et al., Vision, application scenarios, and key technology trends for 6G mobile communications, Sci. China Inf. Sci. 65 (5) (2022) 151301.

[45] Z. Lv, N. Kumar, Software defined solutions for sensors in 6G/IoE, Comput. Commun. 153 (2020) 42–47.

[46] Y. Kim, H. Jeong, J.D. Cho, et al., Construction of a soundscape-based media art exhibition to improve user appreciation experience by using deep neural networks, Electronics 10 (10) (2021) 1170.

[47] C. Ksouri, I. Jemili, M. Mosbah, et al., Towards general Internet of Vehicles networking: routing protocols survey, Concurr. Comput. Pract. Exp. 34 (7) (2022) e5994.

[48] Z. Liu, P. Cui, Y. Dong, et al., MultiSec: a multi-protocol security forwarding mechanism based on programmable data plane, Electronics 11 (15) (2022) 2389.

[49] R. Huang, X. Yang, P. Ajay, Consensus mechanism for software-defined blockchain in internet of things, Internet Things Cyber-Phys. Syst. 3 (2022) 52–60.

[50] Y. Ren, R. Xie, F.R. Yu, et al., Potential identity resolution systems for the industrial Internet of Things: a survey, IEEE Commun. Surv. Tutor. 23 (1) (2020) 391–430.

CHAPTER

An exploration of theory for smart spaces in everyday life: Enriching ambient theory for smart cities

2

H. Patricia McKenna
AmbientEase and the UrbanitiesLab, Victoria, BC, Canada

1 Introduction and motivation

The purpose of this chapter is to address the problem of what would seem to be the underdeveloped theoretical foundation for the notion of smart spaces and, also, where smart spaces fit in the broader smart discourse. Streitz [1] distinguishes between spaces with "system-oriented, importunate smartness" and those that are "people-oriented empowering smartness," where the former space, be it a room, house, vehicle, or city, "would be active (in many cases proactive) and in complete control," and in the latter case "would keep the human in the loop and in control," as in "not at the mercy of an automated system." Streitz [1] describes a smart space or system as one that "makes suggestions and recommendations based on the information collected, but humans still have the final say and make the decision" such that "[t]he space supports and enables smart behavior of people." Through research initiatives, the Urban Theory Lab [2] at the University of Chicago "seeks to contribute to the urgent collective project of imagining new spaces of solidarity and emancipatory forms of urbanization for the flourishing of human and non-human life." As such, this chapter is motivated by the need for exploring understandings of smart spaces in the context of the smartness discourse more broadly and addresses the gap in the research and practice literature for smart spaces theory while being attentive to people and other life forms on the planet. In response to the seeming smart spaces theory gap and associated research challenges, this chapter seeks to: (a) contribute to the research literature for theory and smart spaces; (b) formulate a conceptual framework in support of theory for smart spaces; and (c) propose use of ambient theory for smart cities to accommodate smart spaces more generally. As such, the significance of this work is twofold in that it contributes to the theoretical foundation for smart spaces on the one hand and, on the other hand, shows how ambient theory for smart cities is adaptable for use with smart spaces and smart environments.

Smart Spaces. https://doi.org/10.1016/B978-0-443-13462-3.00006-6

What follows is a background for this work, together with definitions and abbreviations used; a review of the research literature identifying the state of the art for smart spaces and for theory in smart environments; the theorizing of smart spaces; a description of the methodology and analysis used in advancing a solution for theorizing smart spaces; a presentation and discussion of findings including limitations; conclusion and implications; and the outlook going forward as well as future work for research and practice.

2 Background, definitions, and abbreviations

In investigating smart spaces through a series of workshops, it is perhaps worth noting that Frey et al. [3] "broadened [the] scope to smart *environments*." Indeed, a workshop participant, Mark Borkum [3], claims that "there is no such thing as a smart space," arguing that "there are agents or smart objects that can be brought into the space." In addressing the human aspects of smart spaces, Frey et al. [3] state that "[a] key foundation for our views about exploiting spaces ascribed as *smart* is that no space is, or can be, inherently smart" adding that "[our] explorations of the quality of *smartness* led us to conclude that the role of hardware and software is to *confer capability*," and as such, "people are essential for a system to achieve smartness," concluding that "[o]ur view of smartness as *conferred capability* casts technology in a supportive rather than a controlling, or even mediating, role." From a human geography perspective, it is perhaps worth noting that Baude [4] describes the Internet as an extending or an innovating of space. Streitz [1] advanced the notion that "smart spaces make people smarter," also adding that "the smart space is a cooperative space" where "the space functions like a companion supporting users, inhabitants, citizens in a cooperative fashion."

2.1 Definitions

Definitions for key terms used in this work are provided here.

Ambient. McCullough [5] describes the ambient as "an awareness of continuum and continuum of awareness."

Smart environments. According to Sahni et al. [6], smart spaces contribute to the formation of a smart environment "where devices respond to human behavior and needs" and it is worth noting, Lecomte [7] claims that "[d]ue to the increasing embeddedness of pervasive and immersive technologies in the built environment, a new type of spaces known as smart environments emerges" and this is important in that for Lecomte [7], the environment pertains to "the way we relate to the outside world, or more precisely the way we experience reality around us." Accordingly, Lecomte [7] argues that "[b]y merging physical space and digital space, smart environments embody a new type of space known as smart space."

Smart spaces. From a practice perspective, smart spaces are described by Hitachi Vantara [8] as "facilities or public areas outfitted with sensors to collect data that can be used to generate insights" pertaining to environmental conditions and services, while extending more comprehensively to include people interacting with their environment, "blurring the lines between public space and personal space," using the example of smart digital billboards and advertisements, responding to people walking nearby and their cellphones.

Everyday life. The Oxford Reference [9] refers to everyday life as "daily activities in the social world."

2.2 Abbreviations

Abbreviations for key terms used in this work are provided here.

AT. Activity Theory
ANT. Actor Network Theory
ATSC. Ambient Theory for Smart Cities
CSSR. Center for Smart Spaces Research, University of Rochester and the University of California, Irvine
DM. Digital Model
DR. Demand Response
FLEAT. Frictionless Learning Environment + Activity Theory
HITL. Human-In-The-Loop
ISSs. Interactive Smart Spaces
IT. Information Technology
MSS. Metacognitive Smart Spaces
NZEBs. Near-Zero Energy Buildings
SAM. Smart Analog Model
SLNA. Systematic Literature Network Analysis
SSC. Smart Sustainable Cities
TPB. Theory of Planned Behavior
ToC. Theory of Change
UI. User Interface

3 Literature review and state of the art for smart spaces

A review of the research literature for smart spaces is provided in this section in developing the emergent and evolving state of the art. As depicted in Fig. 1, types of smart spaces are said to include smart buildings, smart cities, smart factories, and smart homes [8]. Of note in Fig. 1 are spaces for the "emergent," in acknowledgment that all perspectives may not be covered by this review of the literature, which also includes theory pertaining to smart environments while being attentive to everyday life.

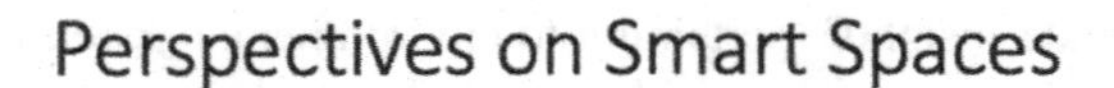

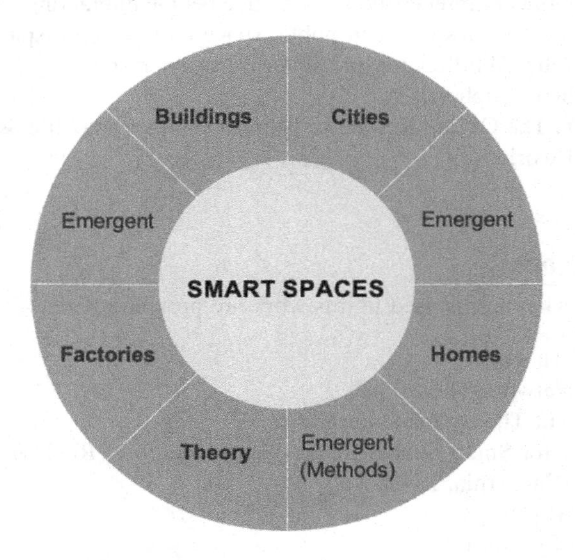

FIG. 1

Perspectives on smart spaces from the research and practice literature.

3.1 Smart spaces

Gilman and Riekki [10] advance a metacognitive approach to understandings of smart spaces that "trigger learning when system performance or user satisfaction needs to be improved." As such, Gilman and Riekki [10] take a metacognitive approach to interaction in smart spaces, contributing a "general framework for metacognitive smart spaces (MSSs)" through which to explore "recognition and adaptation at the meta-level." According to Gilman and Riekki [10], "user-smart space interaction is always dynamic and volatile, evolving with experience" and the MSS framework is designed to improve "experience, user awareness and understanding of system facilities, and overall user acceptance" in support of trustworthiness and reliability. From an architectural perspective, Kim et al. [11] identify the distinguishing features of the smart space as "flexible and evolutionary modes of response to user's needs." Smart space, according to Sahni et al. [6] "does not just mean interconnection of different devices in our surroundings" but additionally is "an environment where the devices respond to human behavior and needs." Vega-Barbas et al. [12] explore interaction patterns for smart spaces, defining interaction as "a mutual or reciprocal action or influence exerted between two or more objects, people, agents, forces, functions, etc." Vega-Barbas et al. [12] provide a classification of interactions, based on the work of Sharp et al. [13], as instruction interactions, conversation interactions, manipulation interactions, and interactions of exploration.

McKenna [14,15] highlights the importance of "the ambient concept for aiding understandings of smart spaces and interactions" supported by a framework that includes urban layers and spaces encompassing infrastructures, experiences, and interactions. For Streitz [1], a smart space "provides status information, advice, guidance and suggestions but does not make the final decisions." Bian [16] introduces a special issue on smart spaces and places for geographers, later released in book form [17], concerned with questions about "how to make spaces and places 'smart', how the 'smartness' affects the way we think" in relation to "spaces and places" as well as "what role geographers play in knowledge production and decision-making in a 'smart' era." From the perspective of practice, WebJunction [18] describes a tool for the creation of smart spaces in the form of community learning in public library spaces in support of a host of activities from design thinking, to prototyping, to collaborating. Cognizant [19] claims that smart spaces are "also known as connected places" and "can range from a building with networked temperature and motion to a vehicle that constantly reports its location, performance and maintenance needs." Of note is the Center for Smart Spaces Research (CSSR) at the Rochester Institute of Technology [20], in collaboration with the University of California, Irvine (UCI), focusing in an integrated way on applications pertaining to health and disease monitoring systems, smart buildings, public spaces and cities, mission critical systems, and transportation, to name a few. McKenna [21] explores the nurturing of theory for smart environments and spaces using the example of ambient theory for smart cities (ATSC).

In summary, Table 1 provides an overview by author and year of perspectives on smart spaces over the time period of 2012–23, highlighting features of smart spaces

Table 1 Perspectives on smart spaces.

Authors	Year	Perspectives on smart spaces
Frey et al.	2012	From smart spaces to smart environments
Gilman and Riekki	2012	Metacognitive approach—recognition and adaptation
Kim et al.	2015	Flexible, evolutionary modes of response to user's needs
Sahni et al.	2018	Devices responding to human behavior and needs
Vega-Barbas et al.	2018	Classification of interaction patterns for smart spaces
McKenna	2019	The ambient and smart spaces
Streitz	2019	Human-computer interaction
Bian	2020	Geography—spaces and places and smartness
WebJunction	2021	Toolkit for creating smart spaces: design, prototype, collaborate
Cognizant	2022	Connected spaces (e.g., buildings, vehicles)
RIT/UCI	2022	Integrated smart spaces research: health, buildings, transportation
McKenna	2023	Nurturing of theory for smart environments and spaces with ATSC

while broadening understandings to smart environments [3]; the metacognitive approach [10]; flexibility and response to user needs [11]; devices designed to be responsive to human behavior and needs [6]; interaction patterns [12], the ambient [15]; human-computer interaction [1]; geography of places, spaces, and smartness [16]; a toolkit for practitioners [18] for design thinking, prototyping, and collaborating; connected spaces as in buildings or vehicles [19]; integrated spaces research for applications pertaining to health, buildings, transportation and much more (RIT/UCI, 2022) [20]; and the nurturing of theory for smart environments and spaces using ambient theory for smart cities [21].

3.1.1 Smart buildings as smart spaces

Kim et al. [11] propose a two-model symbiotic approach to architectural design accommodating a smart analog model (SAM) and a digital model (DM) for "form-finding and prototyping." Hatcher [22] highlights a discussion by Eric Bassier on hyperconvergence as a transformative factor for video surveillance data where the three elements of compute, storage, and networking are brought together into a single IT (information technology) infrastructure for smart buildings and spaces. Rajagopalan et al. [23] provide a business perspective on smarter facilities management "to transform buildings into self-aware, flexible and responsive structures that are adaptable to ever-changing environmental conditions and occupant preferences." Bakker [24] provides an exploration of "how buildings and spaces are designed, built, used, and better understood through technology," highlighting architectural opportunities such as "improved communication, flexibility, well-being, productivity and data collection" from global and interdisciplinary perspectives. Bernstein [25] identifies four changes that need to be made to "remove barriers to collaboration and development" for smart buildings in giving "everyone access to the data, so everyone gets a say in how it's interpreted and used to make work better" through "combining human and device intelligence" for improved collective intelligence in the workplace. Diefenbach et al. [26] advance the notion conceptually of room intelligence (RI) as a personality-based user interface (UI) for smart spaces, "blurring the line between the physical and the digital world." Gianotti et al. [27] explore a human-centered perspective on Internet of Things (IoT)-enhanced interactive smart spaces (ISSs) design using the cases of two schools involving learning and play in a "magic room" and two therapeutic centers. Kolokotsa and Kampelis [28] address demand response (DR) in relation to "near-zero energy buildings (NZEBs), smart communities and microgrids" in support of "significant environmental and economic benefits" where people become actively involved as users of energy. Alsamani et al. [29] contribute to the theoretical foundations of smart spaces with a framework accommodating a bottom-up approach and use design science to develop a small-scale smart space in the form of a smart kiosk (SK) in a smart building space. Tomar et al. [30] provide a work that explores smart buildings in terms of building energy management, renewable energy integration, and grid-interactive management. From a business and practice perspective, Tyson [31] notes that "[a] smart building is presumed to be a technologically advanced building with a strong focus on things like efficiency, occupancy, performance, and reduced cost" but additionally must be "an intelligent place for the people in it," as in "less a building with technology, and much more a place that integrates the right amount of design, experience, operations, and performance intelligence" in support of "human engagement" so as to "foster meaningful human outcomes" such as "simplicity, health and wellness, discovery, learning, and a sense of welcome, ease, or comfort." Gaines [32] describes initiatives in cities involving the designing of buildings in cities for disassembly, the reuse of materials, and the reorganizing of space.

In summary, Table 2 provides an overview by author and year of perspectives on smart buildings as a type of smart space over the time period of 2015–23, highlighting architectural design [11]; hyperconvergence and IT infrastructure [22]; smarter facilities management for business [23]; architectural opportunities in buildings using technology [24]; collaboration and development for collective workplace intelligence [25]; room intelligence blurring the physical and digital [26]; interactive smart space design for school and therapeutic uses [27]; energy demand response in buildings and energy smart communities [28]; small-scale smart space design [29]; energy management for buildings [30]; technologies and designs in buildings that take people into consideration [31]; and rethinking urban design for disassembly, reuse, and the reorganizing of space [32].

Table 2 Perspectives on smart buildings as smart spaces.

Authors	Year	Perspectives on smart buildings as smart spaces
Kim et al.	2015	Architectural design
Hatcher	2019	Hyperconvergence and IT infrastructure
Rajagopalan et al.	2019	Smarter facilities management for business
Bakker	2020	Architectural opportunities using tech for design, creation, use
Bernstein	2020	Data access and improved workplace collaboration and development
Diefenbach et al.	2020	Room intelligence as a user interface weaving physical and digital
Gianotti et al.	2020	Interactive smart spaces (ISSs)—schools and therapeutic centers
Kolokotsa and Kampelis	2020	Energy demand response in smart buildings and communities
Alsamani et al.	2022	Smart kiosk design and implementation
Tomar et al.	2022	Energy and grid management; Renewable energy integration
Tyson	2022	Tech for buildings taking people into consideration
Gaines	2023	Designing for disassembly and reuse and reorganizing of space

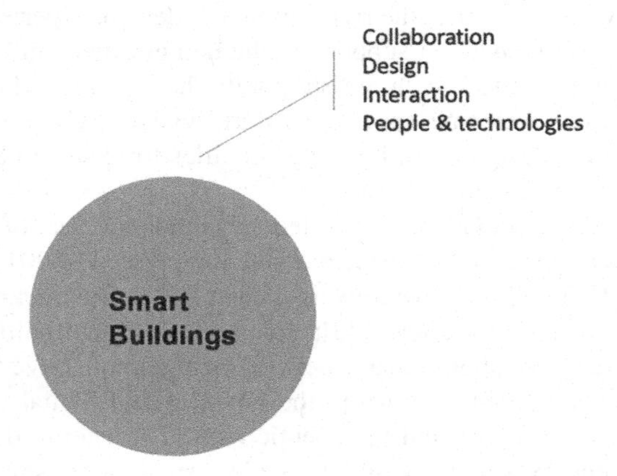

FIG. 2

Perspectives on smart buildings from the research and practice literature.

Visually, Fig. 2 provides an overview of the research literature for the smart buildings dimension of smart spaces, highlighting elements such as collaboration, design, interaction, and the interplay and interactivities of people and technologies.

3.1.2 Smart cities as smart spaces

Streitz [33] calls for "a critical reflection of different manifestations of the 'smart everything' paradigm" based on the need for "reconciling humans and technology" so as "to keep the 'human in the loop' and in control" giving rise to the need for "a citizen-centric design approach for future cities." Caprotti [34] describes two urban spaces as examples of visible renderings of the smart city concept, in the form of the Data Dome in Bristol and the Operations Centre in Glasgow, as "urban experimentation and development" in the United Kingdom. Gomez et al. [35] provide a technology-centric perspective on smart environments, said to include "the smart home, smart health, smart cities, and smart factories" in relation to the Internet of Things (IoT) communication technologies and architectures, while identifying research questions pertaining to "interoperability and standardisation, adaptation and personalisation and entity virtualization" and acknowledging the need for "considering the human aspect." It is worth noting that Streitz [1], in challenging the notion of "the smart everything paradigm," points to the need to "move beyond *'smart-only' cities* towards humane, sociable, and cooperative hybrid cities based on citizen-centered design." Streitz [1] speaks of the notion of "self-aware" cities [33] that are "characterized by how much the city knows about itself and how this is communicated to the city administration and its citizens" and it is in this way that "smart spaces make people smarter" and more engaged. From a geographical perspective, Lynch and Del Casino Jr. [36] explore the notion of intelligence in smart spaces, arguing that it is "multiple, partial, and situated in and in-between spaces, bodies, objects, and technologies" involving many "entanglements" while being "attentive to the … innovations in information processing and to the ways particular intelligences are prioritized while others may be neglected or suppressed." Streitz [1] argues for "people-oriented, empowering smartness" said to be "in the spirit of humanized computing" where the "two trade-offs to be considered" are "human control and automation" on the one hand and "privacy and smartness" on the other, with the proposed solutions of "privacy by design" and "privacy by default." Crampton et al. [37] explore the urban space of festivals using an online survey to discern the attitudes and experiences of festivalgoers toward the rapid introduction of surveillance technologies, revealing the need for "more bottom-up safety measures." Shaw and Sui [38] propose a space-place (splatial) framework (encompassing absolute, relative, relational, and mental space) for geography and geographic information systems researchers for the creative study of "human dynamics in the age of smart technologies." Enlund et al. [39] explore sensors in relation to the production of smart city spaces, pointing to the importance of "the (in) visibility of the data" and "also the (in)visibility of the sensors" since "[d]eploying sensors visibly could contribute to showcasing their usefulness and trigger interest in the data they produce" on the one hand, while on the other hand, such "visibility can be a risk to the proper functioning of sensors and the accuracy of the data they produce." As such, further exploration is encouraged by Enlund et al. [39] focusing, for example, on "what types of lived spaces smart city technologies might produce" and the potential for "sensors to become mediators of needs and wants between different groups of citizens," extending to planners and so on. Komninos et al. [40] employ the notion of "connected intelligence spaces" in "smart city ecosystems" highlighting "physical, social, and digital dimensions" understood as "systems of innovation enabling synergies between human, machine, and collective intelligence" for greater "efficiency and performance" through "innovating rather than optimizing city routines" in support of "a universal architecture of high impact smart city projects." McCarthy et al. [41] describe smart cities in the context of "the vibrant, complex, and human world of urban life." McElroy [42] describes the long and challenging process of repurposing public spaces using the example of the Shipyards District in North Vancouver.

In summary, Table 3 provides an overview by author and year of perspectives on smart cities as smart spaces over the time period of 2017–23, highlighting awareness and design with a focus on people [33]; examples of smart city spaces as experimentation and development [34]; tech-enabled smart environments and the need for incorporating the human aspect [35]; the complex nature of intelligence in smart spaces [36]; people, privacy and smartness [1]; safety and smart festival spaces [37]; study of human dynamics in technology infused smart places and spaces [38]; exploration of smart space production based on the (in)visibility of sensors and data [39]; an ecosystems approach to smart cities design with connected intelligence spaces [40]; urban life of smart cities as vibrant, complex, and human [41]; and the challenges associated with the repurposing of public spaces [42].

Table 3 Perspectives on smart cities as smart spaces.

Authors	Year	Perspectives on smart cities as smart spaces
Streitz	2017	Self-aware cities; Citizen-centric design for future cities
Caprotti	2019	Visual renderings of the SC in urban spaces—experimentation
Gomez et al.	2019	Technology-centric perspective—IoT + human aspect
Lynch and Del Casino Jr	2019	Geographic understandings of spatial dimensions/entanglements
Streitz	2019	Humane, sociable, and cooperative cities—privacy and smartness
Crampton et al.	2020	Rapid move toward smart festivals—surveillance technologies
Shaw and Sui	2020	Splatial framework for understanding smart spaces and places
Enlund et al.	2022	Smart city space production, sensors and data, (in)visibility
Komninos et al.	2022	Connected intelligence spaces: physical, social, digital; ecosystems
McCarthy et al.	2022	Vibrant, complex, and human world
McElroy	2023	Repurposing public spaces

Visually, Fig. 3 provides an overview of the research literature for the smart cities dimension of smart spaces, highlighting awareness; complexity; humaneness; people, connectedness, technologies; and repurposing.

3.1.3 Smart factories as smart spaces

Radziwon et al. [43] provide an overview of visions that exist for smart factories in formulation of a definition as "flexible and adaptive processes that will solve problems arising" in the context of a manufacturing production facility "with dynamic and rapidly changing boundary conditions in a world of increasing complexity" whether "related to automation, understood as a combination of software/hardware or mechanics" in support of "optimization of manufacturing resulting in reduction of unnecessary labour and waste of resources" or "a perspective of collaboration between different industrial and nonindustrial partners" in the form of "dynamic organization" as "smartness." Longo et al. [44] advance the notion of "smart operators" in support of "a human-centered approach" that enhances "capabilities and competencies" in the context of the smart factory and Industry 4.0. Streitz [33] addresses the need for reconciling humans and technology in many contexts including industrial facilities and the need for "keeping the human in the loop." Strozzi et al. [45] provide a literature review of the smart factory concept using a systematic literature network analysis (SLNA), with practical and theoretical implications, along with a definition, citing The Boston Consulting Group [46], as "a production

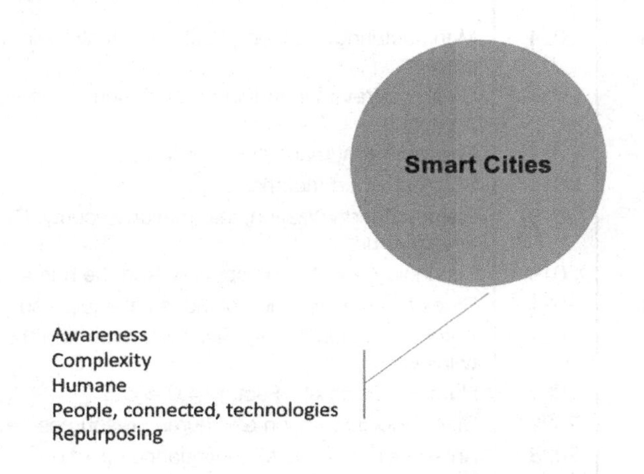

FIG. 3

Perspectives on smart cities from the research and practice literature.

plant where the pillars" of Industry 4.0 "are implemented" including "additive manufacturing, augmented reality, Internet of Things, Big data analytics, autonomous robots, simulation, cyber-security, vertical and horizontal integration and cloud." Herrmann [47] explores the technical components of smart factories in relation to risk, identifying elements such as standardization, information security, information technology (IT) infrastructure availability, and organizational and financial, among others. Gomez et al. [35] provide a technology-centric perspective on smart environments such as smart factories. Cerquitelli et al. [48] address the issue of predictive maintenance of industrial equipment in smart factories in a European context, identifying perspectives and opportunities in support of "manufacturing intelligence." Cooke [49] addresses the "digital twins" issue in the context of Industry 4.0 and the smart factory where "potentially full automation" gives rise to concerns with "the absence of humans." Brunetti et al. [50] use a survey to explore "smart interactive technologies" in "the human-centric factory 5.0" where elements such as sustainability; communication, learning, and knowledge sharing; and the needs of people figure strongly, highlighting resilience and support for critical infrastructure. Sawada et al. [51] address manufacturing, robotics, and service engineering challenges in smart factories by describing activities associated with the development of digital tools for integration into a cyber-physical system along with the development of a course in support of workers learning about relevant digital technologies. Seeking to understand key antecedents in smart factories, Jo [52] found that "personal innovativeness has a significant effect on perceived ease of use, perceived usefulness, commitment to learning, and continuance intention."

Table 4 Perspectives on smart factories as smart spaces.

Authors	Year	Perspectives on factories as smart spaces
Radziwon et al.	2014	Manufacturing; Collaborations-industrial/nonindustrial partners
Longo et al.	2017	Smart operators in Industry 4.0 as human-centered approach
Streitz	2017	Reconciling humans and technology
Strozzi et al.	2017	SLNA of smart factories
Herrmann	2018	Risks—standardization, information security, IT infrastructure
Gomez et al.	2019	Technology-centric perspective and the human aspect
Cerquitelli et al.	2021	Predictive maintenance of industrial equipment
Cooke	2021	Automotive industry process innovation and the digital twins issue
Brunetti et al.	2022	Human-centric AI—Factory 4.0 and 5.0
Sawanda et al.	2022	Digital tool integration and human resources development
Jo	2023	Antecedents of users' continuance intention

In summary, Table 4 provides an overview by author and year of perspectives on smart factories as smart spaces over the time period of 2014–23, highlighting definitions pointing to adaptive, dynamic, flexible, and collaborative [43]; smart operators with capabilities and competencies [44]; keeping humans in the loop [33]; systematic literature network analysis (SLNA) contributing definitions [45]; identification of risk factors [47]; tech-centered approach to smart environments and the need to include the human aspect [35]; predictive maintenance capabilities [48]; full automation and issues with excluding the human from the loop [49]; sustainability, communication, learning, knowledge sharing, and human aspects in factories 5.0 [50]; tools and courses in support of workers and cyber-physical systems [51]; and personal innovativeness [52].

Visually, Fig. 4 provides an overview of the research literature for the smart factories dimension of smart spaces, highlighting collaborations, integrating digital and human resource development, people and technologies, and technical and human factors.

3.1.4 Smart homes as smart spaces

Bal et al. [53] provide a review of some 35 projects or systems in both academia and industry involving collaborative technologies in support of the independent and/or assisted living of seniors in smart homes. Bal et al. [53] note that the smart home concept "may also be referred to as other terms and forms" including "smart space, aware-house, changeable home, attentive house and collaborative ambient intelligence" while among the issues identified are affordability, usability, information, security, and privacy. Solaimani et al. [54] explore the key concepts pertaining to the smart home through a systematic review identifying technological and

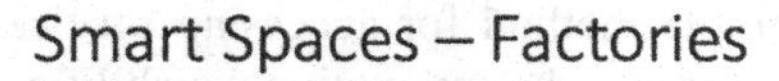

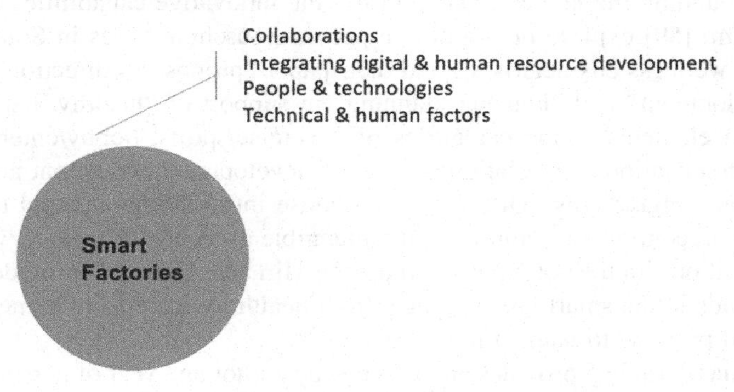

FIG. 4

Perspectives on smart factories from the research and practice literature.

nontechnological dimensions including social-organizational, economical, organizational, legal/legislation-related, and entrepreneurial, using perspectives that are strategic and operational. Wilson et al. [55] provide a systematic analysis of the users of smart homes in terms of three groupings: views—functional, instrumental, socio-technical; uses—prospective users, interactions and decisions, using technologies in the home; and challenges for realizing the smart home—hardware and software, design, and domestication. Acknowledging the need to move from technology-centric solutions to a "user-centric approach for developing smart spaces" from a technology and smart space developer perspective, Sahni et al. [6] explore the challenges and opportunities for smart homes and smart shopping in relation to the four criteria of "type of stakeholders, number of users, dynamicity of smart space, and user's requirement."

While Sahni et al. [6] are attentive to the user and human behavior, they do not seem to consider the need advanced by Streitz [33] to "keep the 'human in-the-loop' and in control." From a health perspective, Helal and Bull [56] advance the notion of smart-ready homes and communities, identifying "a great need to integrate technology with living spaces to provide assistance and independent living" and also "to smarten these spaces for lifelong living" where "the technology and the smart home applications must be flexible, adaptive, and changeable over time." Because "people do not just live at home" but also "in communities" Helal and Bull [56] stress the need for "[l]ooking at the big picture (communities), as well as the small (homes)" with a pathway progressing "beyond smart-ready homes towards smart-ready communities." Focusing on the 65+ population, Loi [57] encourages the use of participatory design approaches in explorations of intelligent systems and ambient computing in support of designing homes "with (instead of for)" people.

Altendeitering and Schimmler [58] describe explorations with "Node-RED," a visual programming language designed to enable people to "easily access data and services within and beyond their smart homes" as well as "create simple, customized programs" through "a simple method for accessing services of smart spaces surrounding the home" in support of "the innovative capabilities of users." Cho and Kim [59] explore one- and two-person household types in South Korea, identifying wellness characteristics and their paper "proposes a direction for smart home development" and "housing planning" in support of "healthy, happy lives" focusing on elements in the categories of "exercise/sports, hobby/entertainment, social communications, occupation/work, self-development/education, and energy conservation." Ingredients that transform a house into a home are said to include "participation, continuous evolution, and intangible aspects" [60], all of which will have implications for the notion of smart homes. Hirt and Allen [61] provide an overview of trends in the smart homes space, from health to exercise to home office to security and privacy, to name a few.

In summary, Table 5 provides an overview by author and year of perspectives on smart homes as smart spaces over the time period of 2011–23.

Table 5 highlights varying terminology (e.g., aware-house, attentive house) and issues (e.g., affordability, usability) [53]; social-organizational, economical, organizational, legal/legislation-related, and entrepreneurial dimensions [54]; views, uses, and challenges [55]; keeping the human in the loop and in control [33]; challenges and opportunities for developers [6]; incorporating the larger community into designs for lifelong living and learning [56]; participatory design for the 65+ population [57]; data access and services in support of user innovation capabilities [58]; wellness characteristics for smart homes in support of exercise, work, learning,

Table 5 Perspectives on smart homes as smart spaces.

Authors	Year	Perspectives on smart homes as smart spaces
Bal et al.	2011	Review of smart homes—collaborative technologies for seniors
Solaimani et al.	2015	Strategic and operational (e.g., social-organizational, etc.)
Wilson et al.	2015	Analysis of users—views, uses, challenges
Streitz	2017	Keeping humans in the loop and in control
Sahni et al.	2018	Challenges and opportunities for smart homes—software developers
Helal and Bull	2019	Smart-ready homes and communities
Loi	2019	Designing with (not for) people 65+
Altendeitering and Schimmler	2020	Visual programming languages for easy access to data and services
Cho and Kim	2022	Wellness categories for smart home development and housing planning
Loi	2022	Home—participation, continuous evolution, intangible aspects
Hirt and Allen	2023	Trends—health to exercise to security and privacy

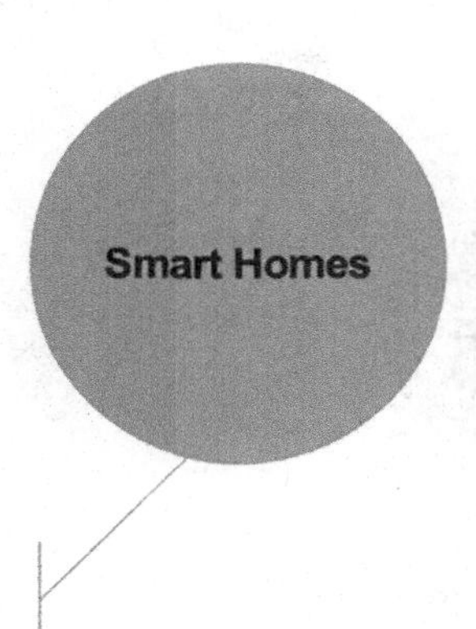

FIG. 5

Perspectives on smart homes from the research and practice literature.

energy conservation and so on [59]; importance of elements such as participation and intangible aspects for making a house a home [60]; and emerging trends, from health to security and privacy [61].

Visually, Fig. 5 provides an overview of the research literature for the smart homes dimension of smart spaces, highlighting collaborative technologies, continuous evolution, health and well-being, intangibles, participation, and people in the loop and in control.

As depicted in Fig. 6, common themes emerging from this review of the research literature for smart spaces, including smart buildings, cities, environments, factories, and homes, pertain to adaptation; the ambient; awareness; collaboration; connection; dynamic, flexible, human-centered designs and approaches; humane, social, and cooperative approaches; keeping people in-the-loop; learning; people and technology interactions; privacy (taking people into consideration, as in privacy by design described by Streitz [1]); recognition; sharing (as in knowledge sharing described by Brunetti et al. [50]); and spaces and places. It is worth noting that collaboration is said to be a synonym for cooperation [62] since, as indicated in Section 3.1.2, Streitz [1] described "the smart space" as "a cooperative space" and as "humane, social, and cooperative." As such, collaboration could serve as a variable for the exploring of smart spaces along with other variables such as connecting, privacy, and sharing.

3.2 Theory and smart environments

Theory accommodating smart environments emerges in the work of Harrison et al. [63], where information technology (IT) foundations and principles for smart cities are identified in terms of being instrumented, interconnected, and intelligent, while

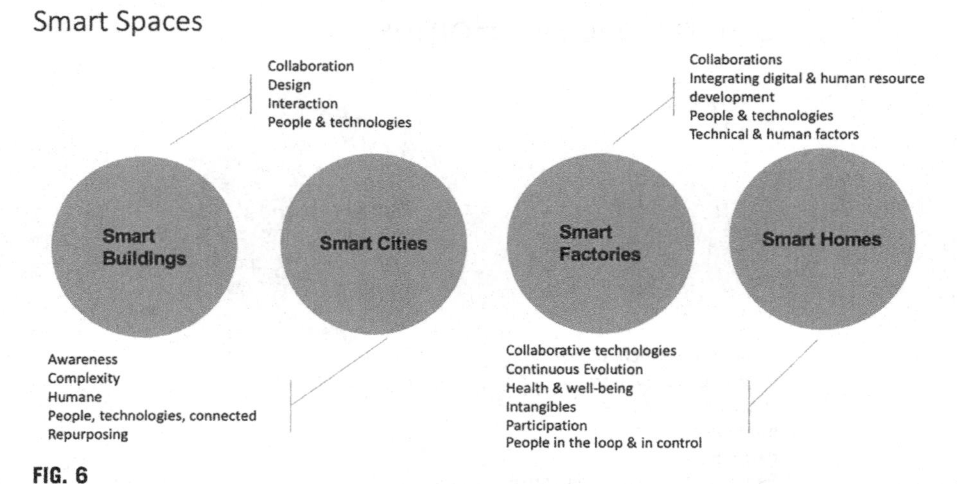

FIG. 6

Perspectives on smart spaces from the research and practice literature.

Harrison and Abbott Donnelly [64] propose a theory of smart cities in response to "a field in want of a good theoretical base" focusing on instrumentation. McKenna and Chauncey [65] propose the use of activity theory (AT) in combination with the frictionless learning environment (FLE) innovation, forming the FLEAT model in support of engagement for "smart teams" in "collaborative real-world problem solving in smart cities," as in "learning cities in action." Dainow [66] employs actor network theory (ANT) to consider the mutual influences of people and digital systems in future smart cities as an "integrated domain," arguing that ANT supports "treating the human and the digital components of the smart city as equal actants within the same environment" while affording analysis that is "essential for a full understanding of the interactions of the human and the digital." Ibrahim et al. [67] propose use of the theory of change (ToC) in support of the transformation toward smart sustainable cities (SSCs) forming a model, encompassing the complex and multidimensional processes involved, to guide the work of "city planners, decision makers, and key stakeholders about how to transform their cities" in response to the "challenges of rapid urbanization." From a knowledge management and e-learning perspective, Petrucco and Ferranti [68] advance the use of activity theory in the design of applications for smart city interventions, shedding light on "interactions."

Soleri's urban effect theory, highlighted by Balsas [69], encompasses the notions of miniaturization, complexity, and duration from an architectural and ecological (arcology) perspective in support of sustainability. McKenna [14,70] advances ambient theory for smart cities "to complement and extend existing theory for smart cities," highlighting awareness and meaningful action involving people and technology interactions, intended for environments and regions more generally and beyond [71,72]. Will-Zocholl [73] explores use of information space theory in evolving workplace environments as a space of social action where all aspects of everyday life are "(re)organized." Enlund et al. [39] draw on the theory of the production of space by Lefebvre [74] in an exploration of sensors for data collection in smart cities and "how the sensors themselves produce smart spaces," where space is said to encompass "material, mental, and social dimensions." Tunji-Olayeni et al. [75] found that while a positive attitude and perceived behavioral control (PBC) "have a significant effect on the intention to adopt green construction," there is a need "from the government, to encourage the widespread adoption of green construction" in the context of South Africa.

In summary, Table 6 provides an overview by author and year of theories and their dimensions pertaining to smart cities, smart environments, and smart spaces spanning the timeframe of 2010–23, highlighting theory for smart environments [63] and the dimensions of information technology (IT) foundations; theory of smart cities [64] as intelligent, instrumented, and interconnected; the frictionless learning environment and activity theory (FLEAT) innovation focusing on collaboration and learning cities in action [65]; actor network theory (2017) for understanding interactions involving the human and digital in smart cities; theory of change [67] and

Table 6 Theories and their dimensions for smart cities, environments, and spaces.

Authors	Year	Theory	Dimensions
Harrison et al.	2010	Theory for smart environments	IT foundations
Harrison and Abbott Donnelly	2011	Theory of smart cities	Intelligent, instrumented, interconnected
McKenna and Chauncey	2014	Frictionless learning environment + activity theory (FLEAT)	Collaboration Learning cities in action
Dainow	2017	Actor network theory	Understanding interactions of the human and digital in SCs
Ibrahim et al.	2017	Theory of Change (ToC)	Transformation to SSC
Petrucco and Ferranti	2017	Activity theory	Design of SC apps; Interactions
Balsas	2020	Urban effect theory (Soleri)	Miniaturization, complexity, duration for sustainability
McKenna	2021	Ambient theory for smart cities	Awareness; Action
Will-Zocholl	2021	Information space theory	Social action
Enlund et al.	2022	Theory of the production of space	Material, mental, and social
Tunji-Olayeni et al.	2023	Theory of planned behavior	Attitude and PBC

transformation to smart sustainable cities; activity theory [68] in support of the design of smart city applications and interactions; Soleri's urban effect theory [69] involving miniaturization, complexity, and duration for sustainability; ambient theory for smart cities [14,70] focusing on awareness and action involving people interacting with technologies; information space theory [73] and social action; theory of the production of space [39] and the material, mental, and social; and theory of planned behavior [75] and the importance of attitude and perceived behavioral control (PBC).

As shown in Fig. 7, a range of theories are being applied to smart spaces, many of which touch on elements of ambient theory for smart cities. Such theories include activity theory in relation to the design of smart city applications and interactions [68]; actor network theory (ANT) in relation to human and digital interactions in smart cities [66]; information spaces theory in relation to workplace environments, social action, and the reorganization of space [73]; theory for smart environments in relation to information technology (IT) foundations and principles for smart cities [63]; theory of smart cities in terms of being intelligent, instrumented, and interconnected [64]; theory of change (ToC) as transformation to smart sustainable cities (SSCs) [67]; theory of planned behavior (TPB) pertaining to attitude and perceived behavioral control [75]; urban effect theory involving miniaturization, complexity, and duration for sustainability [69]; and ambient theory for smart cities involving

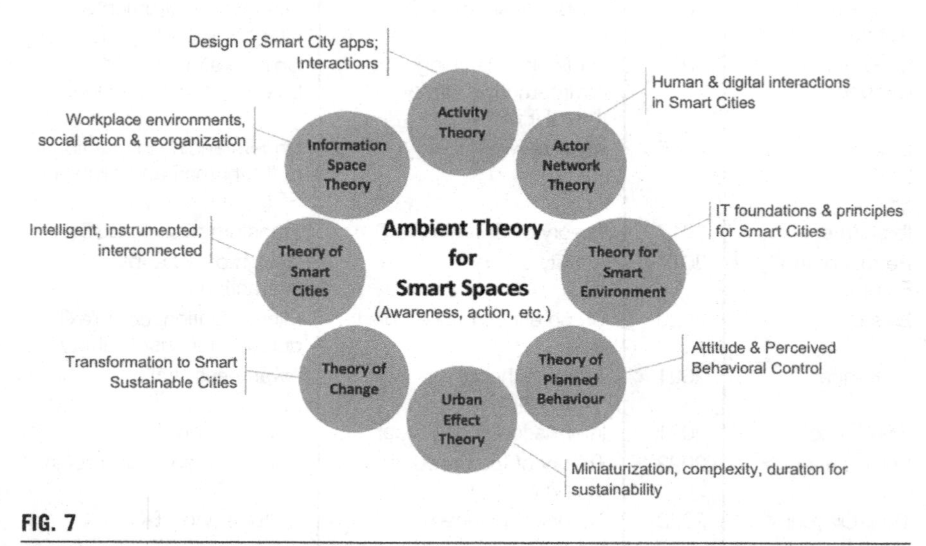

FIG. 7

An overview of emergent theory in support of smart spaces.

awareness in relation to people and technologies in support of adaptive, dynamic, emergent, interactive, and pervasive capabilities as well as involving people meaningfully in actions such as planning, designing, creating, and the like [70].

3.3 Summary and analysis

Based on the literature review for smart spaces and for theory and smart environments, a summary and analysis are presented in Fig. 8 in relation to research gaps, strengths, weaknesses, and the state of the art for the theorizing of smart spaces.

3.3.1 Research gaps

Through descriptions from the research and practice literature of smart spaces in terms of their uses and challenges, a theoretical foundation to guide understandings of smart spaces begins to take form, opening the way for opportunities and challenges. Similarly, while theories have been advanced for smart cities and the urban environment, as well as frameworks for smart spaces, theory development to guide understandings of smart spaces is also beginning to emerge, opening the way for further work.

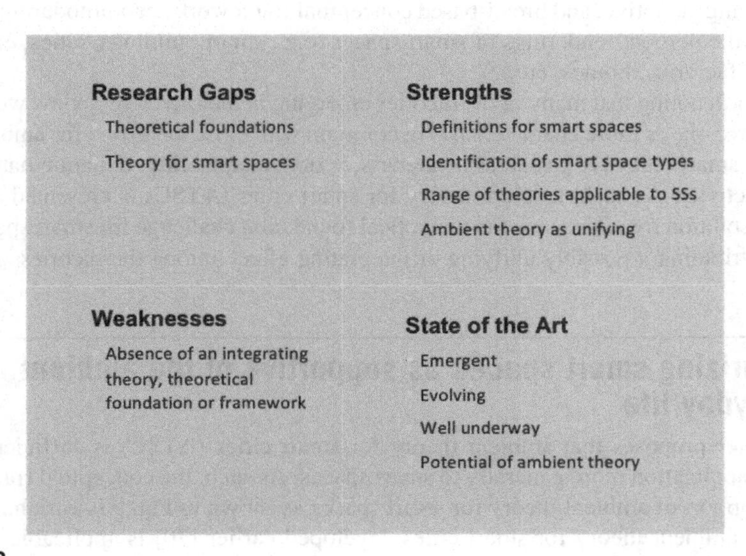

Literature Review for Smart Spaces: Summary & Analysis
Research Gaps
Theoretical foundations
Theory for smart spaces
Strengths
Definitions for smart spaces
Identification of smart space types
Range of theories applicable to SSs
Ambient theory as unifying
Weaknesses
Absence of an integrating theory, theoretical foundation or framework
State of the Art
Emergent
Evolving
Well underway
Potential of ambient theory

FIG. 8

Summary and analysis of the research literature for smart spaces.

3.3.2 Strengths

Among the strengths emerging from the literature review provided is the contribution to understandings of the emergent definitions of smart spaces and the many types of smart spaces identified, including buildings, cities, factories, homes, health, and work environments. Also of note, as shown in Fig. 7, theories have emerged for smart spaces pertaining to activity theory, actor network theory, the ambient, change, contexts, information, the production of space, smart cities, smart environments, and urban effect. Ambient theory would seem to hold potential for providing a unifying effect among theories advanced to date, incorporating as it does elements such as awareness, adaptability, action, and interactions that are evident, to some degree, in the other theories described.

3.3.3 Weaknesses

A key weakness emerging from the literature review provided is the seeming absence of an integrating theory, theoretical foundation, or framework for smart spaces giving rise to the possibility that ambient theory for smart cities may hold encouraging potential.

3.3.4 State of the art

Based on the literature review provided in this chapter, the state of the art regarding the development of a theoretical foundation for smart spaces would seem to be emergent, evolving, and underway. As such, this chapter seeks to further advance the state of the art pertaining to the development of theoretical foundations for smart spaces by proposing the application of ambient theory for smart cities (ATSC), contributing an integrating, adaptive, and broad-based conceptual framework accommodating the various and emergent renderings of smart spaces (e.g., smart buildings, cities, environments, factories, homes, etc.).

It is worth noting that many of the theories emerging in the literature review would seem to have one or more characteristics in common with those identified for ambient theory for smart cities [70], such as awareness, action, adaptability, dynamic nature, and interactivity. As such, ambient theory for smart cities (ATSC) is presented as a proposed solution for addressing the theoretical foundation challenge for smart spaces while contributing a possibly unifying or integrating effect among the theories.

4 Theorizing smart spaces as supportive of the ambient in everyday life

This chapter proposes that ambient theory for smart cities (ATSC) is sufficiently broad for application more generally to smart spaces. As such, the conceptual framework in support of ambient theory for smart spaces as shown in Fig. 9 is enriched to show that ambient theory for smart cities developed earlier [70] is applicable and extensible to smart environments [14,21] and is thus proposed in this work for use with smart spaces. As a solution to the problem of a theoretical foundation

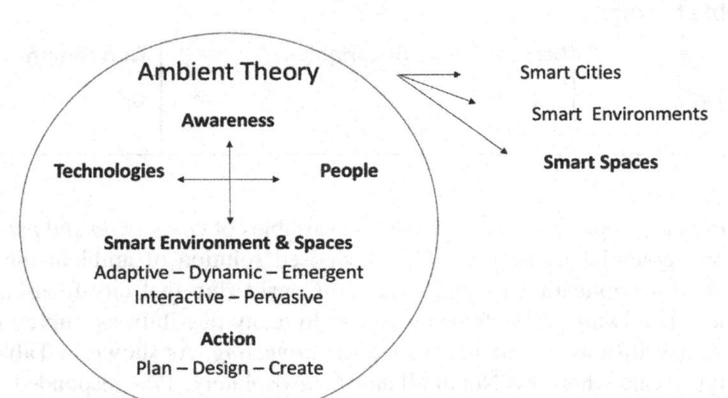

FIG. 9

Conceptual framework in support of ambient theory for smart spaces.

for smart spaces, ambient theory is particularly relevant in terms of considerations for people and technology interactions in smart spaces; the adaptive and dynamic nature of smart spaces; and for meaningful action in support of planning, design, and creative potentials enabled.

5 Use case and analysis

A use case of ambient theory for smart cities (ATSC) is explored in this chapter in terms of application for smart spaces using several variables emerging from the review of the research literature. Survey data are used based on an exploratory case study research approach combined with an explanatory correlational design for the exploration of variables [76].

Using an online space, people were invited to sign up for the study, and demographic data were gathered during this process including age range, gender, location, and self-categorization (e.g., educator, student, community member, etc.). Study participants emerged from cities in Canada, the United States, Europe, and the Middle East. Data collection methods include use of a pretested survey instrument with open-ended and closed questions pertaining to a range of elements including, but not limited to, smart cities, aware technologies, and aware people, along with in-depth interviews guided by a pretested interview protocol. In parallel with the study underlying this work, data were also systematically collected through individual and group discussions across many sectors in a variety of Canadian cities (e.g., Toronto, Vancouver, Greater Victoria). Overall, an analysis was conducted for $n = 79$ consisting of 42% females and 58% males for people ranging in age from their 20s to their 70s.

Table 7 Correlation for assessments of connecting and privacy in smart cities, extensible to smart spaces.

Items	Assessments of variables	Correlation
Connecting Privacy	5 (17%); 7 (83%) 6 (50%); 7 (50%)	0.54

This use case exploration begins with the variables of *connection* and *privacy*, to demonstrate potential usefulness of the proposed solution of ambient theory for smart cities for application to smart spaces. Considering that city-focused social media and other aware technologies give rise to many possibilities, survey participants were asked to assess elements such as *connecting*. As shown in Table 7, on a Likert type scale where 1 = Not at all and 7 = Absolutely, 17% responded at position 5 (Sort of) and 83% responded at the upper end position of 7. When asked to assess the extent to which factors such as *privacy* contribute to increased value for data in smart cities, responses emerge toward the upper end of the scale with 50% at position 6 (Sure) and 50% at position 7.

Using the Real Statistics add-in for Microsoft Excel [77], a Spearman correlation coefficient for ordinal data of 0.54 was found. Correlations in the 0.35–0.65 range are said to be "useful for limited prediction" [78].

Collaborating, an additional element pertaining to city-focused social media and other aware technologies that give rise to many possibilities, was assessed, as shown in Table 8, with 17% of responses at position 5; 17% of responses at position 6; and 66% of responses at the upper end position of 7.

Using the Real Statistics add-in for Microsoft Excel [77], a Spearman correlation coefficient for ordinal data of 0.77 emerged. Correlations in the 0.66–0.85 range are said to be "very good" with "good prediction" potential [78].

Exploring further, *sharing*, an additional element pertaining to city-focused social media and other aware technologies that give rise to many possibilities, was assessed, as shown in Table 9, with 34% responding toward the upper end of the scale at position 6 and 66% responding at position 7.

Using the Real Statistics add-in for Microsoft Excel [77], a Spearman correlation coefficient for ordinal data of 0.63 was found, which is said to be "useful for limited prediction" [78].

Table 8 Correlation for assessments of connecting and collaborating, extensible to smart spaces.

Items	Assessments of variables with aware technologies	Correlation
Connecting Collaborating	5 (17%); 7 (83%) 5 (17%); 6 (17%); 7 (66%)	0.77

Table 9 Correlation for assessments of connecting and sharing, extensible to smart spaces.

Items	Assessments of variables	Correlation
Connecting	5 (17%); 7 (83%)	0.63
Sharing	6 (34%); 7 (66%)	

Qualitatively, learning, which emerged in a variety of ways in the literature review of smart spaces, also emerges in the underlying study for this chapter. For example, an educator speaks in terms of "learning ecologies" in smart cities in support of "discovery, sociability, and connectedness" while another educator and business person speaks in terms of school—city—business collaborations for learning, integrating people and technologies. A community leader speaks of the importance of "citizen/visitor education and awareness" enabled through social media and other aware technologies and the need to "make involvement as intuitive and frictionless as possible." Yet another educator identifies the need for "collaboration projects between public authorities and community groups, where citizens are considered partners" along with the "adoption of technologies, community weaving" and "trust relationships between the authorities and the public." A public administration executive identified the need for "safety" while highlighting key elements of smart cities such as "learning capabilities" and "strategy" as well as "culture."

6 Discussion and findings

The findings of this exploration of the application of ambient theory for smart cities to smart spaces are discussed and presented in terms of results, advantages, limitations, and a summative evaluation; an overview is provided in Fig. 10.

6.1 Results

Results of this work include the identification of variables relevant to smart cities and smart spaces such as *privacy* and *connecting*, and the correlation of these variables with modest findings showing a correlation of 0.54. Exploring further, findings for *connecting* and *sharing* yield a slightly stronger correlation at 0.63 in what could be considered proxies for smart spaces, as in more aware people interacting with social media and aware technologies. Exploring further still, a stronger and more promising correlation is found between *connecting* and *collaborating* at 0.77, suggesting "good prediction" potential, again in what might be considered proxies for smart spaces. Commentary by study participants would seem to support the importance of variables such as connecting and collaborating in smart cities and smart spaces while highlighting related variables such as learning, safety, and trust.

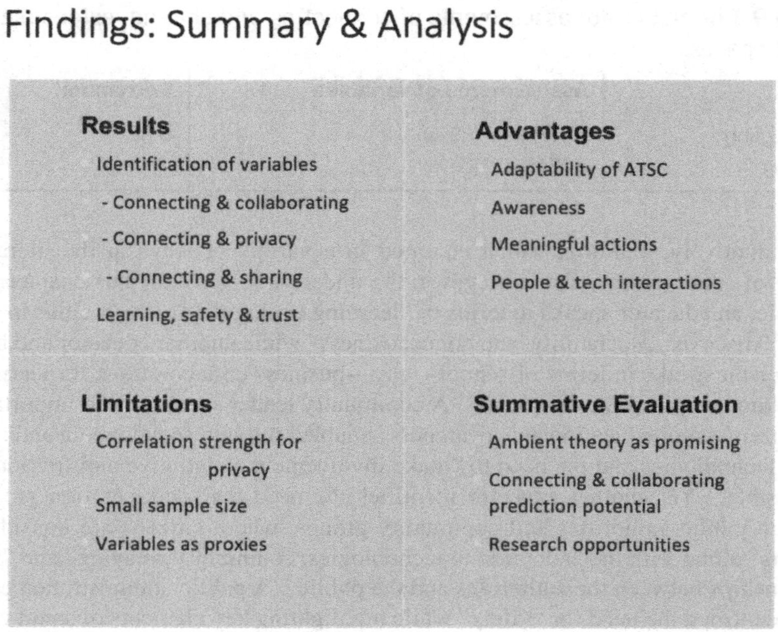

Findings: Summary & Analysis

Results

Identification of variables
- Connecting & collaborating
- Connecting & privacy
- Connecting & sharing

Learning, safety & trust

Advantages

Adaptability of ATSC
Awareness
Meaningful actions
People & tech interactions

Limitations

Correlation strength for privacy
Small sample size
Variables as proxies

Summative Evaluation

Ambient theory as promising
Connecting & collaborating prediction potential
Research opportunities

FIG. 10

Overview of summary and analysis of the findings and discussion for smart spaces.

6.2 Advantages

Among the advantages of the proposed use of ambient theory for smart cities advanced in this chapter is the adaptability of ambient theory for application to smart environments and to smart spaces. Because components of ambient theory for smart cities include awareness, interactions involving people and technologies, and meaningful action, the theory would seem to be particularly relevant for application to smart environments or smart spaces. Thus the conceptual framework in support of ambient theory for smart spaces, as shown in Fig. 9, is intended to assist in understanding actual and potential applications and uses. For example, repurposing public spaces for active and passive uses is described by McElroy [42] using the case of the Shipyards District in North Vancouver, where after many challenging years, considerable collaboration has contributed to the creation of what seems to be a highly regarded waterfront space that takes people and their needs into consideration.

6.3 Limitations

Limitations of this chapter associated with the small sample size underlying the study call for caution in the interpretation of correlation findings. Use of the variables selected as proxies for smart spaces may also be a limitation of this study, opening the way for opportunities going forward for the identification and exploration of other, more direct variables with stronger prediction potentials.

6.4 Summative evaluation

Regarding a summative evaluation of the ambient theory for smart cities application to smart spaces, the performance of this theory would seem to be promising, particularly in relation to the variables of *connecting* and *collaborating*. With further and larger studies, potential exists to learn more about the possibilities for the variables of *connecting* and *sharing* in smart spaces. Similarly, while challenges associated with *privacy* in relation to smart cities are evident, findings in this work present potential opportunities for research going forward that explores privacy in relation to additional variables in smart spaces, such as those of learning, safety, and trust.

7 Conclusions

In conclusion, this chapter contributes to the research literature through providing an exploration of the developing theoretical foundation for smart spaces while further enriching ambient theory for smart cities. This work is significant in that a conceptual framework in support of ambient theory for smart spaces is advanced in the form of ambient theory for smart cities, environments, and spaces, the application of which is explored through selected variables, identified from the research literature. Based on exploratory case study and explanatory correlational findings in this work, an encouraging relationship emerges between *connecting* and *collaborating*, with good prediction potential based on a correlation of 0.77, as proxies for smart spaces, assessed in the context of social media and aware technologies. Other variables explored included *connecting* and *sharing* as well as *connecting* and *privacy* with correlations of 0.63 and 0.54, respectively, both affording "limited prediction" according to Creswell [78], as they are within the range of 0.35–0.65. It should be noted that limitations of the study underlying this chapter, pertaining to small sample size, encourage caution in the interpretation of correlations found, opening the way for other, larger studies going forward.

8 Outlook and future work

Going forward, the outlook for contributions to the theoretical foundations for smart spaces is particularly encouraging in view of the relevance of smart spaces to smart environments, smart cities, and the need for smartness across many domains of study and practice, more generally. For example, the notion by Loi [60] of home as involving "participation, continuous evolution, and intangible aspects" is particularly fitting for the exploration of smart homes as smart spaces going forward. While many potential variables for exploration emerged in the literature review in this chapter, only a few were explored in this work in relation to ambient theory for smart spaces, giving rise to many exploration opportunities going forward for variables pertaining to learning, sustainability, innovation, and complexity, to name a few. This work is also intended to encourage lively discussion, debate, and interpretation in support of the emergent theoretical and evolving foundations of smart spaces.

Five exercises guiding and encouraging research and practice for smart spaces going forward are provided here.

Exercises

1. Identify key characteristics of what you think a smart space should be.
2. Imagine and describe the possibilities of one or more smart spaces in your everyday life.
3. Describe your experience of one or more smart spaces in a city of your choice.
4. Enumerate one or more benefits of smart spaces.
5. Enumerate what you consider the challenges of smart spaces to be.

The audience for this chapter is expected to be researchers and practitioners concerned with evolving understanding and theoretical foundations of smart spaces, in support of more aware, enriching, informed, and meaningful interactions and actions in everyday life.

References

[1] N. Streitz, Beyond 'smart-only' cities: redefining the 'smart-everything' paradigm, J. Ambient. Intell. Humaniz. Comput. 10 (2) (2019) 791–812, https://doi.org/10.1007/s12652-018-0824-1.

[2] UTL, Urban Theory Lab, University of Chicago, Division of Social Sciences, 2022. Retrieved 2 December 2022 from https://urbantheorylab.net.

[3] J.G. Frey, C. Bird, C. Willoughby, Diverse Perceptions of Smart Spaces: Physical, Digital, Personal, Social, and Data Aspects, University of Southampton, UK, 2012.

[4] B. Baude, From digital footprints to urbanity: lost in transduction, in: J. Lévy (Ed.), A Cartographic Turn, Routledge, Lausanne, 2016, pp. 273–297.

[5] M. McCullough, Ambient Commons: Attention in the Age of Embodied Information, The MIT Press, Cambridge, MA, 2013, https://doi.org/10.7551/mitpress/8947.001.0001.

[6] Y. Sahni, J. Cao, J. Shen, Challenges and opportunities in designing smart spaces, in: B. Di Martino, K.C. Li, L. Yang, A. Esposito (Eds.), Internet of Everything. Internet of Things, Springer, Singapore, 2018, https://doi.org/10.1007/978-981-10-5861-5_6.

[7] P. Lecomte, Umwelt as the foundation of an ethics of smart environments, Humanit. Soc. Sci. Commun. 10 (2023) 925, https://doi.org/10.1057/s41599-023-02356-9.

[8] Hitachi Vantara, What are "Smart Spaces?", in: Insights, 2022. Retrieved 5 November 2022 from https://www.hitachivantara.com/en-us/insights/faq/what-are-smart-spaces.html.

[9] Oxford Reference, Everyday life, 2023. Retrieved 31 January 2023 from https://www.oxfordreference.com/display/10.1093/oi/authority.20110803095802730;jsessionid=931E4C492647C3C388D7C84BB95B5FB3.

[10] E. Gilman, J. Riekki, Smart spaces: a metacognitive approach, in: M. Rautianinen, et al. (Eds.), Grid and Pervasive Computing Workshops. GPC 2011, Lecture Notes in

Computer Science, vol. 7096, Springer, Berlin, Heidelberg, 2012, https://doi.org/10.1007/978-3-642-27916-4_17.

[11] D.-Y. Kim, Y. Choe, S.-A. Kim, Implementing a digital model for smart space design: practical and pedagogic issues, Procedia Soc. Behav. Sci. 174 (2015) 3306–3313, https://doi.org/10.1016/j.sbspro.2015.01.998.

[12] M. Vega-Barbas, I. Pau, J.C. Augusto, F. Seoane, Interaction patterns for smart spaces: a confident interaction design solution for pervasive sensitive IoT services, IEEE Access 6 (2018) 1126–1136, https://doi.org/10.1109/ACCESS.2017.2777999.

[13] H. Sharp, Y. Rogers, J. Preece, Interaction Design: Beyond Human-Computer Interaction, vol. 11, Wiley, West Sussex, 2011.

[14] H.P. McKenna, Urban Life and the Ambient in Smart Cities, Learning Cities, and Future Cities, IGI Global, 2023, https://doi.org/10.4018/978-1-6684-4096-4.

[15] H.P. McKenna, Ambient Urbanities as the Intersection Between the IoT and the IoP in Smart Cities, IGI Global, Hershey, PA, 2019.

[16] L. Bian, Introduction: smart spaces and places, Ann. Am. Assoc. Geogr. 110 (2) (2020) 335–338, https://doi.org/10.1080/24694452.2019.1702810.

[17] L. Bian, Smart Spaces and Places, Routledge, 2021. Retrieved 19 July 2022 from https://www.routledge.com/Smart-Spaces-and-Places/Bian/p/book/9780367703547.

[18] WebJunction, Toolkit for creating smart spaces (News blog, 26 January), 2021. Retrieved 30 October 2022 from https://www.webjunction.org/news/webjunction/toolkit-creating-smart-spaces.html.

[19] Cognizant, Smart Spaces, 2022. Retrieved 6 December 2022 from https://www.cognizant.com/us/en/glossary/smart-spaces.

[20] RIT, Center for Smart Spaces Research (Planning), Rochester Institute of Technology (RIT) and University of California, Irvine (UCI), 2022. Retrieved 5 February 2023 from https://www.rit.edu/cssr.

[21] H.P. McKenna, The nurturing of theory for smart environments and spaces: the case of ambient theory for smart cities, in: N. Streitz, S. Konomi (Eds.), Distributed, Ambient and Pervasive Interactions. HCII 2023, Lecture Notes in Computer Science, Springer, Cham, 2023.

[22] J. Hatcher, Hyperconvergence, the cornerstone of future smart buildings? Smart Buildings Magazine, 2019 (2nd December) Retrieved 9 December 2022 from https://smartbuildingsmagazine.com/features/hyperconvergence-the-cornerstone-of-future-smart-buildings.

[23] R. Rajagopalan, T. Hameed, B. Rajamani, B. Sethuraman, Embracing Smarter Facilities Management. Cognizant 20-20 Insights, 2019. Retrieved 5 December 2022 from https://www.cognizant.com/us/en/archives/whitepapers/documents/embracing-smarter-facilities-management-codex2931.pdf.

[24] R. Bakker, Smart Buildings: Technology and the Design of the Built Environment, RIBA Publishing, 2020. Retrieved 22 December 2022 from https://sloanreview.mit.edu/article/getting-smarter-about-smart-buildings/.

[25] E. Bernstein, Getting smarter about smart buildings: Intelligent environments can make the workplace safer *and* improve collaboration, MIT Sloan Manag. Rev. (Fall) (2020).

[26] S. Diefenbach, A. Butz, D. Ullrich, Intelligence comes from within—personality as a UI paradigm for smart spaces, Designs 4 (18) (2020), https://doi.org/10.3390/designs4030018.

[27] M. Gianotti, F. Riccardi, G. Cosentino, F. Garzotto, M. Matera, Modeling interactive smart spaces, in: G. Dobbie, U. Frank, G. Kappel, S.W. Liddle, H.C. Mayr (Eds.), Conceptual Modeling. ER 2020, Lecture Notes in Computer Science, vol. 12400, Springer, Cham, 2020, https://doi.org/10.1007/978-3-030-62522-1_30.

[28] D. Kolokotsa, N. Kampelis, Smart Buildings, Smart Communities and Demand Response, Wiley Telecom, 2020.

[29] B. Alsamani, S. Chatterjee, A. Anjomshoae, P. Ractham, Smart space design—a framework and an IoT prototype implementation, Sustainability 15 (1) (2023) 111, https://doi.org/10.3390/su15010111.

[30] A. Tomar, P.H. Nguyen, S. Mishra, Control of Smart Buildings: An Integration to Grid and Local Energy Communities, Studies in Infrastructure and Control, Springer, 2022.

[31] R. Tyson, From smart buildings to intelligent placemaking, in: Research and Insights Blog, Gensler, 2022. Retrieved 8 December 2022 from https://www.gensler.com/blog/from-smart-buildings-to-intelligent-placemaking.

[32] J. Gaines, The cities built to be reusable: The remnants of most old buildings end up in landfill, but some cities are starting to design them so they can be easily disassembled and repurposed from the start, BBC, 2023 (7 February). Retrieved 27 June 2023 from https://www.bbc.com/future/article/20230207-can-we-design-cities-for-disassembly.

[33] N. Streitz, Reconciling humans and technology: the role of ambient intelligence, Keynote Paper, in: Proceedings of the 2017 Conference on Ambient Intelligence, LNCS 10217, Springer, 2017, pp. 1–16.

[34] F. Caprotti, Spaces of visibility in the smart city: flagship urban spaces and the smart urban imaginary, Urban Stud. 56 (12) (2019) 2465–2479, https://doi.org/10.1177/0042098018798597.

[35] C. Gomez, S. Chessa, A. Fleury, G. Roussos, D. Preuveneers, Internet of things for enabling smart environments: a technology-centric perspective, J. Ambient Intell. Smart Environ. 11 (1) (2019) 23–43, https://doi.org/10.3233/AIS-180509.

[36] C.R. Lynch, V.J. Del Casino Jr., Smart spaces, information processing, and the question of intelligence, Ann. Am. Assoc. Geogr. (2019), https://doi.org/10.1080/24694452.2019.1617103.

[37] J.W. Crampton, K.C. Hoover, H. Smith, S. Graham, J.C. Berbesque, Smart festivals? Security and freedom for well-being in urban smart spaces, Ann. Am. Assoc. Geogr. 110 (2) (2020) 360–370, https://doi.org/10.1080/24694452.2019.1662765.

[38] S.-L. Shaw, D. Sui, Understanding the new human dynamics in smart spaces and places: toward a splatial framework, Ann. Am. Assoc. Geogr. 110 (2) (2020) 339–348, https://doi.org/10.1080/24694452.2019.1631145.

[39] D. Enlund, K. Harrison, R. Ringdahl, A. Börütecene, J. Löwgren, V. Angelakis, The role of sensors in the production of smart city spaces, Big Data Soc. 9 (2) (2022), https://doi.org/10.1177/20539517221110218.

[40] N. Komninos, C. Kakderi, L. Mora, A. Panori, E. Sefertzi, Towards high impact smart cities: a universal architecture based on connected intelligence spaces, J. Knowl. Econ. 13 (2022) 1169–1197, https://doi.org/10.1007/s13132-021-00767-0.

[41] D. McCarthy, H. Field, G. Donnelly, F. McDonald, Smart Cities From Emerging Tech Brew, 2022. Retrieved 5 November 2022 from https://www.emergingtechbrew.com/stories/c/smart-cities.

[42] J. McElroy, What other metro Vancouver municipalities can learn from the Shipyards District, CBC News, 2023 (12 January). Retrieved 6 February 2023 from https://www.cbc.ca/news/canada/british-columbia/shipyards-district-vancouver-municipalities-1.6711147.

[43] A. Radziwon, A. Bilberg, M. Bogers, E.S. Madsen, The smart factory: exploring adaptive and flexible manufacturing solutions, Procedia Eng. 69 (2014) 1184–1190, https://doi.org/10.1016/j.proeng.2014.03.108.

[44] F. Longo, L. Nicoletti, A. Padovano, Smart operators in industry 4.0: a human-centered approach to enhance operators' capabilities and competencies within the new smart

factory context, Comput. Ind. Eng. 113 (2017) 144–159, https://doi.org/10.1016/j.cie.2017.09.016.

[45] F. Strozzi, C. Colicchia, A. Creazza, C. Noè, Literature review on the 'Smart Factory' concept using bibliometric tools, Int. J. Prod. Res. 55 (22) (2017) 6572–6591, https://doi.org/10.1080/00207543.2017.1326643.

[46] The Boston Consulting Group, The Factory of the Future, 2016. Retrieved 13 December 2022 from https://www.bcg.com/publications/2016/leaning-manufacturing-operations-factory-of-future.

[47] F. Herrmann, The smart factory and its risks, Systems 6 (4) (2018) 38, https://doi.org/10.3390/systems6040038.

[48] T. Cerquitelli, N. Nikolakis, N. O'Mahony, E. Macii, M. Ippolito, S. Makris (Eds.), Predictive Maintenance in Smart Factories: Architectures, Methodologies, and Use-Cases, Springer, Singapore, 2021.

[49] P. Cooke, Image and reality: 'digital twins' in smart factory automotive process innovation—critical issues, Reg. Stud. 55 (10–11) (2021) 1630–1641, https://doi.org/10.1080/00343404.2021.1959544.

[50] D. Brunetti, C. Gena, F. Vernero, Smart interactive technologies in the human-centric factory 5.0: a survey, Appl. Sci. 12 (2022) 7965, https://doi.org/10.3390/app12167965.

[51] H. Sawada, Y. Nakabo, Y. Furukawa, N. Ando, T. Okuma, H. Komoto, K. Masui, Digital tools integration and human resources development for smart factories, Int. J. Autom. Technol. 16 (3) (2022) 250–260.

[52] H. Jo, Understanding the key antecedents of users' continuance intention in the context of smart factory, Technol. Anal. Strat. Manag. 35 (2) (2023) 153–166, https://doi.org/10.1080/09537325.2021.1970130.

[53] M. Bal, W. Shen, Q. Hao, H. Xue, Collaborative smart home technologies for senior independent living: a review, in: 15th International Conference on Computer Supported Cooperative Work in Design (CSCWD), 2011, https://doi.org/10.1109/CSCWD.2011.5960116.

[54] S. Solaimani, W. Keijzer-Broers, H. Bouwman, What we do—and don't—know about the smart home: an analysis of the smart home literature, Indoor Built Environ. 24 (3) (2015) 370–383, https://doi.org/10.1177/1420326X13516350.

[55] C. Wilson, T. Hargreaves, R. Hauxwell-Baldwin, Smart homes and their users: a systematic analysis and key challenges, Pers. Ubiquit. Comput. 19 (2015) 463–476, https://doi.org/10.1007/s00779-014-0813-0.

[56] S. Helal, C.N. Bull, From smart homes to smart-ready homes and communities, Dement. Geriatr. Cogn. Disord. 47 (3) (2019) 157–163, https://doi.org/10.1159/000497803.

[57] D. Loi, HOME: exploring the role of ambient computing for older adults, in: M. Antona, C. Stephanidis (Eds.), Universal Access in Human-Computer Interaction. Multimodality and Assistive Environments. HCII 2019, Lecture Notes in Computer Science, vol. 11573, Springer, 2019, https://doi.org/10.1007/978-3-030- 23563-5_39.

[58] M. Altendeitering, S. Schimmler, Data-flow programming for smart homes and other smart spaces, in: IEEE Symposium on Visual Languages and Human-Centric Computing (VL/HCC), 2020, https://doi.org/10.1109/VL/HCC50065.2020.9127268.

[59] M.E. Cho, M.J. Kim, Smart homes supporting the wellness of one or two-person households, Sensors 22 (2022) 7816, https://doi.org/10.3390/s22207816.

[60] D. Loi, Designing the unfinished: a home is not a house, Interaction 29 (3) (2022) 16–18, https://doi.org/10.1145/3529163.

[61] M. Hirt, S. Allen, 10 smart home trends this year, Forbes, 2023 (20 April). Retrieved 28 June 2023 from https://www.forbes.com/home-improvement/internet/smart-home-tech-trends/.

[62] Merriam-Webster, Cooperation, Merriam-Webster Dictionary, 2023 (Online). Retrieved 1 February 2023 from https://www.merriam-webster.com/dictionary/cooperation.

[63] C. Harrison, B. Eckman, R. Hamilton, P. Hartswick, J. Kalagnanam, J. Paraszczak, P. Williams, Foundations for smarter cities, IBM J. Res. Dev. 54 (4) (2010) 1–16.

[64] C. Harrison, I. Abbott Donnelly, A theory of smart cities, in: Proceedings of the 55th Annual Meeting of the ISSS, International Society for Systems Sciences, 2011.

[65] H.P. McKenna, S.A. Chauncey, Taking learning to the city: an exploration of the frictionless learning environment innovation, in: Proceedings of the 6th International Conference on Education and New Learning Technologies (EduLearn14), 2014, pp. 6324–6334.

[66] B. Dainow, Smart city transcendent: understanding the smart city by transcending ontology, Orbit 1 (2017) 1–15.

[67] M. Ibrahim, A. El-Zaart, C. Adams, Theory of change for the transformation towards smart sustainable cities, in: 2017 Sensors Networks Smart and Emerging Technologies (SENSET), 2017, pp. 1–4, https://doi.org/10.1109/SENSET.2017.8125067.

[68] C. Petrucco, C. Ferranti, Design smart city apps using activity theory, Knowl. Manag. E-Learn. Int. J. 9 (4) (2017) 499–511.

[69] C.J.L. Balsas, Paolo Soleri and America's third utopia: the sustainable city-region, J. Urban. Int. Res. Placemaking Urban Sustain. 13 (4) (2020) 410–430, https://doi.org/10.1080/17549175.2020.1726798.

[70] H.P. McKenna, The importance of theory for understanding smart cities: making a case for ambient theory, in: N.A. Streitz, S. Konomi (Eds.), Distributed, Ambient and Pervasive Interactions. HCII 2023, Lecture Notes in Computer Science, vol. 14037, Springer, Cham, 2021, https://doi.org/10.1007/978-3-031-34609-5_8.

[71] M. Batty, Big data, smart cities and city planning, Dialogues Hum. Geogr. 3 (3) (2013) 274–279, https://doi.org/10.1177/2043820613513390. PMID:29472982.

[72] N. Brenner, New Urban Spaces: Urban Theory and the Scale Question, Oxford University Press, New York, 2019.

[73] M. Will-Zocholl, Information spaces(s), in: R. Appel-Meulenbroek, V. Danivska (Eds.), A Handbook of Theories on Designing Alignment Between People and the Office Environment, Routledge, 2021.

[74] H. Lefebvre, The Production of Space, Blackwell, Malden, MA, 1991.

[75] P. Tunji-Olayeni, K. Kajimo-Shakantu, T.O. Ayodele, Factors influencing the intention to adopt green construction: an application of the theory of planned behaviour, Smart Sustain. Built Environ. (2023), https://doi.org/10.1108/SASBE-06-2022-0126.

[76] R.K. Yin, Case Study Research: Design and Methods, sixth ed., Sage, Thousand Oaks, CA, 2017.

[77] C. Zaiontz, Real statistics using excel, 2023. Retrieved from www.real-statistics.com.

[78] J.W. Creswell, Educational Research: Planning, Conducting, and Evaluating Quantitative and Qualitative Research, sixth ed., Pearson, Boston, MA, 2018.

CHAPTER

Data fusion and homogenization

3

Two key aspects for building digital twins of smart spaces

Andres Munoz-Arcentales[a], Javier Conde[a], Álvaro Alonso[a], Joaquín Salvachúa[a], Washington Velasquez[b], and Sonsoles López-Pernas[c]

[a]*Departamento de Ingeniería de Sistemas Telemáticos, Escuela Técnica Superior de Ingenieros de Telecomunicación, Universidad Politécnica de Madrid, Madrid, Spain,* [b]*Escuela Superior Politécnica del Litoral, ESPOL, FIEC, Guayaquil, Ecuador,* [c]*School of Computing, Faculty of Science, Forestry and Technology, University of Eastern Finland, Joensuu, Finland*

1 Introduction

Smart space is any surrounding environment that adapts itself to human behavior and needs by utilizing the data obtained from the interaction between objects and humans [1]. Digital twins (DTs) have emerged as a key concept in the field of smart spaces, enabling the creation of virtual replicas of physical environments. They provide a unique opportunity for creating more efficient and intelligent smart spaces by bringing together data from various sources and processing such data in real time [2]. However, the success of a DT relies heavily on the quality and accuracy of the data used to create it. Since data in smart spaces may stem from multiple sources, data fusion and homogenization are two aspects that play a crucial role in the development of DTs. Data fusion is the process of integrating data from multiple sources to create a more comprehensive and accurate view of the environment [3]. With the advent of internet of things (IoT) devices, there has been an exponential growth in the amount of data generated by different sources in the context of smart spaces. Data fusion allows IoT data to be used to good advantage, to build more accurate DTs. Homogenization, on the other hand, is the process of standardizing data formats and structures to enable seamless integration and processing.

Data fusion is a critical process in the creation of DTs of smart spaces. This process involves several challenges, including dealing with data heterogeneity, managing data volume, and addressing data quality issues [4]. Data heterogeneity refers to the differences in data format, structure, and content between different sources. This can pose

Smart Spaces. https://doi.org/10.1016/B978-0-443-13462-3.00002-9

significant challenges when integrating data from multiple sources. One of the main approaches to addressing data heterogeneity is through the use of data integration techniques, such as ontology-based data integration [5], which enables the integration of data from multiple sources by mapping them to a common ontology. Managing data volume is another critical challenge in data fusion. The growth of IoT devices has led to an exponential increase in the volume of data generated from different sources. This poses challenges in terms of data storage, processing, and analysis. One of the main approaches to managing data volume is through the use of edge computing, which enables data processing to be carried out at the edge of the network, closer to the source of data. Data quality is another critical challenge in data fusion. Data quality issues can arise due to various factors, such as sensor malfunctions, environmental changes, or human errors. Ensuring data quality is crucial for the accuracy and reliability of DTs. One approach to addressing data quality issues is through the use of data cleaning techniques, such as outlier detection and data imputation [6].

Homogenization is another critical aspect of building DTs of smart spaces. The goal of homogenization is to standardize data formats and structures to enable seamless integration and processing [7]. This process involves several challenges, including dealing with data format heterogeneity, managing semantic heterogeneity, and addressing data privacy and security issues. Data format heterogeneity refers to the differences in data formats used by different sources. This can pose challenges when integrating data from multiple sources, as the data may need to be converted into a common format before it can be processed. One approach to addressing data format heterogeneity is through the use of data mapping techniques, which enable the mapping of data from one format to another. Managing semantic heterogeneity is another critical challenge in homogenization. Semantic heterogeneity refers to the differences in the meaning of data between different sources. This can pose challenges when integrating data from multiple sources, as the data may need to be aligned to a common ontology before it can be processed. One approach to addressing semantic heterogeneity is through the use of semantic mapping techniques, which enable the mapping.

This chapter aims to address the challenges of data fusion and homogenization in the implementation of digital twins for smart spaces. By leveraging the technologies from the FIWARE ecosystem, we present a comprehensive approach to combat these challenges and enable the creation of more accurate and valuable digital twins. Our contribution lies in providing a detailed description of the implementation process, including data selection, preprocessing, integration, and interpretation steps. Furthermore, we present a case study on low-emission zones (LEZ), showcasing the practical application of our proposed solution within a smart city context. Through this case study, we demonstrate the replicability of our approach in different scenarios within the smart city environment. Lastly, we discuss the advantages of utilizing FIWARE as the underlying technology, emphasizing how it helps overcome the data fusion and homogenization challenge effectively.

The remainder of this chapter is organized as follows. Section 2 presents a short overview of the background information that is relevant to the topics presented in this chapter. In Section 3, we present the data ingestion challenge in DTs, while in Section 4, we present the problems that affect the building of DTs from data coming from smart spaces and the proposed solution to face them. In Section 5, we present a case study where the solution proposed is applied to the smart city ecosystem focused on the development of a DT for a low-emission zone. Finally, Section 6 presents the discussion and conclusions of this chapter.

2 Background

This section provides background information on the technological ecosystem that is used in the implementation presented in this chapter, namely, the FIWARE ecosystem. **FIWARE** (https://www.fiware.org/) is an open-source framework that offers a set of components known as generic enablers (GEs) to aid in the development and implementation of smart solutions. The framework consists of various modules, each performing specific tasks. The Context Broker GE is a crucial component in any smart solution as it manages, updates, and grants access to context information. In the context of DTs, the Context Broker acts as a liaison between the physical and virtual worlds, facilitating information exchange between the two domains. FIWARE offers a suite of complementary components, structured in three layers, to build around the FIWARE ecosystem. The layers include (1) interface with the IoT, robots, and third-party systems, (2) context data/API management, and (3) processing, analysis, and visualization of context information.

The **Orion Context Broker** (https://github.com/telefonicaid/fiware-orion) is the most widely used version of the Context Broker in FIWARE. It manages context information by implementing a publish-subscribe pattern that provides a Next Generation Service Interface (NGSI) interface. Clients can create, query, and update context elements and subscribe to changes in context information. Other elements interact with Orion through HTTP/HTTPs requests, and only the latest information state is saved. Therefore, other GEs must be used to store the information history.

NGSI defines the way data are modeled and how they are accessed through the FIWARE RESTful Application Programme Interface (API). The FIWARE components are compatible with the NGSIv2 [8] and the linked data version called NGSI-LD [9]. In NGSI, the information is modeled through entities (i.e., representations of physical or logical objects). Each of the context entities has a set of attributes and metadata of attributes to represent its real-life properties. Entities are encoded in JSON (NGSIv2) and JSON-LD (NGSI-LD) formats. NGSI API interactions are made using HTTP/HTTPs through synchronous and asynchronous requests. In the synchronous communication, clients use HTTP methods for creating, updating, deleting, and querying context information. The asynchronous communication is based on the publish/subscribe pattern. In this way, a client subscribes to changes in specific entities, and when a change happens, the client receives a notification with the update.

FIWARE provides the **Draco GE** (https://github.com/ging/fiware-draco) for ingesting data from and to different sources, such as REST APIs, database systems, TCP, or HTTP servers. Draco is a dataflow management system based on Apache NiFi (https://nifi.apache.org/) that supports powerful and scalable directed graphs of data routing, transformation, and system mediation logic. The suite of processors and controllers provided in this GE allows for the conversion of incoming data into NGSI entities and attributes, required for publication in the Context Broker. Draco is also used in the persistence part, with a set of processors in charge of persisting NGSI context data in third-party storage systems, creating a historical view of the data posted in the Context Broker.

In the FIWARE Ecosystem, there are various **GEs that manage authorization and authentication** with regard to users and devices. Every request sent to a FIWARE GE has to be authenticated using a token. The Keyrock GE (https://github.com/ging/fiware-idm) is an Identity Management component based on OAuth 2.0, issuing tokens for previously registered and authenticated subjects. The Wilma PEP Proxy GE (https://github.com/ging/fiware-pep-proxy) acts as a policy enforcement point (PEP), validating tokens to ensure the fulfillment of access control policies based on previously defined roles and permissions for subjects. In complex scenarios, where advanced policy definition is needed, the XACML-based policy decision point (PDP) AuthZForce GE (https://github.com/authzforce/fiware) should be used.

The **FIWARE Smart Data Models** initiative (https://smartdatamodels.org/) provides the necessary tools for defining a common schema that enables the integration of external data sources and using standard data models adopted by the industry. The FIWARE community has already standardized many data models, from different domains with the objective of being portable for different solutions. Nowadays, there are smart models in 13 different domains, including smart cities, smart environment, smart sensoring, smart agrifood, and smart energy, inter alia. Smart Data Models are compatible with both NGSIv2 and NGSI-LD. They are proposed and maintained by the community, and users are able to extend them for their specific case study.

3 Data ingestion challenges in digital twins

The origin of the term "digital twin" is attributed to Michael Grieves and John Vickers, who coined it at the beginning of the 21st century as "a virtual equivalent of a physical product" [10]. The definition of DT has evolved over the years, and although there is no total consensus in the literature regarding its formal definition [11], a DT is characterized by the communication between the physical environment and the virtual environment through a procedure called twinning [12]. The virtual environment captures the current state of the physical object, processes its information, and obtains results that allow changing the physical environment through IoT actuators. Furthermore, the application of artificial intelligence (AI) techniques is compatible with the processing tasks of a DT, providing the DT with predictive capabilities and constituting the so-called intelligent DTs [13].

The concept of smart spaces [14] arose at the same time as the DTs one. A smart space is defined as the physical environment in which smart devices are connected and work together [15]. IoT has become the enabling technology for smart spaces; in fact, the combination of IoT and smart spaces is referred to as IoT-based smart spaces [16]. Some authors propose to divide smart spaces into different domains such as smart cities, smart buildings, smart health, smart industry, or smart transportation [16]. Researchers on smart spaces study their requirements, including physical communication networks (e.g., WiFi, 5G, satellite); types of networks depending on the extension to cover (e.g., personal area networks, local area networks, metropolitan area networks); technologies for capture of information (e.g., IoT protocols, data fusion tools); infrastructure for the deployment of solutions considered smart (e.g., cloud computing, edge computing).

DTs and smart spaces have gained importance in recent years due to the evolution of enabling technologies such as IoT, AI, big data, extended reality, and cloud computing [17]. As a consequence, they are currently applied in a wide variety of fields, including the manufacturing sector [18], the primary sector [19], and the different fields that compound the smart spaces [20].

Although DTs of smart spaces constitute a promising and widespread technology, there are a set of barriers that limit their application that must be addressed. Among these challenges, one of the most relevant is the integration of external data sources in the DT [17]. DTs interact with a multitude of actors, using different protocols, through different formats, with different update periods, with different requirements, etc. As a final result, the DT must integrate all external data sources in a way that is transparent to the rest of the processing components. In other words, one of the challenges of DTs is to carry out data fusion and data homogenization of external data sources.

Regarding the homogenization and fusion of heterogeneous data sources, Lenzereni [21] defines three main components during the integration process: the global schema, the source schema, and the mapping. The global schema is the representation of the information from the point of view of the integrating system. The source schema represents the information in the original data source. The mapping is the specification to transform from one schema to another. The way in which the global schema is created makes it possible to differentiate between two approaches. The first one, known as Local-as-View, is based on a static predefined global schema. In this local proposal, a new mapping is created to adapt the source schema to the global schema when a new data source appears. In contrast, the second approach, known as Global-as-View, consists of extending the global schema dynamically as new data sources appear. The Local-as-View approach requires an initial effort to define the global schema; however, it facilitates the scalability of the system, since the application querying the integrated sources (i.e., the DT) knows the global schema previously. However, this approach is invalid when the environment in which the homogenization is carried out is very dynamic and needs to continuously extend the global schema.

Fusco and Aversano [22] propose a hybrid scheme between the Local-as-View and Global-as-View approaches. Their research is an ontology-based proposal that integrates heterogeneous data sources through a system with mediators and wrappers. Mediators interact with the application, in this case, the DT, through a global and virtual view. Mediators translate the petitions into local views, specific to each data source, and the wrappers communicate with the respective real sources. In contrast, Sreemathy et al. [23] propose the extract, transform, and load (ETL) process for the integration of data sources. In the first phase of the process, called extraction, the connection with the real data sources is established. In the second phase, known as transformation, raw data are transformed into predefined and homogeneous models. In the third phase, called load, the homogenized data are stored in a common repository known as a warehouse. In the solution of Sreemathy et al., the resultant homogenized data sources are stored in the integrator system, so it is necessary to guarantee the consistency of the information. In the proposal of Fusco and Aversano [22], the information is not saved in the integrator system because the mappings are in charge of retrieving the information from the original data sources and adapting them in real time. However, this solution is more complex to implement and is limited to the storage and network capabilities of the real data sources.

There have been proposals for the homogenization of large-scale scenarios. In 2006, Berners-Lee et al. proposed linked data as an enabling technology for the semantic web [24]. The semantic web aims to build a network of interlinked, reusable, machine-readable data that enables self-discovery of information, similar to how humans do it through the traditional web of documents based on HTML documents and links. From the semantic web concept, technologies that facilitate the integration of external data sources have emerged, such as the RDF standard for modeling information through triples, OWL for the definition of ontologies, and SPARQL to perform operations on RDF data sources. Some studies propose the use of linked data and technologies derived from the semantic web to model and homogenize the information of DTs [25]. The work of Jacoby et al. [26] analyzes alternatives for modeling and homogenizing information in DTs, highlighting standards and formats such as SensorThings, Asset Administration Shell, Web of Things, or NGSI-LD. All of them specify the way in which data should be modeled in the DT, how to manage the DT information, and how to implement the twinning process. SensorThings [27] models the DT information as *Things* that receive the information as *DataStreams* (observations) coming from *Sensors*. Additionally, these *Things* perform *Tasks* (actions) through `Actuators`. Asset Administration Shell [28] is focused on industrial environments and integrates data sources as *Assets*. An asset contains properties, operations, and relationships to other assets. Web of Things [29] is a W3C recommendation that defines how physical information should be modeled in virtual environments. The centerpieces of Web of Things are the *Thing Descriptors* composed of *Properties* that define them, *Actions* that they can perform, and *Events* with which they can interact. Finally, NGSI-LD [9] is the standard developed by ETSI for context information modeling, and adopted by the FIWARE initiative. NGSI-LD defines the way of accessing and modeling data through *Entities* with their respective *Properties* and *Relationships* to other *Entities*. The election of one alternative or another will depend on each case study; however, all of them use accessible and interoperable data formats (e.g., JSON, JSON-LD, XML, AutomationML, RDF/XML, OPC-UA) [26].

4 Problem statement and proposed solution

In the previous section, we discussed how data fusion and homogenization have been some of the main barriers when implementing DTs. We also mentioned a set of specifications and standards that aim to define an ecosystem to manage DT data. In addition, some research has proposed tools to ease the development of DTs. Stojanovic et al. [30] proposed the use of the FA3ST toolkit for the implementation of DTs compatible with Industry 4.0 and guaranteeing data sovereignty. Kherbache et al. [31] proposed Eclipse Ditto and Eclipse Hono as tools for easing the connection of physical devices with the rest of the infrastructure in the DT. The FIWARE initiative proposed the Context Broker to manage the context information in the DT [32]. There are also commercial solutions acting as Platforms as a Service for the implementation of DTs such as Azure Digital Twins (https://learn.microsoft.com/azure/digital-twins) or IBM Digital Twin Exchange (https://www.ibm.com/products/digital-twin-exchange). Other research, such as the study carried out by Conde et al. [33], analyzed a complete architecture for the implementation of DTs focused on the information flow. In this chapter, we will study the solution of Conde et al., which proposes a whole FIWARE ecosystem to implement DTs focused on the information flow. Specifically, the feasibility of the Draco GE tool, based on Apache NiFi, for data fusion and data homogenization will be analyzed. Draco follows the ETL approach of Sreemathy et al. [23] limited to simple data transformations. FIWARE Draco is based on the configuration of a set of processors building a graph that enables the integration of data sources into the DT, by homogenizing the information using the FIWARE Smart Data models.

In this proposal, the FIWARE Smart Data Models act as a global schema of the DT. Following Lenzereni's approach [21], the use of global schemes facilitates the integration of new data sources, since the rest of the architecture is transparent to the data fusion and data homogenization process. However, instead of developing a new mapping between the global and local schema, the information is stored in the FIWARE Context Broker and NGSI-LD will be used to manage the context information.

Fig. 1 shows the proposal based on FIWARE Draco for data fusion and data homogenization in DTs. In the following sections, case studies on DTs will be

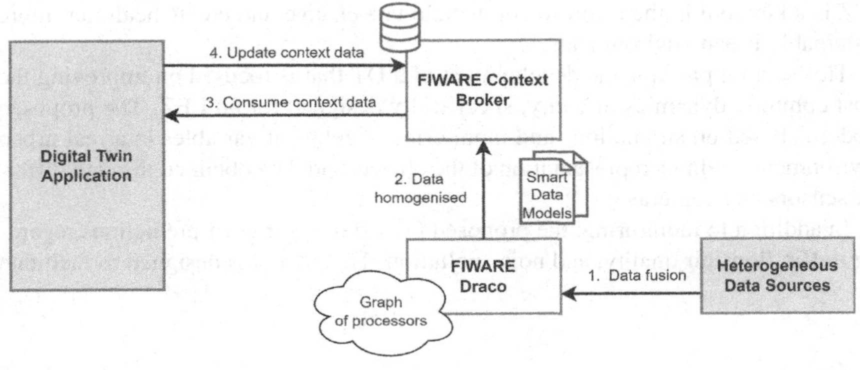

FIG. 1

Data fusion and data homogenization architecture.

presented and analyzed to validate the FIWARE Smart Data Models, NGSI-LD, and FIWARE Draco as a solution to perform data fusion and data homogenization in the development of DTs.

5 Case study

As we have stated in Sections 3 and 4, the heterogeneity and complexity of the data generated by heterogeneous sources pose a significant challenge in building and maintaining DTs. To address this challenge, this chapter presents a solution that leverages a common standard and multiple sources of data to build DTs in smart spaces. Specifically, this section presents one case study that demonstrates how the proposed solution can be applied in the smart city domain. The case study highlights the benefits of the proposed solution in facilitating the design and development of middleware, collecting and homogenizing data, and even having the possibility of consuming and publishing the data from and to Open Data portals. Overall, this chapter describes the contribution to the advancement of DT technology by proposing a solution that can effectively deal with the heterogeneity and complexity of data in smart spaces.

5.1 LEZs: A reference proposal for building a standardized digital twin

A low-emission zone (LEZ) is an area within a city that has specific regulations and restrictions in place to reduce air pollution from vehicles. These zones typically prohibit the most polluting vehicles from entering or driving through the area, and encourage the use of cleaner, more efficient transportation options [34]. This can include public transportation, electric or hybrid vehicles, and cycling or walking. By implementing an LEZ, cities aim to improve air quality, protect public health, and reduce the overall carbon footprint of transportation in the area. Some cities have also implemented congestion charges, which require drivers to pay a fee for entering the LEZ, further incentivizing the use of cleaner transportation options. Overall, an LEZ is a key tool in the effort to combat climate change and create healthier, more sustainable urban environments.

This section presents the development of a DT that is focused on improving the most common dynamics of a city, specifically designed for an LEZ. The proposed model is based on simulations and monitoring of relevant variables in a real urban environment, giving a representation of the objects and data obtained through different sensors and cameras.

In addition to monitoring, the proposed DT offers a series of predictions regarding traffic flow, air quality, and noise pollution. This model is designed to facilitate the assessment of possible modifications to the roads, such as changes in direction, modification of the number of lanes, or pedestrianization, to visualize their potential impact on the dynamics of the city.

To design and test a working prototype, the city of Guadalajara in Spain has been selected, which is to establish an LEZ shortly. The proposed DT monitors and simulates the activities within that zone. The model provides insights into how the different dynamics can be optimized to improve the city's environmental and social conditions.

Overall, the proposed DT provides a comprehensive understanding of the different variables and their interactions in the LEZ, enabling policymakers to make informed decisions to create a sustainable urban environment.

5.1.1 Data infrastructure

The data infrastructure comprises four key stages, namely data ingestion, data processing and modeling, data storing, and data provisioning. The software components that enable these stages are based on FIWARE, an open-source platform that provides GEs for building smart applications.

Data ingestion involves gathering relevant data from various data sources. The data sources use different technologies and protocols to make data available for consumption. Therefore, specific connectors are configured for each source that obtains the data using the defined protocols. After that the data will be available in infrastructure for further processing and analysis.

Data processing and modeling involve transforming the retrieved data into a format that is useful for analysis and visualization. The data sources frequently provide data in different formats and with different fields in each dataset. Hence, data processing involves removing useless information and appropriately tagging, homogenizing, and sorting fields. This is done following a standard data model defined for the specific domain.

Once the data have been modeled and homogenized, it is stored in a database to enable historical registry for further queries and processing. The database is designed to ensure efficient storage and retrieval of data.

In addition to storing data in the database, data consumers can be subscribed to receive real-time data updates. For this purpose, components that manage publish/subscribe mechanisms are required. These components enable data consumers to receive data in real time for further processing and analysis.

5.2 Modeling data

Within the context of the LEMON Zone project, an NGSI-LD general data model has been designed for traffic management in LEZs. This model allows heterogeneous data sources to be integrated into the NGSI-LD standard. As a consequence of data standardization, the information can be easily managed inside the system and it makes the data easy to share and understand by third-party systems. Moreover, as FIWARE Smart Data Models are designed and approved by the industry, they guarantee the quality of the solution.

As is shown in Fig. 2, the centerpiece of the model is the *RestrictedTrafficArea* entity, which allows defining the low-emission traffic zone with generic attributes such as status, location, regulation, etc. A *RestrictedTrafficArea* is composed of RoadSegments entities that manage context information of roads (e.g., allowed vehicle types, maximum speed). RoadSegments information is completed with real-time traffic data modeled as *TrafficFlowObserved* entities. These entities register statistics such as traffic intensity, average vehicle speed, occupancy, etc. The model also allows control of specific Vehicles including information such as its model, physical characteristics, vehicle type, license plate, etc. In addition to managing traffic in the LEZ, it is important to control parking in order to avoid traffic jams. For this purpose, the entities *OffStreetParking* and *OnStreetParking* are offered, which model indoor and outdoor parking areas, respectively. Moreover, the entity *ParkingSpot* allows the modeling of a particular parking spot including information such as its location or status. In a LEZ, it is also necessary to shape the environment. For this purpose, data models are provided to manage weather through *WeatherObserved* entities (by registering parameters such as temperature, precipitation, pressure, wind speed, etc.); noise levels using *NoiseLevelObserved* entities; and air quality through *AirQualityObserved* entities (by managing levels of tropospheric ozone, nitrogen dioxide, sulfur dioxide, etc.) (https://www.fiware.org/smart-data-models).

5.2.1 Data architecture

The proposed data architecture is built upon the utilization of FIWARE GEs to enable both real-time and batch context information delivery to the DT application for the LEZ of cities. This system conforms to the FIWARE standard and is acknowledged in the industry as "Powered by FIWARE" owing to its usage of the Orion Context Broker to handle all context data relating to the LEZ. The Orion Context Broker grants access to data through standard CRUD operations, while also informing all pertinent clients of any updates in context information via its publication and subscription system.

The proposed architecture consists of two GEs, namely Draco and the Context Broker. Draco facilitates the ETL process and storing historical data. On the other hand, the Context Broker serves as a centralized point of access for all stakeholders by exploiting the publish/subscribe features and NGSI-LD.

The flow of data collection and processing is shown in Fig. 3, which illustrates how real-time data are collected via multiple processors using Draco, and then

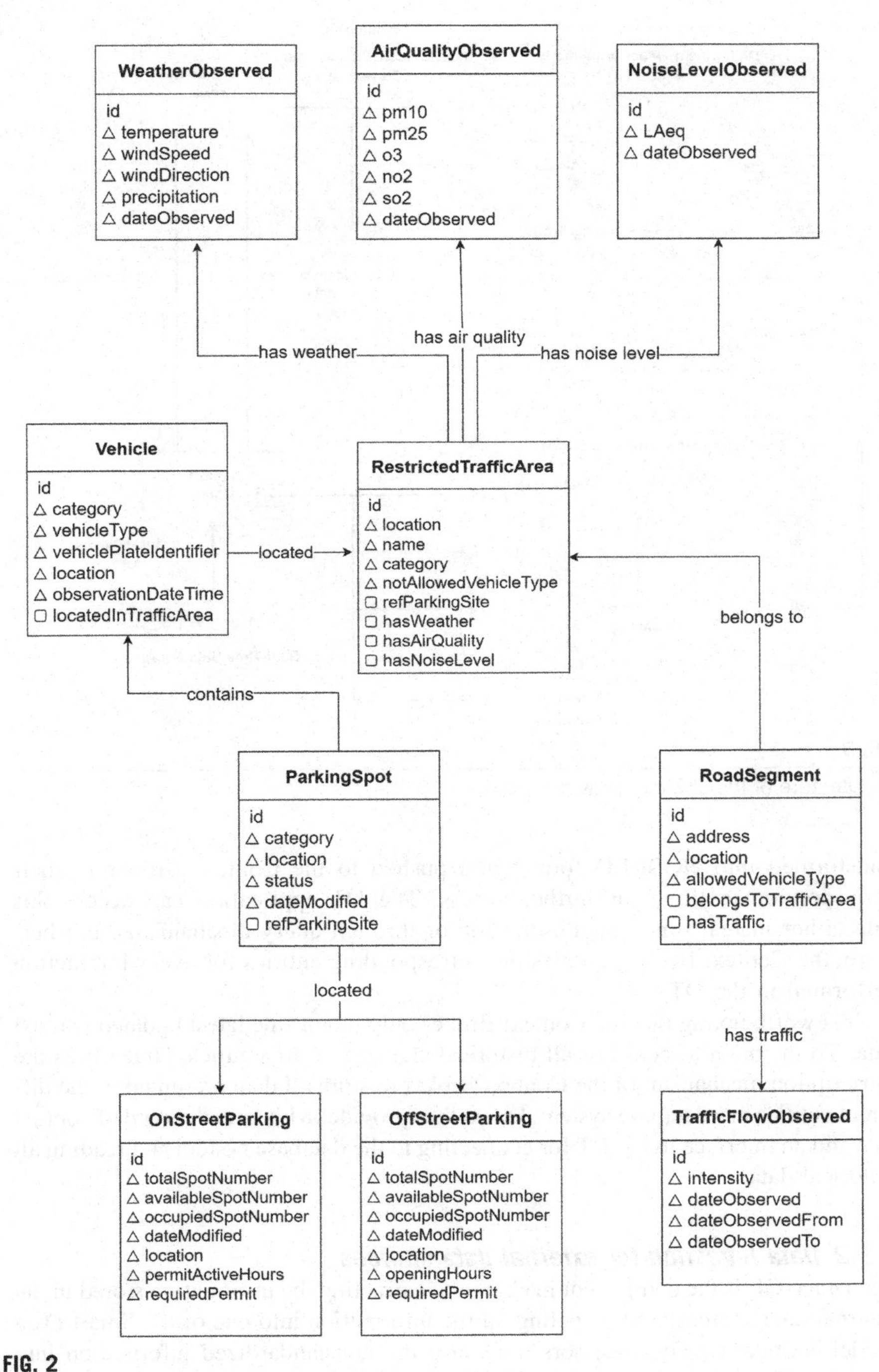

FIG. 2

LEZ Smart Data Model.

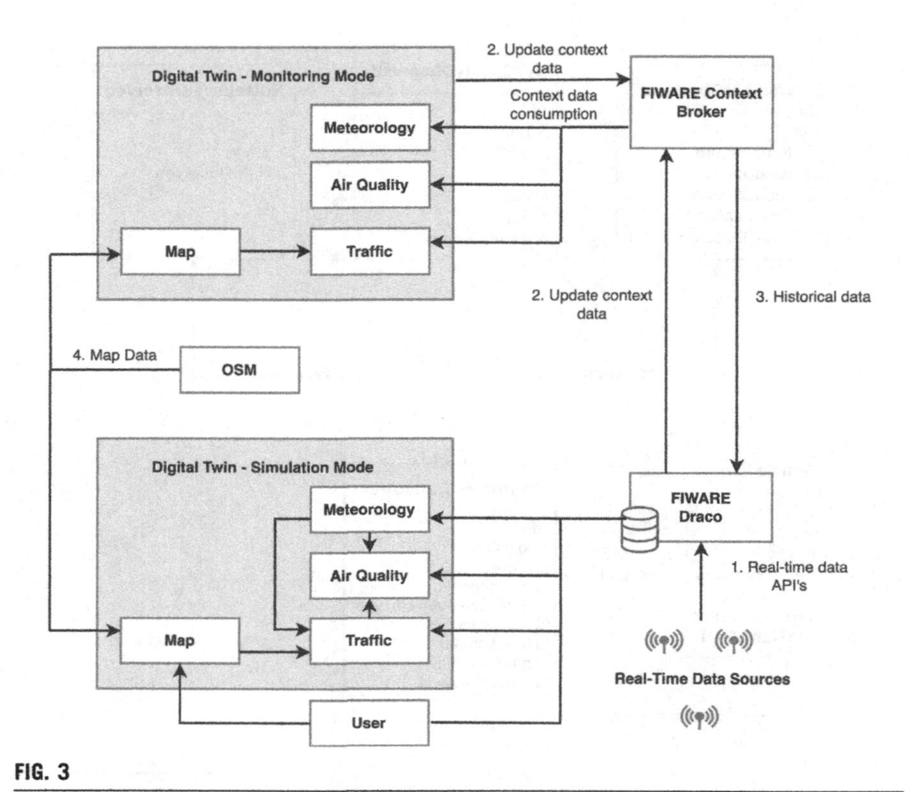

FIG. 3

Architecture of the LEZ digital twin.

transformed into NGSI-LD format and pushed to the Context Broker in their corresponding entities for further access. The DT application can access this data either in real time via subscription or through query mechanisms. Furthermore, the Context Broker updates the corresponding entities for every interaction performed in the DT.

It is worth noting that the Context Broker only retains the latest updated context data. To maintain a record of all historical changes in the entities, Draco uses the subscription mechanism of the Context Broker to store all data presented in the different entities in a database system. Draco thus provides a historical record of context data and an interface to the DT for connecting to the database system and reading all historical data.

5.2.2 Data ingestion for external data sources

The Draco GE is the component in charge of extracting the information stored in the external data sources, and modeling all the information into one of the Smart Data Model entities. Draco processors transform the nonstandardized information into

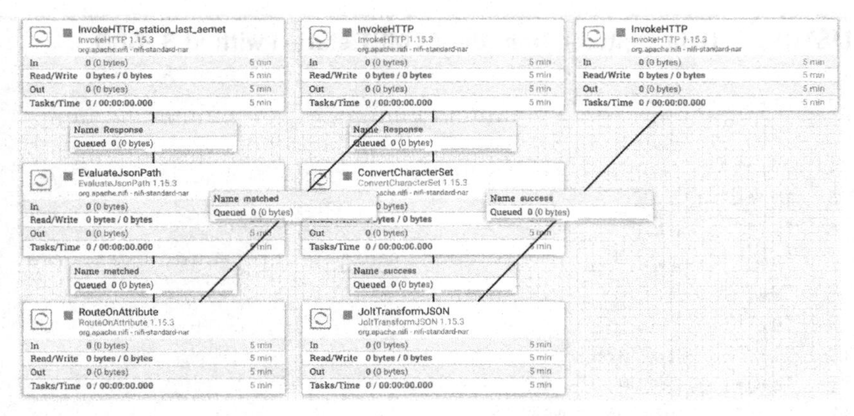

FIG. 4

FIWARE Draco processors flow for collecting data from AEMET station.

NGSI-LD format when new data are received from the data sources. Lastly, Draco publishes the standardized data into the Orion Context Broker.

Concretely, Draco was configured with a set of processors to extract real-time weather information and real-time air quality information. Fig. 4 shows the AEMET template, used to ingest weather data from the Spanish Meteorology Agency (AEMET [https://www.aemet.es/]). In order to get the data, two requests are needed. The first one retrieves the URI required for getting the last measure, or error, if the station is not working. The second one retrieves the data source itself.

The template is divided into these processors:

- InvokeHTTP-station-last-aemet: It sends a GET HTTP request to the weather station.
- EvaluateJsonPath: It extracts the URI to request for new data.
- RouteOnAttribute: It filters valid URIs (the station is operating).
- InvokeHTTP: It sends a GET HTTP request to the retrieved URI.
- ConvertCharacterSet: It converts to UTF-8 character encoding.
- JoltTransformJSON: It transforms the input JSON file into a JSON-LD FIWARE entity. For example, the input JSON.

The main objective of this process is to convert the incoming data into the data format specified by the NGSI-LD standard. Thus the incoming data represented in Listing 1 will be transformed into the NGSILD entity as is shown in Listing 2.

LISTING 1 Data captured from the AEMET station with id 3168D.


```
1 {
2   "idema" :  "3168D",
3   "lon" : -3.149995,
4   "fint" : "2023-01-11T19:00:00",
5   "prec" : 0.0,
6   "alt" : 721.0,
7   "vmax" : 1.5,
8   "vv" : 0.9,
9   "dv" : 106.0,
10  "lat" : 40.630283,
11  "dmax" : 348.0,
12  "ubi" :  "GUADALAJARA",
13  "pres" : 944.8,
14  "hr" : 93.0,
15  "stdvv" : 0.2,
16  "ts" : 5.8,
17  "pres_nmar" : 1030.9,
18  "tamin" : 7.2,
19  "ta" : 7.2,
20  "tamax" : 8.3,
21  "tpr" : 6.1,
22  "vis" : 15.5,
23  "stddv" : 9.0,
24  "inso" : 0.0,
25  "tss5cm" : 8.4,
26  "pacutp" : 0.0,
27  "tss20cm" : 8.1
28 }
```


LISTING 2 Modeled NGSI-LD entity of the AEMET station with id 3168D.


```
1 {
2  "@context":"https://smart-data-models.github.io/dataModel.
   Weather/context.jsonld",
3  "id":"urn:ngsi-ld:WeatherObserved:3168D-1",
4  "type":"WeatherObserved",
5  "address":{
6     "type": "Property",
7     "value":{
8        "addressLocality":"GUADALAJARA",
9        "addressCountry":"ES",
```

```
10        "type":"PostalAddress"
11      }
12  },
13  "atmosphericPressure":{
14      "type":"Property",
15      "value":944.8
16  },
17  "source":{
18      "type":"Property",
19      "value":"http://www.aemet.es"
20  },
21  "dateObserved":{
22      "type":"Property",
23      "value":{
24          "@value":"2023-01-11T19:00:00.00Z",
25          "@type":"DateTime"
26      }
27  },
28  "location":{
29      "type":"GeoProperty",
30      "value":{
31          "type":"Point",
32          "coordinates":[
33              -3.149995,
34              40.630283,
35              721
36          ]
37      }
38  },
39  "precipitation":{
40      "type":"Property",
41      "value":0
42  },
43  "windSpeed":{
44      "type":"Property",
45      "value":0.9
46  },
47  "windSpeedMax":{
48      "type":"Property",
49      "value":1.5
50  },
```

Continued

LISTING 2 Modeled NGSI-LD entity of the AEMET station with id 3168D—cont'd

```
51  "windDirection":{
52     "type":"Property",
53     "value":106
54  },
55  "windDirectionMax":{
56     "type":"Property",
57     "value":348
58  },
59  "relativeHumidity":{
60     "type":"Property",
61     "value":0.93
62  },
63  "temperature":{
64     "type":"Property",
65     "value":7.2
66  },
67  "temperatureMin":{
68     "type":"Property",
69     "value":7.2
70  },
71  "temperatureMax":{
72     "type":"Property",
73     "value":8.3
74  },
75  "stationCode":{
76     "type":"Property",
77     "value":"3168D"
78  }
79 }
```

A similar procedure is applied to the real-time air quality data source. In this case, the data are obtained from the Guadalajara Council. Fig. 5 summarizes the set of processors used:

- InvokeHTTP: Sends a HTTP GET petition to the Guadalajara Council.
- UpdateAttribute: Transforms the time data information to the ISO8601 format.
- JoltTransformJSON: It transforms the input JSON file into a JSON-LD FIWARE entity.

According to the aforementioned statement, it becomes apparent that the procedure employed bears a striking resemblance to the AEMET case. As a result, both the

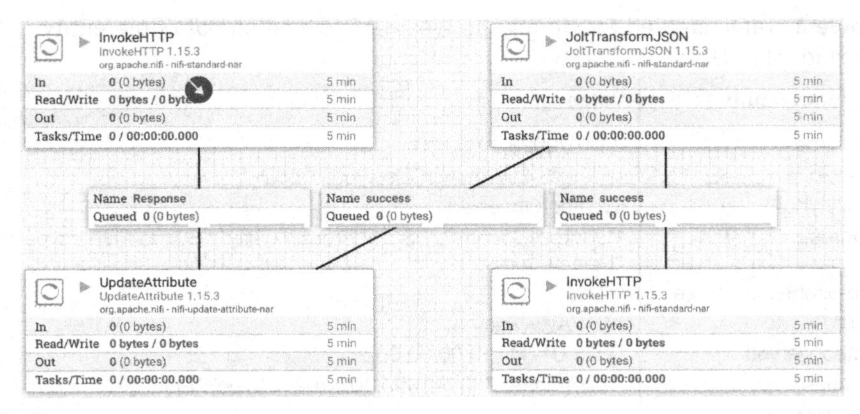

FIG. 5

FIWARE Draco processors flow for collecting AirQuality data.

input and output data utilized for the AirQuality modeling are documented in Listings A.1 and A.2 correspondingly.

5.3 Data provisioning to client applications

The application for DTs has the capability to receive data through two primary interfaces, which are facilitated by the FIWARE architecture. On one hand, the application uses the REST API provided by the CB for querying the different entities in order to retrieve the data. In this particular deployment, the DT requests the context broker for weather data, specifically wind speed, pressure, and other related information, provided by AEMET. An example of a query made to the Context Broker to request weather information for the station with id 3168D in Guadalajara City is shown in Listing 3.

LISTING 3 HTTP request to the Context Broker.

```
1 curl -G -iX GET 'http://broker.org/ngsi-ld/v1/entities/urn:
  ngsi-ld:WeatherObserved:3168D-1'\
2 -H 'Link: <https://smart-data-models.github.io/dataModel.
  Weather/context.jsonld>;'
```

On the other hand, the second interface available in this scenario for retrieving historical data of the changes that occurred in the entities of the CB is via a direct query to a relational database. In this implementation, data are stored in a PostgreSQL database. The structure of each created table adheres to the NGSI specification for recording context data. For instance, for the *WeatherObservation* entity, a table is created with the name historical.weatherobserved, and every attribute of that entity is mapped as a column in that table; finally a row with the value of the attributes is inserted as a record, as is represented in Table 1, which shows an example of a *WeatherObservation* entity stored in the PostgreSQL.

Table 1 Table created for storing the WeatherObservation data from entity with id 3168D.

Column name	Value
recvtime	11/01/2023 19:00:00
entityid	urn:ngsi -ld:WeatherObserved:3168D-1
entitytype	WeatherObserved
address	{"addressCountry":"ES","addressLocality":"Guadalajara","type": "PostalAddress"}
atmosphericpressure	944.8
source	http://www.aemet.es
dateobserved	2023-01-11T19:00:00.00Z
location	{"coordinates":{[}-3.1499,40.6302{]},"type":"Point"}
precipitation	0
windspeed	0.9
windspeedmax	1.5
winddirection	106
winddirectionmax	348
relativehumidity	0.93
temperature	7.2
temperaturemin	7.2
temperaturemax	8.3
stationcode	3168D

5.4 Client application

The purpose of this section is to present a general overview of the DT application created for providing a digital representation of an LEZ within a city. To accomplish this, we obtained a detailed 3D map of the area using OpenStreetMap (OSM) and then generated 3D representations of various features such as buildings, parks, squares, and parking spaces. Moving vehicles were also simulated within the environment, and the entry and exit points of the LEZ were clearly marked. Additionally, the layout of roads, paths, bicycle lanes, and pedestrian streets were created, and traffic control cameras were placed throughout the environment. To ensure that air quality and meteorological data were incorporated, the application was connected with the relevant data providers. The application also included configuration for parking areas, traffic signs, and an alert system. Unity was used as the engine to represent the data and objects retrieved by the data architecture presented in Section 5.2.1.

The main view of the application is shown in Fig. 6, where multiple layers of the city are visualized with various textures and colors, representing the type of object, layer of information, and sensors. The base layer of the map includes terrain types ranging from green spaces to buildings and parking places, where buildings are automatically generated from OSMs floorplans. Cameras are

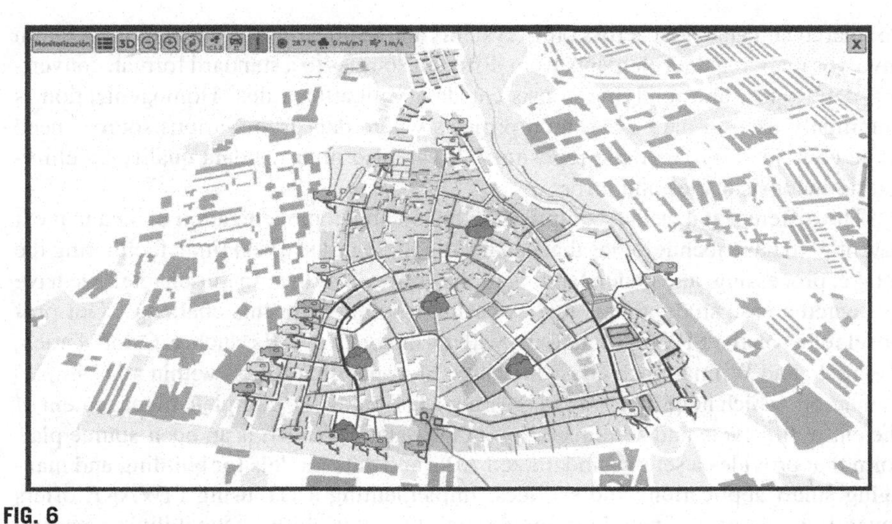

FIG. 6

Digital twin client application view.

displayed in green and red, indicating their active and inactive status, respectively, while traffic and weather data are visualized through the clouds and cameras. The top menu of the application allows the user to filter and highlight various aspects of the data.

Moreover, the application provides real values for most of its variables. Users can observe the actual number of vehicles within a zone, categorized according to the type of vehicle and emission label. Apart from traffic data, the application displays contamination values obtained from the sensors and weather values such as temperature, wind speed, and rain flow.

6 Discussion and conclusions

DTs use data from sensors, information systems, machine learning, and other technologies to create a digital model that mimics the physical world. DTs are used in various industries, such as manufacturing, health care, and urban planning to simulate and optimize operations, monitor performance, and predict outcomes. In the context of smart spaces, DTs allow for real-time monitoring and analysis of physical systems, providing valuable insights into their behavior and performance. This information can be used to optimize energy consumption, reduce operational costs, and improve the overall sustainability of the built environment.

The field of action of a DT is vast, so the data can come from different information sources to improve the physical object's capacity. However, data fusion and homogenization are needed to transform data from various sources or formats to one single standardized format. This process ensures that data can be easily compared and analyzed across other systems and applications. Data homogenization involves mapping data elements from different sources to a standard format, converting data types, and applying consistent data validation rules. Homogenization is commonly used in data integration projects where data from various sources need to be combined into a single repository. It can also improve data quality by eliminating inconsistencies and errors.

The different stages that comprise the fusion and homogenization of data make it essential to have technologies that can handle these tasks in real time, facilitating the intake, processing, and understanding of the entire data flow. In this chapter, we have presented a case study aimed at showcasing a solution for this challenge. Our proposal relies on the FIWARE platform with its majority of GEs, such as Orion, Draco, Keyrock, and Wilma. Each GE is in charge of a specific process within a DT implementation, which helps not only in creating a DT but in the complete management of the entire life cycle and administration of the data. FIWARE is an open-source platform that provides a set of standardized APIs and data models for building and managing smart applications and services. Implementing a DT using FIWARE offers several advantages, including interoperability, scalability, flexibility, security, cost-effectiveness, and community support. First, FIWARE provides standardized APIs and data models, based on NGSI, that enable interoperability between different systems, allowing DTs to seamlessly integrate with other smart applications and services. Moreover, the FIWARE ecosystem of tools is designed to be highly scalable, making it ideal for implementing DTs that need to process large amounts of data and support a large number of users. It also provides a modular architecture that allows developers to build and deploy DTs in a flexible and customizable way, according to their specific requirements. FIWARE provides robust security features, including authentication, authorization, and encryption, ensuring that DTs and the data they process are secure and protected. Moreover, FIWARE is open source and free to use, making it a cost-effective solution for implementing DTs compared to proprietary alternatives. Lastly, FIWARE has a large and active community of developers and users, providing support and resources for building and deploying DTs.

Appendix: AirQuality input and output data

LISTING A.1 Air Quality raw data.


```
1
2 {
3   "2023-01-11T19:00:00.000Z":{
4       "aqi_SO2":1.02,
5       "aqi_PM10":1.8416666667,
6       "aqi_O3":1.2,
```

```
7        "aqi_NO2":1.9,
8        "modelled_SO2":0,
9        "modelled_PM10":0,
10       "modelled_O3":0,
11       "modelled_NO2":0,
12       "val_SO2":2.0,
13       "val_PM10":16.8333333333,
14       "val_O3":10.0,
15       "val_NO2":36.0,
16       "culprit":"NO2",
17       "aqi":1.9
18    }
19 }
```

LISTING A.2 Air Quality NGSI-LD formatted output.

```
1 {
2   "id":"urn:ngsi-ld:AirQualityObserved:Guadalajara:ES1537A",
3   "type":"AirQualityObserved",
4   "source":{
5      "value":"https://www.eea.europa.eu",
6      "type":"Property"
7   },
8   "location":{
9      "type":"GeoProperty",
10     "value":{
11        "coordinates":[
12           -3.1716,
13           40.6298
14        ],
15        "type":"Point"
16     }
17  },
18  "address":{
19     "value":{
20        "addressLocality":"Guadalajara",
21        "type":"PostalAddress",
22        "addressCountry":"ES"
23     },
24     "type":"Property"
25  },
```

Continued

LISTING A.2 Air Quality NGSI-LD formatted output—cont'd

```
26   "dateObserved":{
27      "type":"Property",
28      "value":"2023-01-11T19:00:00.000Z"
29   },
30   "@context":[
31      "https://uri.etsi.org/ngsi-ld/v1/ngsi-ld-core-context.
        jsonld",
32      "https://smart-data-models.github.io/dataModel.Environment
        /context.jsonld"
33   ],
34   "so2":{
35      "value":2.0,
36      "type":"Property",
37      "unitCode":"GQ"
38   },
39   "pm10":{
40      "value":16.8333333333,
41      "type":"Property",
42      "unitCode":"GQ"
43   },
44   "o3":{
45      "value":10.0,
46      "type":"Property",
47      "unitCode":"GQ"
48   },
49   "no2":{
50      "value":36.0,
51      "type":"Property",
52      "unitCode":"GQ"
53   }
54 }
```

References

[1] Y. Sahni, J. Cao, J. Shen, Challenges and opportunities in designing smart spaces, in: Internet of Things, Springer, Singapore, 2017, pp. 131–152, https://doi.org/10.1007/978-981-10-5861-5_6.

[2] C. Zhang, W. Xu, J. Liu, Z. Liu, Z. Zhou, D.T. Pham, Digital twin-enabled reconfigurable modeling for smart manufacturing systems, Int. J. Comput. Integr. Manuf. 34 (7–8) (2021) 709–733.

[3] X. Qin, Y. Gu, Data fusion in the Internet of Things, Procedia Eng. 15 (2011) 3023–3026.

[4] E. Rahm, H.H. Do, et al., Data cleaning: problems and current approaches, IEEE Data Eng. Bull. 23 (4) (2000) 3–13.

[5] A. Halevy, A. Rajaraman, J. Ordille, Data integration: the teenage years, in: Proceedings of the 32nd International Conference on Very Large Data Bases, 2006, pp. 9–16.

[6] F. Ridzuan, W.M.N.W. Zainon, A review on data cleansing methods for big data, Procedia Comput. Sci. 161 (2019) 731–738.

[7] G.C. Doubell, A.H. Basson, K. Kruger, P.D.F. Conradie, A digital twin system to integrate data silos in railway infrastructure, in: Service Oriented, Holonic and Multi-Agent Manufacturing Systems for Industry of the Future: Proceedings of SOHOMA 2022, Springer, 2023, pp. 142–153.

[8] J.J. Hierro, M. Reyes, K. Zangelin, I. Arias, C. Romero, A.J. López, M. Capdeville, G. Privat, S. García, M. Bauer, FIWARE-NGSI v2 Specification, FIWARE, 2018. https://fiware.github.io/specifications/ngsiv2/stable/.

[9] Context Information Management (CIM); NGSI-LD API, European Telecommunications Standards Institute, 2022.

[10] M. Grieves, Digital twin: manufacturing excellence through virtual factory replication, White Paper 1 (2014) 1–7.

[11] M. Singh, E. Fuenmayor, E.P. Hinchy, Y. Qiao, N. Murray, D. Devine, Digital twin: origin to future, Appl. Syst. Innov. 4 (2) (2021) 36. https://doi.org/10.3390/asi4020036.

[12] D. Jones, C. Snider, A. Nassehi, J. Yon, B. Hicks, Characterising the Digital Twin: a systematic literature review, CIRP J. Manuf. Sci. Technol. 29 (2020) 36–52.

[13] B.A. Talkhestani, T. Jung, B. Lindemann, N. Sahlab, N. Jazdi, W. Schloegl, M. Weyrich, An architecture of an intelligent digital twin in a cyber-physical production system, at - Automatisierungstechnik 67 (9) (2019) 762–782.

[14] L. Rosenthal, V. Stanford, NIST smart space: pervasive computing initiative, in: Proceedings IEEE 9th International Workshops on Enabling Technologies: Infrastructure for Collaborative Enterprises (WET ICE 2000), 2000, pp. 6–11.

[15] M.-O. Pahl, G. Carle, G. Klinker, Distributed smart space orchestration, in: 2016 IEEE/IFIP Network Operations and Management Symposium, 2016, pp. 979–984.

[16] E. Ahmed, I. Yaqoob, A. Gani, M. Imran, M. Guizani, Internet-of-things-based smart environments: state of the art, taxonomy, and open research challenges, IEEE Wirel. Commun. 23 (5) (2016) 10–16.

[17] M. Attaran, B.G. Celik, Digital twin: benefits, use cases, challenges, and opportunities, Decis. Anal. J. 6 (2023) 100165.

[18] L. Lattanzi, R. Raffaeli, M. Peruzzini, M. Pellicciari, Digital twin for smart manufacturing: a review of concepts towards a practical industrial implementation, Int. J. Comput. Integr. Manuf. 34 (6) (2021) 567–597.

[19] A. Nasirahmadi, O. Hensel, Toward the next generation of digitalization in agriculture based on digital twin paradigm, Sensors 22 (2) (2022) 498. https://doi.org/10.3390/s22020498.

[20] F.J. Villanueva, O. Acena, N.J. Dorado, R. Cantarero, J.F. Bermejo, A. Rubio, On building support of digital twin concept for smart spaces, in: G. Fortino, F.Y. Wan, A. Nurnberger, D. Kaber, R. Falcone, D. Mendonca, A. Guerrieri (Eds.), Proceedings of the 2020 IEEE International Conference on Human-machine Systems (ICHMS), IEEE, 2020, pp. 66–69. 1st IEEE International Conference on Human-Machine Systems (ICHMS), ELECTR NETWORK, September 7–9, 2020.

[21] M. Lenzerini, Data integration: a theoretical perspective, in: Proceedings of the ACM SIGACT-SIGMOD-SIGART Symposium on Principles of Database Systems, 2002, pp. 233–246.
[22] G. Fusco, L. Aversano, An approach for semantic integration of heterogeneous data sources, PeerJ Comput. Sci. 6 (2020) e254.
[23] J. Sreemathy, K. Naveen Durai, E. Lakshmi Priya, R. Deebika, K. Suganthi, P.T. Aisshwarya, Data integration and ETL: a theoretical perspective, in: 2021 7th International Conference on Advanced Computing and Communication Systems (ICACCS), vol. 1, 2021, pp. 1655–1660.
[24] T. Berners-Lee, J. Hendler, O. Lassila, The Semantic Web: a new form of Web content that is meaningful to computers will unleash a revolution of new possibilities, Sci. Am. (2001) 34–43.
[25] J. Conde, A. Munoz-Arcentales, A. Alonso, G. Huecas, J. Salvachúa, Collaboration of digital twins through linked open data: architecture with FIWARE as enabling technology, IT Prof. 24 (6) (2022) 41–46.
[26] M. Jacoby, T. Usländer, Digital twin and internet of things–current standards landscape, Appl. Sci. 10 (18) (2020) 6519. https://doi.org/10.3390/app10186519.
[27] OGC SensorThings API, Open Geospatial Consortium, 2019. https://www.ogc.org/standard/sensorthings/.
[28] The Asset Administration Shell: Implementing digital twins for use in Industrie 4.0, Plattform Industrie 4.0, 2022. https://www.plattform-i40.de/IP/Redaktion/EN/Downloads/Publikation/Details_of_the_Asset_Administration_Shell_Part1_V3.html.
[29] V. Charpenay, T. Kamiya, M. McCool, S. Käbisch, M. Kovatsch, Web of Things (WoT) Thing Description, W3C, 2023. https://www.w3.org/TR/wot-thing-description11/.
[30] L. Stojanovic, T. Usländer, F. Volz, C. Weißenbacher, J. Müller, M. Jacoby, T. Bischoff, Methodology and tools for digital twin management; the FA3ST approach, IoT 2 (4) (2021) 717–740.
[31] M. Kherbache, M. Maimour, E. Rondeau, Digital twin network for the IIoT using eclipse ditto and hono, IFAC Paper 55 (8) (2022) 37–42.
[32] A. Abella, A. Alonso, M. Bauer, J. Conde, L. Frost, F.L. Gall, B. Orihuela, G. Privat, J. Salvachua, G. Tropea, K. Zangelin, FIWARE for Digital Twins, FIWARE Foundation e. V., 2021.
[33] J. Conde, A. Munoz-Arcentales, A. Alonso, S. López-Pernas, J. Salvachúa, Modeling digital twin data and architecture: a building guide with FIWARE as enabling technology, IEEE Internet Comput. 26 (3) (2022) 7–14.
[34] J. Tarriño-Ortiz, J. Gómez, J.A. Soria-Lara, J.M. Vassallo, Analyzing the impact of Low Emission Zones on modal shift, Sustain. Cities Soc. 77 (2022) 103562.

CHAPTER

Secured digital-twin data service for the Internet of smart things

4

Mian Qian[a], Cheng Qian[a], Adamu Hussaini[a], Guobin Xu[b], Weixian Liao[a], and Wei Yu[a]

[a]*Department of Computer and Information Sciences, Towson University, Towson, MD, United States,* [b]*Department of Computer Science, Morgan State University, Baltimore, MD, United States*

1 Introduction

With the advance of sensing, communication, computing, and machine learning technology, the Internet of Things (IoT) has continued to evolve significantly and has impacted people's daily lives, empowering a variety of smart-world systems in energy, transportation, manufacturing, healthcare, city infrastructure, and others [1–14]. On the consumer side, large tech companies such as Amazon and Google released their voice assistants in 2016. Industrial 4.0, named Industrial IoT (IIoT), has enabled numerous industrial processes automatically, including the automobile, healthcare, manufacturing, and transportation sectors [4]. Those smart systems leverage IoT technology to enable real-time monitoring capability and identify potential security risks and issues. Meanwhile, due to the development of the IoT, the IoT concept has been growing rapidly (from things for intranets to worldwide deployed devices). Nonetheless, integrating large, connected networks, such as smart cities with smart transportation and other resources, requires a more extensive communication and network infrastructure to manage the massive number of devices and data. To enable the IoT in various smart systems, the IoT system must evolve from a limited and closed network system to a global information communication network infrastructure that connects various devices ("things") belonging to different application and management domains.

However, maintaining security in such a complex IoT digital world is complicated. Due to increasing reliance on IoT, most operations are performed remotely, which has led to the exponential growth of security issues such as unauthorized access, malware, denial of service (DoS), social engineering, false data injection, timing disruption attacks, etc. [15–24]. For example, in 2021, approximately one billion IoT-related attacks were reported [25,26]. Similarly, in October 2016, the DDoS attack launched by Mirai malware affected 145,000 IoT devices, which is the second

Smart Spaces. https://doi.org/10.1016/B978-0-443-13462-3.00019-4

most significant world-level attack. Thus, identifying threats that have occurred in the past, or determining potential risks to protect IoT systems from cyberattacks, is an urgent issue that must be considered. In our prior research, we used a three-layer architecture to taxonomize threats to a typical IoT system to gain a thorough understanding of these threats in IoT systems [27]. The three layers are the perception layer (also called the "things layer"), the network layer (mainly used for network connection), and the application layer.

There are various related tutorials, surveys, and research studies in IoT security [5,28–30]. For instance, researchers in Refs. [31,32], employed artificial intelligence (AI) technologies, especially the deep neural network (DNN), to monitor the IoT network systems and other systems. Researchers in Ref. [33] proposed a two-stage AI-based intrusion detection for software defined (SD) IoT networks. In their work, the first stage is called the "feature selection stage," which picks up features of the network flow to detect abnormal behaviors. The second stage is denoted as the "flow classification stage," which leverages a common ML algorithm, random forest, to classify the network behaviors into five classes. Both stages work together to detect intrusions through AI self-learning. In addition, Kumar et al. [34] proposed an IoT security framework that employs ML/DL (machine learning/deep learning) algorithms against DoS/DDoS attacks. Their experimental results revealed that deep learning achieved 99.5% and machine learning 99.9% accuracy, respectively. Chaabouni et al. [35] extensively surveyed ML algorithms using network intrusion detection system (NIDS) techniques for IoT systems, which focused more on monitoring network performance. Their research found that ML could achieve a significant success rate regarding the security and privacy of IoT. Research in Ref. [36] used ML to detect cyber-physical attacks on a wireless 3D printer. According to their study, the accuracy rate was extremely high. Nonetheless, adversaries can use fake data to train the ML model to affect the classification performance of an ML model [5]. For example, Wang et al. [37] used IR light (an invisible light) to implement an I-Can-See-the-Light Attack (ICSL attack) on Tesla Model 3 (an autonomous vehicle with IoT devices) to confirm that current autonomous vehicles treat invisible light as normal light, which leads to a wrong response. Likewise, Tariq et al. [38] used fake data and created two novel databases to simulate the Deepfake Impersonation (DI) attacks on popular face recognition platforms such as Naver, Microsoft, and Amazon (AWS). Using ML or other techniques directly on the physical system cannot guarantee the security of the IoT system in all cases. In this regard, we can leverage DT to perform a comprehensive evaluation of the virtualized physical system to carry out detection without affecting the original system.

Generally speaking, DT is a replica representing a virtual copy of physical objects, processes, and equipment. It makes it possible to analyze, predict, and optimize operations using historical and collected real-time data [39]. DT is becoming more popular because it can perform sophisticated simulations and optimizations, especially in process automation [40]. In this chapter, we propose using DT as a security solution (protection and defense) infrastructure on top of IoT systems to improve the security of the systems. Compared with other security tools, DT can be leveraged to detect potential threats and simulate the reaction as physical IoT assets to assist users with incident response evaluation. Furthermore, DT can consolidate different applications, operating systems, and platforms into one data-oriented service platform so that monitoring and control of the IoT system can be effectively supported.

In this chapter, our main contributions are summarized as follows:

- We propose a taxonomy of potential threats to IoT systems according to the three-layer architecture [27]. The three layers of a typical IoT system are the perception layer, also called the "things layer," the network layer, and the application layer. We choose common types of threats for each layer to explain why IoT systems are vulnerable to these threats and the limitations or problems they face in solving the issues by themselves.
- We provide some case studies to study effective countermeasures by using DT to protect the IoT system based on the aforementioned three-layer architecture. Similar to the taxonomy for threats, we choose some common types of threats for each layer. We identify countermeasures based on our literature reviews to show that using DT is a suitable and effective way to defend against investigated threats.
- We design a taxonomy for categorizing the problem space of leveraging DT to secure IoT systems concerning use, scope, and objectives and show some examples based on the existing research in our defined problem space. Furthermore, we use real-world SSH brute-force and DDoS attacks as case studies to demonstrate the efficacy of using DT as defense infrastructure to secure IoT systems. In order to characterize the problem space of using DT to protect IoT systems, we propose a three-dimensional framework. In our framework, we consider three factors: security usage (monitoring, attack detection, and threat prediction), security scope (physical components, DT for processes, or DT for the whole system), and security objectives (confidentiality, integrity, and availability).
- We discuss future study trends on how to develop a secured DT system. Since DT is a virtual version of an IoT system, it relies on IoT devices to collect, distribute, and analyze data from physical objects. Any threat to the IoT system can affect the DT deployed in the system. In addition to using DT to protect the IoT system, we need to consider how to protect DT by designing different mechanisms: for instance, using multiple DTs to establish a self-corrective environment, using suitable ML on top of DT to detect threats, or protecting the communication between DT and IoT system.

The remainder of this chapter is organized as follows: In Section 2, we introduce the three-layer architecture of IoT systems and investigate a taxonomy of potential threats to IoT systems. In Section 3, we introduce the security requirements in IoT and show some examples of mitigating attacks using DT. In Section 4, we design a taxonomy to define the problem space of applying DT to secure IoT systems from the aspect of security use, scope, and objectives. In Section 5, we use real threats in IoT systems as case studies to demonstrate how DT can protect IoT systems (e.g., mitigating the DDoS attack via machine learning techniques). In Section 6, we discuss future directions for developing secured DT systems. Finally, we conclude the chapter in Section 7.

2 Layer structure of IoT systems and threats

In this section, we first illustrate the layer structure of IoT systems and then discuss various threats associated with different layers.

2.1 Layer structure of IoT systems

IoT systems have grown dramatically in the past few years and have been widely used in the manufacturing, transportation, agriculture, healthcare, and utility management domains. A typical IoT system consists of three layers [27]: the perception layer, the network layer, and the application layer, as shown in Fig. 1. (i) *Perception Layer:* It is also often referred to as the sensing layer or the things layer, which includes all the things (sensors, edge devices, actuators, smart objects). (ii) *Network Layer:* It sometimes is known as the "transmission layer," linking the application and perception layers. Typically, it uses wired or wireless communication networks to carry and transmit the data collected by sensors in the perception layer. Additionally, it establishes connections among various networks, smart devices, and network devices. Unfortunately, it makes the new attack surface and develops new vulnerabilities in the system. (iii) *Application Layer:* It offers services to different applications as needed by the sensory data. Every application that makes use of IoT technology or in which IoT is implemented is defined by the application layer. It specifies IoT applications, including smart homes, smart cities, and smart health. In particular, in the context of smart homes and smart offices, the application layer faces many security problems. This is particularly true for smart devices with weak computing capabilities and limited storage space.

2.2 Potential threats to IoT system

Recently, an increasing number of threats have targeted IoT systems. For instance, in 2017, one of the leading global medical device manufacturing companies, St. Jude Medical, Inc., recalled 465,000 pacemakers due to their vulnerability to IoT threats [41]. Also in 2017, thousands of baby heart monitors were hacked. In addition, attacks against car systems, such as the Jeep attack, can allow an adversary to remotely control the engine system [42]. In the following, we briefly describe potential security threats to IoT systems based on the three-layer architecture.

2.2.1 Threats to perception layer

The perception layer consists of sensors and smart devices that provide services such as location, data recording, security, etc. The following are typical security risks related to the perception layer: replay attacks, eavesdropping, timing attacks, and node capture, among others [43,44]. For instance, some IoT devices are vulnerable because they lack the necessary built-in security controls to defend against threats. The adversary could try to compromise a node through malicious code by physically capturing sensors and extracting all information, including crucial details from their

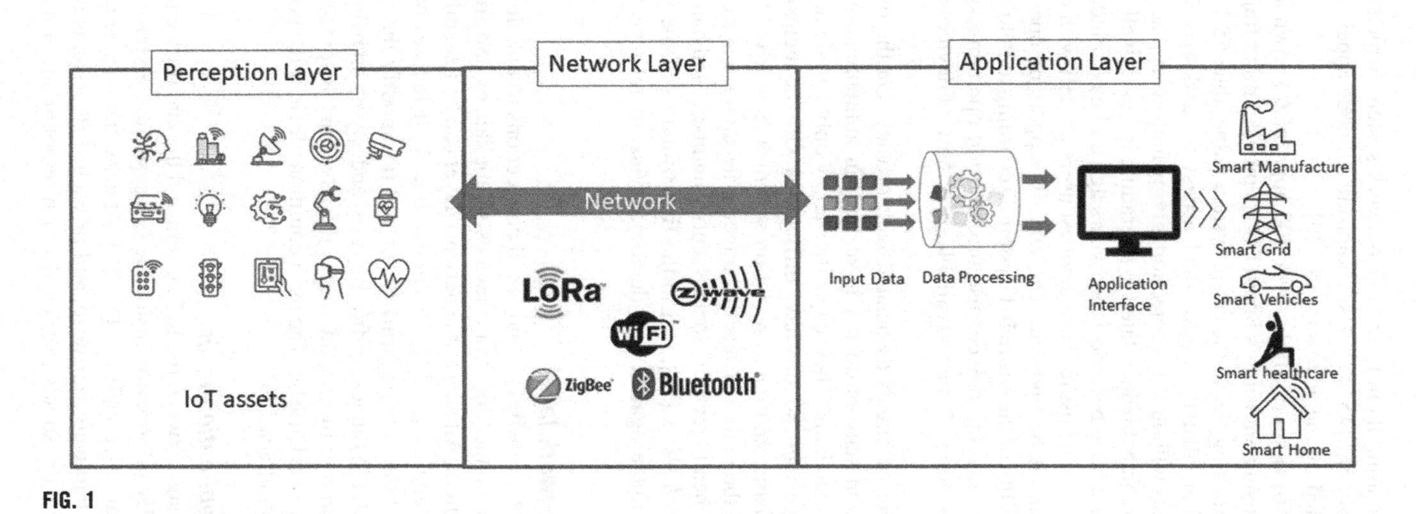

FIG. 1

Three-layer architecture for IoT system.

memories, then propagating them to the IoT network system, which can be further used to launch other attacks [45–47]. The compromised node appears as part of the system but is controlled by the adversary [48,49].

Using Samsung [50] as an example, over 500 apps, also known as SmartApps, have been developed using Samsung's SmartThings programming frame. Numerous security flaws in SmartThings leave the entire system vulnerable: 55% of SmartApps requested device actions that they never had access to, and 42% of SmartApps responded with requests without any protection, giving hackers full access to the host device. There are also some threats related to upgrading issues as well. In this regard, some IoT devices could have been designed years ago and have outdated operating systems. In order to continue using these devices, users are left with outdated operating systems that cannot be updated. The end of support of operating systems increases the vulnerability of the entire IoT system. According to Palo Alto Networks [51], 83% of medical imaging devices use unsupported operating systems. When security patches for Windows 7 are unavailable, over 56% of devices are vulnerable to cyberattacks.

There are a number of threats associated with privacy. On the one hand, most low-power consumption sensors do not have built-in authentication and energy-saving encryption mechanisms. Their capacities allow only certain and limited feature functions to be executed, so they have difficulties with the inclusion of security controls and mechanisms, and data protection schemes. Sensitive data leakage is another issue. One of the main reasons that adversaries attack a system is to obtain sensitive data, such as health records, social security numbers, and bank information. According to *Irish Tech News* [52], just in the first 6 months of the year 2021, 1.5 billion attacks were made against IoT devices, trying to retrieve sensitive data from them.

2.2.2 Threats to network layer

The network layer includes wireless and wired connections to link devices. In addition, it consists of the connection from devices to appliances and from devices to servers or edge computers. Attacks to the network layer can be generally categorized as passive and active attacks [53,54]: (i) *Passive attack:* It focuses on "eavesdropping" or "monitoring" the network traffic, but will not modify the network itself. Man-in-the-middle (MITM) attack, traffic analysis, and spying all belong to this type of attack. (ii) *Active attack:* In such an attack, the adversary leverages the attack to make changes to the network traffic. The most common traffic is IP spoofing, MITM, DoS attack, and DDoS attack [55].

2.2.3 Threats to application layer

The application layer sits between the IoT system and the end user, enabling the end user to access or modify information inside the IoT system (e.g., false data injection attacks to affect critical functionalities) [56–59]. Attacks in this layer can be grouped into three categories: hardware, software, and privacy. IoT systems use the remote control to get or send data, so adversaries also use remote control techniques, like

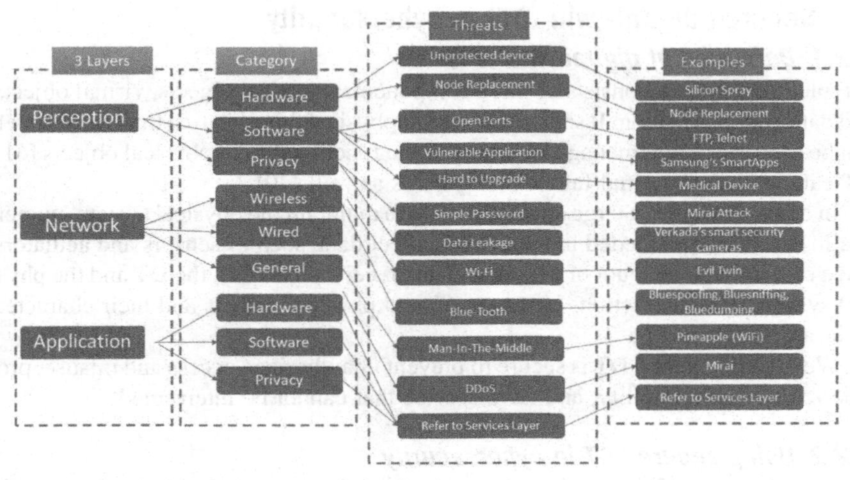

FIG. 2

Threats in IoT system.

port scanning and IP spoofing, to look for computer systems that are vulnerable to attack. Therefore attacks such as man-in-the-middle (MITM), malicious code attacks, and cross-site scripting are representative attacks in this layer. In addition to these widely used attacks on IoT systems, other attacks, such as IoT ransomware and shadow IoT, are also becoming common. Based on our review of the potential attacks on three layers, we propose a framework as shown in Fig. 2 to demonstrate the relationship between IoT layers and security threats.

3 Secured digital-twin for IoT system

3.1 Needs for cyber defense

Recently, the attack surface has rapidly expanded due to the rapid adoption of IoT devices, edge computing, and next-generation wireless networks, offering adversaries more opportunities to target devices. As a result, companies have begun to develop tools to protect their digital assets in cyberspace. However, cybersecurity professionals face workload overflow with the continuing growth of cyberattacks and more tools developed for cyber defense. According to the State of Security report for the year 2021 [60], IT professionals need to handle an average of 11,000 alerts per day. Due to the shortage of labor resources, many warnings are ignored or bypassed, which leaves potential risks in the IoT system. Thus we need a platform that can consolidate all the different devices into one operating system, automatically detect threats, and simulate solutions before implementing them on the physical network. These systems will improve the efficiency of cyber protection, strengthen the entire IoT system, and reduce resource usage during the solution implementation.

3.2 Secured digital-twin (DT) in cybersecurity

3.2.1 Definition of digital twin (DT)

A typical DT system consists of three components: physical objects, virtual objects, and data communication. It shows the state of physical things in real time. DT models can be used to predict, manage, or optimize the functionality of physical objects [61]. DT can address abnormal failures in systems as well [40].

In order to acquire a large amount of data to maintain the physical system, numerous IoT devices are needed in the DT to collect data, such as sensors and actuators, and a considerable amount of data is sent and received between the DT and the physical system. The interactions between DT and physical objects and their characteristics are shown in Fig. 3.

We assume that the DT is secure to prevent unauthorized access and misuse, provide reasonable reliability, and have service that cannot be interrupted.

3.2.2 Using secured DT in cybersecurity

A typical secured DT system is not a carbon "copy" version of its physical device. In addition to the real-time reflection, it has two distinct characteristics as follows: (i) *Interaction:* Data must be transmitted between physical and DT all the time. (ii) *Iteration:* DT not only shows the situation of the physical object, but also needs to be able to modify or control the physical objects. Therefore, a number of research studies [62–67] have been turning to DT to assist with incident response, such as security simulations, threat detection, and response evaluations.

Over the last decades, DT has undergone various developments. It has been implemented in different areas of human endeavors, such as the healthcare field, the transportation sector, the building industry, the manufacturing sector, and the

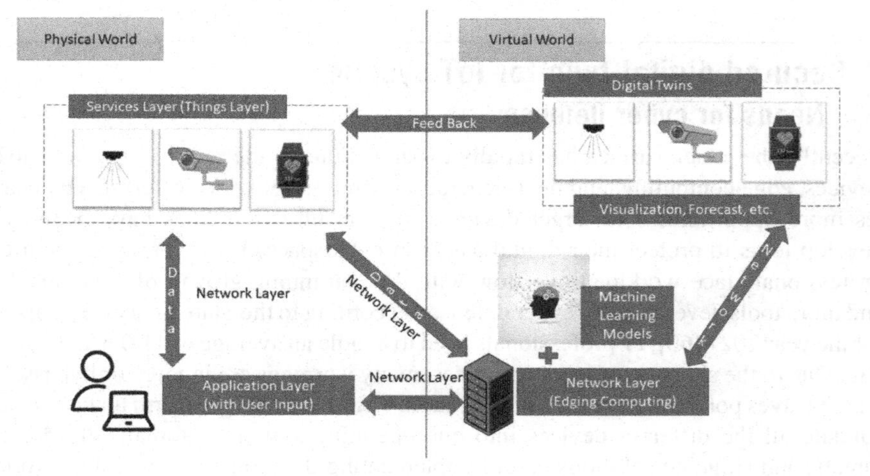

FIG. 3

Relationship between DT and physical objects.

aviation industry, among others. However, when it is operational, little is achieved regarding its protection and privacy.

In our recent research [40], we classified different threats to DT layer by layer. Then, we examined how to use other supporting technologies (e.g., simulation, blockchain technology [68], modeling, intrusion detection method [17], and emulation [69,70]) to secure DT-based CPS and provided a workable defensive mechanism called the Secured DT Development Life Cycle (SDTDLC). Similarly, when studying the DT of wireless systems, Khan et al. [71] categorized threats to DT into the DT (edge/cloud twin) layer and the physical interaction layer. Some of the protections suggested are encryption of twin control information, on-demand virtualized security functions for the twin layer, and continued motoring of twin control information as security solutions for the twin layer. At the same time, novel forensics to analyze the attacks, a lightweight devices authentication protocol, and encryption for wireless security were provided as tools for protecting the physical interaction layer.

According to research conducted by Maillet-Contoz et al. [72], the identified security weaknesses of the IoT systems concern end-device security. However, testing the integration of the security layer on various IoT systems is difficult due to the heterogeneity of devices and protocols. Thus DT was proposed to make it possible to quickly integrate, test, and validate the devices' security to provide end-to-end protection of IoT systems. Karaarslan and Babiker [73] investigated several security threats and challenges at the DT layer, data transmission channels, and ML models. In addition, potential preventative methods (hashing, network resiliency, antimalware, algorithm robustness enhancement, etc.) were provided.

Based on the security demands of the DT for industrial IoT, Feng et al. [74] employed game theory to characterize the interaction process between adversaries and defenders in the network. Additionally, they proposed a targeted allocation strategy for network security protection resources in a DT environment. The approach adopted to analyze network security vulnerabilities through the lens of game theory offers new insights into how to recognize and efficiently remedy security flaws in DT-based networks.

3.3 Building DT for IoT systems

The development of DT for IoT systems will offer data gathering, processing, visualization, and device management platforms with high connectivity. This makes it possible for IoT solutions to be developed, managed, and scaled quickly. It will also increase performance, offer scalability, and provide fault tolerance.

Recall that adversaries against IoT systems use all types of cyberattacking tools and mechanisms, including man-in-the-middle (MITM) attacks, privilege escalation attacks, brute-force attacks, malware, and DoS/DDoS attacks. One of the known DDoS attacks is Mirai malware [75]. When ARC-based smart devices are infected with the malware, a network of remotely controlled devices, which are called "bots" or "zombies," is created. As a result, DDoS attacks are frequently carried out via this network of bots. In early October 2016, Mirai's developer released the source code.

According to reports from various locations [75–77], 164 countries and more than 300,000 IoT devices are affected by the Mirai malware, especially CCTV cameras.

One of the pertinent research questions that needs to be asked is whether DT can be used to protect the IoT system. In the following, using a three-layer IoT architecture [27], we perform literature reviews to gain insight into applying the DT security approach to a complicated IoT system.

3.3.1 Perception layer

Since some smart objects (e.g., RFID tags, sensors, and healthcare devices) in IoT systems have a limited computing capacity, restricted energy, and small memory storage, lightweight encryption algorithms are widely used in the IoT systems. However, these low-energy solutions are vulnerable to different attacks. Unprotected devices, or devices that use simple or no passwords, that have unused open ports, or vulnerable applications installed on the devices, are common security issues that IoT assets face. Using DT to duplicate physical devices can help the designer identify IoT assets with associated weaknesses in the perception layer.

Several recent studies used DT to detect abnormal performance from IoT assets to identify potential threats to physical devices. For example, two DTs were used in a DT for an additive manufacturing study [78]. One is called the DT Core, which is the expected result that the physical assets are required to achieve, similar to a target or a baseline needed to be met for the whole system. The other one is DT Function, which continuously connects with physical IoT assets to change with various input data from physical devices. In their study, DT Core acts like a discrete, static DT system, while the DT Function serves as a continuous change system. First, the DT Function checks with the DT Core to identify potential assets whose performance is far from the baseline (DT Core). Then, the DT Function controls activities of the physical IoT devices, such as restarting or shutting down the device, to bring the performance back to normal.

Chhetri et al. [79] proposed the use of IoT sensors on the side channel of the manufacturing line to build a real-time DT product. They compared the DT product with the physical system to capture abnormal situations such as cyber-physical parts failure, the physical domain as raw materials input, and overall environment changes. Their study concluded that such a method could detect side-channel emissions, for instance, vibration, acoustic, or magnetic data, to avoid information leakages through the side channel.

These studies use the DT as the baseline or real-time reflection to compare with another DT or physical system to find any deviation, and then to identify abnormal behavior of the manufacturing production line. DT can be considered an authentication tool for IoT assets. Most IoT assets do not have authentication features. Thus adding authentication feature functions on their DTs, such as password, fingerprint, or two-factor authentication, allows users to reach the physical IoT assets only if they can pass the authentications from the DT side. In this way, IoT assets can be protected by their DT without adding extra investment on physical protection methods, such as locks, safety cameras, etc.

3.3.2 Network layer

IoT sensors utilize various network protocols for remote access, such as WiFi, Bluetooth, ZigBee, RFID, and Z-Wave. However, easy-to-access means easy to be hacked, especially for the ad-hoc protocol used in IoT systems, which consists of multiple connected devices, making all the network routes complicated and complex, and making it difficult to detect malicious attacks through remote access protocols.

One of the solutions to secure the network layer is to add IDS/IPS in the demilitarized zone (DMZ) to observe the network system. However, this solution might slow down the transmission speed, and the failure rate needs to be considered. Another solution is to identify the trusted devices. Only trusted devices will be added and allowed to transfer data. This solution requires human beings to monitor the latest network system over time to ensure the accuracy of the trusted devices, and the devices need to be free of malicious code. Recently, block-chain-based end-to-end communication protection has also been studied.

In order to evaluate the security performance of the network settings or simulate the traffic that occurs in the network, numerous studies have been developed using DT to simulate the network environment. For example, Baert et al. [64] used DT to set up a Bluetooth mesh to monitor the IoT networking system. With traditional methods to monitor the network management system, such as using tools to measure the bandwidth, or examining the node's interface, or installing applications on devices to obtain the network performance, the time required to measure the performance is relatively high, and these common methods have trouble providing comprehensive views of network performance. By using the DT to simulate the Bluetooth mesh, more accurate and complete insights into the network can be provided, which can be used to improve the network settings quickly and effectively.

Hong et al. [66] conducted research on using DT on data center networks, which collect and distribute a large volume of data transmitted among IoT devices. Considering various equipment from diverse vendors with different configuration interfaces, using vendor-specific CLIs (command-line interfaces) could not satisfy the complex and multifunction services that the data center network requires. By using DT to model the entire network system, the network designer can simulate the deployment among various devices with a unified DT model, identify potential issues, and efficiently retrieve the data for the users. Instead of manually typing in the CLIs, the DT can send data to a unified application and then automatically collect data, manage the data distribution among servers, and send out alerts to IoT devices through a centralized method.

In addition to using DT to simulate the IoT network settings, we can use DT to simulate potential attacks from the network to find out weaknesses of the whole network settings, and then look for solutions to improve the network performance. Table 1 describes some examples of attacks in the network layer and countermeasures to use DT technologies to mitigate the threats.

Table 1 Network layer threats and countermeasures.

Common attack	Description	Type of threat	Countermeasures
MITM attacks	Man-in-the-middle attacks can be passive attacks that eavesdrop on the communication between hosts or active attacks that capture the information and modify or intercept the data.	Passive/active	Solution 1: Blockchain [80] the communication between DT and physical devices. Solution 2: Use DT to set up a trusted network system [81]. Only trusted devices can be used to transmit data.
DoS/DDoS attack	Denial of service (DoS) is an attack to flood a server with network packets. A DDoS attack means a group of affected computers targets a single system with a DoS attack.	Passive/active	Solution 1: Use DT to examine the trusted devices [81]. Only trusted devices will be allowed to transmit data. Solution 2: Adding IDS/IPS [82–84] with DDoS detection with DT to reduce the chances of DoS/DDoS attack.
Sinkhole attack	A compromised node tries to attract network traffic by a fake routing update. It can be used to launch other attacks such as selective forwarding attacks, dropping or altering routing information, or even acknowledgment spoofing attacks	Active attack	Solution 1: Using DT to identify trusted route solutions [85] to select only trusted routes to transmit the data. Solution 2: Using IDS [81,82,85] in the DT to simulate the process to filter out potentially compromised devices.

3.3.3 Application layer

In a typical DT frame, DT normally uses cloud services to store the input data and then transfer the data to the associated application. The application layer is the place where the user can interface with the computer and directly control the DT. Thus snooping the password, installing malware to computers or applications, opening the backdoor, or even using ransomware to lock the computer become the methods that adversaries use to attack the whole DT system. Table 2 lists a few typical instances of attacks on the application layer along with the protective measures that are taken.

Just as Shi et al. [63] mentioned, DT and SVE (simulation, virtualization, and emulation) are useful technologies to duplicate physical systems. Among these four technologies, DT is a relatively new concept. Compared to simulation, which only focuses on statically duplicating the physical system, DT can replicate the physical system flexibly and continuously. Meanwhile, DT can be involved in

Table 2 Application layer threats and countermeasures.

Common attack	Description	Countermeasures
Malware	Malware means malicious software, a file or code that explores, steals, or breaks the computer operating system, applications, etc. Malware comes in different variants, such as computer viruses, worms, ransomware, keyloggers, Trojan horses, spyware, rootkits, and mobile malware [86,87]. Mirai is one representative IoT malware that appeared in 2016 and is used to attack IoT systems.	Solution 1: Add IDS or IPS or antivirus software in computer systems to detect well known malware. Then use DT to communicate/control the smart devices. Solution 2: Add a gateway between IoT devices and servers [88] to prevent malware from impacting the server and then affecting other connected devices (can be applied to the edge computer on the service layer side). Also, use a DT to monitor the process and ensure that everything is on track.
Password leaked	Password might be stolen by installed malicious software such as key logger, or be too simple and guessed by a brute-force attack, or make use of the default username as a password or use the same password for a long time, increasing the chances of being leaked.	Solution 1: Encrypt data from device to device [81,88] to avoid data leakage. The communication between DT and IoT assets needs to be encrypted. Solution 2: Use DT to encrypt the message from DT to the IoT device and monitor the performance of the IoT device. If abnormal activities occur on the physical device, it will either upgrade the device or even drop the device.
Phishing emails	Using email as a medium to fool users into clicking a fake link to download viruses into the computer. Email is an inexpensive and easy method to distribute viruses. Therefore, to spread viruses, phishing emails are very common.	Solution 1: Continue to educate users not to click linkage in suspicious emails. Solution 2: Install antivirus to filter out known phishing emails from edge computing in the red services layer or at the DT edge computer side [88,89]. Considering the response speed, installing it at the application layer is better. We can add an additional edge computer to the DT side to filter out potential risks before moving there.

the whole life cycle of the real system. The benefits of using DT to simulate physical systems include but are not limited to the following: (i) interacting with the physical part in "real-time," (ii) duplicating the physical system with high-fidelity instead of just "copying" the system without any flexibility, (iii) being used in the whole life cycle instead of only the research and development phase, and (iv) the open source used in DT, such as eclipse ditto, can reduce the development cost for the simulation area.

4 Taxonomy for DT-based IoT security

In the previous section, we reviewed how to use DT to secure IoT systems based on IoT's three layers: the perception layer, the network layer, and the application layer. We now design a three-dimensional framework to categorize the problem space of leveraging DT to secure IoT systems. In our framework, we consider the following three aspects: security use (monitoring, attack detection, and threat prediction), security scope (physical components, DT for process, or DT for overall system), and security objectives (confidentiality, integrity, and availability). In the following, we describe these three dimensions in detail.

4.1 Security use

During the past decade, the study of DT has grown enormously. Existing research efforts emphasize three unique values of using a DT, which are visualization, simulation, and prediction.

4.1.1 Visualization

DT is created to "represent" real-world entities and processes in a digital virtual world. It uses software programming, geolocation, big data-driven methods, and others to duplicate physical assets. Using Tesla as an example [90], Tesla is using the DT application "Thinkwik" to reflect any car produced by Tesla. Any real-time issues with Tesla motors will be recorded and then fixed by downloading over-the-air (OTA) software updates.

4.1.2 Simulation

Simulation tools are widely used in DT technology to reflect the impact on real-world processes or assets. Using smart manufacturing as an example, in a smart manufacturing system, numerous cyber-physical production systems are used in the production line with various materials, and at different times to launch the production line. Using DT reflects the physical process or products and demonstrates the input conditions to the manufacturer's production line. Users can use DT to estimate outputs and compare them with real situations, to identify potential outlier data through the production time. Baert et al. [64] used DT to the Bluetooth mesh in a data center to simulate network settings and then compared results with real-world's data to identify abnormal situations.

4.1.3 Prediction

Compared to other simulation tools, DT not only simulates real-world settings, but also can predict "future" things that might occur to physical devices. For example, ABB [91] used DT in marine DC grid systems, in which a DT was developed based on the initial settings of a physical device. Since the grid systems would be dynamically changed by the wind, wave, and current, several DT models were developed to analyze the impact on the grid system. DT not only recorded the benchmark (initial settings) of the grid system, but also provided analytical output by a modified or slightly different system. In this way, potential threats to the grid system can be identified.

Based on offered visualization, simulation, and prediction capabilities, DT can be used to conduct monitoring, attack detection, and threat prediction.

4.1.4 Monitoring

DT's virtualization of the physical assets enables its capability to monitor the IoT system. Based on different ways to design a DT, monitoring can be implemented at a remote access domain such as the cloud or on-site within the same building with an IoT system. For example, Tesla's DT "Thinkwik" [90] is implemented in the cloud to monitor all vehicles produced by Tesla. In the Rogage et al. [92] study, DT was used to monitor on-site large infrastructure projects, such as plant construction, to avoid potential risks that might hurt human beings. ABB is using on-shore DT [91] in their ABB Marine & Ports business to monitor the fleet operations centers.

4.1.5 Attack detection

Attack detection means continuously monitoring and identifying abnormal activities within IoT systems. Once an attack is detected, the detection system will send alerts to the administration systems to schedule defense resources to contain the attack. Three types of intrusion detection methods are widely used in IoT systems: signature-based, anomaly-based, and hybrid attack detection [93]. The signature-based detection method detects attacks by comparing them with the recorded history of existing attacks. The signature-based method's benefits are fast response and high accuracy. However, zero-day malware cannot be detected by signature-based systems because it is not archived in history records. Anomaly-based intrusion detection uses the pattern of the threats to identify threats. Using anomaly-based, also called behavior-based. Detection can identify zero-day attacks, but it relies on the models to analyze the behavior patterns, which might cause a low accuracy rate or slow response [94]. Hybrid intrusion detection combines signature-based and anomaly-based schemes to recognize threats.

DT's virtualization and simulation can be used in the IoT attack detection, especially in the hybrid detection. When DT is used to detect attacks, at least two statuses of DTs are required. One is used as a baseline for IoT systems to represent the status before attacks. The other one is real-time DT that either uses simulation to reflect things that changed after the attack or uses virtualization to represent the latest status after the attacks. Attacks or anomaly activities can be easily detected by comparing the baseline with the real-time DTs. For example, Yigit et al. [95] considered the scenario in which one DT's status duplicated the physical settings of the core network settings and the other DT status used a Yet Another Next Generation (YANG) model, automated feature selection (AutoFS) module, and ML to simulate new variable input factors to the network system. After comparing two models, a DDoS attack can be effectively identified for IoT systems. Likewise, Varghese et al. [96] designed a DT-based intrusion detection system for industrial control systems. In their proposed method, the DT module within the IDS system is used to collect data from IoT assets, plus the simulator modules, to simulate 23 types of well-known attacks, such as network DoS attacks, command injection attacks, and naive measurement modification, and to simulate the impact and output for the system.

4.1.6 Threat prediction

DT can be used to predict potential threats to the IoT system. Similar to simulation, multiple DT statuses are used to predict potential threats. Previously, we introduced ABB's DT in marine DC grid systems [91], which used a DT module to predict potential threats to marine DC grid systems. In the Yuan et al. study [97], DT was used to assist the federated learning model to cluster island-sensitive healthcare information. In their protocol, DT is used to create virtual models for federated learning with virtual client selection. Different clustered centers cluster data based on different priorities. Different DTs are used to reflect this priority, ensure that nothing is missed or skipped during the clustering process, and then share the overall results with the global DT for information sharing. In their model, multiple DTs can work together to predict any potential risks raised by federated learning itself to avoid information loss. Likewise, Qiao et al. [98] proposed a five-dimensional DT model to combine design parameters, process data, manufacturing process data, and the feedback process into their DT predict model. From a security point of view, the DT model of the tool system can monitor processes, detect anomalies, send alarms, and use various variables to predict potential risks to securely monitor and defend the system.

4.2 Security scope

DT can be classified by the scope for which it was applied, meaning the security target, from a specified component or physical assets to the production process of an overall system, or even multisystem level. For instance, General Electric (GE) [99] categorized its DT portfolio into three levels: asset DT, process DT, and network DT. According to GE, their asset DT works with specified physical assets, and the network DT is used to monitor the real-time changes to the grid. The DT creates models to predict the impacts on its process and then optimizes the processes. From a security perspective, we can classify DT into three scopes: DT for physical assets, DT for network, and DT for the overall system. Fig. 4 shows the classification in DT scope and Table 3 presents the classification based on DT scope that includes some examples of research efforts.

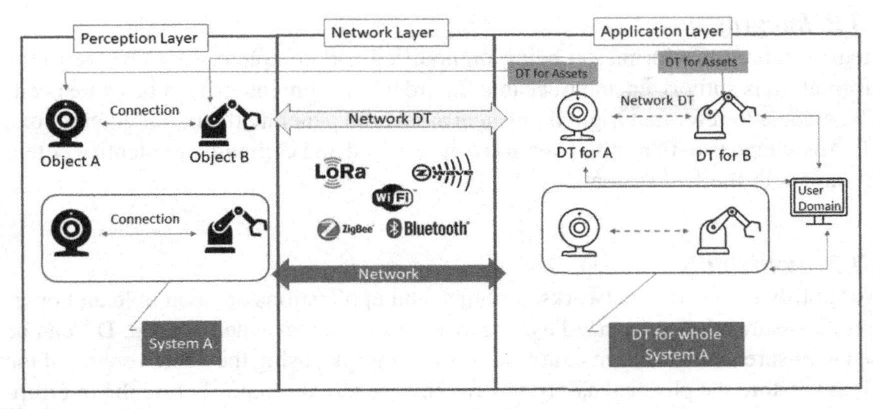

FIG. 4

IoT scope.

Table 3 DT with three levels of scopes.

DT scope	DT examples	Purpose
Physical components	1. DT for manufacturing side channel [79] 2. DT for large on-site infrastructure [92] 3. DT "Thinkwik" used by Tesla to monitor the engine [90]	Virtualization, simulation
Network	1. DT for Bluetooth Mesh [64] 2. DT for data center network [66] 3. Yigit's YANG model with DT [95]	Virtualization, simulation, prediction
Overall system	1. DT for additive manufacturing [78] 2. DT to set up trusted devices [81] 3. DT for marine grid system [91]	Virtualization, simulation, prediction

4.3 Security objectives

Confidentiality, integrity, and availability, also known as the CIA triad, are used in our model to consider the security objectives.

4.3.1 Confidentiality

Confidentiality means that the information is accessible only by authorized individuals. Recalling the Yuan et al. study [97], the federated learning model can categorize information into sensitive and nonsensitive groups. For sensitive data, it is kept in the organization, and stored in an "island" way to ensure only authorized users have access to the data. DT is used to assist the federated learning model to cluster island-sensitive healthcare information to ensure that nothing is missed or skipped during the clustering process. Another mechanism to ensure confidentiality is authentication. For example, Lopez et al. [100] studied the DT used in the smart grid system, which added security policies to provide authentication.

4.3.2 Integrity

Integrity refers to information being unmodified and complete. Any change to the information is authorized and traceable. In order to ensure integrity, a baseline needs to be created to refer to. Original physical assets can generate the initial version for a DT. Any changes to that initial version can be saved and compared to identify potential threats to the IoT system.

4.3.3 Availability

Availability means that networks, systems, and applications are available and operating. It assures that authorized users can access resources when needed. DT can be used to ensure availability in many ways. For example, using the latest version of the DT can restore the physical assets and systems back to the status before the interruption. DT can be used to monitor the network systems to ensure everything runs properly as well. Furthermore, DT is chosen to assist IDS to detect DDoS attacks to secure the IoT system. Table 4 demonstrates the classification based on the CIA triad, and Fig. 5 shows an example of mapping the existing works using our designed 3D framework with respect to DT use, security scope, and security objectives.

5 Proof of concept

We now choose two representative threats, SSH port brute-force attacks and DDoS attacks, to provide proof of the concept and to demonstrate how DT can be used to secure IoT systems. Note that we assume the DT is in place and is used to monitor the security of IoT systems.

Table 4 Digital twins for CIA triad.

Attack type	Attack example	DT application example
Confidentiality	Brute-force attack	**1.** DT for authorization, such as DT for federated learning model [97]. **2.** DT with blockchain to protect communication [40]
Integrity	Man-in-the-middle attack	**1.** DT for the federated learning model [97] **2.** DT to set up the trusted device [81] **3.** DT for additive manufacturing with DT Core and DT Function [78]
Availability	DDoS attack	**1.** DT for IDS [82,95] **2.** DT for additive manufacturing with DT Core and DT Function. Once DT Function is far away from DT Core, DT Core will correct the data from DT Function [78].

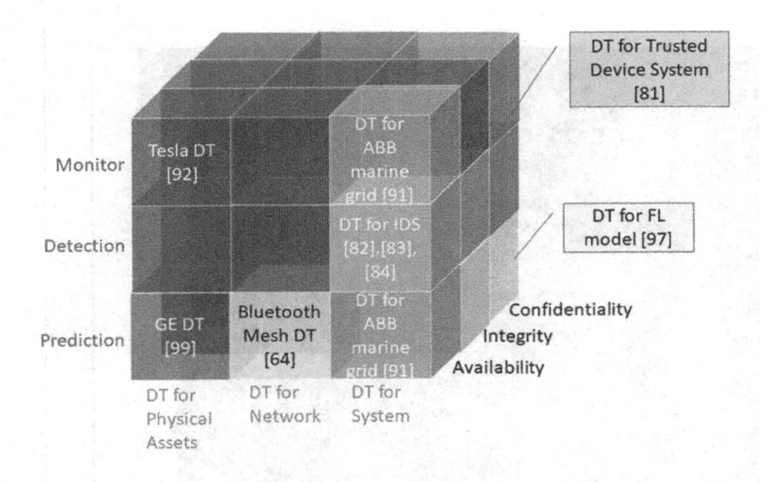

FIG. 5

Problem space for DT-based IoT security.

5.1 Using DT to handle SSH port brute-force attacks

Since Linux/Unix-based operating systems are widely used in IoT devices, such ports are vulnerable to SSH port brute-force attacks. This attack focuses on guessing user login credentials and breaking into the system. Figs. 6 and 7 demonstrate SSH brute-force attacks worldwide.

Since the attack can target all users running Linux/Unix-based systems with the SSH port open on the public network, we can use the following mechanisms to defend against the attack [16]: (i) *Set Up Access Control:* Linux/Unix-based systems all have built-in firewalls. In this case, we can configure firewall rules to defend

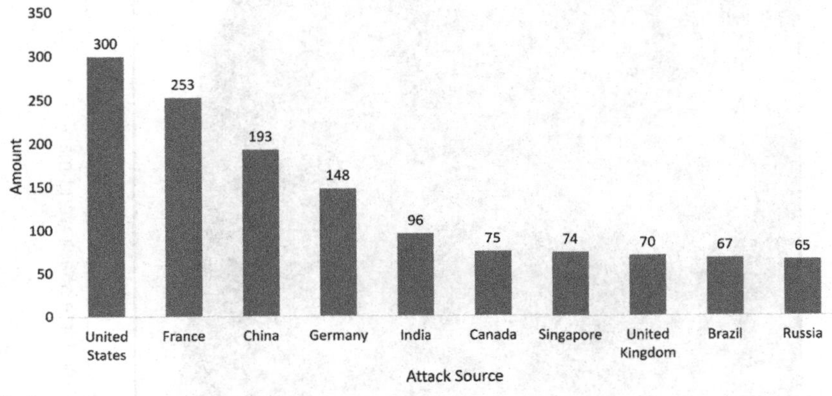

FIG. 6

SSH brute-force attack.

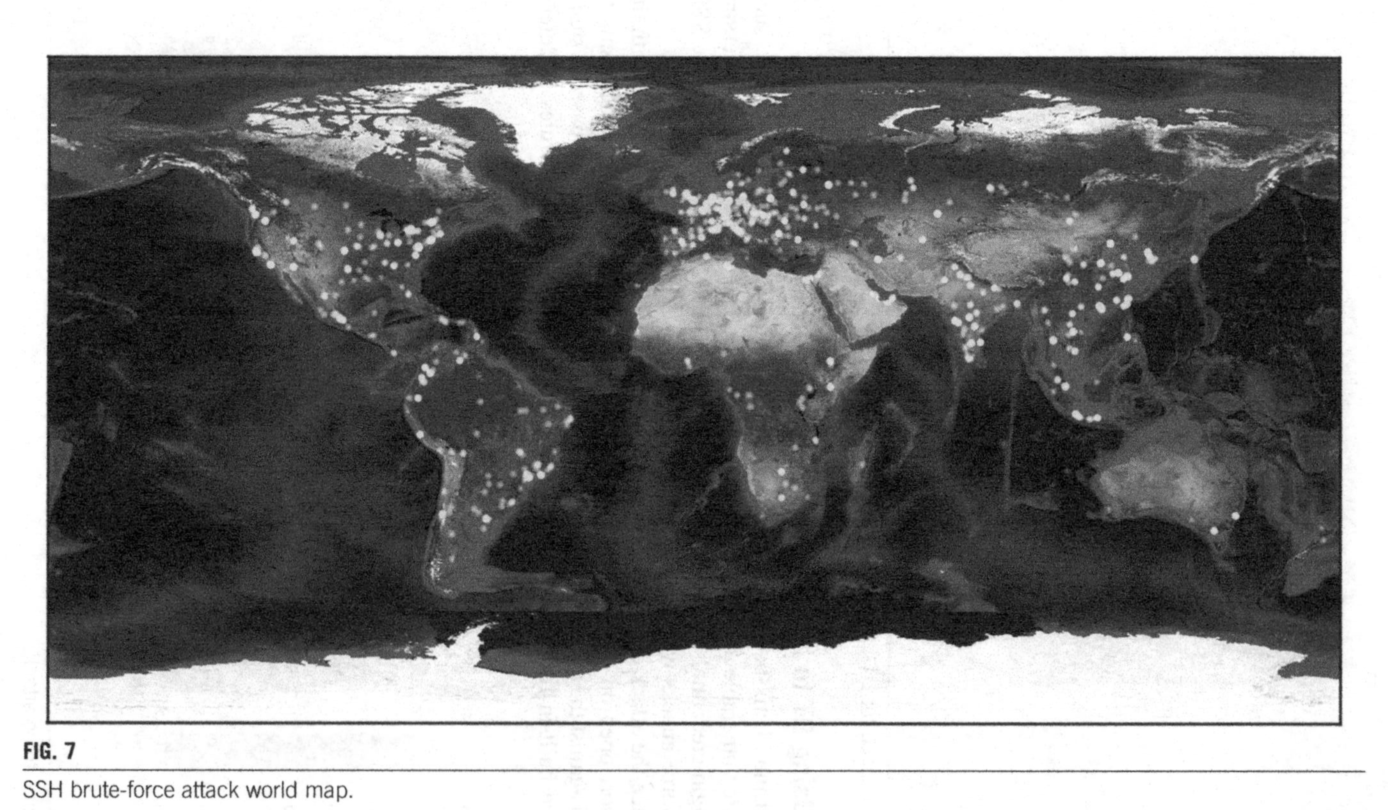

FIG. 7

SSH brute-force attack world map.

against attacks. For example, we allow specific users to access specific ports with the command "iptables -A INPUT -p tcp -s 192.168.0.2 – dport 11211 -j ACCEPT". This command can make the user whose IP is 192.168.0.2 access port 11211, and if there are no other rules, other users cannot access port 11211. Intrusion detection systems can be used as defensive strategies against the SSH brute-force attack, including fail2ban. (ii) *Change The Default Port:* Note that changing the default port is the easiest way to solve this problem. Although it is still possible for an adversary to find the target device's SSH port via a port scan, it can reduce SSH port attacks by scripted bots. *URPF, often called Unicast Reverse Path Forwarding:* URPF provides the essential function of the router, which can deal with network attacks based on IP spoofing attacks. After the router receives the data packet, it will check the validity of the source IP address. If it is a legitimate address, the router will forward it. Alternatively, it will simply drop packets.

We consider utilizing DT with current defense methods to protect the IoT system. DT builds a cyber replica of an IoT system and model, meaning how many devices exist in the network with SSH ports open. Then, based on this model, DT can notify the IDS system or a firewall to mitigate the impact.

5.2 DT with ML-based detection

On top of DT, ML can also protect IoT systems. The DDoS attack is one of the common attacks in the network layer. Compared to SSH port brute-force attacks, DDoS attacks scan the vulnerability of target devices as well. Nonetheless, the DDoS attack targets a limited resource of an Internet entity, such as the host and network [55]. Once an excessive number of requests consume resources, the end host will be preoccupied with responding to these requests and will be unable to process the other legitimate requests. For example, Mozi malware, which first appeared in 2019, affected about 438,000 hosts (routers and cameras), accounting for 89% of all IoT attacks discovered by IBM in 2020 [29].

To deal with DDoS attacks, we consider deploying IDS/IPS gateways between IoT devices and servers to detect malware. In addition, the gateway can be deployed to monitor malicious activities that target IoT devices. In this case, we can obtain the evaluation result from the DT end (e.g., extracting the state of network activities concerning attacks). In this way, by aggregating malicious activities from different gateways, the administrators can have a global view of ongoing and historical attacks toward IoT systems.

In addition, there are several ways to mitigate DDoS attacks based on DT. For instance, we could include a trust device feature function on top of the DT so that traffic from vulnerable IoT devices can be filtered or vulnerable devices may be removed from the system by enforced access control policies. Furthermore, we could maintain a list to include all IoT devices that must be verified for vulnerability. Once these devices connect to the DT, the DT will scan the devices by examining their security state based on data collected through the DT. Once trusted devices are identified, they will be marked and put into the trusted list. If the device is identified as nontrust, it may be isolated from the IoT system or disabled. It is worth noting that such a mechanism needs to maintain the list dynamically. By utilizing DT's security scanning, some devices that were previously marked as trusted can become malicious.

We can engage ML models based on DT to filter out affected devices before DT sends data to applications. ML-based techniques have become viable in detecting DDoS attacks. For example, ML approaches have been applied in IDS or IPS systems to detect suspicious network threats [5]. SDN (Software-Defined Networking) is a platform that provides efficient and dynamic management of physical systems. In this case, it can become an identical way to evaluate the adversarial impacts via the DT-based ML model. As a result, we used data from Kaggle, one well-known site for ML dataset sharing, to retrieve raw data for our evaluations as an example. Notably, we used the *"DDoS SDN dataset"* [101], consisting of 104,345 (63,561 benign and 40,784 malicious) requests and 23 columns of features with the binary ground truth label (0 as malicious and 1 as benign). We used the Scikit-learn package from Python to implement classification algorithms (random forest, decision tree, support vector machine, and logistic regression) to determine whether a request is normal (benign) or malicious. Fig. 8 shows the relationship between IP address

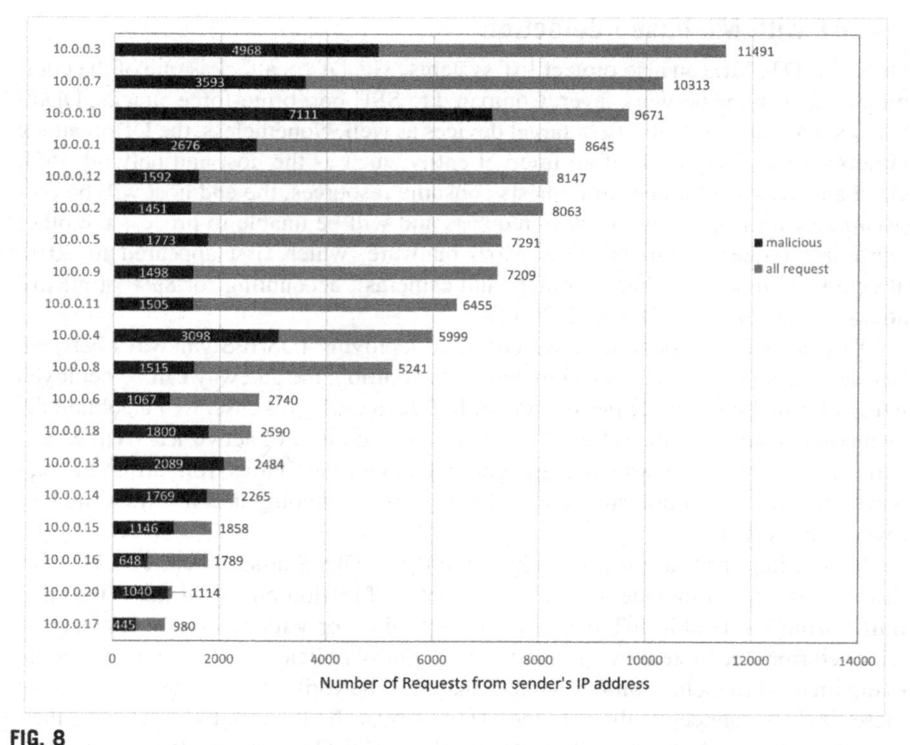

FIG. 8

Requests with IP address.

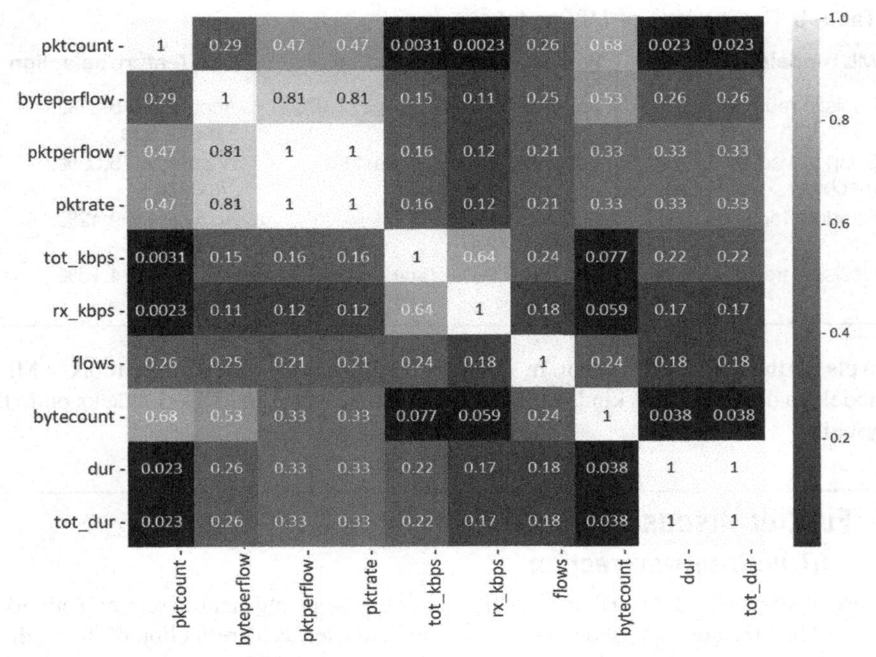

FIG. 9

Heatmap.

and request quantities. The *blue bar (light gray in print version)* represents the total requests, and the *red bar (dark black in print version)* represents the malicious ones. We also generate a heatmap, as shown in Fig. 9, to illustrate the relationship among characteristics, such as packet duration, byte counts, etc.

In order to evaluate the performance of the ML models, we use the following metrics: (i) *True positive (TP):* this refers to the probability that detected attacks are real attacks, (ii) *False positive (FP):* this refers to the probability that the system mistakenly marks a nonattack as an attack, (iii) *False negative (FN):* this refers to the probability that, when there is an attack, the system does not detect it, and (iv) *True negative (TN):* this refers to the probability that when there is no attack, the system correctly recognizes it as a benign one.

Note that Tonkal et al. [84] used 14 feature weights between 1.11 and 17.87 for different factors. Factors such as source IP address, number of packets, and destination IP have the highest weights because these are the most critical elements in the DDoS attack. In our example, we use these feature weights for these critical features and compare the accuracy and time information between the model without features and the model with feature weighting, as shown in Table 5.

According to results from Table 5, we conclude that integrating the proper ML model into DT can be used to detect DDoS attacks to improve the overall security

Table 5 Comparison of ML models.

ML models	Without feature selection	With feature selection
Logistic regression	Accuracy: 76.64%, Time: 7.97 s	Accuracy: 75.21%, Time: 2.68 s
Support vector machine	Accuracy: 78.4%, Time: N/A	Accuracy: 75.65%, Time: N/A
Random forest	Accuracy: 99.99%, Time: 28.42 s	Accuracy: 99.42%, Time: 24.85 s
Decision tree	Accuracy: 98%, Time: 227 s	Accuracy: 94.19%, Time: 72.80 s

levels for the whole IoT system. In summary, DT can work with security tools or ML models to detect various kinds of threats, such as malware and DDoS attacks on IoT systems.

6 Further discussion

6.1 DT with self-correction

Most of the studies on DT assume threats come from physical assets or network areas. The DTs are supposed to be secured and are used as a "reflection of the truth" to detect anomalous behavior [19,67,102,103]. However, DT relies on IoT technology to collect, distribute, and analyze data from physical objects, solely on sensor data from the perception layer to assist the DT in creating a digital copy of the physical system. As a result, adversaries can use attack strategies such as a DoS attack or self-propagated malware to subvert the sensor or gateway in this instance. Additionally, DT cannot access the available physical system once the gateway or sensor has been compromised. Disconnection might occur any time between physical assets and DT. In addition, propagation delay can cause the data in DT to go out of data instead of in real-time reflection.

An anomaly detection framework is proposed to identify anomalies from DT itself [104]. The framework uses the Gaussian mixture model to compare the results between the plant and the DT. Once anomalies are identified, the model will detect the differences and correct the DT accordingly. This study used the physical plant as a baseline for DT to follow. This provides one example of having a DT that self-corrects its mistakes and detects problems in physical plants. Under this expectation, more builds are needed that have multiple DTs to focus on different aspects of the physical IoT assets, to provide complete protection to both DT and IoT systems.

6.2 Mounting suitable ML model on DT systems

DT provides real-time communication with physical devices and uses a unified interface for IoT assets. Nonetheless, DT with ML models to detect malware and DDoS attack can be cost-effective [82,84,88,105–107]. Meanwhile, DT is a suitable platform to test and evaluate the performance of the integrated ML models. With the various advantages of DT and ML in terms of network performance improvement in IoT, challenges in DT and ML shall be considered. For example, before we deploy ML models, we need to obtain valuable, unbiased, and accurate data to train the models. Thus proper features for ML shall be carefully reviewed to ensure the expected results. Furthermore, in order to improve the efficiency of data transmission in the network, data size must be minimized, while the bandwidth of the network system must be maximized. Fragmentation technology is widely used, and headers of the packet data will be used for the receiver side to reorder the data. Once a suitable ML model is selected, with huge data collected from IoT devices, how to reduce the overhead size of the data and how to process the data with low latency and high reliability are other challenging issues in ML and DT that need to be addressed as well.

6.3 Privacy-preserving data collection and analysis

The DT systems contain massive amounts of privacy-sensitive data, such as medical records, autonomous vehicle sensing data, and real-time operational data. As a result, it is critical to build a robust, secure, and privacy-preserving system for the digital communication between DT and IoT assets. For example, new technologies, such as blockchain [40,68], can be used on top of DT to protect the message through hashing methods during communication. This method provides traceability and ensures no intervention from a third party. Note that knowing how to conduct the authentication procedure on DTs and protect the communication between DT and IoT devices has become a daunting task in this scenario. Other privacy-preserving communication, data publishing, and analysis approaches should be considered as well [53,108,109].

7 Final remarks

In this chapter, we classified a number of IoT system threats using the three-layer architecture. Then, for each layer, we described the reason why IoT systems are susceptible to these threats and the difficulties they encounter while attempting to resolve problems alone. The IoT refers to the overall system of interconnected devices and the technology that enables communication between those devices and the cloud/edge within those devices. In order to characterize the problem space of using DT to protect IoT systems, we proposed a three-dimensional framework that considers three factors: security usage (monitoring, attack detection, and threat prediction), security scope (physical components, DT for process, and DT for the whole system), and security objectives (confidentiality, integrity, and availability). Additionally, using a dataset obtained from the Kaggle dataset repository, SSH brute-force and DDoS attacks were carried out as two case studies to demonstrate the usefulness of engaging DT as a security architecture to protect IoT devices. Finally, we identified research areas that need further investigation, such as privacy-preserving data collection and analysis, DT with self-correction, and mounting suitable ML models on DT systems.

References

[1] J. Lin, W. Yu, N. Zhang, X. Yang, H. Zhang, W. Zhao, A survey on internet of things: architecture, enabling technologies, security and privacy, and applications, IEEE Internet Things J. 4 (5) (2017) 1125–1142.
[2] J.A. Stankovic, Research directions for the internet of things, IEEE Internet Things J. 1 (1) (2014) 3–9.
[3] W. Yu, F. Liang, X. He, W.G. Hatcher, C. Lu, J. Lin, X. Yang, A survey on the edge computing for the internet of things, IEEE Access 6 (2018) 6900–6919.
[4] H. Xu, W. Yu, D. Griffith, N. Golmie, A survey on industrial internet of things: a cyber-physical systems perspective, IEEE Access 6 (2018) 78238–78259.
[5] F. Liang, W.G. Hatcher, W. Liao, W. Gao, W. Yu, Machine learning for security and the internet of things: the good, the bad, and the ugly, IEEE Access 7 (2019) 158126–158147.
[6] M. Mohammadi, A. Al-Fuqaha, S. Sorour, M. Guizani, Deep learning for IoT big data and streaming analytics: a survey, IEEE Commun. Surv. Tutor. 20 (4) (2018) 2923–2960.
[7] Y. Sun, H. Song, A.J. Jara, R. Bie, Internet of things and big data analytics for smart and connected communities, IEEE Access 4 (2016) 766–773.
[8] J. Lin, W. Yu, X. Yang, Q. Yang, X. Fu, W. Zhao, A novel dynamic en-route decision real-time route guidance scheme in intelligent transportation systems, in: 2015 IEEE 35th International Conference on Distributed Computing Systems, 2015, pp. 61–72.
[9] N.Y. Philip, J.J.P.C. Rodrigues, H. Wang, S.J. Fong, J. Chen, Internet of things for in-home health monitoring systems: current advances, challenges and future directions, IEEE J. Sel. Areas Commun. 39 (2) (2021) 300–310.
[10] J. Wang, N. Varshney, C. Gentile, S. Blandino, J. Chuang, N. Golmie, Integrated sensing and communication: enabling techniques, applications, tools and data sets, standardization, and future directions, IEEE Internet Things J. 9 (23) (2022) 23416–23440.
[11] H. Xu, X. Liu, W. Yu, D. Griffith, N. Golmie, Reinforcement learning-based control and networking co-design for industrial internet of things, IEEE J. Sel. Areas Commun. 38 (5) (2020) 885–898.
[12] Z. Lv, L. Qiao, Analysis of healthcare big data, Futur. Gener. Comput. Syst. 109 (2020) 103–110. [Online]. Available: https://www.sciencedirect.com/science/article/pii/S0167739X20304829.
[13] Z. Lv, B. Hu, H. Lv, Infrastructure monitoring and operation for smart cities based on IoT system, IEEE Trans. Ind. Inform. 16 (3) (2020) 1957–1962.
[14] R. Du, P. Santi, M. Xiao, A.V. Vasilakos, C. Fischione, The sensable city: a survey on the deployment and management for smart city monitoring, IEEE Commun. Surv. Tutor. 21 (2) (2019) 1533–1560.
[15] M.A. Baballe, A. Hussaini, M.I. Bello, U.S. Musa, Online attacks types of data breach and cyber-attack prevention methods, Curr. Trends Inf. Technol. 12 (2) (2022) 21–26p.
[16] X. Liu, C. Qian, W.G. Hatcher, H. Xu, W. Liao, W. Yu, Secure internet of things (IoT)-based smart-world critical infrastructures: survey, case study and research opportunities, IEEE Access 7 (2019) 79523–79544.
[17] Z. Chen, G. Xu, V. Mahalingam, L. Ge, J. Nguyen, W. Yu, C. Lu, A cloud computing based network monitoring and threat detection system for critical infrastructures, Big Data Res. 3 (2016) 10–23. special Issue on Big Data from Networking Perspective.

[Online]. Available: https://www.sciencedirect.com/science/article/pii/S2214579615000520.
[18] X. Yang, X. Zhang, J. Lin, W. Yu, P. Zhao, A Gaussian-mixture model based detection scheme against data integrity attacks in the smart grid, in: 2016 25th International Conference on Computer Communication and Networks (ICCCN), 2016, pp. 1–9.
[19] G. Grieco, G.L. Grinblat, L. Uzal, S. Rawat, J. Feist, L. Mounier, Toward large-scale vulnerability discovery using machine learning, in: Proceedings of the Sixth ACM Conference on Data and Application Security and Privacy, Association for Computing Machinery, New York, NY, 2016, pp. 85–96.
[20] Q. Yang, D. Li, W. Yu, Y. Liu, D. An, X. Yang, J. Lin, Toward data integrity attacks against optimal power flow in smart grid, IEEE Internet Things J. 4 (5) (2017) 1726–1738.
[21] X. Yang, J. Lin, W. Yu, P.-M. Moulema, X. Fu, W. Zhao, A novel en-route filtering scheme against false data injection attacks in cyber-physical networked systems, IEEE Trans. Comput. 64 (1) (2015) 4–18.
[22] S. Kim, K.-J. Park, C. Lu, A survey on network security for cyber–physical systems: from threats to resilient design, IEEE Commun. Surv. Tutor. 24 (3) (2022) 1534–1573.
[23] Q. Yang, D. An, W. Yu, On time desynchronization attack against IEEE 1588 protocol in power grid systems, in: 2013 IEEE Energytech, 2013, pp. 1–5.
[24] J. Lin, W. Yu, N. Zhang, X. Yang, L. Ge, Data integrity attacks against dynamic route guidance in transportation-based cyber-physical systems: modeling, analysis, and defense, IEEE Trans. Veh. Technol. 67 (9) (2018) 8738–8753.
[25] N. Sun, J. Zhang, P. Rimba, S. Gao, L.Y. Zhang, Y. Xiang, Data-driven cybersecurity incident prediction: a survey, IEEE Commun. Surv. Tutor. 21 (2) (2018) 1744–1772.
[26] S.C. Sethuraman, V. Vijayakumar, S. Walczak, Cyber attacks on healthcare devices using unmanned aerial vehicles, J. Med. Syst. 44 (1) (2020) 29.
[27] H. Xu, F. Liang, W. Yu, Internet of things: architecture, key applications, and security impacts, in: Encyclopedia of Wireless Networks, Springer, 2020, pp. 672–681.
[28] I. Butun, P. Österberg, H. Song, Security of the internet of things: vulnerabilities, attacks, and countermeasures, IEEE Commun. Surv. Tutor. 22 (1) (2020) 616–644.
[29] J.G. Almaraz-Rivera, J.A. Perez-Diaz, J.A. Cantoral-Ceballos, Transport and application layer DDoS attacks detection to IoT devices by using machine learning and deep learning models, Sensors 22 (9) (2022) 3367.
[30] Z. Lv, Y. Han, A.K. Singh, G. Manogaran, H. Lv, Trustworthiness in industrial IoT systems based on artificial intelligence, IEEE Trans. Ind. Inform. 17 (2) (2021) 1496–1504.
[31] I. Zakariyya, H. Kalutarage, M.O. Al-Kadri, Robust, effective and resource efficient deep neural network for intrusion detection in IoT networks, in: Proceedings of the 8th ACM on Cyber-Physical System Security Workshop, Association for Computing Machinery, 2022, pp. 41–51.
[32] W.G. Hatcher, W. Yu, A survey of deep learning: platforms, applications and emerging research trends, IEEE Access 6 (2018) 24411–24432.
[33] J. Li, Z. Zhao, R. Li, H. Zhang, Ai-based two-stage intrusion detection for software defined IoT networks, IEEE Internet Things J. 6 (2) (2019) 2093–2102.
[34] P. Kumar, H. Bagga, B.S. Netam, V. Uduthalapally, Sad-IoT: security analysis of DDoS attacks in IoT networks, Wirel. Pers. Commun. 122 (1) (2022) 87–108.
[35] N. Chaabouni, M. Mosbah, A. Zemmari, C. Sauvignac, P. Faruki, Network intrusion detection for IoT security based on learning techniques, IEEE Commun. Surv. Tutor. 21 (3) (2019) 2671–2701.

[36] M. Wu, Z. Song, Y.B. Moon, Detecting cyber-physical attacks in cybermanufacturing systems with machine learning methods, J. Intell. Manuf. 30 (2019) 1111–1123.
[37] W. Wang, Y. Yao, X. Liu, X. Li, P. Hao, T. Zhu, I can see the light: attacks on autonomous vehicles using invisible lights, in: 2021 ACM SIGSAC Conference on Computer and Communications Security, 2021, pp. 1930–1944.
[38] S. Tariq, S. Jeon, S.S. Woo, Am I a real or fake celebrity? evaluating face recognition and verification APIs under deepfake impersonation attack, in: Proceedings of the ACM Web Conference 2022, Association for Computing Machinery, 2022, pp. 512–523.
[39] C. Qian, X. Liu, C. Ripley, M. Qian, F. Liang, W. Yu, Digital twin—cyber replica of physical things: architecture, applications and future research directions, Future Internet 14 (2) (2022) 64.
[40] A. Hussaini, C. Qian, W. Liao, W. Yu, A taxonomy of security and defense mechanisms in digital twins-based cyber-physical systems, in: 2022 IEEE International Conferences on Smart Data, IEEE, 2022, pp. 597–604.
[41] I. Arghire, St. Jude Medical Recalls 465,000 Pacemakers Over Security Vulnerabilities, 2017, 08. [Online]. Available: https://www.securityweek.com/st-jude-medical-recalls-465000-pacemakers-over-security-vulnerabilities/.
[42] A. Greenberg, Hackers Remotely Kill a Jeep on the Highway—With Me in It, 2015, 07. [Online]. Available: https://www.wired.com/2015/07/hackers-remotely-kill-jeep-highway/.
[43] X. Wang, S. Chellappan, W. Gu, W. Yu, D. Xuan, Search-based physical attacks in sensor networks, in: Proceedings. 14th International Conference on Computer Communications and Networks, 2005. ICCCN 2005, 2005, pp. 489–496.
[44] X. Yang, X. He, W. Yu, J. Lin, R. Li, Q. Yang, H. Song, Towards a low-cost remote memory attestation for the smart grid, Sensors 15 (8) (2015) 20799–20824. [Online]. Available: https://www.mdpi.com/1424-8220/15/8/20799.
[45] X. Wang, W. Yu, A. Champion, X. Fu, D. Xuan, Detecting worms via mining dynamic program execution, in: 2007 Third International Conference on Security and Privacy in Communications Networks and the Workshops—SecureComm 2007, 2007, pp. 412–421.
[46] W. Yu, C. Boyer, S. Chellappan, D. Xuan, Peer-to-peer system-based active worm attacks: modeling and analysis, in: IEEE International Conference on Communications, 2005. ICC 2005, vol. 1, 2005, pp. 295–300.
[47] W. Yu, X. Wang, X. Fu, D. Xuan, W. Zhao, An invisible localization attack to internet threat monitors, IEEE Trans. Parallel Distrib. Syst. 20 (11) (2009) 1611–1625.
[48] V. Hassija, V. Chamola, V. Saxena, D. Jain, P. Goyal, B. Sikdar, A survey on IoT security: application areas, security threats, and solution architectures, IEEE Access 7 (2019) 82721–82743.
[49] J.-P.A. Yaacoub, O. Salman, H.N. Noura, N. Kaaniche, A. Chehab, M. Malli, Cyber-physical systems security: limitations, issues and future trends, Microprocess. Microsyst. 77 (2020) 103201.
[50] A. Mangino, M.S. Pour, E. Bou-Harb, Internet-scale insecurity of consumer internet of things: an empirical measurements perspective, ACM Trans. Manag. Inf. Syst. 11 (4) (2020) 1–24.
[51] B. Jovanovic, Internet of Things Statistics for 2022—Taking Things Apart, 2022, [Online]. Available: https://dataprot.net/statistics/iot-statistics/.
[52] S. Leonard, The Most Vulnerable IoT Devices: Think Before You Buy, 2021, [Online]. Available: https://irishtechnews.ie/the-mostvulnerable-iot-devices/.

[53] M. Yang, J. Luo, Z. Ling, X. Fu, W. Yu, De-anonymizing and countermeasures in anonymous communication networks, IEEE Commun. Mag. 53 (4) (2015) 60–66.
[54] R. Pries, W. Yu, X. Fu, W. Zhao, A new replay attack against anonymous communication networks, in: 2008 IEEE International Conference on Communications, 2008, pp. 1578–1582.
[55] J. Mirkovic, P. Reiher, A taxonomy of DDoS attack and DDoS defense mechanisms, SIGCOMM Comput. Commun. Rev. 34 (2) (2004) 39–53, https://doi.org/10.1145/997150.997156. [Online]. Available:.
[56] Q. Yang, J. Yang, W. Yu, D. An, N. Zhang, W. Zhao, On false data-injection attacks against power system state estimation: modeling and countermeasures, IEEE Trans. Parallel Distrib. Syst. 25 (3) (2014) 717–729.
[57] J. Lin, W. Yu, X. Yang, G. Xu, W. Zhao, On false data injection attacks against distributed energy routing in smart grid, in: 2012 IEEE/ACM Third International Conference on Cyber-Physical Systems, 2012, pp. 183–192.
[58] Q. Yang, D. An, R. Min, W. Yu, X. Yang, W. Zhao, On optimal PMU placement-based defense against data integrity attacks in smart grid, IEEE Trans. Inf. Forensics Secur. 12 (7) (2017) 1735–1750.
[59] J. Lin, W. Yu, X. Yang, Towards multistep electricity prices in smart grid electricity markets, IEEE Trans. Parallel Distrib. Syst. 27 (1) (2016) 286–302.
[60] Forrester Study: The 2020 State of Security Operations, 2021, 2. [Online]. Available: https://www.paloguard.com/datasheets/forresterthe-2020-state-of-security.pdf.
[61] M. Eckhart, A. Ekelhart, Digital twins for cyber-physical systems security: state of the art and outlook, in: Security and Quality in Cyber-Physical Systems Engineering, Springer, 2019, pp. 383–412.
[62] M. Dietz, M. Vielberth, G. Pernul, Integrating digital twin security simulations in the security operations center, in: Proceedings of the 15th International Conference on Availability, Reliability and Security, Association for Computing Machinery, New York, NY, 2020, pp. 1–9.
[63] L. Shi, S. Krishnan, S. Wen, Study cybersecurity of cyber physical system in the virtual environment: a survey and new direction, in: Proceedings of the 2022 Australasian Computer Science Week, Association for Computing Machinery, New York, NY, 2022.
[64] M. Baert, E. De Poorter, J. Hoebeke, A digital communication twin for performance prediction and management of bluetooth mesh networks, in: Proceedings of the 17th ACM Symposium on QoS and Security for Wireless and Mobile Networks, Association for Computing Machinery, New York, NY, 2021.
[65] Y. Li, Q. Liu, A comprehensive review study of cyber-attacks and cyber security; emerging trends and recent developments, Energy Rep. 7 (2021) 8176–8186.
[66] H. Hong, Q. Wu, F. Dong, W. Song, R. Sun, T. Han, C. Zhou, H. Yang, Netgraph: an intelligent operated digital twin platform for data center networks, in: Proceedings of the ACM SIGCOMM 2021 Workshop on Network-Application Integration, Association for Computing Machinery, New York, NY, 2021, pp. 26–32.
[67] J. Corral-Acero, F. Margara, M. Marciniak, C. Rodero, F. Loncaric, Y. Feng, A. Gilbert, J.F. Fernandes, H.A. Bukhari, A. Wajdan, et al., The 'digital twin' to enable the vision of precision cardiology, Eur. Heart J. 41 (48) (2020) 4556–4564.
[68] W. Gao, W.G. Hatcher, W. Yu, A survey of blockchain: techniques, applications, and challenges, in: 2018 27th International Conference on Computer Communication and Networks (ICCCN), 2018, pp. 1–11.

[69] P. Moulema, W. Yu, D. Griffith, N. Golmie, On effectiveness of smart grid applications using co-simulation, in: 2015 24th International Conference on Computer Communication and Networks (ICCCN), 2015, pp. 1–8.
[70] W. Gao, J.H. Nguyen, W. Yu, C. Lu, D.T. Ku, W.G. Hatcher, Toward emulation-based performance assessment of constrained application protocol in dynamic networks, IEEE Internet Things J. 4 (5) (2017) 1597–1610.
[71] L.U. Khan, Z. Han, W. Saad, E. Hossain, M. Guizani, C.S. Hong, Digital twin of wireless systems: overview, taxonomy, challenges, and opportunities, IEEE Commun. Surv. Tutor. (2022).
[72] L. Maillet-Contoz, E. Michel, M.D. Nava, P.-E. Brun, K. Leprêtre, G. Massot, End-to-end security validation of IoT systems based on digital twins of end-devices, in: 2020 Global Internet of Things Summit (GIoTS), 2020, pp. 1–6.
[73] E. Karaarslan, M. Babiker, Digital twin security threats and countermeasures: an introduction, in: 2021 International Conference on Information Security and Cryptology (ISCTURKEY), 2021, pp. 7–11.
[74] H. Feng, D. Chen, H. Lv, Z. Lv, Game theory in network security for digital twins in industry, Digit. Commun. Netw. (2023).
[75] A.N. Desai, IoT Devices Remain Highly Vulnerable as a Billion 'Smart' Electronics Were Attacked in 2021, 2022, [Online]. Available: https://www.neowin.net/news/iot-devices-remain-highly-vulnerableas-a-billion-039smart039-electronics-were-attacked-in-2021/.
[76] G. Lin, S. Wen, Q.-L. Han, J. Zhang, Y. Xiang, Software vulnerability detection using deep neural networks: a survey, Proc. IEEE 108 (10) (2020) 1825–1848.
[77] S. Liu, M. Dibaei, Y. Tai, C. Chen, J. Zhang, Y. Xiang, Cyber vulnerability intelligence for internet of things binary, IEEE Trans. Ind. Inform. 16 (3) (2019) 2154–2163.
[78] E.C. Balta, D.M. Tilbury, K. Barton, A digital twin framework for performance monitoring and anomaly detection in fused deposition modeling, in: 2019 IEEE 15th International Conference on Automation Science and Engineering (CASE), IEEE, 2019, pp. 823–829.
[79] S.R. Chhetri, S. Faezi, A. Canedo, M.A.A. Faruque, Quilt: quality inference from living digital twins in IoT-enabled manufacturing systems, in: Proceedings of the International Conference on Internet of Things Design and Implementation, 2019, pp. 237–248.
[80] D.M. Mendez Mena, B. Yang, Blockchain-based whitelisting for consumer IoT devices and home networks, in: Proceedings of the 19th Annual SIG Conference on Information Technology Education, Association for Computing Machinery, New York, NY, 2018, pp. 7–12.
[81] N. Djedjig, D. Tandjaoui, F. Medjek, I. Romdhani, Trust-aware and cooperative routing protocol for IoT security, J. Inf. Secur. Appl. 52 (2020) 102467.
[82] N.Z. Bawany, J.A. Shamsi, K. Salah, DDoS attack detection and mitigation using SDN: methods, practices, and solutions, Arab. J. Sci. Eng. 42 (2017) 425–441.
[83] L.E.S. Jaramillo, Malware detection and mitigation techniques: lessons learned from Mirai DDoS attack, J. Inf. Syst. Eng. Manag. 3 (3) (2018) 19.
[84] Ö. Tonkal, H. Polat, E. Başaran, Z. Cömert, R. Kocaoğlu, Machine learning approach equipped with neighbourhood component analysis for DDoS attack detection in software-defined networking, Electronics 10 (11) (2021) 1227.
[85] D. Airehrour, J. Gutierrez, S.K. Ray, Secure routing for internet of things: a survey, J. Netw. Comput. Appl. 66 (2016) 198–213.

[86] A. Hussaini, B. Zahran, A. Ali-Gombe, Object allocation pattern as an indicator for maliciousness-an exploratory analysis, in: Proceedings of the Eleventh ACM Conference on Data and Application Security and Privacy, 2021, pp. 313–315.
[87] E. Nwaibeh, S. Kamara, S. Oladejo, H. Adamu, Epidemiological model of computer malware prevalence and control, J. Niger. Assoc. Math. Phys. 49 (1) (2019) 133–140.
[88] S.-Y. Hwang, J.-N. Kim, A malware distribution simulator for the verification of network threat prevention tools, Sensors 21 (21) (2021) 6983.
[89] H. Musa, B. Modi, I.A. Adamu, A.A. Aminu, H. Adamu, Y. Ajiya, A comparative analysis of different feature set on the performance of different algorithms in phishing website detection, Int. J. Artif. Intell. Appl. 10 (3) (2019).
[90] D. Piromalis, A. Kantaros, Digital twins in the automotive industry: the road toward physical-digital convergence, Appl. Syst. Innov. 5 (4) (2022). [Online]. Available: https://www.mdpi.com/2571-5577/5/4/65.
[91] J. Nowak, M. Stakkeland, Your Systems May Be Optimized but Digital Twins Could Learn to Do It Better, 2019, 05. [Online]. Available: https://new.abb.com/news/detail/24663/your-systems-maybe-optimized-but-digital-twins-could-learn-to-do-it-better.
[92] K. Rogage, E. Mahamedi, I. Brilakis, M. Kassem, Beyond digital shadows: digital twin used for monitoring earthwork operation in large infrastructure projects, AI Civ. Eng. 2730-5392, 1 (2022), https://doi.org/10.1007/s43503-022-00009-5. [Online]. Available:.
[93] W. Yu, D. Griffith, L. Ge, S. Bhattarai, N. Golmie, An integrated detection system against false data injection attacks in the smart grid, Secur. Commun. Netw. 8 (2) (2015) 91–109. [Online]. Available: https://onlinelibrary.wiley.com/doi/abs/10.1002/sec.957.
[94] W. Yu, H. Zhang, L. Ge, R. Hardy, On behavior-based detection of malware on android platform, in: 2013 IEEE Global Communications Conference (GLOBECOM), 2013, pp. 814–819.
[95] Y. Yigit, B. Bal, A. Karameseoglu, T.Q. Duong, B. Canberk, Digital twin-enabled intelligent DDoS detection mechanism for autonomous core networks, IEEE Commun. Stand. Mag. 6 (3) (2022) 38–44.
[96] S.A. Varghese, A.D. Ghadim, A. Balador, Z. Alimadadi, P. Papadimitratos, Digital twin-based intrusion detection for industrial control systems, in: 2022 IEEE International Conference on Pervasive Computing and Communications Workshops and Other Affiliated Events (PerCom Workshops), IEEE, 2022, pp. 611–617.
[97] X. Yuan, J. Zhang, J. Luo, J. Chen, Z. Shi, M. Qin, An efficient digital twin assisted clustered federated learning algorithm for disease prediction, in: 2022 IEEE 95th Vehicular Technology Conference:(VTC2022-Spring), IEEE, 2022, pp. 1–6.
[98] Q. Qiao, J. Wang, L. Ye, R.X. Gao, Digital twin for machining tool condition prediction, Procedia CIRP 81 (2019) 1388–1393.
[99] F. Tao, J. Cheng, Q. Qi, M. Zhang, H. Zhang, F. Sui, Digital twin-driven product design, manufacturing and service with big data, Int. J. Adv. Manuf. Technol. 94 (9) (2018) 3563–3576.
[100] J. Lopez, J.E. Rubio, C. Alcaraz, Digital twins for intelligent authorization in the b5g-enabled smart grid, IEEE Wirel. Commun. 28 (2) (2021) 48–55.
[101] A. Kezin, DDoS SDN Dataset, 2021, [Online]. Available: https://www.kaggle.com/datasets/aikenkazin/ddos-sdn-dataset.
[102] F. Mohamed, J. Abdeslam, E.B. Lahcen, Towards new approach to enhance learning based on internet of things and virtual reality, in: Proceedings of the International

Conference on Learning and Optimization Algorithms: Theory and Applications, Association for Computing Machinery, New York, NY, 2018, pp. 1–5.

[103] R. Coulter, Q.-L. Han, L. Pan, J. Zhang, Y. Xiang, Data-driven cyber security in perspective—intelligent traffic analysis, IEEE Trans. Cybern. 50 (7) (2019) 3081–3093.

[104] C. Gao, H. Park, A. Easwaran, An anomaly detection framework for digital twin driven cyber-physical systems, in: Proceedings of the ACM/IEEE 12th International Conference on Cyber-Physical Systems, Association for Computing Machinery, New York, NY, 2021, pp. 44–54.

[105] M. Li, H. Zhou, Y. Qin, Two-stage intelligent model for detecting malicious DDoS behavior, Sensors 22 (7) (2022) 2532.

[106] A.K. Pathak, S. Saguna, K. Mitra, C. Åhlund, Anomaly detection using machine learning to discover sensor tampering in IoT systems, in: ICC 2021-IEEE International Conference on Communications, IEEE, 2021, pp. 1–6.

[107] L. Erhan, M. Ndubuaku, M. Di Mauro, W. Song, M. Chen, G. Fortino, O. Bagdasar, A. Liotta, Smart anomaly detection in sensor systems: a multi-perspective review, Inf. Fusion 67 (2021) 64–79.

[108] X. Yang, T. Wang, X. Ren, W. Yu, Survey on improving data utility in differentially private sequential data publishing, IEEE Trans. Big Data 7 (4) (2021) 729–749.

[109] W. Gao, W. Yu, F. Liang, W.G. Hatcher, C. Lu, Privacy-preserving auction for big data trading using homomorphic encryption, IEEE Trans. Network Sci. Eng. 7 (2) (2020) 776–791.

CHAPTER

5 Creating environmentally conscious products and environments with smart materials

Aditi Maheshwari and Andreea Danielescu
Accenture Labs, San Francisco, CA, United States

1 Introduction

Rapid industrialization, growing demand for goods and services, and population growth over the past century fueled the linear economy model—or the take-make-waste model—where resources and energy were extracted from nature, used to produce goods and services, consumed, and ultimately thrown away [1]. However, as we look to the future, we realize that this production and consumption system is not adequate to address the needs of our society. Studies show that we now require the equivalent of 1.7 Earths to replenish the resources that have been consumed and absorb the pollution that has been generated thus far. On top of that, resource use is predicted to double by 2050, which means that if we continue to produce and consume at this rate, three planet Earths will be needed by 2050 to sustain human life [2].

This linear economy that has depleted our natural resources and caused the global climate crisis is no longer sustainable—and the global economy has found it imperative that we adopt more environmentally conscious approaches to producing products and services. As advocated by William McDonough and Michael Braungart in their book *Cradle to Cradle: Remaking the Way We Make Things*, products should be designed for circularity, that is, made in a way that they can be easily disassembled and their components reused or recycled at the end of their useful lives. This can be achieved by using renewable nature-derived materials, reducing the amount of waste generated during production, and reusing or recycling materials whenever possible. In addition to these, bioresorbable and biodegradable materials that completely dissolve or disintegrate back into the natural environment as harmless substances will also play a crucial role in advancing product sustainability. Together, these circular manufacturing techniques can help to reduce the environmental footprint of products and promote sustainable development by minimizing waste and maximizing the use of resources.

Smart Spaces. https://doi.org/10.1016/B978-0-443-13462-3.00001-7

Replacing rare metals, plastics, and petroleum-based products with sustainable alternatives and redesigning products with fewer materials for easier disassembly and recycling require material innovation. New materials will drive the future of sustainable, smart products and cities. However, today, we all realize how far away from that we are—with e-waste, plastics, and clothing, among other items, piling up in landfills. Despite this reality, demand for smarter and more feature-rich products and devices continues to increase, accentuating the need for materials and methods that enable a continued growth in intelligence at the edge without an increase in hazardous waste. To that end, in this chapter we will explore how technologies of the 4th Industrial Revolution [3], including smart materials, additive manufacturing, and printed electronics, can be used to create low environmental footprint systems that are functional and interactive, yet drastically reduce the negative impact on the environment. These technologies have the potential to revolutionize the way we create products and interact with our environment, and they have a significant role to play in sustainable future cities.

In addition to product sustainability, environmentalists, researchers, and corporations are also looking to enable "resilience" in our natural and built environments, i.e., the ability to resist change or recover from disturbance in a way that preserves the essence of a system's structure and function. Mass deforestation, floods, wildfires, etc., are just a few examples of disturbances that have been induced by human activities and now pose a threat to our ecosystems. The pervasiveness of low-power and intelligent sensor networks, IoT-enabled devices, and self-sustaining energy harvesting devices have now made it possible to monitor, safeguard, and conserve our natural and urban environments. Combined with smart materials, these technologies hold the potential to sense environmental anomalies for corrective action, revive native biodiversity, enrich depleted soil with nutrients, reforest burnt land, and allow for easier management of agricultural and rangeland systems [4].

So, what makes materials "smart"? Traditionally, smart materials are defined as materials that can sense and react to external stimuli such as pressure, temperature, magnetism, etc., in a visible and tangible way, through a change in their molecular structure. Consider nitinol, a shape memory alloy used in arterial stents for minimally invasive surgery. After deformation, nitinol can revert to its original shape on heating. It is inserted in the patient's artery in compressed form, where it gradually reexpands using heat from the body to keep the artery open. More recently, however, the term "smart materials" can also refer to the approach of directly combining specialized materials and electronics to create material systems that are responsive. Examples of such systems include conductive threads used in e-textiles, conductive inks for printed electronics, and 4D printing—a renovation of 3D printing that allows for dynamic and shape changing 3D prints [5]. What unifies all of these materials and material systems under the umbrella term of "smart materials" is that their behavior and properties can be controlled in predictable, repeatable, and useful ways. Smart materials can enable calmer and more seamless designs [6]. The Jacquard jacket by Google and Levi's is one example of how conductive threads were used to create gesture-responsive clothing that seamlessly connects to our smartphones for distraction-free biking [7]. Besides simplifying design, smart materials also reduce the number of parts needed, reduce weight, offer new form factors, and make designs more robust.

Recently, there has been an increased focus on utilizing smart materials and additive manufacturing techniques in a variety of sustainability-oriented applications, including transient and biodegradable electronics, edible electronics, and designing for reusability. From the use of engineered living materials for constructing self-repairing buildings [8] and other forms of bioengineered urban infrastructure, to photocatalytic self-cleaning capabilities on textiles and other surfaces [9], smart material systems are poised to play a crucial role in enabling sustainable, autonomous, and smarter products and environments. Through a series of R&D projects from our lab, we discuss the potential of smart material innovations in creating intelligent and environmentally conscious:

(a) Consumer Goods and Products.
(b) Apparel and E-Textiles.
(c) Natural and Built Environments.

2 Consumer goods, devices, and products

Redesigning everyday goods and products for circularity requires a thorough analysis of the production and consumption behavior at each stage of the value chain to inform design and material decisions. From raw material substitution to diverting waste out of landfills, material innovations play an important role in circular product design. Within the materials and electronics communities, **transient electronics** is a rapidly growing field of research that aims to develop electronic products that are not only functional but also ephemeral. This field of research has gained significant attention in recent years due to the growing concern for the environmental impact of electronic waste. These electronics are made from biodegradable materials, such as soluble polymers, starch, wax, cellulose, or chitin, and are typically broken down by natural processes within a few months or years. This makes them ideal for use in applications where disposable electronics are needed, such as medical devices, environmental sensors, and agricultural applications.

Transient electronics can help to reduce waste and protect the environment by avoiding the accumulation of electronic waste. They can be used to create products that are sustainable in other ways. For example, biodegradable electronics can be used in agriculture to monitor crop growth and soil health. These devices can be left in the soil after harvest and will dissolve into the environment, preventing large amounts of electronic waste from being generated by the agriculture industry. In the field of biomedicine, transient electronics can be used to create medical devices that are designed to be implanted into the body and then dissolve over time. This can reduce the risk of infection and other complications, while also allowing the body to heal more naturally. In the field of security and military applications, transient electronics can be used to create devices that are designed to be intentionally destroyed in the event of a security breach.

In the following paragraphs, we discuss a few examples of material innovations that have enabled functional transient systems, without compromising on environmental sustainability.

2.1 A home-compostable heating patch made of leaf skeletons for use in packaging and beyond

The majority of "self-heating" packages on the market today rely on one-shot exothermic reactions and lack important features such as reusability or temperature controllability. They also create waste in the form of single-use packages. To overcome these challenges, our team at Accenture Labs along with researchers from University of California, Berkeley leveraged existing organic structures that are readily available in nature—in this case, leaf skeletons—to create a new system capable of being electrically controlled as a portable heater [10]. As leaves grow on trees, their veins branch out in a fractal pattern to cover relatively large areas, naturally accomplishing patterns similar to those in conventional Joule heaters. The new system utilizes leaf skeletons coated with chitosan (a biomaterial derived from shrimp shells) and a commercially available, highly conductive, water-based silver ink certified to be nontoxic (Circuit Scribe). While the leaf skeleton provides the structural foundation of the heater, the chitosan helps provide stability and prevents the concentration of the electrical current at any point on the leaf that would lead to burning of the structure (Fig. 1).

FIG. 1

Leaf skeletons coated with chitosan and silver nanowires.

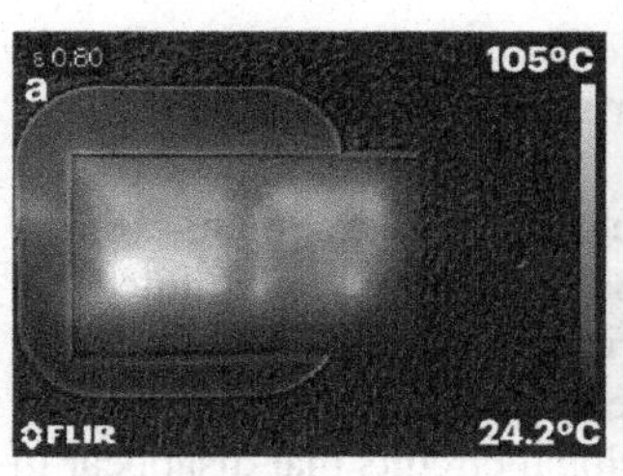

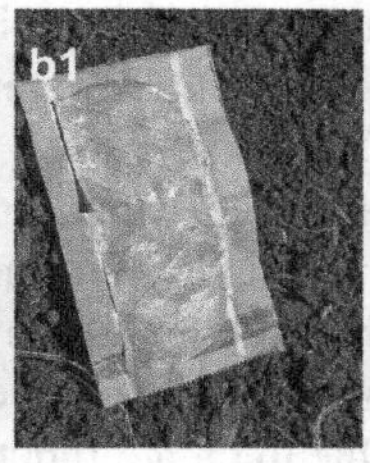

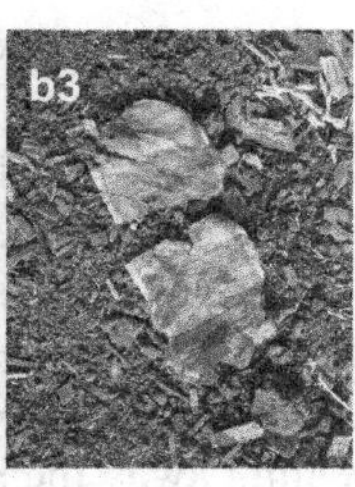

FIG. 2

(A) Infrared image of package placed onto charging mat. (B1–3) The same packaging turns into compost in under 60 days.

To heat, the packaging relies on wireless inductive charging—which is becoming readily available these days, with even some mobile phones capable of charging other electronics. The system can wirelessly heat to >70°C, is flexible, lightweight, low-cost, and reusable, and maintains its functionality over long periods of heating and multiple power cycles. Furthermore, the use of organic materials makes the heater fully biodegradable in at-home composting conditions over the course of 60 days (Fig. 2).

The technology can be used on-the-go to heat food and beverages when embedded in paper packaging or coffee sleeves and can even be used in milk cartons for in-pack pasteurization. It can also be used as a biodegradable heater for wax strips, hair masks, or body oils. Other applications include activating shape morphing materials [5,11,12] to enable more efficient manufacturing and shipping and create tangible user interfaces, activating thermally expandable microspheres [13] to protect arbitrarily shaped contents in a package, on-body heat therapy patches, and thermo-responsive clothing. Ultimately, this technology provides increased convenience without the environmental impact when disposed of and the concept can be applied to a wide set of applications for heating and sensing.

2.2 Passive moisture and temperature sensing with self-destroying circuits

The biodegradable heater demonstrates a technology that was designed with a responsible end-of-life for the product in mind, causing minimal environmental impact. But the concept of sustainable design can be taken a step further to create sensors and devices where end-of-life provides their primary function. The same material properties that make them sustainable, and are often considered detrimental to durability, can instead provide the device's core function. This is the fundamental principle driving the field of transient electronics.

Transient electronic materials provide a way for devices to exist for as long as they're useful, then dissolve, transform, or otherwise disappear in a sustainable and environmentally responsible way. Along with researchers from Georgia Institute of Technology, we designed a set of transient electronic materials that can be used in a wide set of applications, including dissolvable water sensors and melting temperature sensors [14].

To create water-dissolvable moisture sensors, we utilized a hydrographic film, a commonly used printing and transferring medium for making patterns and printed circuits on 3D surfaces and objects. The hydrographic film is composed of a PVA printing layer and a PET backing. By inkjet printing conductive traces on the PVA layer, and peeling away the backing, we were able to create functional circuits that disappear in water over time. By further placing the PVA circuit over a hydrogel layer, we demonstrated a water-leak sensor (Fig. 3). On attaching this sensor to a pipe, the hydrogel will slow down the leak by absorbing the water. As the hydrogel absorbs the water and expands, the physical deformation will trigger a resistance change in the circuit. Gradually, the water will pass through the hydrogel and dissolve the PVA sheet, increasing the trace resistance and providing a clear signal that water is present.

Besides water, heat is another common trigger for the destruction of transient electronics. In another experiment, we transferred inkjet-printed silver nanoparticle traces to a beeswax substrate using a print-transferring technique to create highly conductive circuitry on wax. The most exciting aspect of this approach is that the conductivity of the traces remains relatively unchanged over a large temperature range, and a dramatic increase in resistance is seen only when the temperature threshold has been met. Sensitivity of the sensor can be tuned by controlling the shape and dimensions of the printed traces as well as the material properties of the beeswax substrate. Using serpentine traces with this approach relieves the local stress concentration and makes the circuit more stable, increasing the endurance of the device before its cut-off point. Changing the circuit's path can therefore allow a designer to control the time it takes for a device to be destroyed. This application harnesses

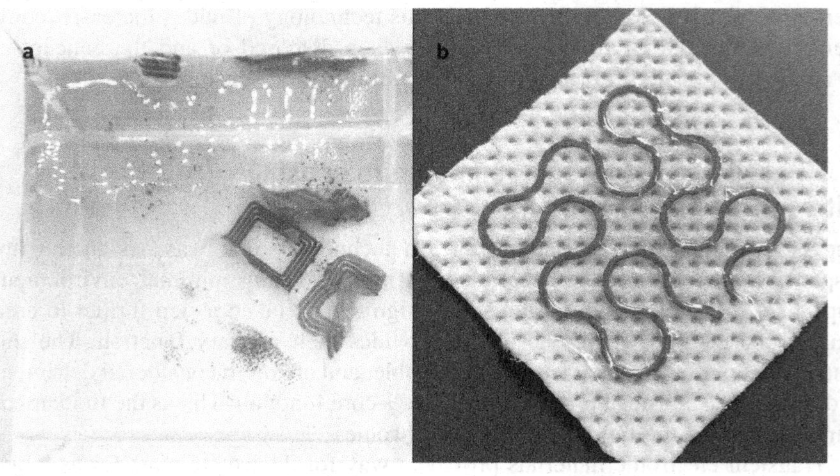

FIG. 3

(A) Inkjet printed conductive traces on PVA. (B) PVA-hydrogel combined water leakage sensor.

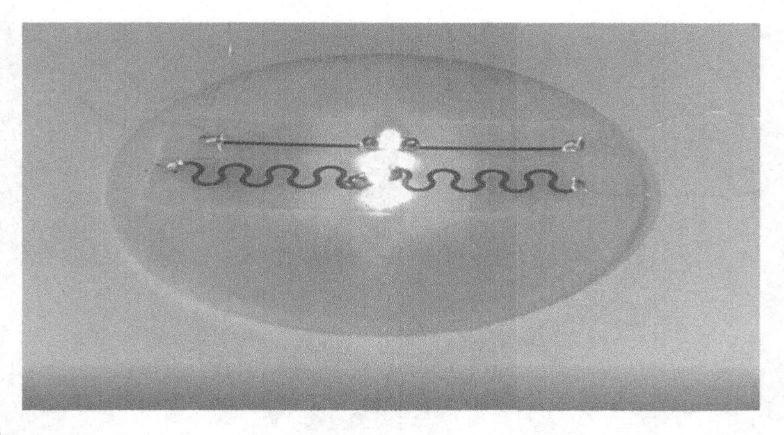

FIG. 4

Conductive circuitry on beeswax for temperature-sensitive circuits.

sustainable materials susceptible to different conditions to achieve sensing, for example creating a sensor that melts to detect a temperature threshold. The wax can either biodegrade or be recaptured and formed into a new sensor after it has melted (Fig. 4).

To enable interactivity and communication from the transient electronics to mobile phones, smart home systems, etc. we created a chipless radio frequency ID (RFID) tag. This passive radio frequency (RF) system does not require any power or additional electronics to function. The basic operating principle of RFID is that an external antenna transmits a radio signal to a tag and measures the reflected signals. Most commonly, the RFID tag controls the shape of the reflected waveform using a tiny chip. However, unique antenna patterns based on time- or frequency-domain reflectometry can also change the reflected waveform and eventually eliminate the need for a chip [15]. By doing so, RFID tags can be printed easily using sustainable conductive materials that are easily broken down without a significant environmental footprint when discarded. Among various tag data encoding methods, we chose a hybrid coding technique that changes the magnitude of the resonance peaks in the radar cross section (RCS) and designed C-shape metallic strip resonators that could resonate between 2.5 GHz to 7.5 GHz based on Vena et al.'s primitives [16], which allowed us to encode 22.9 bits in a dimension of 2×4 cm through absence/presence and frequency shift coding techniques. Ultimately, this implementation of a chipless RFID system allowed us to passively communicate through various fully decomposable transient sensor systems.

2.3 Data sanitation with edible electronics

Edible electronics, a subcategory of transient electronics, are devices that can be safely ingested and broken down in the body. These devices are made from edible materials, such as edible inks, gold or silver foil, etc., and are designed for applications such as medical sensors, dietary tracking devices, and drug delivery

FIG. 5

Conductive circuitry on chocolate for data sanitation purposes.

systems, as well as entertainment and interactivity. Edible electronics are particularly relevant to security and data sanitation applications, and for creating ingestible and bioresorbable sensors for health monitoring. To demonstrate an application of edible electronics for data sanitation, we combined edible gold leaf as an electrode material with chocolate. The gold-patterned chocolate was created by laser cutting the contour of the circuit design on a gold leaf placed on a PET substrate. After obtaining the desired gold leaf pattern, the gold circuit was transfer printed onto chocolate using commercial sugar paper as the transfer medium. Using the chipless RFID technique described earlier, the gold-patterned chocolate bar created in our lab can contain identifiable information, e.g., access control for users to obtain temporary access to areas, objects, or events, such as a concert, and then can be eaten and destroyed, when no longer needed (Fig. 5).

3 Smart apparel and textiles

The fashion industry is one of the largest contributors to waste and pollution in the world. Fast fashion is damaging. Instead of encouraging a reduction in production, it encourages consumer spending through continuously evolving styles and by creating clothes that are not meant to last. The fabrics don't break down, resulting in landfills full of clothing that isn't reused or recycled. One way to prevent this is by embedding additional functionality in our clothing, such as the ability to monitor health, or change patterns or colors, or self-adjust to increase wearer comfort, so that consumers are disincentivized to throw clothing away frequently. These functions are precisely what e-textiles enable. *E*-textiles are textile materials that are embedded with electronic components, such as sensors or actuators, to perform specific functions. Aside from making our clothes smarter in seamless and unobtrusive ways, e-textiles allow us to embed reusability in garments, upcycle waste fabrics, and reduce the amount of waste going into landfills.

While e-textiles extend the useful life of our garments and encourage a mindset shift towards reusability, their production and ultimate disposal can also create waste. To create e-textiles sustainably, technologists need to "design for disassembly"—either by allowing individual components to be retrieved from the e-textile system for recycling and reuse, or by designing with transient electronic systems to begin with. Embedding electronics otherwise would cause even fabrics that do break down to be even less eco-friendly.

3.1 Sustainable and multifunctional textile sensors and actuators

One unique way to create circular e-textiles is by upcycling waste fabrics by carbonizing them to create environmentally friendly and flexible textile sensors. In a recent set of experiments at Accenture Labs and MIT, waste cotton fabrics were treated at high temperatures (750–900°C) in an inert environment in a tube furnace, and then cooled down, to transform the fabric into a flexible and conductive material. The carbonization shrinks the cotton fabric and changes its color to black. The conductive cotton fabric can be optionally encapsulated in a biodegradable material like chitosan, natural rubber latex (NRL), or biodegradable Ecoflex to improve its performance, durability, and washability, and used as an active element for applications in wearable devices or interactive textiles. For example, carbonized textiles can be used to create flexible thermometers or pressure sensors, which can be integrated into clothing to monitor body temperature or physical activity (Fig. 6).

The active element can be used to create a variety of sensors for heat or fire detection, airflow sensing, touch, strain, or proximity sensing, or can even be used as a heating element. At their end-of-life, they can be home composted, or reused by grinding into a powder, and coating on top of another waste fabric to create a fresh sensor. Therefore these sensors can be created using a circular manufacturing process, enabling the transformation from fast fashion to smart sensors in clothing, or creating new ways of medical health monitoring (Fig. 7).

These biodegradable sensors can also be embedded into shirts to create functional garments. For example, we demonstrated the use of the sensors as haptic transfer devices to facilitate remote passive interactions. When you miss your family member, friend, or partner, you can touch the sensor sewn into your shirt, which then warms up the sensor that's attached to theirs, communicating to them that you're thinking about them. As another example, the sensors can be used in motion capture suits for immersive experiences in virtual environments, where they can enable you to connect with your digital avatar in the Metaverse for gaming, sports, and other applications (Fig. 8).

Not only can we embed biodegradable or recyclable sensors and actuators on our clothes, but we can also customize how our outfits look using smart materials like

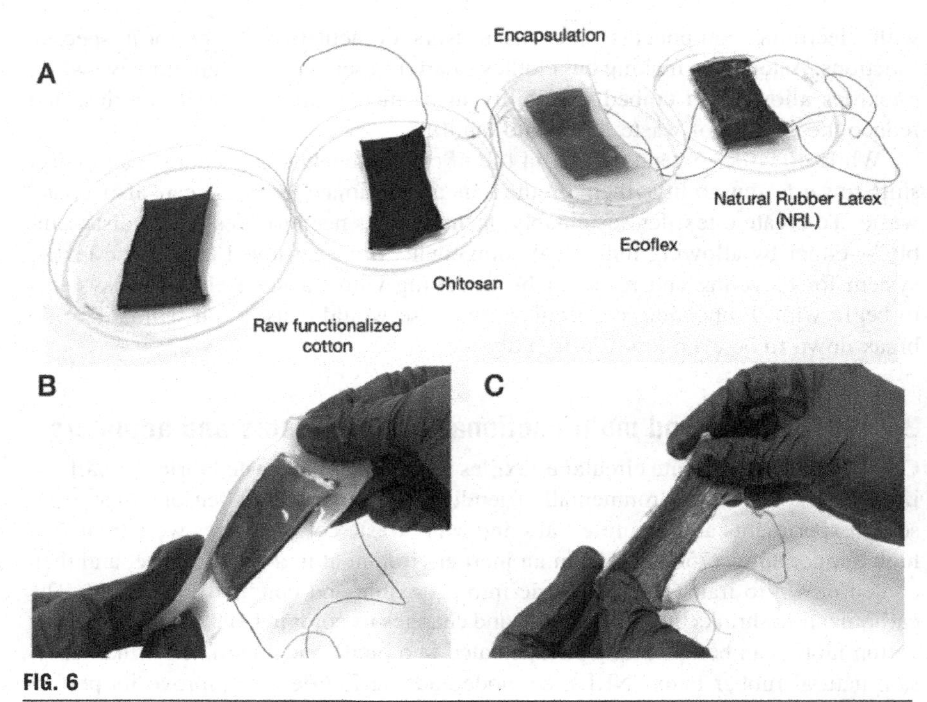

FIG. 6

(A) Biodegradable, stretchable, and flexible sensors made of waste cotton encapsulated in a variety of material. (B) Stretchability of Ecoflex coated sensors. (C) Stretchability of NRL coated sensors.

thermochromic pigments and shape morphing fibers, enabling more functionality within the same piece of garment to encourage a sustainable lifestyle where "less is more." Chromorphous [17] is one such example, which demonstrates the use of thermochromic threads into clothing. These threads change color based on temperature, allowing the garment to not only change appearances, but also to indicate heat build-up around the body, i.e., when the wearer is too hot or too cold. Similarly, researchers at MIT [18] created self-ventilating clothing using bacteria as the actuation medium, to create more energy-efficient garments, where the clothing self-adjusts to the wearer's body temperature instead of the wearer having to rely on heating or cooling systems. In both these systems, the same piece of clothing morphs into different styles, colors, patterns, and even adjusts to environmental conditions and external temperature, thereby allowing the wearer to purchase fewer clothing items and mitigate waste (Fig. 9).

3.2 Energy harvesting and storage through textiles

E-textiles have the potential to build on the ubiquity and scalability of textiles and textile manufacturing to create new experiences for individuals and provide comfortable, long-term monitoring of vitals for medical patients, remote field workers, and

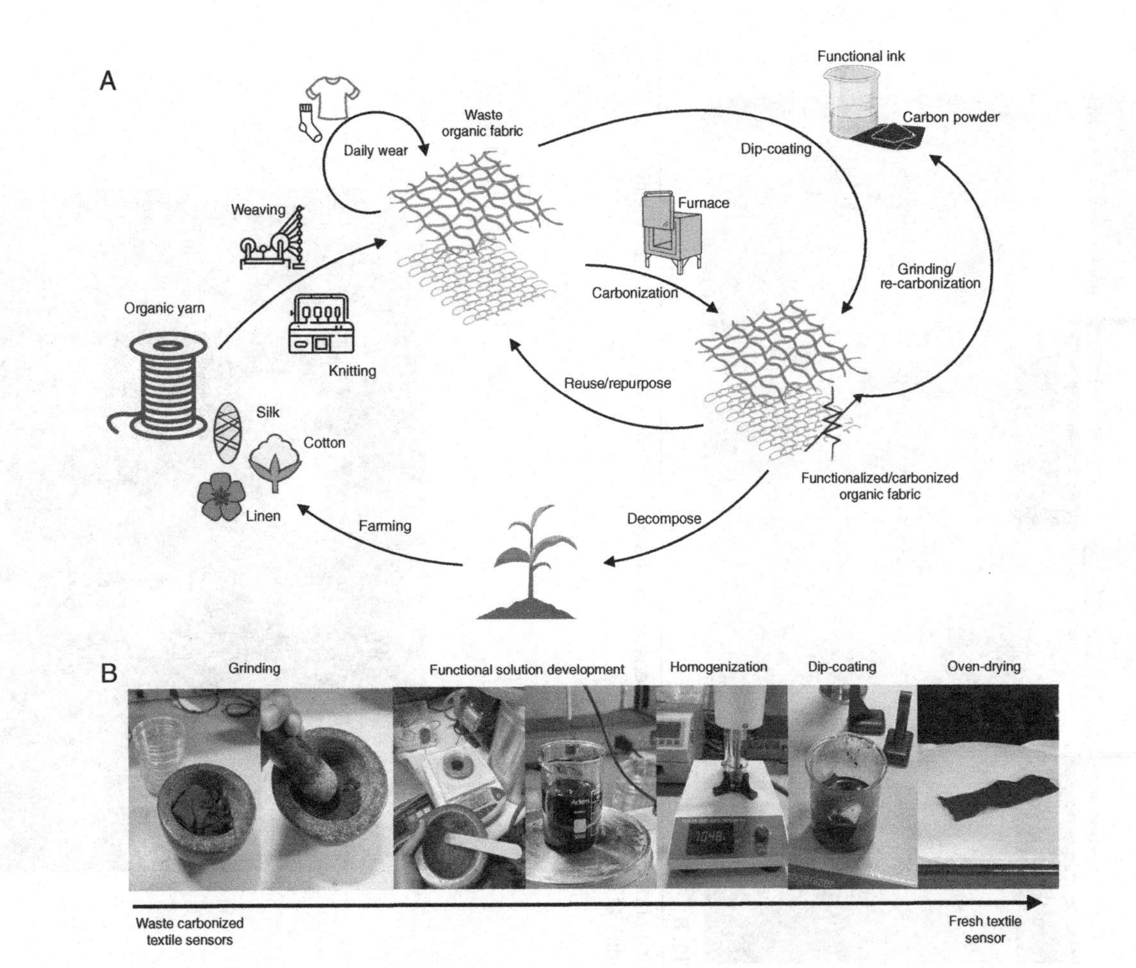

FIG. 7

(A) Life Cycle of circular textile sensors. (B) Process showing end-of-life processing to create fresh textile sensors out of used ones.

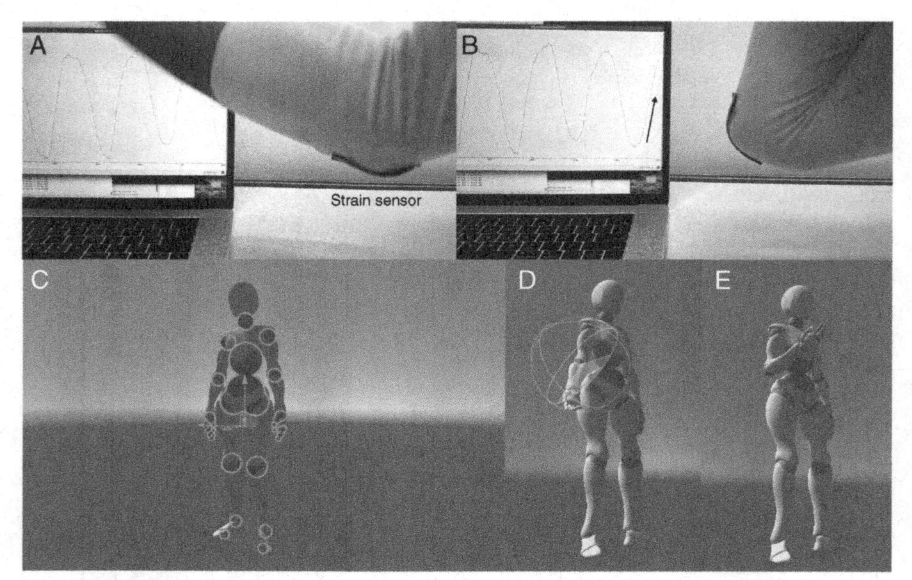

FIG. 8

Use of biodegradable textile sensors for strain and motion sensing in virtual reality applications.

FIG. 9

Biologic—Self-ventilating clothing using living nanoactuators by Lining Yao at MIT Media Lab.

the self-tracking community. All textile sensors, biodegradable or otherwise, need to be powered using an energy source. Today we use batteries to power our wearables, which are bulky and heavy, require nonrenewable metals to be produced, are uncomfortable to wear for long periods of time, and generate hazardous waste in landfills. The realization of a pure-textile system, without the use of any hard electronics, requires the development of flexible and manufacturable energy storage that provides sufficient power for long-term operation while also considering issues such as durability, sustainability, and safety. To eliminate batteries and the e-waste associated with them, we worked with researchers at Drexel University to create a fully fabric-based textile wearable that embeds power storage directly into the fabric and has no rigid components [19]. This allows for soft integrated structures that are comfortable and form fitting in everyday use.

Using a novel 2D material called MXene, our research team created a textile-based energy storage device fabricated using conductive additive-free (MXene) $Ti_3C_2T_x$ paint that produces enough power to operate a standard microcontroller for over 90 min and is expected to be able to power optimized low power electronics for up to a day. MXene provides a scalable and sustainable material for manufacturing e-textile systems, and our MXene-based textile supercapacitors are high capacity and fast charging, with the ability to directly integrate with fabrics through sewing or knitting techniques (Fig. 10).

Our textile energy storage device can be combined with energy harvesting technology along with passive or active textile communications using conductive threads to create fully self-powered fabric-based textiles. Energy harvesting can be enabled using common and nonhazardous oxides, as well as by using MXene harvesters [20,21].

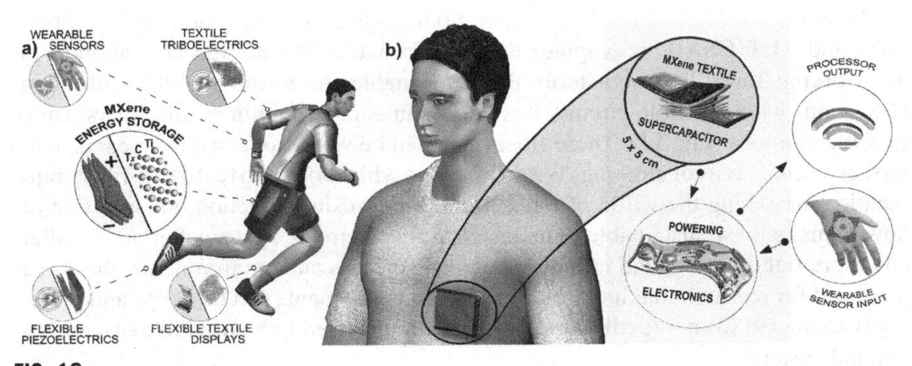

FIG. 10

(A) Applications of MXene-based textile energy storage. (B) Textile integrated with MXene yarn-based supercapacitor.

3.3 Textile intelligence through programmable fibers and novel computational substrates

In addition to developing textile-based sensors, actuators, power management, and communication systems, there is a need to embed digital system capabilities and processing units within the textile to realize intelligence within fabric-based wearables. So far, e-textile fibers have been analog, with the ability to carry continuous electronic signals rather than discrete, digital ones. However, in a recent development, MIT researchers created the first programmable fiber with digital capabilities, able to sense, store, analyze, and infer activity digitally, after being sewn into a shirt [22]. By placing hundreds of square silicon microscale digital chips into a polymer preform and precisely controlling the flow, the researchers were able to create a polymer fiber with continuous electrical connection between the chips over a length of tens of meters. The researchers demonstrated the ability to write, store, and read information on their digital fiber, including a 767-kilobit full-color short movie and a 0.48-megabyte music file that can be stored for 2 months without power. Finally, the research team also embedded predictive AI intelligence within the textile by including a neural network of 1650 connections within the fiber memory. By training the neural network with 270 min of data, the team got it to predict the minute-by-minute activity of the shirt's wearer with 96% accuracy.

Another example of in-fiber electronics is the integration of semiconducting diodes of high quality in thermally drawn fibers created by MIT researchers, a technology that is now commercially developed by AFFOA (Advanced Functional Fabrics of America) [23]. The researchers added miniaturized LEDs and photosensing diodes along with copper wires to a polymer preform, which, when heated in a furnace during the fiber-drawing process, resulted in a long fiber with the diodes lined up along its center and connected by the copper wires [24]. These optoelectronic fibers can be woven into washable and intelligent textiles that expand the fundamental capabilities of fabrics to encompass communications, lighting, physiological monitoring, and more.

To build on this vision of intelligent textile systems, researchers at Accenture Labs and MIT CSAIL (Computer Science & Artificial Intelligence Laboratory) are working towards design tools that can enable the development of ultrathin (1.5–2 mm wide) flexible circuits that can be integrated within textile fibers using existing methods (Fig. 11). These fibers can then be woven or sewn into seams, collars, or similar areas of clothing. We do note that while efforts to realize in-fiber computation are taking the world of e-textiles in a promising direction, the question of how to make these sustainable remains open. As electronics get smaller and smaller and more tightly integrated into our everyday surfaces and environments, there is a dire need for research that looks into how these components can be easily and effectively extracted from e-textiles and e-fibers and upcycled to be used in new computational systems.

The applications of smart materials in sustainable consumer goods and fashion are wide ranging. The fast fashion you buy for a party could be transformed into your

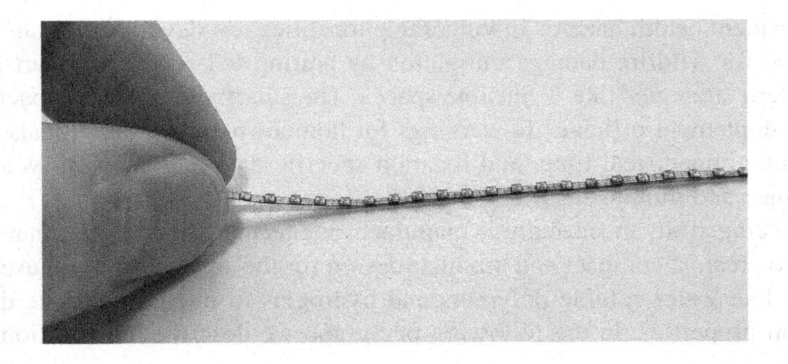

FIG. 11

Ultrathin electronics encased in fibers for integration into fabrics by Accenture Labs and MIT CSAIL researchers.

smart running shirt when it goes out of style, the self-heating coffee cup becomes fertilizer for your garden, or your ticket to a concert becomes your snack before the show. Smart materials provide us with the ability to design our consumables to have function after their usable lifespan, designing for the full life cycle of the objects. In the next section, we discuss how these technologies can support the creation, restoration, and maintenance of resilient natural and urban environments.

4 Natural and built environments

Our natural and built environments are facing unprecedented dangers—from large scale deforestation to massive wildfires. As interest within the tech community expands beyond urban settings, novel tools and devices are being developed to support more sustainable interactions with natural environments and inform conservation action. In the following text we provide examples of how smart materials are facilitating the development of the next generation of technologies that support resilient natural and built environments.

4.1 IoT smart probes for wildfire resilience in urban environments

Despite rising concerns around wildfire damage in wildland-urban interfaces, most homeowners do not take proactive steps to mitigate wildfire dangers. Government mandates require the creation of defensible spaces around houses in high-risk areas, but homeowners continue to perceive defensible spaces as a contingency, or an optional preventative measure rather than a strategic action.

Our team at Accenture Labs sought to develop technological solutions to tackle the raging wildfire problem across the United States and reduce its effects on urban communities. To enable the everyday person to create an actively managed and monitored fire-safe space around their home that mitigates the risk of wildfire damage and consequent health hazards in vulnerable localities, we developed a state-of-the-art method for wildfire damage mitigation by pairing IoT-enabled “smart probes” with modern strategies like defensible spaces. The smart probe is a compact device that can supplement official fire warnings for homeowners and individuals by providing fine-grained, real-time, and location-specific data and enabling year-round maintenance and monitoring.

We leveraged advanced additive manufacturing techniques like multimaterial 3D printing to create a compact and modular design for the smart probe and used novel materials like water soluble polymers and hydrogels to integrate unique drought-prevention properties. In the following paragraph we describe the functionality of our probe:

The smart probe allows for precise monitoring and proactive maintenance of defensible spaces through three distinct functionalities:

- Early fire warnings through continuous environmental observation
- Autonomous protection of the homeowner’s property
- Proactive upkeep reminders for the homeowner

The smart probe is equipped with solar-powered sensors that detect precise environmental parameters and map the air conditions around the home. The probe acts as an early detection device that sounds the alarm and alerts the homeowner when high-risk situations arise, or when ignition is detected nearby. Additionally, subsoil moisture sensors remind the homeowner to water the land when drought-like conditions are detected and can connect to water sprinkler systems for automated land upkeep. By watering only when needed, the probes enable more efficient water use on a massive scale. To enable autonomous maintenance of defensible spaces, the probes are fitted with dissolvable spikes or “nose cones” that deliver nontoxic moisture-retaining hydrogels to vegetation on the homeowner’s property. Well-hydrated defensible spaces allow firewise plants to stay healthy and reduce the risk of vegetation catching on fire. Finally, fire suppressant caps attached to the probe release fire extinguishing chemicals when they contact flames, acting as the first line of defense against ground fire (Fig. 12).

Each probe is registered with a LoRaWAN mesh network, a global, decentralized way of sending small amounts of data from IoT devices. LoRaWAN connectivity can enable a powerful network of interconnected probes that easily analyze geographically distributed data with near real-time insights and also allows the probes to act as an extension to homeowners’ existing smart home technology.

The probe is modular and adaptable, and features can be added or removed depending on the needs of the property. For example, in areas further from home, the probe might be fitted with the sensor insert to detect when environmental conditions pose a fire risk. In areas closer to the home or near vegetation, probes might be fitted with the suppressant insert and the hydrogel spike to counteract small fires or droughts.

The smart probe proves that viable fire sensing, monitoring, and proactive maintenance can be done collectively through a singular network-integrated and compact device. The precise temporal and spatial data collected by networks of these probes

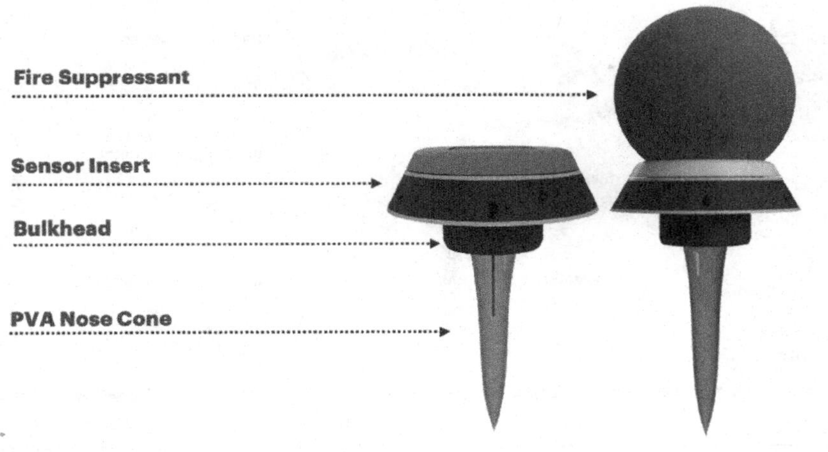

FIG. 12

Components of the smart probe.

can be used at the local homeowner level, the larger neighborhood level, and even the state-wide level for fire tracking and prevention. Data collected via homeowners and commercial land managers can be used to reduce fire risk and damage while the collection of larger-scale network data can be used in predictive analytics to promote community wildfire resilience in the future. Enabling homeowners and individuals with tools like the smart probe can raise awareness around wildfire safety, improve the accuracy of fire prediction systems, and democratize personal wildfire preparation, ultimately creating resilient urban environments.

4.2 E-seeds for scalable and rapid reforestation of natural and seminatural environments

Reforestation of land, especially after wildfire damage, is expensive, laborious, and unsafe for plantation crews. An alternative is aerial seeding, which is a technique for sowing seeds by spraying them from an airplane, helicopter, or drone. It is invaluable in remote, steep, or hazardous areas in which soils need to be rejuvenated for agriculture or for land regeneration. However, the process is often inefficient and requires seed application rates to be increased manyfold to ensure reasonable seed establishment rates [4].

To address these limitations, researchers at Accenture Labs and the Morphing Matter Lab at Carnegie Mellon University developed e-seeds: biomimetic seed carriers that can bury themselves in the soil to facilitate targeted aerial reforestation, soil enrichment, and soil sensing at scale (Fig. 13).

The e-seed is a biomimetic shape-changing interface inspired by the morphology and dispersion abilities of the naturally occurring erodium seed. It is a wood-based

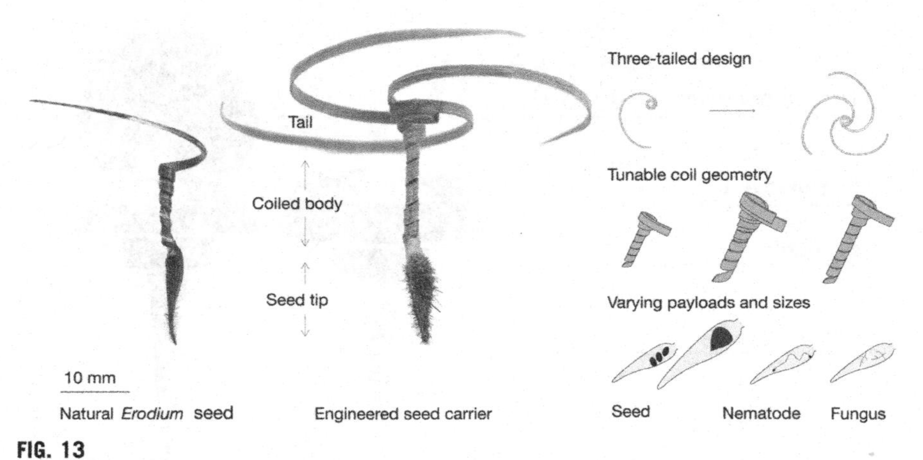

FIG. 13

Design concept of the biomimetic e-seed carrier.

payload carrier that harnesses moisture fluctuations in its environment to self-drill and self-anchor in the soil. It comprises a "tip" that can encapsulate a payload (e.g., seeds, nutrients, microorganisms, etc.), a helical "body" that can self-drill, and one or more "tails" that allow the e-seed to orient itself upright.

To optimize the e-seed for different payloads and environments, Luo et al. developed a lightweight parametric computational design tool to control the size, shape, curvature, and torsion of the e-seed [25]. Besides planting, e-seeds can also be integrated with miniaturized sensors or biodegradable stimuli-sensitive inks for in-situ detection of key environmental parameters, with the possibility of aerial deployment in difficult or remote locations (Fig. 14).

As demonstrated in a recent *Nature* article [26], the researchers showed how design optimization of the e-seed allowed the biodegradable seed carrier to not only plant seeds much more efficiently and autonomously in the soil, but also deliver soil-enriching microbes like insect-parasitic nematodes near the roots of the plants to prevent them from rot and infestation.

Bioinspired innovations like the e-seed that were engineered and optimized to solve the challenge of efficient, large-scale, aerial seeding is a prime example of problem-led engineering and highlights the potential for natural materials to be highly controllable and functional, as well as sustainable.

4.3 Self-cleaning surfaces for public health and safety

Public health and safety have emerged as key priorities in urban environments ever since the global COVID-19 pandemic. Although COVID is spread through air particles, rather than on surfaces, contaminated surfaces are still known to spread bacteria and viruses. Ensuring the cleanliness and disinfection of public surfaces can help prevent future pandemics. Today, most high touch textile or fabric surfaces used

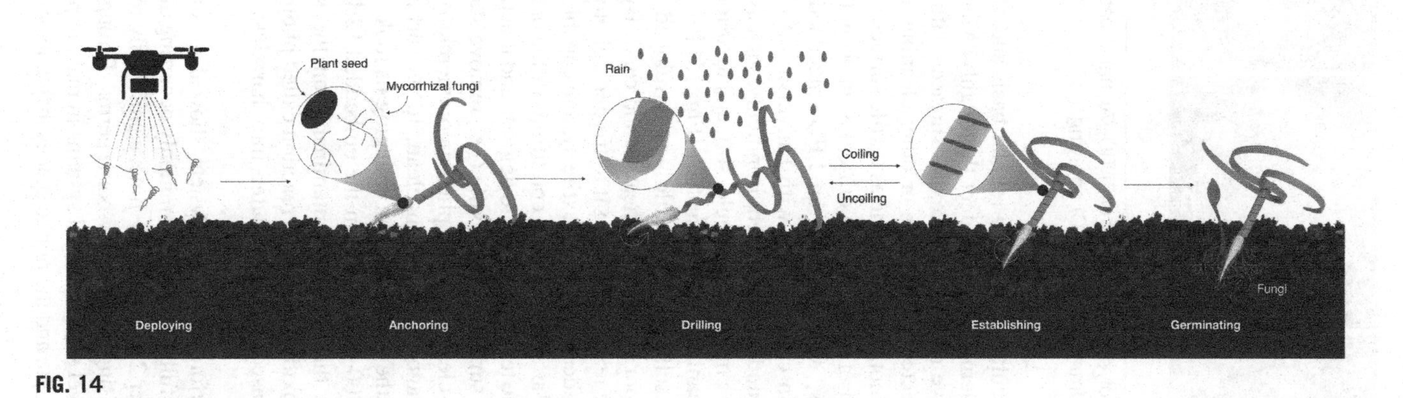

FIG. 14

E-seeds as a tool for mass reforestation and soil enrichment.

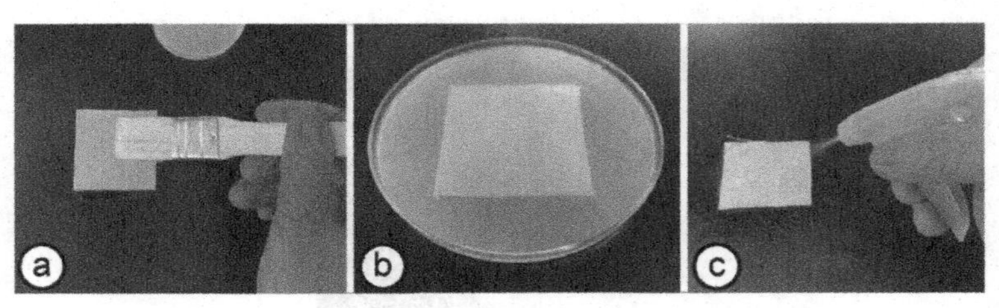

FIG. 15

Process of coating textile (polyester) swatches with TiO_2/Ag nanocoating using three different methods: (A) painting, (B) immersion, and (C) spraying.

in public spaces are difficult to clean on a regular basis and often require the use of environmentally toxic chemicals to disinfect thoroughly. Many public spaces (like buses, hospitals, etc.) have replaced soft textile surfaces with rigid plastic surfaces that can be easily disinfected with a wipe or other cleaning agent. However, these plastic surfaces are often not as comfortable as textile surfaces, disincentivizing their use in spaces where the user experience is critical, such as in theaters, or in cars. To combat these challenges, we developed an LED-embedded, photocatalytic self-cleaning textile system in collaboration with University of Colorado Boulder that can be seamlessly integrated into public environments, such as in automotive or theater seating, allowing for a more comfortable, soft, and luxurious experience without compromising on the cleanliness of the surface (Fig. 15) [9].

Our LED-embedded self-cleaning textile was developed for use in high-touch surfaces in low light indoor spaces (i.e., theaters and cars). By embedding the light source (LEDs) within the textile and using it to trigger a photocatalytic reaction, we were able to successfully design a novel textile system capable of self-cleaning continuously, irrespective of ambient lighting conditions or the presence of humans. We use "clean" to describe the textile's ability to degrade and inactivate any organic contaminants present on the surface of the textile (i.e., remove stains and kill bacteria). To make our textile self-cleaning, we applied a commercially available photocatalytic titanium dioxide nanocoating, Lumaclean, to its surface. This nanocoating adheres to the surface of the textile without changing its look or feel. When exposed to visible or UV light (200–500 nm) from the embedded LEDs, the photocatalytic nanocoating reacts with the organic contaminants on the surface of the textile through a reduction and oxidation (redox) reaction called photocatalysis, thus breaking down the carbon-based organic matter into harmless by-products such as CO_2 and H_2O (Fig. 16).

Antibacterial tests performed on our system show that our light integrated self-cleaning system is able to fully inactivate commonly occurring disease-causing bacteria on public surfaces in under 2 h of exposure time, significantly reducing the chances of public health outbreaks arising from surface-borne germs. Through this project, we demonstrated the utility of novel smart material systems in making public spaces safer, and building trust between people and the public spaces they interact with. By reducing the

FIG. 16

Self cleaning textile System with integrated LEDs and photocatalytic nanocoating for high touch public surfaces.

need to use chemical disinfectants frequently on surfaces, we also show how embedding autonomy and intelligence in public infrastructure can allow for sustainable management of urban spaces.

5 Conclusion

Embedding "smarts" into everyday products sustainably requires deeper investigation into novel materials that do away with the e-waste problem. As an example, current PCBs are made from epoxy resin and fiberglass and require energy-intensive methods of shredding and incineration to be processed when thrown away. To tackle this issue, Jiva Materials developed the Soluboard, a PCB laminate alternative made of natural fiber and other biodegradable ingredients capable of disintegrating in warm water. This allows the natural fibers to be composted, the remaining solution to be disposed of using standard domestic wastewater systems, and the electronic components to be removed for reprocessing [27].

When used to create sustainable smart spaces and products, the potential for smart materials is limitless—from fully degradable insulin pumps to IoT sensors that can be dissolved leaving behind the multipurpose chips for reuse, to batteries that can be discarded in a backyard compost or used for garden fertilizer. The technologies and concepts highlighted in this chapter introduce methodologies to create components for smart, sustainable products while decreasing their environmental impact.

We note that research on methods to embed intelligence into everyday spaces is ever evolving. A novel research area ripe for exploration is the incorporation of human-like, dynamic intelligence in intelligent systems. As pioneers of this area of research, Cortical Labs is working on growing organic human neural networks across silicon chips within a simulated environment that sends information to the neurons about their environment, with positive or negative feedback [28,29]. The silicon chip has a set of pins that send electrical impulses into the neural structure, and receive impulses back in return. The organic-digital intelligence uses lower power than traditional electronic systems, is self-programming and infinitely flexible, and has more intuition, insight, and creativity than AI, similar to human intelligence.

Ultimately, to realize products like sustainable electronics, eco-conscious e-textiles, and bio-inspired and bio-integrated sensors and actuators for ecological restoration and urban resilience, there is a need for continued research in new materials and bio- and nature-inspired techniques to create complex components that can be assembled into low impact designs. We encourage technologists, designers, and researchers to build on the concepts introduced in our research and leverage a materials-first product design process that prioritizes the use of biological structures, natural and biodegradable materials, and principles of circularity to create new applications and designs that tackle today's most pressing environmental challenges.

References

[1] P. Lacy, J. Long, W. Spindler, The Circular Economy Handbook: Realizing the Circular Advantage, first ed., Palgrave Macmillan, 2020.

[2] The 2050 Criteria: Guide to Responsible Investment in Agricultural, Forest, and Seafood Commodities, World Wildlife Fund, 2012.

[3] P. Lacy, D. Waughray, Harnessing the Fourth Industrial Revolution for the Circular Economy Consumer Electronics and Plastics Packaging, World Economic Forum, 2018.

[4] A. Maheshwari, A. Kumar Aggarwal, A. Danielescu, Designing tools and interfaces for ecological restoration: an investigation into the opportunities and constraints for technological interventions, in: Proceedings of the 2022 CHI Conference on Human Factors in Computing Systems. CHI '22, Association for Computing Machinery, 2022, pp. 1–17, https://doi.org/10.1145/3491102.3517664.

[5] B. An, Y. Tao, J. Gu, et al., Thermorph: democratizing 4D printing of self-folding materials and interfaces, in: Proceedings of the 2018 CHI Conference on Human Factors in Computing Systems. CHI '18, Association for Computing Machinery, 2018, pp. 1–12, https://doi.org/10.1145/3173574.3173834.

[6] A. Tugui, Calm technologies in a multimedia world, Ubiquity (March) (2004) 1, https://doi.org/10.1145/985616.985617.

[7] I. Poupyrev, A Smarter Wardrobe With Jacquard by Google, Google, 2019. https://blog.google/products/atap/smarter-wardrobe-jacquard-google/. (Accessed 14 July 2023).

[8] L.M. Gonzalez, Beyond the Brick: Collaborations With a Sensing Microbial System in the Built Environment, Thesis, Massachusetts Institute of Technology, 2022. https://dspace.mit.edu/handle/1721.1/144850. (Accessed 14 July 2023).
[9] F. Bell, A. Hong, A. Danielescu, et al., Self-deStaining textiles: designing interactive systems with fabric, stains and light, in: Proceedings of the 2021 CHI Conference on Human Factors in Computing Systems. CHI '21, Association for Computing Machinery, 2021, pp. 1–12, https://doi.org/10.1145/3411764.3445155.
[10] K.W. Song, A. Maheshwari, E.M. Gallo, A. Danielescu, E. Paulos, Towards decomposable interactive systems: design of a backyard-degradable wireless heating interface, in: Proceedings of the 2022 CHI Conference on Human Factors in Computing Systems. CHI '22, Association for Computing Machinery, 2022, pp. 1–12, https://doi.org/10.1145/3491102.3502007.
[11] J. Gu, D.E. Breen, J. Hu, et al., Geodesy: self-rising 2.5D tiles by printing along 2D geodesic closed path, in: Proceedings of the 2019 CHI Conference on Human Factors in Computing Systems. CHI '19, Association for Computing Machinery, 2019, pp. 1–10, https://doi.org/10.1145/3290605.3300267.
[12] G. Wang, Y. Tao, O.B. Capunaman, H. Yang, L. Yao, A-line: 4D printing morphing linear composite structures, in: Proceedings of the 2019 CHI Conference on Human Factors in Computing Systems. CHI '19, Association for Computing Machinery, 2019, pp. 1–12, https://doi.org/10.1145/3290605.3300656.
[13] H. Kaimoto, J. Yamaoka, S. Nakamaru, Y. Kawahara, Y. Kakehi, ExpandFab: fabricating objects expanding and changing shape with heat, in: Proceedings of the Fourteenth International Conference on Tangible, Embedded, and Embodied Interaction. TEI '20, Association for Computing Machinery, 2020, pp. 153–164, https://doi.org/10.1145/3374920.3374949.
[14] T. Cheng, T. Tabb, J.W. Park, et al., Functional destruction: utilizing sustainable materials' physical transiency for electronics applications, in: Proceedings of the 2023 CHI Conference on Human Factors in Computing Systems. CHI '23, Association for Computing Machinery, 2023, pp. 1–16, https://doi.org/10.1145/3544548.3580811.
[15] C. Herrojo, F. Paredes, J. Mata-Contreras, F. Martín, Chipless-RFID: a review and recent developments, Sensors 19 (15) (2019) 3385, https://doi.org/10.3390/s19153385.
[16] A. Vena, E. Perret, S. Tedjini, Chipless RFID tag using hybrid coding technique, IEEE Trans. Microw. Theory Tech. 59 (12) (2011) 3356–3364, https://doi.org/10.1109/TMTT.2011.2171001.
[17] ChroMorphous, A New Fabric Experience, ChroMorphous, 2023. http://chromorphous.com/. (Accessed 14 July 2023).
[18] bioLogic, 2023. https://tangible.media.mit.edu/project/biologic. (Accessed 14 July 2023).
[19] A. Inman, T. Hryhorchuk, L. Bi, et al., Wearable energy storage with MXene textile supercapacitors for real world use, J. Mater. Chem. A 11 (7) (2023) 3514–3523, https://doi.org/10.1039/D2TA08995E.
[20] Y. Wang, T. Guo, Z. Tian, K. Bibi, Y.Z. Zhang, H.N. Alshareef, MXenes for energy harvesting, Adv. Mater. 34 (21) (2022) 2108560, https://doi.org/10.1002/adma.202108560.
[21] D. Patron, K. Jost, A. Cook, et al., Knitted wireless power harvesting and storage, in: The Fiber Society 2014 Fall Meeting and Technical Conference, 2014. https://drexel.edu/functional-fabrics/research/projects/wi-fi-power-harvesting/. (Accessed 16 July 2023).
[22] G. Loke, T. Khudiyev, B. Wang, et al., Digital electronics in fibres enable fabric-based machine-learning inference, Nat. Commun. 12 (1) (2021) 3317, https://doi.org/10.1038/s41467-021-23628-5.

[23] AFFOA, Technology, 2023. https://affoa.org/technology/. (Accessed 16 July 2023).
[24] M. Rein, V.D. Favrod, C. Hou, et al., Diode fibres for fabric-based optical communications, Nature 560 (7717) (2018) 214–218, https://doi.org/10.1038/s41586-018-0390-x.
[25] D. Luo, J. Gu, F. Qin, G. Wang, L. Yao, E-seed: shape-changing interfaces that self drill, in: Proceedings of the 33rd Annual ACM Symposium on User Interface Software and Technology. UIST '20, Association for Computing Machinery, 2020, pp. 45–57, https://doi.org/10.1145/3379337.3415855.
[26] D. Luo, A. Maheshwari, A. Danielescu, et al., Autonomous self-burying seed carriers for aerial seeding, Nature 614 (7948) (2023) 463–470, https://doi.org/10.1038/s41586-022-05656-3.
[27] Jiva Materials, The World's First Fully Recyclable PCB Substrate, Jiva Materials, 2023. https://www.jivamaterials.com/. (Accessed 16 July 2023).
[28] B.J. Kagan, A.C. Kitchen, N.T. Tran, et al., In vitro neurons learn and exhibit sentience when embodied in a simulated game-world, Neuron 110 (23) (2022) 3952–3969.e8, https://doi.org/10.1016/j.neuron.2022.09.001.
[29] Cortical Labs, DishBrain Intelligence, 2023. https://corticallabs.com/. (Accessed 16 July 2023).

CHAPTER

IoT interoperability enhances smart and healthy living

6

Andreas Andreou[a], Constandinos X. Mavromoustakis[a], and Evangelos K. Markakis[b]

[a]*Department of Computer Science, University of Nicosia and University of Nicosia Research Foundation, Nicosia, Cyprus,* [b]*Department of Electrical and Computer Engineering, Hellenic Mediterranean University, Heraklion, Crete, Greece*

1 Introduction

Older adults tend to have several concurrent medical conditions, requiring a number of prescribed medications to help control their chronic diseases. Thus there is a need for a personalized approach to the safe and effective use of these medications to ensure the best possible results. Specifically, in this chapter we focus on patients over 65 with diabetes and high blood pressure who must be monitored to avoid hyper/hypoglycemic episodes or hypertension. Medication can be adjusted after the healthcare provider reviews their health parameters by remote monitoring. The goal is to optimize and personalize the treatment accordingly and in real time, to prevent potentially harmful health conditions.

The IoT network of connected devices and sensors allows end users to control and monitor environments in a simple yet powerful way. But while the IoT's potential is vast, getting the devices to work together seamlessly is a significant challenge. For the IoT to live up to its full potential, there needs to be a high level of interoperability between devices and systems. Interoperability is the ability of different methods, services, and applications to interact, allowing free flow of data among them, so that when other devices are connected, the acquired data can be used to control and monitor activities. However, due to the lack of standardization in the industry, it is often tricky for devices from different manufacturers to collaborate.

Manufacturers are turning to different technologies to overcome this challenge to ensure interoperability. These include open-source protocols such as MQ Telemetry Transport (MQTT) and Advanced Message Queuing Protocol (AMQP) and standards developed by organizations like the AllSeen Alliance. These protocols and standards provide developers with a common language for communicating with two connected devices. It helps to reduce development costs and time to market, resulting in faster innovation. Furthermore, cloud-based platforms are also helping to boost interoperability. With them, developers can connect their products to

Smart Spaces. https://doi.org/10.1016/B978-0-443-13462-3.00011-X

different types of systems without developing an interface for each one. It allows free flow of data between systems, platforms, and devices, creating a truly connected ecosystem. Ultimately, the goal of interoperability is to make the IoT more accessible. By making it easier for different devices to work together, we can unlock the power of the IoT and create more intelligent, efficient, and secure systems [1]. With the right tools and technologies in place, interoperability will help bring the future of the IoT within reach.

Smart and healthy living is essential for individuals and communities, as it increasingly incorporates technology and innovations that have the potential to make life easier, healthier, and more efficient. Healthy living encompasses lifestyle practices that promote overall well-being, such as consuming nutritious food, exercising regularly, and getting restful sleep. Integrating smart and healthy living into everyday habits is beneficial for many reasons. By using intelligent technologies, we can reduce our energy, fuel, and water usage, reducing our carbon footprint and saving money in the long run. Meanwhile, focusing on healthy habits promotes increased productivity and improved mental awareness, leading to a better quality of life (QoL).

A few critical factors exist for older adults looking to pursue smart and healthy living. First, smart technology investments should be made to make their home more energy-efficient and automated. Some key ideas include adding insulation, repairing air leaks to prevent energy loss, and installing solar panels, energy-efficient appliances, and light-emitting diode (LED) systems. Second is a focus on healthy lifestyle choices by eating nutritious meals, taking frequent breaks to stretch and relax, and getting adequate rest [2]. Additionally, it is important to spend time outdoors, to get fresh air and sunlight, and to develop healthy routines such as creating to-do lists and setting achievable goals. Humans can maximize their potential and improve their QoL by approaching their lives from a smart and nutritional perspective. In addition, healthy living for older adults can now involve real-time monitoring by healthcare providers, who can intervene any time with guidelines, to give medication, or in situations of emergency.

1.1 Motivation and contribution

The high rates of population aging motivate investigation of solutions to maintain the healthy living of seniors. Smart and healthy sustainability for older adults requires predicting and preventing unhealthy situations. Continuous monitoring of physiological parameters such as blood glucose, blood pressure, heart rate, weight, etc. is of paramount importance in long-term chronic diseases such as diabetes and hypertension. However, the acquired data should be provided for evaluation through analytics and interpretation by healthcare providers, so they can prevent hyper/hypoglycemic events in patients with diabetes or diagnose hypertension episodes.

Motivated by the aforementioned, we analyze the acquired data to help predict clinical deterioration and establish dynamic thresholds for control of vitals. Also, intending to assist older individuals with self-management of their conditions, we have enabled the predictions to be displayed to the user. Therefore the harmonized dataset is analyzed on the basis of older adults' normal intervals for clinical parameters. Healthcare professionals can then provide personalized advice and treatment interventions based on the analytic findings. In addition, the findings will allow clinicians to act earlier to reduce the risks of possibly harmful situations and improve patients' disease control.

Anomalies may occur due to various abnormal activities in standard processes. They are classified as point anomaly, contextual anomaly, and collective anomaly. A point anomaly occurs when a data value behaves differently from all other data of the same type in the dataset, called an outlier. It is a common anomaly type. Contextual anomalies are found in datasets that include behavioral and contextual features. Our contribution was based on identifying abnormalities within the datasets. The aim was to predict the occurrence of hyperglycemic or hypoglycemic episodes in older adults suffering from diabetes and the occurrence of hypertension. However, the underlying and major goal was to enable remote monitoring of older adults by healthcare providers. Integrating IoT devices allows healthcare providers to be more vigilant and connect with patients proactively.

1.2 Related works

The authors of Ref. [3] applied a systematic survey regarding IoT frameworks for smart city applications, resulting in IoT development and defining the term "smart city." Khan, utilizing a modified deep convolutional neural network, proposed an IoT framework to assess heart disease more accurately [4]. Based on the same orientation, Andreou et al. proposed an IoT framework for age-friendly communities, aiming to prevent situations similar to COVID-19 [5,6]. Energy management was investigated by Prakash et al. and Andreou et al. within the same context, designing a system to completely replace power outage in a region with partial load shedding in a controlled manner according to the consumer's preference [7,8]. Also, Wirtz, Weyerer, and Schichtel introduced an IoT framework integrated with government, consolidating a smart ecosystem [9].

Although we are on the cusp of the smart cities era, there are various challenges to securing the development of a robust intelligent ecosystem [10]. Those challenges have been investigated by the authors in [11] where they identify and prioritize innovative engineering challenges through the best-worst method. Milosevic et al. [12] interpreted the substantial barriers for smart city development using a fuzzy multi-criteria decision support system to focus on the areas in danger of terror attacks. Intuitionistic fuzzy sets were used by Goala et al. to present an innovative aggregation operation for the same context [13]. Notwithstanding the preceding, to optimize the smart operations of a smart city's framework, governments need to equip the infrastructure with innovative technologies that have efficient methods and techniques for the safety and well-being of its inhabitants [14].

Anomaly detection is identifying data and patterns that do not fit in an acquired dataset [15]. Defining anomaly detection more accurately is finding patterns that occur during normal actions and exhibit abnormal behavior within a dataset [16]. According to Agrawal and Agrawal, the patterns can be considered as anomalies or outliers [17]. Real-time anomaly detection within datasets is necessary to identify possible malfunctions in a minimum time and to produce solutions [18]. Anomaly detection is a structure that autonomously determines the dependencies between the data and analyzes them by monitoring them continuously, rather than the detection of exceeding a predetermined limit of a value in the dataset [19]. Anomaly detection systems are theoretically based on solid foundations and support fast detection, easy maintenance, and reusability for small, medium, or large-scale problems that may arise in production systems. In this way, it ensures that the models planned to be developed are subjected to early testing processes. Classification processes in anomaly detection systems can be used in supervised and unsupervised methods [20].

1.3 Methodology

The data flow is displayed in Fig. 1. As can be seen, the data is driven from the clinical devices (blood pressure monitor, weight scales, pulse oximeter, and blood glucometer) via Bluetooth to an application on the participant's smartphone or tablet device. The patients will be able to view their clinical parameters as well as alerts triggered by the system in anomalous circumstances. Also, reminders and medication lists will be available on end users' screens. Then the data will be uploaded

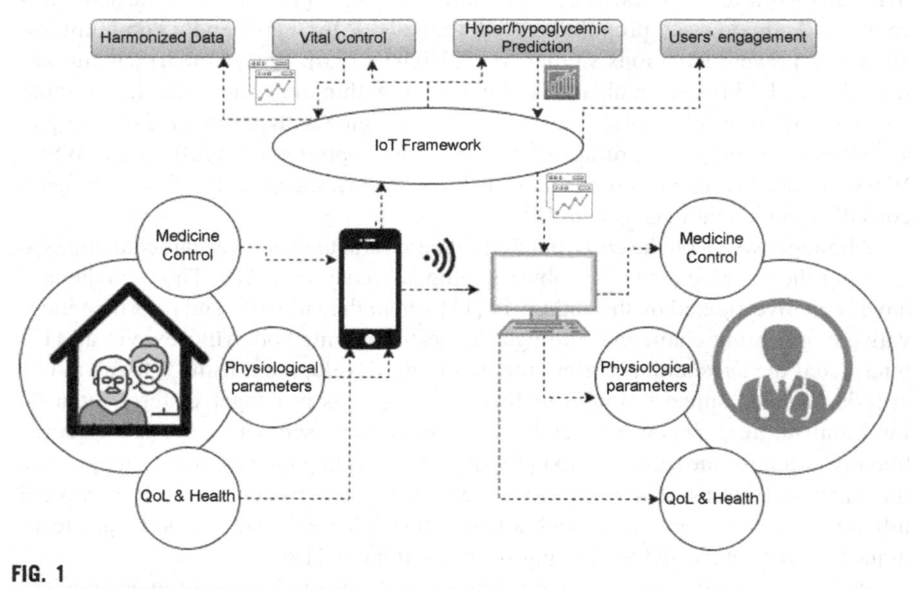

FIG. 1

Data flow.

through Wi-Fi to the associated digital solution and stored on the server. Finally, data will be transferred from the digital solutions via Wi-Fi to the data lake where analysis will be performed. Finally, the interpretation of the data will be driven back to the servers and displayed through a browser-based dashboard to enable doctors to track adherence.

2 Proposed framework

Aiming to address and improve deficiencies in medication adherence and treatments of older individuals living at home with diabetes, we propose integrating an IoT framework in which the healthcare provider can remotely monitor the most important physiological parameters of the patient. The anticipated benefits of this highly personalized approach are to help diabetes patients self-monitor their health conditions, physiological parameters, and medication compliance. The personalized approach advocated in this study and practiced as a pilot activity considers the early identification of hyper/hypoglycemic episodes that lead to worsening of symptoms and, in some cases, provides the opportunity to adjust medications and treatments to deliver a safer and more effective use of medication in-home, thus improving the quality of life of the care recipients and reducing rehospitalization occurrences.

2.1 Overview

Fig. 2 illustrates the architecture of the proposed framework. As can be seen, the input will be data driven from the patients. Acquired data include physiological parameters like heart rate, blood pressure, blood glucose, etc., which will be driven through Bluetooth to the application installed on patients' smartphones or tablets. Patients will have access to their data as well as medication list reminders and alerts in case of abnormal acquired values. The data will then be driven through Wi-Fi to the data lake, where the analytics take place. Throughout analysis, we will procure the interpretation of the collected data and forecast the occurrence of hypertension or hypo-/hyperglycemic episodes. The results from the evaluation will be available through the app to a dashboard that can be run on any browser. Hence, the healthcare provider will be able to interpret the assessment from the analytics evaluation and act accordingly, either by changing the patient's medication or sending an alert.

2.2 Outlier detection

Abnormal vitals lead to outliers in the dataset that the system should identify. Therefore we identify them by clustering methods. More specifically, for data mining, we utilized the k-means^{++} algorithmic technique, proposed by Arthur and Vassilvitskii, which can select the initial data values for the k-means clustering algorithm [21,22]. The k-means clustering method divides n observations into k clusters in which each observation belongs to the cluster with the adjoining mean, serving as a prototype of

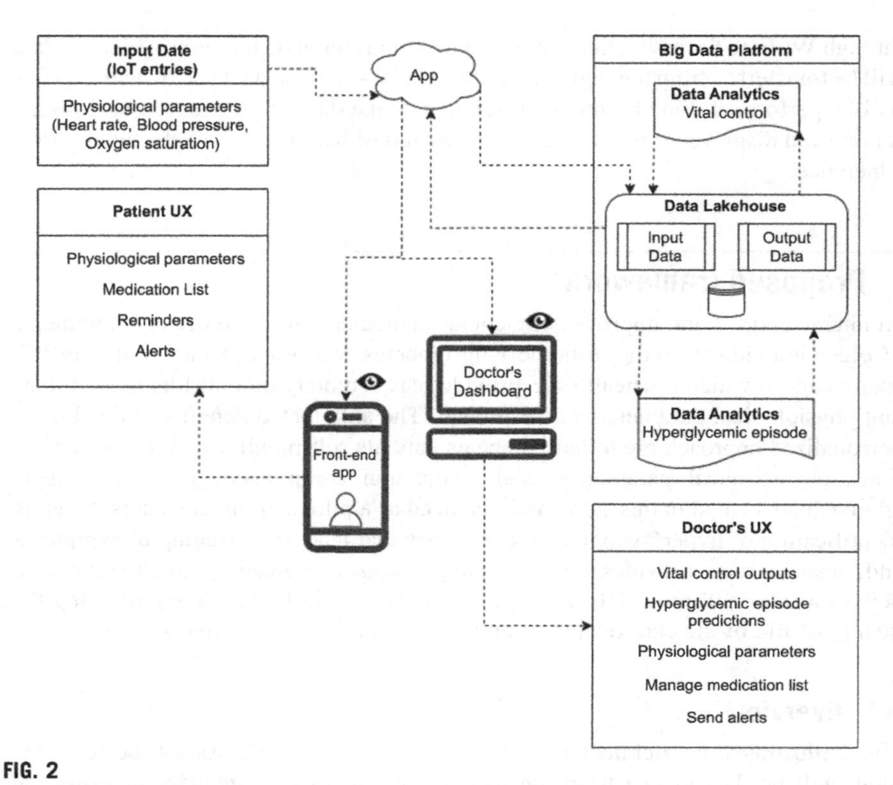

FIG. 2

System architecture.

the cluster results to Voronoi diagrams. Also, the most used fuzzy clustering algorithm, the fuzzy C-means clustering algorithm [23], was deployed to force border point data among two clusters to fall randomly into a cluster, as this clustering technique can fragment n clusters with every data point in the dataset belonging to every cluster to a certain degree.

For the k-means clustering method, the goal is to find the objective function by Eq. (1), where x is a point from the set of observations $(x_1, x_2, \ldots, x_n)$, $k \leq n$ is the number of clusters, and μ_j is the mean of the k sets $S = (S_1, S_2, \ldots, S_k)$. More precisely, the objective function minimizes the intracluster variance. We use the fuzzy C-means clustering method for the best result of an overlapped dataset. It enables the inclusion of data values in more than one cluster, and each data point could be included in the clusters with a different degree of membership. Hence, it is achievable by the depreciation of $g(x)$, in Eq. (2), where the additional factor w_j^m indicates the degree of membership for $x \in S_j$. In addition, to overcome the problem that the means μ_j determined by the k-means method could be subjectively improper in contradiction to the optimal clustering for $f(x)$, we deployed the k-means^{++} algorithm.

$$f(x) = \arg min_S \sum_{j=1}^{k} \sum_{x \in S_j} \|x - \mu_j\|^2 \tag{1}$$

$$g(x) = \arg min_S \sum_{j=1}^{k} \sum_{x \in S_j} w_j^m \|x - \mu_j\|^2, m \geq 1 \tag{2}$$

However, before proceeding with the algorithmic technique, we first need to normalize the distance of each data point x by Eq. (3), where x_{min} and x_{max} are the minimum and the maximum Euclidean distance among the distances of the data points from the means μ_j, respectively.

$$x_{normal} = \frac{x - x_{min}}{x_{max} - x_{min}} \tag{3}$$

We utilized the silhouette coefficient to distinguish the best clusters and applied a weighting technique to determine our preference for these clusters [24]. The weight $W(x)$ is calculated by Eq. (4), where $p(x_i)$ is the probability of a point being included in a cluster, and m denotes the cluster centers after implementing the fuzzy C-means clustering method.

$$W(x) = \sum_{i=1}^{m} \sqrt[w_i]{\prod_{i=1}^{m} [p(x_i)]^{w_i}} \tag{4}$$

Then we investigate the inclusion of each point in the corresponding cluster center using a threshold ϑ and the maximum of the weights $W_{max}(x)$. Thus if the point fulfills the inequality (5), it is considered an outlier, as it has a limited chance of being included in a cluster with legitimate data points.

$$W_{max}(x) \leq \frac{1 + \vartheta m}{m} \tag{5}$$

Algorithm 1 presents the pseudocode of the proposed method that generates the outliers. Initially, the system reads the dataset, and we set the threshold denoted by ϑ. We also input the required values of a and b, which indicate the percentages of the total data comprising the maximum Euclidean distances from the corresponding cluster centers to be removed after steps 2 and 6, respectively. Then we calculate the clusters and their mean values. To this point, we need to investigate whether the Euclidean distances of each point from the corresponding mean are greater than $a\%$ of the maximum Euclidean distance. If so, we recall them as maximum points x_{max} and the remaining points establish new clusters with new means, where we reinvestigate whether the Euclidean distances of each remaining point from the corresponding mean is greater than the $b\%$ of the maximum Euclidean distance. Therefore if the number of elements denoted by $\text{N}(x)$ is less than the mean number of elements denoted by $\overline{\text{N}}(x)$, we form m cluster centers and apply Eq. (3) for normalization. Finally, after the iterations for the three aforementioned clustering methods, we calculate the weights by Eq. (4) and examine the outlier condition (5) that indicates whether the data point should be labeled as an outlier.

Algorithm 1

Input: Dataset, ϑ, a, b
Output: Outliers
1: Calculate the cluster centres k and their corresponding means μ_j
2: *if* $\|x-\mu_j\| > \frac{a\|x-\mu_j\|_{max}}{100}$ *do*
3: $x_{max} \leftarrow x$
4: $\mu_j' \leftarrow \mu_j$
5: $k' \leftarrow k$
6: *if* $\left\|x_{max}-\mu_j'\right\| > \frac{b\|x-\mu_j\|_{max}}{100}$ *do*
7: $k' \leftarrow k + k_{new}$
8: *if* $\mathrm{N}(x) < \overline{\mathrm{N}}(x)$ *do*
9: $\mu_j'' \leftarrow \mu_j$
10: $m \leftarrow k'$
11: $x_{normal} \leftarrow$ (3)
12: *end if*
13: *end if*
14: *end if*
15: **Repeat**
16: $W(x) \leftarrow$ (4)
17: *if* $W_{max}(x) \leq \frac{1+\vartheta m}{m}$ *then*
18: x: *outlier*
19: *else*
20: x: *data*
21: *end if*
22: **End**

3 Experimental results

We experimentally collected data from older adults to examine their interaction with the technological components and determine if the data acquisition ran appropriately. After the implementation of the process for several days, as presented in Figs. 3 and 4, the data from blood pressure and heart rate, respectively, are illustrated on the graph, and the outliers can be quickly identified. In addition, by curve fitting and forecasting the obtained curve, we can determine anomalous situations, like hypertension or heart palpitations.

In Fig. 3, two cases can be seen where the blood pressure exceeded the required levels. *Yellow dots (light gray in print version)* determine those cases. In those cases, the healthcare provider can intervene and indicate the appropriate medications or alerts to the corresponding patient. Also, we can see a *red dot (dark gray in print version)* in Fig. 4, which presents a patient's heart rate over time. Therefore, as can be seen from the level of the *red dot (dark gray in print version)*, the healthcare provider needs to be triggered by an alarm that the patient's situation is not normal.

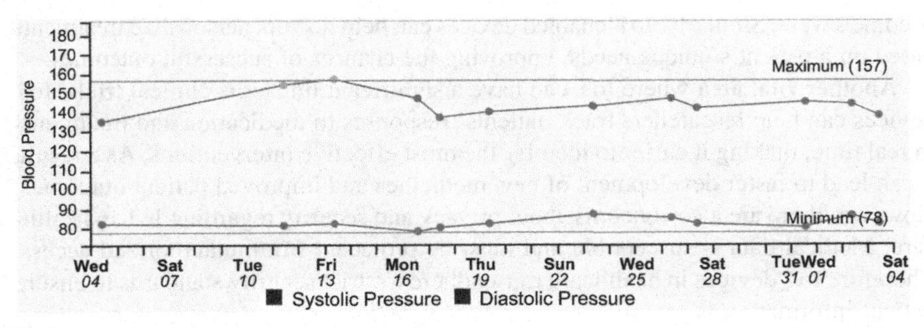

FIG. 3

Blood pressure.

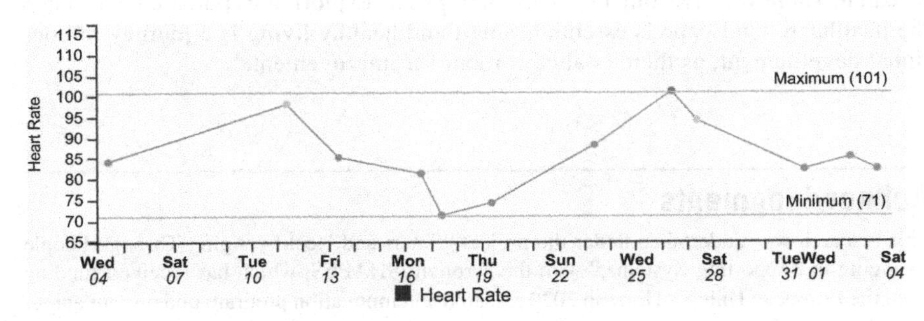

FIG. 4

Heart rate.

4 Conclusion and discussion

The Internet of Things (IoT) has significantly advanced in several sectors, including healthcare. IoT technology, combined with medical devices, sensors, and wearables, is becoming a game-changer in the healthcare industry. The IoT can revolutionize various aspects of healthcare, such as patient monitoring, diagnosis, treatment, clinical trials, and medication development. One of the primary advantages of using IoT in healthcare is the ability to monitor patients remotely. Current research proves that it benefits patients with chronic diseases or who require constant monitoring. For instance, connected devices can monitor a patient's vital signs, such as heart rate, blood pressure, and oxygen levels, and transmit this data to healthcare providers in real time. It allows physicians to identify potential health issues before they escalate and provide timely interventions. IoT technology can also improve patient outcomes by facilitating early diagnosis, faster treatment, and better disease management. For example, wearable devices, such as smartwatches, can collect data on a person's physical activity, sleep patterns, and other vital signs. This information can be used to detect early warning signs of illnesses and intervene before they become severe. Similarly, IoT-enabled devices can help doctors personalize treatments based on a patient's unique needs, improving the chances of successful outcomes.

Another vital area where IoT can have a significant impact is clinical trials. IoT devices can help researchers track patients' responses to medication and treatments in real time, making it easier to identify the most effective interventions. As a result, it can lead to faster development of new medicines and improved patient outcomes. However, there are also concerns about privacy and security regarding IoT in healthcare. Medical data is susceptible and must be protected from unauthorized access. Therefore IoT devices in healthcare must adhere to strict security standards to ensure patient information is secure.

In conclusion, IoT technology offers a tremendous opportunity to transform the healthcare industry. With the proper infrastructure, policies, and regulations, IoT can improve patient outcomes, reduce healthcare costs, and drive innovation in medical research. Therefore investing in IoT technology and exploring its potential to reshape the healthcare landscape is essential. Smart and healthy living is a journey of continual development, as there is always room for improvement.

Acknowledgments

This research was undertaken under the project "Smart and Healthy Aging Through People Engaging in Supporting Systems," with the acronym SHAPES, which has received funding from the European Union's Horizon 2020 research and innovation program under grant agreement No. 857159.

References

[1] A. Andreas, C.X. Mavromoustakis, G. Mastorakis, S. Mumtaz, J.M. Batalla, E. Pallis, Modified machine learning technique for curve fitting on regression models for COVID-19 projections, in: IEEE 25th International Workshop on Computer Aided Modeling and Design of Communication Links and Networks, 2020.

[2] A. Andreou, C.X. Mavromoustakis, G. Mastorakis, E. Pallis, N. Magaia, E.K. Markakis, Intelligently reduce transportation's energy consumption, in: Intelligent Technologies for Internet of Vehicles, Springer, 2021, pp. 293–309.

[3] J.J.P. Abadía, C. Walther, A. Osman, K. Smarsly, A systematic survey of Internet of Things frameworks for smart city applications, Sustain. Cities Soc. 83 (2022).

[4] M.A. Khan, An IoT framework for heart disease prediction based on MDCNN classifier, Access 8 (2020) 34717–34727.

[5] A. Andreou, C.X. Mavromoustakis, G. Mastorakis, J.M.B.a.E. Pallis, Evaluation of the COVID-19 era by using machine learning and interpretation of confidential dataset, Electronics 10 (23) (2021).

[6] A. Andreas, C.X. Mavromoustakis, G. Mastorakis, J.M. Bata, J.N. Sahalos, E. Pallis, E. Markakis, IoT cloud-based framework using of smart integration to control the spread of COVID-19, in: IEEE International Conference on Communications, 2021.

[7] P. Pawar, P. Vittal, Design and development of advanced smart energy management system integrated with IoT framework in smart grid environment, J. Energy Storage 25 (2019).

[8] A. Andreas, C.X. Mavromoustakis, G. Mastorakis, J.M. Batalla, E. Pallis, Energy conservation by using the integration of distributed energy system in smart vehicles, in: IEEE 7th International Energy Conference, Riga, Latvia, 2022.

[9] B.W. Wirtz, J.C. Weyerer, F.T. Schichtel, An integrative public IoT framework for smart government, Gov. Inf. Q. 36 (2) (2019) 333–345.

[10] A. Andreas, C.X. Mavromoustakis, G. Mastorakis, J.M. Batalla, J.N. Sahalos, E. Pallis, E. Markakis, Enhancement of COVID-19 detection by unravelling its structure and selecting the optimal attributes, in: IEEE Global Communications Conference, 2022.

[11] M.I. Khan, S. Khan, U. Khan, A. Haleem, Modeling the Big Data challenges in context of smart cities—an integrated fuzzy ISM-DEMATEL approach, Int. J. Build. Pathol. Adapt. 41 (2021) 422–453.

[12] M.R. Milosevic, D.M. Milosevic, A.D. Stanojevic, D.M. Stevic, D.J. Simjanovic, Fuzzy and interval AHP approaches in sustainable management for the architectural heritage in smart cities, Mathematics 9 (4) (2021).

[13] S. Goala, D. Prakash, P. Dutta, P. Talukdar, K.D. Verma, G. Palai, A decision support system for surveillance of smart cities via a novel aggregation operator on intuitionistic fuzzy sets, Multimed. Tools Appl. 21 (2022).

[14] T. Abbate, F. Cesaroni, M.C. Cinici, M. Villari, Business models for developing smart cities. A fuzzy set qualitative comparative analysis of an IoT platform, Technol. Forecast. Soc. Chang. 142 (2019) 183–193.

[15] V.Q. Nguyen, L. van Ma, J. Kim, LSTM-based anomaly detection on big data for smart factory monitoring, J. Digit. Contents Soc. 19 (4) (2018) 789–799.

[16] A. Andreas, C.X. Mavromoustakis, G. Mastorakis, J.M. Batalla, M. Mukherjee, E. Pallis, Internal virus detection framework based on IoT semantic interoperability, in: IEEE International Conference on Communications, Seoul, Korea, Republic of, 2022.

[17] S. Agrawal, J. Agrawal, Survey on anomaly detection using data mining techniques, Comput. Sci. 60 (1) (2015) 708–713.

[18] G. Hwang, S. Kang, A.J. Dweekat, J. Park, T.W. Chang, An IoT data anomaly response model for smart factory performance measurement, Int. J. Ind. Eng. Theory Appl. Pract. 25 (5) (2018) 702–718.

[19] S. Schriegel, A. Maier, O. Niggemann, Big data and machine learning for the smart factory—solutions for condition monitoring, diagnosis and optimization, in: Industrial Internet of Things, Springer, 2017, pp. 473–485.

[20] A. Andreas, C.X. Mavromoustakis, G. Mastorakis, D.-T. Do, J.M. Batalla, E. Pallis, E.K. Markakis, Towards an optimized security approach to IoT devices with confidential healthcare data exchange, Multimed. Tools Appl. 80 (2021) 31435–31449.

[21] D. Arthur, S. Vassilvitskii, k-means++: the advantages of careful seeding, in: 8th Annual ACM-SIAM Symposium on Discrete Algorithms., Philadelphia, PA, USA, 2007.

[22] A. Likas, N. Vlassis, J.J. Verbeek, The global k-means clustering algorithm, Pattern Recogn. 36 (2) (2003) 451–461.

[23] J.C. Bezdek, R. Ehrlich, W. Full, FCM: the fuzzy c-means clustering algorithm, Comput. Geosci. 10 (2–3) (1984) 191–203.

[24] A.M. Bagirov, R.M. Aliguliyev, N. Sultanova, Finding compact and well-separated clusters: clustering using silhouette coefficients, Pattern Recogn. 135 (2023).

CHAPTER

Ecosystem of smart spaces: An overview review

7

Emeka Ndaguba[a] and Christopher Arukwe[b]
[a]*Department of Building & Human Settlements Development, Faculty of Engineering, the Built Environment and Technology, Nelson Mandela University, Port Elizabeth, South Africa,*
[b]*Greyspot Consult Research Unit, Pretoria, South Africa*

1 Introduction

This chapter discusses smart spaces as an emerging research niche area and attempts to develop the ecosystem of the sector and the capabilities it holds. More conveniently, smart spaces are not delinked from smart cities, smart technologies, and smart management in establishing a culture for smart realities in understanding smart spaces. The ambivalence of smart spaces demonstrates the paucity of research on Shared Spaces in urban affairs and built environments. Nonetheless, certain principles make the bond between smart spaces and smart cities and technologies common, particularly when the focus is on agility, sensors, and multifunctionality of the shared space to bring about increased empowerment, connectivity, and interaction. The ineffectiveness or deficiencies witnessed by workers conducting their tasks in silos have led to unquantifiable institutional losses that have made the need for smart spaces a revolutionary proposition or venture [1].

A smart space is a physical environment that uses advanced technologies such as IoT, big data, and AI to improve the efficiency, comfort, and security of the space [2,3]. Smart spaces normally include a variety of different environments such as homes, offices, buildings, cities, and campuses. These spaces are designed to be more connected, responsive, and automated, allowing for real-time monitoring and control of various systems and infrastructure [1]. Smart spaces encompass a range of different aspects that contribute to their ongoing development and advancement.

Improved energy efficiency is one of the key benefits of smart spaces [4,5]. Smart spaces can use technology such as smart thermostats, appliances, and lighting systems to optimize energy usage and reduce costs [6]. Salimi and Hammad [7] argue that smart HVAC systems can also be used to adjust the temperature and humidity levels based on occupancy, weather, and time of day, improving comfort and reducing energy consumption [8].

Smart Spaces. https://doi.org/10.1016/B978-0-443-13462-3.00010-8

Another important aspect of smart spaces is the ability to improve security and enhance performance [9]. They can use such technologies as cameras, alarms, and door and window sensors to improve security and deter burglaries. Smart access control systems can also be used to control access to the space, allowing for better monitoring of who enters and exits the space [10].

Smart spaces also improve the overall perception or experience of the users [7]. Such technologies as online portals, mobile apps, and chatbots to provide users with easy access to various services and information are frequently used in smart spaces, to improve indoor positioning systems that enable users to obtain real-time feeds based on their location-based information and directions.

The hallmarks of smart spaces are their integration and compatibility with other systems such as smart homes, smart buildings, and smart cities, which allows for more holistic control and monitoring of various systems [6,11]. The perception that smart spaces could use technologies such as IoT, big data, and analytics to collect and analyze data from various systems and infrastructure to make data-driven decisions that improve the overall functioning of the space serves to offset the various faults identified with smart spaces. However, despite their potential benefits, several criticisms and critiques of smart spaces should also be considered.

Criticisms of smart spaces can be grouped into four main categories: (i) exacerbation of socioeconomic disparities, (ii) privacy and surveillance concerns, (iii) transparency and accountability concerns, and (iv) issues of standardization and compatibility of devices and technologies [12–14]. With the technology still in its early stages, challenges with interoperability among devices and systems can limit the potential benefits of smart spaces. Additionally, the collection and handling of personal data raises concerns about privacy and surveillance. There is also a lack of accountability and transparency in smart space operation, which can make it difficult for individuals to understand how these systems work or to hold those responsible accountable. To address these issues, it is important to ensure that the benefits of smart spaces are accessible to all, particularly low-income communities, and that privacy and security are protected through proper data management and accountability.

Overall, smart spaces are a good example of how technology can be integrated into our built environment to improve efficiency, comfort, and security. By implementing smart space technologies, we can meet the needs of users and improve the overall functioning of the space, even though smart spaces are a complex and dynamic concept that encompasses different aspects of technology, infrastructure, and urban design. While they can bring many benefits to individuals and communities, it is important to consider the criticisms that have been raised, in order to ensure that smart spaces are developed in a way that is equitable, transparent, and accountable. This research on smart space management contributes to our collective understanding of the importance of user-centric design in the creation of intelligent spaces that offer a favorable user experience. This dialogue pertains to the significance of intuitive user interfaces, personalized services, and convenient access to information. The research prioritized user needs and preferences, thereby fostering the creation of intelligent environments that are tailored to specific requirements and augment user contentment. Having an environment tailored to the needs of the community will bring about reduced waste, consumption, motion, and energy in the community and increase productivity and output, due to more efficient and judicious use of resources.

In the following discussions, this chapter can be broken into three topics: technology, place, and management. However, before we engage with this approach, we must first establish the meaning of the ecosystem of smart spaces to enable the research to flow easily.

2 Ecosystem of smart spaces

The ecosystem of smart spaces is made up of a variety of different components that, working together, create seamless interoperability and an integrated environment [15]. Nonetheless, there are seven identifiable components of a smart space ecosystem: sensors, standards and protocols, actuators, network infrastructure, analytics and AI, user interfaces, and cloud-based services. Sensors typically are devices that collect data from the environment, such as temperature, humidity, light levels, and occupancy. Actuators are devices that control various systems within the smart space, such as lighting, HVAC, and security systems. Network infrastructure includes the hardware and software that connect all the devices and systems within the smart space, allowing them to communicate with each other. Analytics and AI are the technologies that analyze the data collected by the sensors and make decisions based on that data. User interfaces are the ways that users interact with the smart space, such as through a mobile app or a voice-controlled assistant. Cloud-based services are services that are hosted on remote servers and provide various functionalities, such as data storage, data analysis, and management of smart space devices. Standards and protocols are the rules and guidelines that ensure that all the devices and systems within the smart space can communicate with each other effectively. In more detail, the ecosystem of smart spaces is further diagnosed to understand the nature of smart spaces, their benefits, and their disadvantages.

2.1 Sensors

Sensors are a critical component of the smart space ecosystem for urban sustainability, as they are responsible for collecting data from the environment [16,17]. Different types of sensors exist and their usefulness differs significantly, according to Dong et al. [18], Tan et al. [19] and Attar et al. [20]. Some of the sensor types are defined as follows:

- *Environmental sensors*: These sensors can measure various environmental factors, such as temperature, humidity, light levels, and air quality. These sensors

can be used to control HVAC systems, lighting, and other environmental systems to create a comfortable and energy-efficient environment.
- *Occupancy sensors*: These sensors can detect the presence of people in a space and can be used to control lighting and HVAC systems based on occupancy.
- *Location sensors*: These sensors can determine the location of people and objects within a space, and can be used for navigation, wayfinding, and asset tracking.
- *Motion sensors*: These sensors can detect motion and can be used for security and surveillance, as well as for controlling lighting and other systems.
- *Image sensors*: These sensors can capture images and video, and can be used for security and surveillance, as well as for monitoring traffic and other activity in a space.
- *Proximity sensors*: These sensors can detect the presence of nearby objects and can be used for navigation, wayfinding, and other applications.
- *Sound sensors*: These sensors can measure sound levels and can be used for monitoring noise pollution, as well as for controlling sound systems.
- *Chemical sensors*: These sensors can measure the presence of chemicals or pollutants in the air and can be used to monitor air quality and safety [21–23].

These sensors tend to communicate with each other and with other devices and systems in the smart space, through the network infrastructure.

2.2 Actuators

Domb [24] argues that actuators are devices that control various systems within the smart space, allowing the system to respond to the data collected by sensors [25]. Some examples of actuators or actuator sensors used in smart spaces, according to Bai and Huang [26], Sciuto and Nacci [27], and Domingues et al. [28], include:

- *Lighting actuators*: These actuators can control the brightness and color of lights in a space and can be used to adjust lighting levels based on occupancy, daylight, or other factors.
- *HVAC actuators*: These actuators can control heating, ventilation, and air conditioning systems, and can be used to adjust temperature and humidity levels based on occupancy and environmental sensors.
- *Motor actuators*: These actuators can control motors and other mechanical devices, such as window blinds, and can be used to adjust the position of devices based on sensor data.
- *Security actuators*: These actuators can control security systems such as cameras, alarms, and locks, and can be used to activate or deactivate security systems based on sensor data.
- *Audio actuators*: These actuators can control sound systems, and can be used to adjust volume levels, play music or other audio, or provide announcements based on sensor data.

- *Robotics actuators*: These actuators are used to control robotic devices such as drones, robots, and autonomous vehicles, which can be used for security, delivery, cleaning, or other purposes [29–31].

The data collected by sensors is analyzed and processed by the smart space management system, which can use the data to make decisions, control systems, and perform tasks automatically. Actuators can be controlled remotely or by using preset rules, making it possible for the smart space to respond in real time to changes in the environment.

2.3 Network infrastructure

The network infrastructure is a critical component of the smart space ecosystem [32], as it connects all the devices, sensors, actuators, and systems within the smart space, allowing them to communicate with each other [33]. Some examples of network infrastructure used in smart spaces include:

- *Wired networks*: These networks use cables to connect devices and systems, and can provide high-speed, reliable data transfer [34]. They can be used to connect devices within a building or campus, such as lighting and HVAC systems, and can also connect to remote servers for data storage and analysis [35].
- *Wireless networks*: These networks use radio frequencies to connect devices and systems and can be used to connect devices within a building or campus, as well as to connect devices that move around, such as smartphones, laptops, and vehicles [36].
- *Gateways*: These are devices that connect different networks and protocols, allowing devices and systems to communicate with each other [37]. For example, a gateway can connect a wired building automation system to a wireless IoT network.
- *Cloud-based services*: These services are hosted on remote servers and provide various functionalities such as data storage, data analysis, and management of smart space devices [38]. This allows for remote monitoring and control of smart spaces.
- *Standards and protocols*: These are the rules and guidelines that ensure that all the devices and systems within the smart space can communicate with each other effectively. Examples of standards and protocols used in smart spaces include Zigbee, Z-Wave, BACnet, and MQTT [39].

All these components of the network infrastructure work together to create a seamless and integrated environment for smart spaces. The data collected by sensors and sent to the management system is transmitted through the network infrastructure to the backend servers, where it is analyzed and processed. This enables the smart space to respond in real time to changes in the environment, and to make decisions and control systems automatically.

2.4 Analytics and AI

Analytics and AI are important components of the smart space ecosystem, as they are responsible for analyzing the data collected by sensors and making decisions based on that data.

- *Analytics*: Analytics refers to the process of collecting, storing, and analyzing data to uncover insights and make data-driven decisions [40]. In smart spaces, analytics can be used to identify patterns, trends, and anomalies in sensor data, which can help to optimize resource usage, reduce energy consumption, and improve overall operational efficiency [41].
- *Artificial Intelligence (AI)*: AI refers to the simulation of human intelligence in machines, which can be used to make decisions, control systems, and perform tasks automatically [42]. In smart spaces, AI can be used to control systems and make decisions based on sensor data, such as adjusting lighting and HVAC based on occupancy, or optimizing energy usage based on weather forecasts [43].
  - *Machine Learning (ML)*: Machine learning is a subset of AI that allows the system to learn from data and improve performance over time without being explicitly programmed [44]. In smart spaces, ML algorithms can be used to identify patterns and anomalies in sensor data, and to make predictions about future events [45].
  - *Natural Language Processing (NLP)*: NLP is a subset of AI that allows the system to understand and respond to natural language inputs, such as speech or text [46]. In smart spaces, NLP can be used to enable voice-controlled interfaces, such as virtual assistants, which can help users to interact with the smart space more easily [47].
- *Predictive maintenance*: With the use of machine learning algorithms, the system can predict when a device or equipment will fail and schedule maintenance before it happens [48]. This can prevent downtime, increase efficiency, and save costs in the long run.

Overall, analytics and AI are key components of the smart space ecosystem, as they allow the system to analyze sensor data and make decisions automatically, which can help to optimize resource usage, reduce energy consumption, and improve overall operational efficiency.

2.5 Cloud-based services

Cloud-based services are an important component of the smart space ecosystem, as they provide various functionalities such as data storage, data analysis, and management of smart space devices [49]. These services are hosted on remote servers, which allows for remote monitoring and control of smart spaces.

- *Data storage*: Cloud-based services provide a secure and scalable way to store the data collected by sensors in smart spaces [50]. This data can then be accessed and analyzed from anywhere, enabling real-time monitoring and control of the smart space.

- *Data analysis*: Cloud-based services provide powerful computing resources and analytics tools that can be used to analyze the data collected by sensors in smart spaces [51,52]. This can include machine learning algorithms, rule-based systems, and decision-making engines, which can help to identify patterns, trends, and anomalies in sensor data.
- *Management of smart space devices*: Cloud-based services can be used to remotely manage and control devices and systems within the smart space [53]. This can include configuring devices, monitoring their status, and controlling them through a web interface.
- *Software-as-a-Service (SaaS)*: Cloud-based services can provide software applications as a service, which can be accessed and used over the internet [54]. This enables smart spaces to use software applications such as energy management systems, security systems, and other smart space management solutions.
- *Infrastructure-as-a-Service (IaaS)*: Cloud-based services can provide infrastructure such as servers, storage, and networks as a service [55]. This enables smart spaces to use and manage their own infrastructure, which can help to reduce costs and increase flexibility.
- *Platform-as-a-Service (PaaS)*: Cloud-based services can provide a platform on which developers can build, run, and manage applications and services [56]. This enables smart spaces to develop and deploy their own applications and services, which can help to increase functionality and customizability.

Cloud-based services can help to reduce the costs and complexity of managing and maintaining smart spaces and can provide scalability and flexibility. They also allow for remote access and control, which can be useful for monitoring and maintaining smart spaces.

2.6 Standards and protocols

Al-Turjman [57] argues that standards and interoperable communication protocols are important components of the smart space ecosystem having different energy-efficient (EE) metrics, as they ensure that all the devices and systems within the smart space can communicate with each other effectively.

- *Zigbee*: Zigbee is a wireless communication protocol used for IoT devices. It is designed for low-power, low-data rate applications, and is commonly used in smart home and building automation systems [58].
- *Z-Wave*: Z-Wave is another wireless communication protocol used for IoT devices. It is designed for low-power, low-data rate applications and is also commonly used in smart home and building automation systems [59].

- *BACnet*: BACnet is a communication protocol used for building automation and control systems [60]. It is designed for use in HVAC, lighting, and other building systems, and is widely used in commercial buildings.
- *MQTT*: MQTT is a lightweight communication protocol used for IoT devices. It is designed for low-power, low-data rate applications, and is commonly used in smart home and building automation systems [61].
- *IP*: Internet Protocol (IP) is the communication protocol used for the Internet, and it is used in many smart spaces to connect devices and systems to the Internet and to each other.
- *CoAP*: Constrained Application Protocol (CoAP) is a protocol that is similar to HTTP and is used in IoT devices that have limited resources.

These standards and protocols allow devices and systems within the smart space to communicate with each other and with remote servers for data storage and analysis. This enables the smart space to respond in real time to changes in the environment, and to make decisions and control systems automatically. Using standards and protocols ensures that the smart space is interoperable, meaning different devices and systems from different vendors can work together seamlessly [22].

2.7 User interfaces

User interfaces are an important component of the smart space ecosystem, as they provide a way for users to interact with the smart space and control its various systems [62,63]. Some examples of user interfaces used in smart spaces are:

- *Web-based interfaces*: These interfaces can be accessed through a web browser and can be used to control devices and systems within the smart space, such as lighting and HVAC systems [64]. They can also be used to access real-time data and analytics, such as energy usage and occupancy patterns.
- *Mobile apps*: These are mobile applications that can be downloaded on smartphones or tablets and can be used to control devices and systems within the smart space [65]. They can also be used to access real-time data and analytics.
- *Voice-controlled interfaces*: These interfaces allow users to control devices and systems within the smart space using natural language commands, such as "turn off the lights" or "set the temperature to 72 degrees" [66]. They can also be used to access real-time data and analytics.
- *Touchscreen interfaces*: These interfaces can be built into walls or other surfaces and can be used to control devices and systems within the smart space, such as lighting and HVAC systems [67]. They can also be used to access real-time data and analytics.
- *Physical buttons and controls*: These interfaces can be built into devices and systems within the smart space and can be used to control them directly, such as turning lights on or off, or adjusting the temperature [68].
- *Augmented Reality (AR) interfaces*: These interfaces can use AR technology to provide an interactive and immersive experience for the users [69]. This can be used for navigation, training, or providing information about the smart space.

- *Virtual Reality (VR) interfaces*: These interfaces can use VR technology to provide an immersive and interactive experience for the users [70]. This can be used for simulation, training, or providing information about the smart space.

User interfaces are designed to provide a seamless and intuitive experience for the users, making it easy for them to interact with and control the smart space. It is also important for the interfaces to be accessible to people with different abilities and to be responsive to changes in the environment [71].

All these components work together to create a smart space that can automatically adjust to the needs of the people who use it (see Table 1), resulting in a more comfortable, efficient, and sustainable environment.

Table 1 Integrated environment for smart space ecosystem.

Component	Description	Role in smart spaces
Wired networks	Networks that use cables to connect devices and systems and provide high-speed, reliable data transfer	Connect devices within a building or campus and connect to remote servers for data storage and analysis
Wireless networks	Networks that use radio frequencies to connect devices and systems	Connect devices within a building or campus and connect devices that move around, such as smartphones, laptops, and vehicles
Gateways	Devices that connect different networks and protocols	Allow devices and systems to communicate with each other, for example, a wired building automation system to a wireless IoT network
Cloud-based services	Services hosted on remote servers that provide functionalities such as data storage, data analysis, and device management	Allow for remote monitoring and control of smart spaces
Standards and protocols	Rules and guidelines that ensure effective communication among all devices and systems within the smart space	Examples include Zigbee, Z-Wave, BACnet, and MQTT
Devices and sensors	Devices and sensors used to collect data and communicate with the management system	Examples include lighting, HVAC, smartphones, laptops, and vehicles
Management system	System that processes data and makes decisions	Provide functionalities such as data storage, data analysis, monitoring, and control
Data flow and processing	Flow of data from devices to the management system and process of data analysis and decision-making	Allows smart spaces to respond in real time to changes in the environment and to make decisions and control systems automatically

There are three major ways through which the ecosystem of smart spaces can be further explored, demonstrating its strengths and weaknesses. In this study, we identified three critical components: place (or location), technology, and management.

Place is crucial because no initiative exists in a vacuum; rather, it must exist within a location, and the people in those places signal the affirmation or rejection of the implementation of smart contracts. When the place and people issues are sorted out, the next task is to consider the nature of the technology to invest in and implement that is favorable to all. Thereafter, the discussion must shift towards the management and enforcement of zoning protocols for smart initiatives.

3 Smart cities (place)

In recent times, smart cities are cited in urban environments that use advanced technologies and digital solutions to improve the efficiency, sustainability, and liveability of the city. They are usually located in urban enclaves incorporating advanced technologies and integrating multifaceted systems.

This can include using technology to optimize city services such as transportation, energy, and public safety, as well as using technology to improve the quality of life for residents.

A smart city is an urban area that incorporates advanced technologies, such as the Internet of Things (IoT), big data, and artificial intelligence (AI), to improve the efficiency, sustainability, and livability of the city. One of the key features of smart cities is the use of IoT devices to collect data on various aspects of urban life. These devices can be used to monitor things like traffic flow, air and water quality, and energy usage. This data is then analyzed to identify patterns and trends that can be used to optimize the performance of the city. For example, data on traffic flow can be used to optimize traffic lights and reduce congestion, while data on energy usage can be used to improve the efficiency of buildings and reduce energy costs.

Another important feature of smart cities is the use of big data and AI to analyze and process the large amounts of data generated by IoT devices. These technologies can help to identify patterns and trends that may not be immediately obvious to human operators, such as predicting machine failure before it occurs. They can also be used to create predictive models that can help to optimize the performance of the city.

Smart cities also often incorporate advanced visualization and control systems, allowing managers to monitor and control various urban systems remotely. This can help to reduce downtime, improve efficiency, and increase overall performance.

However, smart cities also raise concerns about privacy and surveillance. Smart cities often involve the collection of large amounts of data on individuals, which can be used for a variety of purposes. This raises concerns about how this data is collected, stored, and used, as well as who has access to it. There is also a concern that the data collected could be used for surveillance or for profiling individuals.

Thus, while smart cities have the potential to bring significant benefits, they also raise concerns about privacy and surveillance. It is crucial to address these concerns in the planning and implementation of smart cities, in order to ensure that they are developed in a way that is equitable, transparent, and accountable.

Some key features of smart cities include:

- *Connectivity*: Smart cities use advanced networks and technologies, such as IoT and 5G, to connect devices and systems throughout the city.
- *Data collection and analysis*: Smart cities use sensors and other technologies to collect data on various aspects of the city, such as traffic, weather, and energy usage. This data is then analyzed to optimize city services and make data-driven decisions.
- *Smart infrastructure*: Smart cities use technology to improve the efficiency and sustainability of infrastructure such as transportation, buildings, and energy systems.
- *Citizen engagement*: Smart cities use technology to engage citizens and involve them in the decision-making process. Examples include online platforms for reporting issues, providing feedback, and participating in community planning.
- *Public safety*: Smart cities use technology to enhance public safety, including surveillance systems, emergency response systems, and predictive policing.

Smart cities can help to improve the quality of life for residents, reduce environmental impact, and make cities more efficient and sustainable. They also have the potential to create new economic opportunities, attract investment, and improve urban services. However, it's important to consider the ethical and social implications of smart cities, such as data privacy, cybersecurity, and equitable access to technology.

3.1 Smart cities prioritization framework

In answering the question of what makes a city a smart city, Hannah Williams argues that:

> *A smart city is the re-development of an area or city using Information and Communication Technologies (ICT) to enhance the performance and quality of urban services such as energy, connectivity, transportation, utilities, and others. A smart city is developed when 'smart' technologies are deployed to change the nature and economics of the surrounding infrastructure [72].*

The success or failure of a smart city lies in its people, the environment, and the nature of its governance framework (see Fig. 1). The people, governance, and environment are the fundamental or the primary basis for consideration of readiness of a smart or intelligent city. Where the first three issues are resolved, then the secondary discourse of livability, economy, and mobility can be engaged, and later the tertiary basis, which includes sustainability, community assets, education, service delivery, data, and transport, including digital infrastructure, waste management innovation,

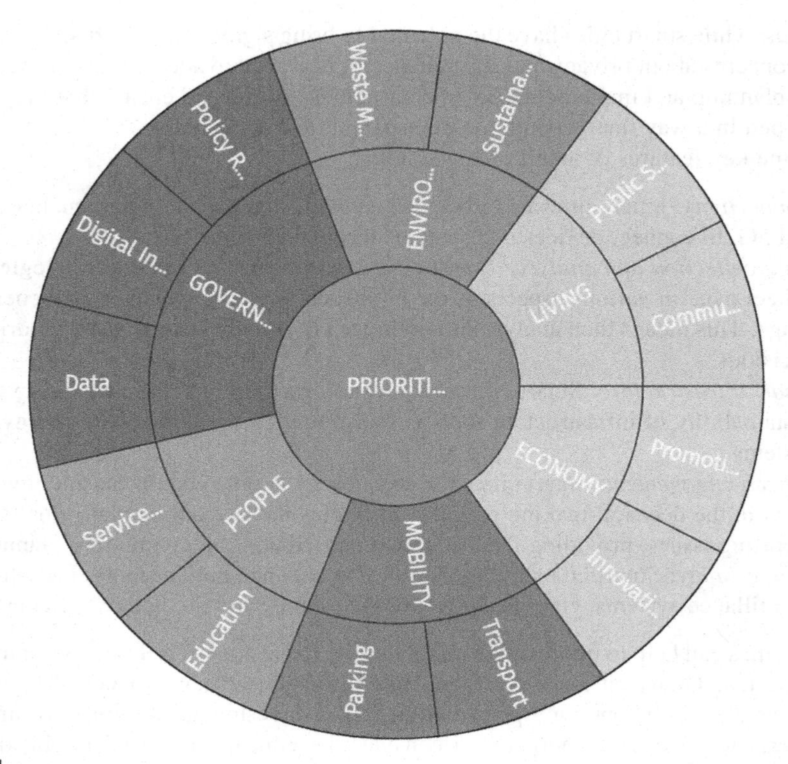

FIG. 1

Smart cities prioritization framework.

promotion, and policy and resourcing (see Fig. 1). This shows that the smart city prioritization framework has three levels instead of two. Features of smart cities can also include smart traffic management, smart energy management, and public safety systems. According to Buzzageek [72], there are varieties of technologies and systems in smart cities such as:

- *Smart transportation*: Smart transportation systems can include things like traffic management systems, intelligent transportation systems, and connected vehicles, which can be used to optimize traffic flow, reduce congestion, and improve public transportation. For example, traffic management systems can use data from cameras and sensors to adjust traffic light timings, reduce traffic congestion, and improve traffic flow. Intelligent transportation systems can provide real-time information to drivers about traffic conditions and alternative routes. Connected vehicles can communicate with each other and with infrastructure to improve safety, reduce congestion, and optimize traffic flow.
- *Smart energy*: Smart energy systems can include things like smart grids, renewable energy systems, and energy-efficient buildings, which can be used to

optimize energy usage and reduce costs. For example, smart grids can use data from sensors to monitor energy usage and adjust supply to meet demand in real time. Renewable energy systems can use data from weather forecasts to optimize energy production and reduce costs. Energy-efficient buildings can use data from sensors to monitor energy usage and adjust HVAC and lighting systems to reduce energy consumption.

- *Smart environment*: Smart environment systems can include things like air quality monitoring systems, waste management systems, and water management systems, which can be used to improve the overall livability of the city. For example, air quality monitoring systems can use data from sensors to monitor air quality and provide real-time information to citizens. Waste management systems can use data from sensors to optimize collection and disposal of waste. Water management systems can use data from sensors to monitor water usage and detect leaks, reducing water waste.
- *Smart public services*: Smart public services can include things like emergency response systems, public safety systems, and e-governance systems, which can be used to improve the overall efficiency of the city. For example, emergency response systems can use data from cameras and sensors to identify and respond to emergency situations. Public safety systems can use data from cameras and sensors to improve surveillance and deter crime. E-governance systems can use data from sensors and citizen engagement platforms to improve transparency, participation, and engagement of citizens in city decision-making.
- *Smart citizen engagement*: Smart citizen engagement systems can include things like mobile applications, social media platforms, and citizen portals, which can be used to increase transparency, participation, and engagement of citizens in city decision-making. For example, mobile applications can provide citizens with real-time information about transportation, traffic, and city services. Social media platforms can be used to engage citizens in discussions about city issues and gather feedback. Citizen portals can provide citizens with easy access to city services, information, and data.
- *Smart data and analytics*: Smart data and analytics systems can include things like sensors, big data platforms, and analytics tools, which can be used to collect and analyze data from various city systems and infrastructure, in order to make data-driven decisions that improve the overall functioning of the city. For example, sensors can be used to collect data on traffic, weather, energy usage, and air quality. Big data platforms can be used to store, process, and analyze this data. Analytics tools can be used to extract insights from this data and make data-driven decisions that improve the overall functioning of the city [73–76].

As suggested earlier, there are seven models for understanding the smart space: smart homes, smart offices, smart factories, smart buildings, smart cities, smart campuses, and smart communities.

There are five major benefits accruing to each of the smart agents from the use of these smart agents: (i) advanced technologies such as IoT, big data, and AI to improve the efficiency, comfort, and security of the home [73], (ii) ability to improve energy efficiency and reduce cost [74,75], (iii) improving security and enabling ease of monitoring owners property [76], (iv) improving the overall experience of residents/occupants, and (v) integrated systems.

3.2 Benefits of smart city technologies in smart spaces

Smart city technologies offer several benefits to cities, including:

- *Improved efficiency and cost savings*: Smart city technologies can help cities to optimize resource usage, reduce energy consumption, and improve overall operational efficiency.
- *Better quality of life for citizens*: Smart city technologies can improve citizen services and provide access to real-time information on city services, transportation, and public safety.
- *Increased sustainability*: Smart city technologies can help cities to reduce their environmental footprint and promote sustainable practices, such as reducing energy consumption and promoting sustainable transportation.
- *Enhanced public safety and emergency response*: Smart city technologies can improve public safety and emergency response by providing real-time information on crime and emergency situations, and by making it easier for first responders to access and share critical information.
- *Improved mobility*: Smart city technologies can improve transportation and mobility by providing real-time information on traffic and public transportation, and by making it easier for citizens to find and use alternative modes of transportation.
- *Better decision-making*: Smart city technologies can provide city officials with real-time data and insights, allowing them to make better-informed decisions on issues such as resource allocation, infrastructure development, and public services.
- *Economic development*: Smart city technologies can help to attract businesses, entrepreneurs, and investors, creating jobs and promoting economic growth.

Overall, smart city technologies can help cities to become more livable, sustainable, and efficient, and can improve the quality of life for their citizens. However, some potential drawbacks must also be considered. It is important to weigh the pros and cons carefully and to take steps to mitigate potential risks and negative impacts.

4 Smart technologies

As mentioned earlier, several models exist for understanding smart spaces; they include but are not limited to smart homes, smart offices, smart buildings, smart cities, smart factories, smart campuses, and smart communities. Research on smart homes, smart offices, smart buildings, and smart cities is more plentiful compared to smart factories, smart campuses, and smart communities. Hence, the literature concentrates more on the known than the unknown.

4.1 Smart homes

These are residential spaces that use technology to improve energy efficiency, comfort, and security. Smart home systems can be controlled remotely and can include features such as automated lighting and temperature control, and home security systems [73]. These technologies allow for the collection and analysis of data from various home systems and infrastructure, such as lighting, HVAC, and security, in order to make data-driven decisions that improve the overall functioning of the home (see Fig. 2).

Smart homes typically include a variety of devices such as:

- *Smart thermostats*: These devices can be used to control the temperature of the home and can be programmed to automatically adjust the temperature based on the homeowner's schedule and preferences. This can help to reduce energy consumption and improve comfort.
- *Smart lighting*: Smart lighting systems can be controlled remotely and can be programmed to turn lights on and off at specific times or based on occupancy. This can help to reduce energy consumption and improve security.
- *Smart appliances*: Smart appliances such as refrigerators, washing machines, and ovens can be controlled remotely, allowing the homeowner to monitor and control the energy usage of these devices.

FIG. 2
Smart home system.
Source: AI Runway APP 2023.

- *Smart security*: Smart home security systems can include features such as door and window sensors, cameras, and alarms that can be controlled remotely through a smartphone or other device. This can help to improve security and provide peace of mind for homeowners.
- *Smart entertainment*: Smart home entertainment systems can include features such as smart TVs, speakers, and streaming devices that can be controlled remotely through a smartphone or other device [74,75,77].

Overall, smart homes can also be integrated with smart home assistants such as Amazon Alexa or Google Home, allowing for voice control of home devices, and can be connected to other home automation systems like Z-Wave and Zigbee [78].

4.2 Smart offices

According to Berawi et al. [79], smart offices are workspaces that use advanced technologies such as IoT, big data, and AI to improve the efficiency, comfort, and productivity of the office.

4.2.1 Open smart office

These are commercial spaces that use technology to improve energy efficiency, comfort, and productivity. Smart office systems can be controlled remotely and can include features such as automated lighting and temperature control and meeting room management.

Smart offices typically include a variety of devices such as:

- *Smart lighting systems*: These systems can be controlled remotely and can be programmed to turn lights on and off based on occupancy and natural light levels, reducing energy consumption and costs.
- *Smart HVAC systems*: These systems can be controlled remotely and can be programmed to adjust the temperature and humidity levels based on occupancy, weather, and time of day, reducing energy consumption and increasing comfort.
- *Smart access control systems*: These systems can be used to control access to the office, allowing building managers to monitor who enters and exits, and to restrict access to certain areas for security purposes.
- *Smart security systems*: These systems can include cameras, alarms, and door and window sensors that can be monitored remotely and can help to improve security and deter burglaries.
- *Smart meeting room management*: These systems can allow employees to schedule and reserve meeting rooms and can be integrated with other systems such as video conferencing and lighting to improve the efficiency and productivity of meetings.
- *Smart energy management systems*: These systems can monitor and analyze the energy usage of the building and provide real-time data on energy consumption, costs, and carbon emissions, allowing building managers to identify opportunities for energy savings [8,30,43].

Overall, smart offices can help to improve energy efficiency, comfort, and productivity, while also providing building managers with more convenient control over their office spaces.

4.3 Smart buildings

These are buildings that use technology to improve energy efficiency, comfort, and security. Smart buildings are increasingly being integrated into smart city projects (see Fig. 3). Older buildings can be retrofitted with sensors and new buildings can be constructed with sensor capabilities that provide real-time space management and enhanced public safety. Additionally, these sensors can be used to monitor the structural health of buildings by detecting wear and tear, and notifying officials when repairs are needed. Citizens can also contribute by notifying officials of necessary repairs through smart city applications. Furthermore, sensors can detect leaks in water mains and other pipe systems, improving the efficiency of public workers and reducing costs. In this way, smart buildings contribute to the sustainability of the city's infrastructure and overall quality of life for citizens (adapted from [80,81]).

Many functions of the smart building ecosystem can be controlled remotely and can include features such as automated lighting and temperature control and building management systems (Fig. 4).

Smart buildings typically include a variety of devices, such as:

- *Smart HVAC systems*: These systems can be controlled remotely and can be programmed to adjust the temperature and humidity levels based on

FIG. 3

Smart office structure.

Source: AI Runway APP 2023.

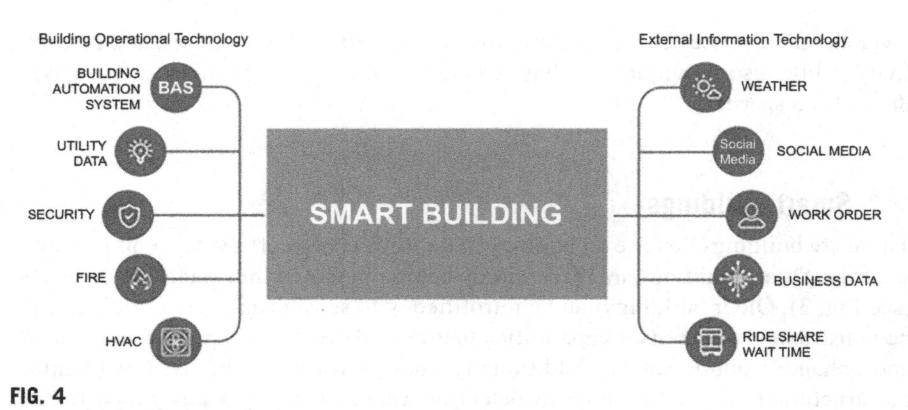

FIG. 4

Smart building ecosystem.

occupancy, weather, and time of day, reducing energy consumption and increasing comfort.

- *Smart lighting systems*: These systems can be controlled remotely and can be programmed to turn lights on and off based on occupancy and natural light levels, reducing energy consumption and costs.
- *Smart access control systems*: These systems can be used to control access to the building, allowing building managers to monitor who enters and exits, and to restrict access to certain areas for security purposes.
- *Smart security systems*: These systems can include cameras, alarms, and door and window sensors that can be monitored remotely and can help to improve security and deter burglaries.
- *Smart elevators*: Smart elevator systems can be programmed to optimize usage, reducing waiting times and energy consumption.
- *Smart energy management systems*: These systems can monitor and analyze the energy usage of the building and provide real-time data on energy consumption, costs, and carbon emissions, allowing building managers to identify opportunities for energy savings.
- *Building Automation Systems (BAS)*: These systems integrate and control multiple building systems such as HVAC, lighting, access control, and security. They can also provide real-time monitoring, analytics, and reporting for building management.
- Smart buildings can also be integrated with building management systems, allowing for remote control and monitoring of building systems.

Overall, smart buildings can help to improve energy efficiency, comfort, and security, while also providing building managers with more convenient control over their buildings.

5 Smart spaces management (management)

As previously discussed, the term smart space refers to a physical environment equipped with various sensors, devices, and systems that can collect and analyze data and can use this information to improve the functionality and efficiency of the space. Examples of smart spaces include smart homes, smart buildings, and smart cities. These spaces employ technologies such as the Internet of Things (IoT), artificial intelligence (AI), and machine learning to gather and analyze data, which is then used to make decisions, control systems, and perform tasks automatically. The goal of smart spaces is to improve the quality of life for the people who use them and make the spaces more efficient and sustainable.

Smart space management, the third critical component through which we are exploring the ecosystem of smart spaces, also involves the use of the advanced technologies already discussed—IoT, big data, and AI—to optimize the functionality of smart spaces [82]. IoT devices and sensors are used to collect data from various systems and infrastructure within the space, and this data is analyzed using big data and AI algorithms to identify patterns, trends, and anomalies, enabling data-driven decision-making. According to Chen et al. [83], this can include predictive maintenance, automatic fault detection, and energy-efficient scheduling of systems.

Building automation systems (BASs) are used to integrate and control multiple building systems, providing real-time monitoring, analytics, and reporting of the space [84]. Cloud-based services are used for data storage, analysis, and remote control and monitoring of the space. User interfaces such as mobile apps, online portals, and chatbots are used to provide easy access to various services and information within the space. These user interfaces can be designed to integrate with the various systems and infrastructure within the space, providing a seamless user experience [85]. Additionally, standards and protocols are important for the interoperability of different smart devices and systems within the space, ensuring that they can work together smoothly. According to Lee et al. [80], smart spaces can be considered as systems of connected devices that exchange data and interact with one another; this can include a variety of systems and infrastructure such as lighting, heating, ventilation, air conditioning (HVAC), access control, and security. This allows for real-time monitoring, analytics, and reporting of the space, giving managers the ability to quickly identify and address any issues that may arise [81].

As mentioned, smart space management also includes the use of cloud-based services to store and analyze data, as well as to provide remote control and monitoring of the space. This allows managers to access and control various systems and infrastructure within the space from a centralized location [86], crucial for optimizing and improving the overall performance of the space and providing users with a better experience.

According to Ndaguba et al. [84], smart space management offers several benefits, including:

- *Improved efficiency*: By using data-driven decision-making and real-time monitoring, smart space management enables better energy and resource management, leading to increased efficiency and cost savings.
- *Enhanced user experience*: Smart space management allows for personalized services, easy access to information, and improved safety and security, resulting in a better user experience.
- *Better sustainability*: By optimizing energy consumption, reducing waste, and promoting sustainable practices, smart space management can help promote environmental sustainability.
- *Increased productivity*: Smart space management can create a more comfortable and productive work environment, leading to increased productivity and employee satisfaction.
- *Improved safety and security*: Through real-time monitoring and analytics, smart space management can quickly detect and address safety and security issues, ensuring a safer environment for users [87–89].

Overall, smart space management can lead to a more efficient, sustainable, productive, and user-friendly smart space.

6 Contribution of this research to the field of smart spaces

Investigation pertaining to smart space management contributes to the comprehension of how sophisticated technologies can be employed to optimize the functionality and performance of smart spaces. This chapter offers valuable insights regarding the fundamental constituents, advantages, and obstacles that are linked to proficient management of smart environments. Through an examination of the utilization of Internet of Things (IoT), big data, and artificial intelligence (AI) in the realm of smart space management, this study provides significant insights and direction for practitioners, policymakers, and scholars engaged in this field.

The research in question, which is an attempt to provide an overview of smart spaces, has made distinct and precise contributions, as follows:

- The study puts forth an all-encompassing conceptual framework that delineates the fundamental constituents and technologies involved in smart space management. This exposition offers a comprehensive perspective on the integration of Internet of Things (IoT), big data, and artificial intelligence (AI) to optimize the performance of smart spaces.
- The study underscores the potential advantages of smart space management, encompassing improved efficiency, enriched user engagement, superior sustainability, better output, and enhanced safety and security. Through the process of quantification and explanation, the research underscores the beneficial effects that smart space management can confer on individuals, organizations, and the environment.

- The present study delves into the intricacies and nuances of intelligent space management, taking into account the various challenges and factors that come into play during the implementation phase. The discourse pertains to a multitude of concerns, including but not limited to apprehensions regarding privacy, potential security vulnerabilities, high costs, complicated operational procedures, and overreliance on technological infrastructure. Through the identification of these considerations, the research offers invaluable insights for practitioners and decision-makers alike, enabling them to effectively mitigate potential risks and optimize the benefits associated with smart spaces.
- The scholarly literature underscores the significance of incorporating diverse technological modalities, including but not limited to the Internet of Things, large-scale data analytics, artificial intelligence, and cloud-based solutions, within the context of smart space management. The statement underscores the significance of interoperability standards and protocols in guaranteeing uninterrupted communication and cooperation among heterogeneous systems and devices within a smart environment.
- The importance of user-centric design in the creation of smart spaces that offer a favorable user experience is emphasized by the research. The discourse pertains to the significance of intuitive user interfaces, personalized services, and convenient access to information. The research endeavors to prioritize user needs and preferences, thereby fostering the creation of smart spaces tailored to specific requirements and that heighten user satisfaction.

In its entirety, the research serves to enhance the body of knowledge in the realm of smart space management through the provision of a thorough exposition of its fundamental constituents, advantages, obstacles, and implementation factors. The chapter provides pragmatic observations and suggestions that may assist interested partners in proficiently administering and improving smart spaces, with the aim of enhancing efficiency, sustainability, and user satisfaction.

7 Conclusion

Smart spaces are physical environments that are equipped with various sensors, devices, and systems that can collect and analyze data and use this information to improve the functionality and efficiency of the space. Smart spaces offer many benefits, such as improved efficiency, enhanced comfort and convenience, increased safety and security, better communication and collaboration, and increased sustainability (see Table 1 for a high-level overview of the key components of the ecosystem of smart spaces). However, it is important to note that smart spaces also have potential drawbacks such as privacy concerns, security risks, high costs, complexity, and dependence on technology. Therefore it is important to weigh the pros and cons carefully and to take steps to mitigate potential risks and negative impacts when implementing smart spaces. In this sense, smart spaces have the potential to improve the quality of life for the people who use them and make the spaces more efficient and sustainable.

Research on smart spaces has revealed several key findings related to their implementation and benefits.

- One study by Yigitcanlar et al. [90] found that smart cities, a type of smart space, can enhance sustainability by improving energy efficiency, reducing waste, and increasing the use of renewable resources. Another study by Farooq [91] found that smart homes can improve quality of life by providing greater convenience, comfort, and security for occupants.
- In terms of implementation, a study by Lee et al. [80] identified the need for open standards and interoperability in smart spaces, to allow different devices and systems to work together seamlessly. Another study by Sikdar et al. [78] highlighted the importance of user-centric design in creating smart spaces that are easy to use and provide a positive user experience.
- The use of advanced technologies such as IoT, big data, and AI has also been shown to have significant benefits for smart spaces. According to Chen et al. [83], these technologies enable real-time monitoring, data analysis, and predictive maintenance, leading to improved efficiency, reduced costs, and increased safety.

Research on smart spaces has highlighted their potential to transform the way we live, work, and interact with our environments. By providing greater efficiency, sustainability, and user experience, smart spaces have the potential to improve quality of life and create a more connected, intelligent world [90,91].

Smart spaces have evolved significantly over the past few years, thanks to the rapid advancements in technology. According to Zhang et al. [82], today smart spaces refer to environments that are integrated with advanced technologies such as the Internet of Things (IoT), big data, and artificial intelligence (AI) to optimize their functionality. The concept of smart spaces started with the development of smart homes, where various home appliances were connected to a network and controlled through a central device. This has now expanded to include smart buildings, smart cities, and smart factories, among others.

Smart spaces are now used to manage a wide range of systems and infrastructure, including lighting, heating, ventilation, air conditioning, security, and energy management. As noted by Varshney and Sarker [88], the use of sensors and data analytics has enabled real-time monitoring, data-driven decision-making, and better resource management.

In recent years, the adoption of smart space technologies has increased significantly, driven by the need for improved efficiency, sustainability, safety, and user experience. As noted by Cho and Choi [92], smart spaces have also become more accessible and affordable, with the development of open standards and the growth of cloud-based services.

In conclusion, as smart spaces continue to evolve, they are expected to become even more integrated and intelligent, with the ability to adapt to changing user needs and preferences. This is expected to transform the way we live, work, and interact with our environments, leading to a more efficient, sustainable, and connected world [82].

References

[1] G.D. Markman, T.L. Waldron, P.T. Gianiodis, M.I. Espina, E pluribus unum: Impact entrepreneurship as a solution to grand challenges, Acad. Manag. Perspect. 33 (4) (2019) 371–382.

[2] Y. Sahni, J. Cao, J. Shen, Challenges and opportunities in designing smart spaces, in: Internet of Everything: Algorithms, Methodologies, Technologies and Perspectives, Springer, 2018, pp. 131–152.

[3] Y.C. Huang, K.Y. Wu, Y.T. Liu, Future home design: an emotional communication channel approach to smart space, Personal Ubiquitous Comput. 17 (2013) 1281–1293.

[4] M. Casini, Smart Buildings: Advanced Materials and Nanotechnology to Improve Energy-efficiency and Environmental Performance, Woodhead Publishing, 2016.

[5] P.S. Moura, G.L. López, J.I. Moreno, A.T. De Almeida, The role of Smart Grids to foster energy efficiency, Energ. Effic. 6 (2013) 621–639.

[6] S. Balandin, H. Waris, Key properties in the development of smart spaces, in: Universal Access in Human-Computer Interaction. Intelligent and Ubiquitous Interaction Environments: 5th International Conference, UAHCI 2009, Held as Part of HCI International 2009, San Diego, CA, USA, July 19-24, 2009. Proceedings, Part II 5, Springer, Berlin Heidelberg, 2009, pp. 3–12.

[7] S. Salimi, A. Hammad, A generalized inhomogeneous Markov chain occupancy model for open-plan offices using Real Time Locating System data, in: Proceedings of the 16th IBPSA Conference Rome, Italy, Sept. 2–4, 2019. https:/doi.org/10.26868/25222708.2019.210196.

[8] P. Carreira, A.A. Costa, V. Mansur, A. Arsénio, Can HVAC really learn from users? A simulation-based study on the effectiveness of voting for comfort and energy use optimization, Sustain. Cities Soc. 41 (2018) 275–285.

[9] S. El Jaouhari, A. Bouabdallah, A.A. Corici, SDN-based security management of multiple WoT smart spaces, J. Ambient Intell. Humanized Comput. 12 (2021) 9081–9096.

[10] A. Evesti, J. Suomalainen, E. Ovaska, Architecture and knowledge-driven self-adaptive security in smart space, Computers 2 (1) (2013) 34–66.

[11] G. Halegoua, Smart Cities, MIT Press, 2020.

[12] M. Bilandzic, M. Foth, Libraries as coworking spaces: understanding user motivations and perceived barriers to social learning, Library Hi Tech 31 (2) (2013) 254–273.

[13] K. Kourtit, P. Nijkamp, Impact of cultural "ambiance" on the spatial distribution of creative professions: A modeling study on the Netherlands, Int. Region. Sci. Rev. 41 (1) (2018) 103–128.

[14] E.D. Madyatmadja, N.A.R. Noverya, A.B. Surbakti, Feature and application in smart campus: a systematic literature review, in: In 2021 International Conference on Information Management and Technology (ICIMTech), 1, IEEE, 2021, pp. 358–363.

[15] A. Pliatsios, C. Goumopoulos, K. Kotis, A review on iot frameworks supporting multi-level interoperability—the semantic social network of things framework, Int. J. Adv. Internet Technol 13 (1) (2020) 46–64.

[16] H. El Alami, A. Najid, ECH: An enhanced clustering hierarchy approach to maximize lifetime of wireless sensor networks, IEEE Access 7 (2019) 107142–107153.
[17] G. Han, Z. Tang, Y. He, J. Jiang, J.A. Ansere, District partition-based data collection algorithm with event dynamic competition in underwater acoustic sensor networks, IEEE Trans. Industr. Inform. 15 (10) (2019) 5755–5764.
[18] B. Dong, B. Andrews, K.P. Lam, M. Höynck, R. Zhang, Y.S. Chiou, D. Benitez, An information technology enabled sustainability test-bed (ITEST) for occupancy detection through an environmental sensing network, Energ. Buildings 42 (7) (2010) 1038–1046.
[19] S.Y. Tan, M. Jacoby, H. Saha, A. Florita, G. Henze, S. Sarkar, Multimodal sensor fusion framework for residential building occupancy detection, Energ. Buildings 258 (2022) 111828.
[20] R. Attar, E. Hailemariam, S. Breslav, A. Khan, G. Kurtenbach, Sensor-enabled cubicles for occupant-centric capture of building performance data, Ashrae Trans. 117 (2) (2011).
[21] L.M. Dang, K. Min, H. Wang, M.J. Piran, C.H. Lee, H. Moon, Sensor-based and vision-based human activity recognition: A comprehensive survey, Pattern Recognit. 108 (2020) 107561.
[22] B. Dong, V. Prakash, F. Feng, Z. O'Neill, A review of smart building sensing system for better indoor environment control, Energ. Buildings 199 (2019) 29–46.
[23] T. Roitsch, L. Cabrera-Bosquet, A. Fournier, K. Ghamkhar, J. Jiménez-Berni, F. Pinto, E.S. Ober, New sensors and data-driven approaches—A path to next generation phenomics, Plant Sci. 282 (2019) 2–10.
[24] M. Domb, Smart home systems based on internet of things, in: Internet of Things (IoT) for Automated and Smart Applications, IntechOpen, 2019.
[25] M. Mazzara, I. Afanasyev, S.R. Sarangi, S. Distefano, V. Kumar, M. Ahmad, A reference architecture for smart and software-defined buildings, in: 2019 IEEE International Conference on Smart Computing (SMARTCOMP), IEEE, 2019, pp. 167–172.
[26] Z.Y. Bai, X.Y. Huang, Design and implementation of a cyber physical system for building smart living spaces, Int. J. Distrib. Sensor Networks 8 (5) (2012) 764186.
[27] D. Sciuto, A.A. Nacci, On how to design smart energy-efficient buildings, in: 2014 12th IEEE International Conference on Embedded and Ubiquitous Computing, IEEE, 2014, pp. 205–208.
[28] P. Domingues, P. Carreira, R. Vieira, W. Kastner, Building automation systems: concepts and technology review, Comput. Standards Interf. 45 (2016) 1–12.
[29] W. Yan, Z. Wang, H. Wang, W. Wang, J. Li, X. Gui, Survey on recent smart gateways for smart home: systems, technologies, and challenges, Trans. Emerg. Telecommun. Technol. 33 (6) (2022) e4067.
[30] A. Pandharipande, M. Zhao, E. Frimout, Connected indoor lighting based applications in a building IoT ecosystem, IEEE Internet Things Mag. 2 (1) (2019) 22–26.
[31] J. Milošević, A. Teixeira, K.H. Johansson, H. Sandberg, Actuator security indices based on perfect undetectability: Computation, robustness, and sensor placement, IEEE Trans. Autom. Control 65 (9) (2020) 3816–3831.
[32] O. Vermesan, P. Friess (Eds.), Internet of Things: Converging Technologies for Smart Environments and Integrated Ecosystems, River Publishers, 2013.
[33] J. Yun, I.Y. Ahn, S.C. Choi, J. Kim, TTEO (Things Talk to Each Other): Programming smart spaces based on IoT systems, Sensors 16 (4) (2016) 467.
[34] H. Gao, C. Liu, Y. Li, X. Yang, V2VR: reliable hybrid-network-oriented V2V data transmission and routing considering RSUs and connectivity probability, IEEE Trans. Intell. Transp. Syst. 22 (6) (2020) 3533–3546.

[35] D.D. Clark, K.T. Pogran, D.P. Reed, An introduction to local area networks, Proc. IEEE 66 (11) (1978) 1497–1517.
[36] R.O. LaMaire, A. Krishna, P. Bhagwat, J. Panian, Wireless LANs and mobile networking: standards and future directions, IEEE Commun. Mag. 34 (8) (1996) 86–94.
[37] M. Starsinic, System architecture challenges in the home M2M network, in: 2010 IEEE Long Island Systems, Applications and Technology Conference, IEEE, 2010, pp. 1–7.
[38] J. Cubo, A. Nieto, E. Pimentel, A cloud-based Internet of Things platform for ambient assisted living, Sensors 14 (8) (2014) 14070–14105.
[39] A. Ali, G.A. Shah, M.O. Farooq, U. Ghani, Technologies and challenges in developing machine-to-machine applications: A survey, J. Network Comput. Appl. 83 (2017) 124–139.
[40] K. Vassakis, E. Petrakis, I. Kopanakis, Big data analytics: Applications, prospects and challenges, in: Mobile Big Data: A Roadmap from Models to Technologies, 2018, pp. 3–20.
[41] H. Janetzko, F. Stoffel, S. Mittelstädt, D.A. Keim, Anomaly detection for visual analytics of power consumption data, Comput. Graph. 38 (2014) 27–37.
[42] M. Haenlein, A. Kaplan, A brief history of artificial intelligence: on the past, present, and future of artificial intelligence, Calif. Manage. Rev. 61 (4) (2019) 5–14.
[43] T.A. Nguyen, M. Aiello, Energy intelligent buildings based on user activity: a survey, Energ. Buildings 56 (2013) 244–257.
[44] R. Saravanan, P. Sujatha, A state of art techniques on machine learning algorithms: a perspective of supervised learning approaches in data classification, in: 2018 Second International Conference on Intelligent Computing and Control Systems (ICICCS), IEEE, 2018, pp. 945–949.
[45] H. Maciel, G.S. Ramos, A.L. Aquino, A sensor network solution to detect occupation in smart spaces in the presence of anomalous readings, IEEE Sensors Lett. 5 (7) (2021) 1–4.
[46] A. Torfi, R.A. Shirvani, Y. Keneshloo, N. Tavaf, E.A. Fox, Natural language processing advancements by deep learning: a survey, 2020. arXiv preprint arXiv:2003.01200.
[47] V.T. Hayashi, W.V. Ruggiero, Hands-free authentication for virtual assistants with trusted IoT device and machine learning, Sensors 22 (4) (2022) 1325.
[48] Y. Bouabdallaoui, Z. Lafhaj, P. Yim, L. Ducoulombier, B. Bennadji, Predictive maintenance in building facilities: A machine learning-based approach, Sensors 21 (4) (2021) 1044.
[49] J. Mineraud, O. Mazhelis, X. Su, S. Tarkoma, A gap analysis of Internet-of-Things platforms, Comput. Commun. 89 (2016) 5–16.
[50] A. Kallel, M. Rekik, M. Khemakhem, IoT-fog-cloud based architecture for smart systems: Prototypes of autism and COVID-19 monitoring systems, Software: Pract. Exp. 51 (1) (2021) 91–116.
[51] A. Gurtov, M. Liyanage, D. Korzun, Secure communication and data processing challenges in the Industrial Internet, Baltic J. Modern Comput. 4 (4) (2016) 1058–1073.
[52] Z. Khan, A. Anjum, S.L. Kiani, Cloud based big data analytics for smart future cities, in: In 2013 IEEE/ACM 6th International Conference on Utility and Cloud Computing, IEEE, 2013, pp. 381–386.
[53] T. Cádrik, P. Takáč, J. Ondo, P. Sinčák, M. Mach, F. Jakab, F. Cavallo, M. Bonaccorsi, Cloud-based robots and intelligent space teleoperation tools, in: Robot Intelligence Technology and Applications 4: Results from the 4th International Conference on Robot Intelligence Technology and Applications, Springer International Publishing, 2017, pp. 599–610.

[54] R. Rai, S. Mehfuz, G. Sahoo, Efficient migration of application to clouds: analysis and comparison, GSTF J. Comput. (JoC) 3 (2013) 1–5.
[55] S. Bhardwaj, L. Jain, S. Jain, Cloud computing: A study of infrastructure as a service (IAAS), Int. J. Eng. Inform. Technol. 2 (1) (2010) 60–63.
[56] S. Pastore, The platform as a service (paas) cloud model: Opportunity or complexity for a web developer? Int. J. Comput. Appl. 81 (18) (2013) 29–37.
[57] F. Al-Turjman, Smart-city medium access for smart mobility applications in Internet of Things, Trans. Emerg. Telecommun. Technol. 33 (8) (2022) e3723.
[58] T. Elarabi, V. Deep, C.K. Rai, Design and simulation of state-of-art ZigBee transmitter for IoT wireless devices, in: 2015 IEEE International Symposium on Signal Processing and Information Technology (ISSPIT), IEEE, 2015, pp. 297–300.
[59] S. Al-Sarawi, M. Anbar, K. Alieyan, M. Alzubaidi, Internet of Things (IoT) communication protocols, in: 2017 8th International Conference on Information Technology (ICIT), IEEE, 2017, pp. 685–690.
[60] H.M. Newman, Bacnet: The Global Standard for Building Automation and Control Networks, Momentum Press, 2013.
[61] S. Kraijak, P. Tuwanut, A survey on internet of things architecture, protocols, possible applications, security, privacy, real-world implementation and future trends, in: In 2015 IEEE 16th International Conference on Communication Technology (ICCT), IEEE, 2015, pp. 26–31.
[62] M. Dragone, J. Saunders, K. Dautenhahn, On the integration of adaptive and interactive robotic smart spaces, Paladyn, J. Behav. Robot. 6 (1) (2015) 000010151520150009.
[63] E. Gilman, O. Davidyuk, X. Su, J. Riekki, Towards interactive smart spaces, J. Amb. Intell. Smart Environ. 5 (1) (2013) 5–22.
[64] R. Piyare, Internet of things: ubiquitous home control and monitoring system using android based smart phone, Int. J. Internet Things 2 (1) (2013) 5–11.
[65] D.G. Korzun, Internet of things meets mobile health systems in smart spaces: an overview, in: Internet of Things and Big Data Technologies for Next Generation Healthcare, Springer, 2017, pp. 111–129.
[66] B. Subirana, H. Levinson, F. Hueto, P. Rajasekaran, A. Gaidis, E. Tarragó, P. Oliveira-Soens, The MIT Voice Name System, 2022. arXiv preprint arXiv:2204.09657.
[67] P. Joddrell, A.J. Astell, Studies involving people with dementia and touchscreen technology: a literature review, JMIR Rehab. Assistive Technol. 3 (2) (2016) e5788.
[68] D. Kern, A. Schmidt, Design space for driver-based automotive user interfaces, in: Proceedings of the 1st International Conference on Automotive User Interfaces and Interactive Vehicular Applications, 2009, pp. 3–10.
[69] M. Billinghurst, Augmented reality in education, New Horizons Learn. 12 (5) (2002) 1–5.
[70] A. Cannavò, C. Demartini, L. Morra, F. Lamberti, Immersive virtual reality-based interfaces for character animation, IEEE Access 7 (2019) 125463–125480.
[71] M.E. Cho, M.J. Kim, Characterizing the interaction design in healthy smart home devices for the elderly, Indoor Built Environ. 23 (1) (2014) 141–149.
[72] Buzzageek, 2023. How Smart is a Smart City. Buzzageek, Retrieved from https:/www.buzzageek.com.au/how-smart-is-a-smart-city, on the 11 Feb. 2023.
[73] F.K. Aldrich, Smart homes: past, present and future, in: Inside the Smart Home, Springer London, London, 2003, pp. 17–39.

[74] C. Stolojescu-Crisan, C. Crisan, B.P. Butunoi, An IoT-based smart home automation system, Sensors 21 (11) (2021) 3784.
[75] M. Moniruzzaman, S. Khezr, A. Yassine, R. Benlamri, Blockchain for smart homes: Review of current trends and research challenges, Comput. Electr. Eng. 83 (2020) 106585.
[76] W. Li, T. Yigitcanlar, I. Erol, A. Liu, Motivations, barriers and risks of smart home adoption: From systematic literature review to conceptual framework, Energy Res. Soc. Sci. 80 (2021) 102211.
[77] A.A. Dashtaki, M. Khaki, M. Zand, M.A. Nasab, P. Sanjeevikumar, T. Samavat, M.A. Nasab, B. Khan, A day ahead electrical appliance planning of residential units in a smart home network using ITS-BF algorithm, Int. Trans. Electr. Energy Syst. 2022 (2022), https://doi.org/10.1155/2022/2549887 (Article ID 2549887; 13 pp).
[78] A.K. Sikder, L. Babun, Z.B. Celik, A. Acar, H. Aksu, P. McDaniel, E. Kirda, A.S. Uluagac, Kratos: Multi-user multi-device-aware access control system for the smart home, in: Proceedings of the 13th ACM Conference on Security and Privacy in Wireless and Mobile Networks, 2020, pp. 1–12.
[79] M.A. Berawi, A.A. Kim, F. Naomi, V. Basten, P. Miraj, L.A. Medal, M. Sari, Designing a smart integrated workspace to improve building energy efficiency: an Indonesian case study, Int. J. Constr. Manag. 23 (3) (2023) 410–422.
[80] H. Lee, S. Choi, D. Kim, K. Lee, Smart space: An overview, Int. J. Precis. Eng. Manuf. Green Technol. 6 (6) (2019) 1189–1199.
[81] Y. Jin, L. Xie, X. Ding, Design of smart campus based on internet of things and building automation system, in: Proceedings of the 5th International Conference on Control and Robotics Engineering (ICCRE 2020), 2020, pp. 237–242.
[82] Y. Zhang, Y. Zhang, H. Chen, Smart space management: A review of the state-of-the-art, Build. Environ. 198 (2021) 107880, https:/doi.org/10.1016/j.buildenv.2021.107880.
[83] J. Chen, L. Gao, D. Zhan, Y. Wang, X. Wang, Smart space energy management based on big data and deep reinforcement learning, IEEE Trans. Smart Grid 12 (2) (2021) 1626–1635.
[84] E. Ndaguba, J. Cilliers, S. Ghosh, S. Herath, E.T. Mussi, Operability of smart spaces in urban environments: a systematic review on enhancing functionality and user experience, Sensors 23 (15) (2023) 6938.
[85] M.R. Haque, S. Rubya, An overview of chatbot-based mobile mental health apps: insights from app description and user reviews, JMIR Mhealth Uhealth 11 (1) (2023) e44838.
[86] J. Yu, G. Li, L. Zhang, B. Dong, An energy-efficient and cost-effective scheduling algorithm for smart homes based on cloud computing, IEEE Internet Things J. 8 (6) (2021) 4566–4575.
[87] A.M. Rahmani, T.N. Gia, B. Negash, A. Anzanpour, I. Azimi, M. Jiang, P. Liljeberg, Exploiting smart e-Health gateways at the edge of healthcare Internet-of-Things: a fog computing approach, Future Gen. Computer Syst. 78 (2018) 641–658.
[88] K.R. Varshney, S. Sarker, Smart building and facility management: a review of smart sensors, energy management systems, and building automation systems, ACM Trans. Manag. Inf. Syst. 10 (1) (2019) 1–29, https:/doi.org/10.1145/3326486.
[89] D. Wang, Y. Hu, Y. Wen, Smart building energy management: a review, Renew. Sustain. Energy Rev. 123 (2020) 109761, https:/doi.org/10.1016/j.rser.2020.109761.

[90] T. Yigitcanlar, M. Foth, M. Kamruzzaman, Towards post-anthropocentric cities: Reconceptualizing smart cities to evade urban ecocide, J. Urban Technol. 26 (2) (2019) 147–152.
[91] A. Farooq, Smart Product line model for fire brigade service to assist civilians, Int. J. Sci. Eng. Res. 9 (4) (2018).
[92] Y. Cho, A. Choi, Application of affordance factors for user-centered smart homes: a case study approach, Sustainability 12 (7) (2020) 3053.

CHAPTER

Exploring the role of IoT for sustainable enhancement in smart spaces

G. Sree Lakshmi[a] and S. Lakshminarayana[b]

[a]*EEE Department, CVR College of Engineering, Hyderabad, Telangana, India,* [b]*Department of CSSE, College of Engineering, Andhra University, Visakhapatnam, Andhra Pradesh, India*

1 Introduction

The Internet of Things (IoT) has emerged as a revolutionary force, transforming to interact with the world around us in the quickly changing context of contemporary technology. The IoT is a huge network of interconnected devices and sensors that interact and share data via the Internet. This technology enables these devices to acquire, process, and act on information in real time. The concept of "smart spaces" is one of the IoT's most promising applications, opening up a wealth of potential.

Smart spaces are a broad category that includes a variety of settings, such as smart cities, smart buildings, smart homes, and more, that use IoT technologies to streamline operations, increase productivity, and enhance the general quality of life for residents. The goal of sustainability, or the idea of meeting present demands without sacrificing the capacity of future generations to meet their own, lies at the core of this theory. Smart spaces are buildings or public spaces that have sensors installed to collect data that can be used to learn more about their surroundings, the services they offer, and how their residents interact with them. These insights from smart spaces can be gleaned from historical data as well as data collected in real time, which can then be applied to enhance operations, user experiences, and safety [1–3].

Smart environments should be defined in a way that goes beyond just the technology that makes them possible. A concept concerning a person's relationship with her surroundings is referred to as a "smart space" or "smart environment." Smart digital billboards enable outdoor marketers to target specific pedestrians with advertisements as they pass by (automated systems update ads based on the real-time stream of data from nearby cell phones), gathering information about onlookers and, in the process, bringing a sense of personality to what was once a largely anonymous public space. As waste reduction and a smaller carbon footprint become more important, sustainability is increasingly on the boardroom agenda. Diverse

Smart Spaces. https://doi.org/10.1016/B978-0-443-13462-3.00013-3

technology, data, hardware, experiences, and rules are just a few examples of the different components that make up buildings and spaces. These components can integrate and synchronize. About 40% of all greenhouse gas emissions are produced by these elements. People can be more productive in a connected, sustainable, accessible, and secure setting by making the space "smart."

Without the development of the Internet of Things (IoT), the cloud, artificial intelligence (AI), machine learning (ML), big data, and processing capacity, smart spaces would have slow growth. This technical convergence is crucial to aggregate the incoming data in a building. The data is then published for essential business leaders to use in making critical decisions that will affect the overall optimization of the building, its assets, and its occupants after it has been normalized into a standard communication protocol. As sensors, processors, and intelligent software can be installed in any place to make it smarter, many smart spaces have advanced in sophistication, and new uses are constantly being embraced. Our immediate surroundings have advanced, as have areas that may not often be on our minds, but still help us in the background [4–6].

IoT devices are one of the most accurate and best ways to collect and verify data, since they are automated and are not subject to human error, and this is how the IoT supports and drives a sustainability culture through smart spaces. The Internet of Things (IoT) transforms a structure or environment into a smart place by removing the need for humans to gather, interpret, and translate data from machines, air quality, temperatures, or human aspects. After the information is gathered, it can be analyzed, visualized, or merged with AI and ML. IoT is unquestionably the most cost- and accuracy-effective collection of instruments for gathering sustainability data as compared to other options, such as manual data inputs. It works well as an enabler when used with the cloud. Significant obstacles do, however, come along with its immense potential. To guarantee that the advantages of the IoT for sustainability are fully realized, security and privacy issues around IoT devices and data processing, as well as the ethical implications of data collecting and utilization, must be addressed [7,8].

This chapter explains the important role of IoT in promoting long-term improvements in smart spaces. With severe global concerns such as climate change, resource depletion, and environmental degradation, there is an urgent need to investigate new solutions that promote ecological balance and social well-being. The IoT is used as a tool to revolutionize resource management, energy efficiency, waste reduction, and mobility, among other things, by integrating it into smart places. Also, the revolutionary impact of the IoT in the creation of sustainable smart places is covered. By understanding the opportunities and challenges, one can see a path to a more resilient, environmentally sustainable, and egalitarian future. As technology continues to influence our environment, embracing the IoT for sustainable development becomes not only an option, but a requirement for fostering a better planet for future generations.

2 Technologies used to create insights into smart spaces

A widely used framework categorizes smart spaces into three distinct environments that interact as one: a virtual computing environment, the physical environment, and the human environment. This framework bridges our conceptual understanding of smart spaces and the physical technology that makes them a reality.

a. **Virtual Computing Environment:** The virtual computing environment layer gives smart devices access to private network services or the Internet, which enables them to connect to other components of the distributed systems that run the smart space environment.
b. **Physical Environment:** The most diverse layer of smart spaces is the physical environment layer, which also comprises the embedded sensors, microprocessors, tracking tags, and other physical components of the smart space.
c. **Human Environment:** This layer is filled with personal smart devices, including pacemakers, wearable smart devices, and cell phones that people carry around with them. This means that humans can develop smart space settings using cell towers, cell networks, and smartphones to create a virtual, physical, and human smart environment that can be considered a large area of smart space.

The approach to smart space settings with several layers (virtual, physical, and human layers) allows us to group smart enabling technologies according to how they are used.

3 Different kinds of smart spaces

This section discusses several kinds of smart spaces in existence today.

Smart Buildings: Several characteristics of smart houses are also used in smart buildings, just on a larger scale. Smart buildings use smart spaces to connect automated building systems that monitor lighting, heating, cooling, security and access, parking structures, water and power meters, fire systems, boiler or chiller plants, elevators, and more, depending on the building. While a hospital with medical supplies and equipment may use smart RFID technology to track, secure, and replenish its inventory as needed, an office building may be made smarter to maintain unit occupancy and reduce costs [9]. These will have a capability to exchange, interact with the nearby similar buildings.

Smart Cities: Smart cities are metropolitan areas with smart space technology installed for governance. Tens of thousands of individuals could benefit from the enhanced efficiency of smart cities. Already, clever initiatives are improving the lives of citizens. Some cities use smart spaces to improve the digital inclusion of their residents by digitizing public records and making them accessible online. Many cities have also launched programs to put and conceal antennas on light poles to provide their residents with digital access [10].

To convey alarms and to provide the administrators of these spaces with insights, smart spaces must collect, transmit, store, manage, and analyze data from environment sensor systems.

Smart Homes: Smart homes enable smart spaces to connect numerous home appliances and systems, enhancing our living spaces' efficiency and comfort. Home automation may now take care of home systems that manage lighting, air conditioning, entertainment, and even security. A number of products, including smart refrigerators and ovens, are now intelligent. There is even a name for this: domotics [11,12].

Smart Factories: Smart factories are nothing new, as many manufacturing sectors served as the petri dish in which "smart" technology was developed. The connected smart factory is currently undergoing a paradigm change toward a digital supply network where numerous factories and suppliers are networked and smaller units can make decisions based on system-wide data. An automated supply chain produces a flexible and effective operation with less downtime and greater adaptability to changing needs, which is important given the size of globally synchronized supply chains [13,14].

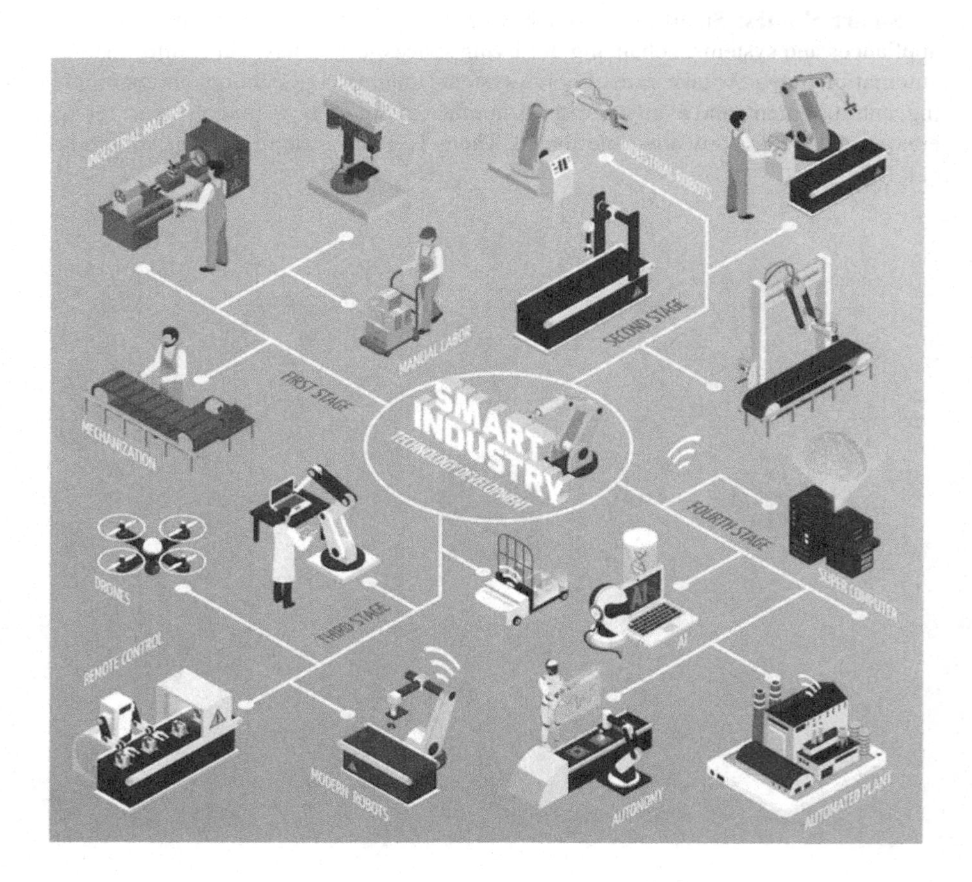

4 Types of sensors used in smart spaces

The following are the main types of sensors employed in smart spaces:

1. Temperature Sensors
2. Humidity Sensors
3. Motion/Occupancy Sensors
4. Contact Sensors
5. Gas/Air Quality Sensors
6. Electrical Current Monitoring Sensors
7. Other Types of Sensors

4.1 Temperature sensors

Temperature sensors measure heat to identify temperature changes. Since the introduction of the Internet of Things, they have been used for various purposes in addition to controlling heating and cooling systems.

As an illustration, many devices used in manufacturing and computing are temperature-sensitive and must be shielded from overheating. Businesses may automate HVAC controls to maintain perfect conditions and automatically spot failure or faults using smart temperature sensors [15].

Improper control of temperature can be dangerous, and temperature is essential for people's comfort. For example, it is the duty of everyone in charge of a commercial space or who rents out a property to lower the danger of *Legionella* bacteria exposure. This bacteria can spread in both hot and cold water systems and can thrive in any area where the water temperature is between 20°C and 45°C.

There are four types of temperature sensors:

a. **Semiconductor-based Sensors:** These identical diodes, mounted on an integrated circuit, use temperature-sensitive voltage compared to current conditions to track temperature changes.
b. **Thermocouple:** As the name implies, this comprises two wires constructed of different metals and positioned at various sites. The difference in voltage between the two points indicates a change in temperature.
c. **Resistance Temperature Detector:** A ceramic or glass core is wrapped in a film or wire, and the change in resistance of the element with temperature is used to determine the temperature. They can be the priciest sensors but also tend to be the most accurate.
d. **Negative Temperature Coefficient Thermistor:** Resistance is strong at low temperatures but rapidly decreases as temperature rises, precisely and quickly reflecting changes.

4.2 Humidity sensors

The quantity of water vapor in the atmosphere is known as humidity, sometimes known as relative humidity. Similar to how some temperatures are difficult for many devices to handle, humid environments can also be problematic. Condensation in the atmosphere leads to corrosion in several types of machinery.

Maintaining optimal conditions is made possible by humidity sensors, allowing quick reactions to changes. These sensors are used to regulate the HVAC and AC systems in residences and workplaces. Any facility that needs a moisture-free environment uses them, including factories, hospitals, museums, greenhouses, and weather stations [16].

There are three common types of humidity sensors:

a. **Capacitive:** The sensor, which has two electrodes and a porous dielectric material in the center, measures humidity using water vapor; when the vapor contacts the electrodes, a voltage shift occurs.

b. **Resistive:** While less sensitive than capacitive, these sensors work on a similar principle and gauge relative humidity by electrical change. The change in resistance on the electrodes is measured using ions found in salts.
c. **Thermal:** Based on the humidity of the air around them, two matching thermal sensors transmit electricity. The difference between them measures the humidity reading. One is covered in dry nitrogen and the other monitors ambient air.

4.3 Motion/occupancy sensors

Motion sensors detect physical movement in a specific region, such as that of a person, animal, or item, and convert that data into an electric signal. Motion detection has been employed in the security sector for many years to warn businesses of intruders. These can be found in everyday items like hand dryers, automatic doors, and toilet flushes. Additionally, they can be used to automate building functions like heating and lighting based on whether an area is used or not, assisting in reducing energy consumption and operating expenses.

However, a new application has been discovered for these sensors, to assist organizations in understanding how rooms and areas are utilized. Occupancy sensors let businesses determine which areas are used the most or which desks or meeting rooms are available at any one time by instantly detecting the presence of people or items. Efficiency in space use can result in significant cost savings and productivity gains in large organizations.

Motion and occupancy sensors work by sensing infrared energy or by emitting radio waves, ultrasonic waves, or both and measuring the reflection of those waves off a moving object. Passive infrared (PIR) motion sensors for underdesk use are available. This little wireless device has two slots constructed of an infrared-sensitive substance. Both slots pick up an equal quantity of ambient infrared radiation while the sensor is not in use. One-half of the sensor detects movement before the other when a person enters the field of view. The sensor learns that someone is there due to this difference in radiation levels between the two slots [17,18].

There are three main types of motion detector sensors:

Motion Sensors or Passive Infrared (PIR) Sensor: These sensors function by sensing the heat that humans release. This sensor will detect motion and provide an alert to the presence of people when one enters its field of view. The sensor keeps track of an area and can provide updates on occupancy levels. These sensor types are GDPR (General Data Protection Regulation) and privacy-compliant options because they don't retain or send any photos or personal data. PIR sensors are available in various configurations; some are mounted on walls or ceilings, while others are placed behind desks. They are a discreet, low-maintenance, and cost-effective choice and are also simple to install.

a. **Desk occupancy Sensors:** These adhere to a desk's underside. Only motion within a 180-degree radius is detected by the PIR sensor because it has a cowl covering only half of the sensor. This makes it very accurate, only identifying a person under a desk and not anyone strolling behind or to the side of it. When combined with a narrow-angle lens, this further improves its accuracy.

b. **Table occupancy Sensors**: These attach to the underside of a table and have a narrow-angle lens, just like desk sensors. This PIR sensor, however, can detect persons seated around a table because it senses motion in a 360-degree circle. The detection range is 0.5 m based on the height of an average table.
c. **Room occupancy sensors** are ceiling-mounted devices with a 360-degree field of vision and a wide-angle lens for person detection. The predicted detection range is 5 m based on a normal ceiling height of 2.5 m and a 64-degree detection angle.
d. **Cubicle occupancy sensors:** These sensors detect a 180-degree view, just like desk sensors. They may be mounted on a cubicle wall or ceiling to identify persons inside that area, to prevent people from wandering through. They are excellent for restrooms or conference rooms.

Time-of-flight sensors: This type of sensor sends out an infrared light beam that bounces off objects and returns to the sensor; the time it takes for the beam to arrive at the sensor allows for precise distance measurement, making it a useful entry/exit sensor. The sensor uses these measures to detect whether the person is approaching or moving away from the sensor.

a. People-flow sensors: These allow for the live monitoring of any single-occupant doorway's traffic flow by detecting persons and the direction of movement. Because they are bidirectional, you can count the people in space by counting those who are walking toward or away from the sensor. Given that these sensors are not cameras and that they only detect movement rather than faces, they are consistent with GDPR and privacy laws.

Infrared array sensors: These sensors measure the temperature as a person or object moves closer to or farther away from the sensor, allowing detection of moving or static objects, temperature distribution, thermal images, and moving directions. The field of view widens as the distance from the sensor grows, while the apparent angular size gets smaller. The sensor can accurately detect temperature and form as the distance between it and the object increases [19].

a. People counter and movement sensors: Find out how many people are in a space, where in the space they are, and how they are moving around. These sensors are GDPR and privacy compliant because they only detect human movement.

4.4 Contact sensors

As well as being called position, status, and building monitoring sensors, contact sensors have other names. A door, window, or other similar device can easily be detected as open or closed using contact sensors.

These sensors are divided into two parts, one of which is fixed to the door or window and the other to the frame. They are touching (indicating that the door or window is closed), and when they have been pushed apart, the two components use magnetic fields (as the door or window is opened).

Knowing what is going on around a facility at any given time is useful for various reasons, including safety, security, and energy efficiency. The current status of all the doors and windows in the building can be viewed, including the doors on refrigerators, cabinets, and cupboards, utilizing contact sensors for building monitoring. Building controls can be automated based on live occupancy, to automatically identify open or broken windows, unlocked doors or cabinets, a presence in a room, or unlocked doors or cabinets [20].

4.5 Gas/air-quality sensors

Gas sensors are used to track changes in air quality and to identify different gases. In the manufacturing, pharmaceutical, petrochemical, and mining industries, they are used to monitor hazardous gases, identify toxic or combustible gases, and monitor air quality. Depending on the application, these sensors can be used to monitor gases such as carbon dioxide, carbon monoxide, hydrogen, nitrogen oxide, oxygen, or air pollution.

While safety is a major concern in many applications, the consequences of poor air quality aren't necessarily severe or even that obvious. Rising carbon dioxide levels can cause stale, stuffy air in today's well-insulated houses, as well as problems like fatigue and headaches. Together with productivity, air quality can impact people's comfort and well-being. It should come as no surprise that more companies are employing environmental monitoring to maintain temperature and air quality, given that employers have to offer a safe working environment.

There are three common types of air-quality sensors:

a. **Oxygen:** This electrochemical sensor can detect any gas that may be electrochemically reduced or oxidized.
b. **Carbon Monoxide:** This sensor functions similarly to the oxygen sensor and is likewise an electrochemical sensor.
c. **Carbon Dioxide:** A carbon dioxide detector passes an infrared beam via a light tube to determine how much of the beam's energy is still there.

4.6 Electrical current monitoring sensors

Electrical current (CT) sensors track energy use in real time at the machine, zone, or circuit level. There are two basic applications for knowing how much energy is being utilized. First is energy savings, by identifying where energy is consumed and where energy is wasted. Moreover, assets can be programmed to turn off automatically when not in use. Second, if typical operating conditions can be identified, the sensor can alert as to when a piece of equipment isn't running as it should. For instance, a motor may have been overworked if its operating current is higher than typical. This realization allows one to plan maintenance rather than spending money on unnecessary inspections. Unscheduled downtime can be minimized by immediately addressing any possible issues [21,22].

4.6.1 Types of CT sensors

a. **Split Core:** This type is perfect for current arrangements since they can be opened up and placed around a conductor.
b. **Hall Effect/DC:** These sensors measure the changing voltage when an object is placed in a magnetic field using a technique known as the Hall effect. They can measure both AC and DC current. They might have an open loop or a closed loop. While closed loops give quick reaction and little temperature drift, open loops are small, inexpensive, and accurate.
c. **Rogowski coils:** These flexible current transformers are simple to install. The conductor is wrapped in a small coil, which is then snapped shut.
d. **Solid core:** These sensors are complete loops with no means of opening, making them ideal for new installations. They are recognized for being extremely accurate.

4.7 Other types of sensors

A number of other types of sensors can be employed in smart spaces, as follows:

a. **Optical Sensors:** Optical sensors are used to measure electromagnetic energy, such as light and electricity. They are used to track temperature, light, radiation, electric and magnetic fields, and other variables in sectors like healthcare, energy, and communications.
b. **Proximity Sensors:** Similar to motion sensors, proximity sensors can both detect an object's presence and gauge its proximity. Reverse parking sensors in automobiles are one of the best-known applications.
c. **Pressure Sensors:** Similar to machine monitoring, pressure sensors identify changes in pressure and notify the system administrator. In addition to being helpful for water and heating systems, this is also important in manufacturing.
d. **Water-quality Sensors:** Environmental management uses water-quality sensors to assess the concentrations of chemicals, ions, organic substances, suspended particles, and pH in water.
e. **Chemical Sensors:** Chemicals can be found in the air or in the water using chemical sensors. They are employed to monitor industrial operations, examine the quality of the air and water in urban areas, and find radioactive items, toxic compounds, and explosives.
f. **Smoke Sensors:** Smoke sensors measure the amount of gases and particulates in the air. Although they have existed for some time, the growth of IoT has allowed these sensors to alert consumers to issues right away.
g. **Level Sensors:** The level of fluids, liquids, or other substances in an open or closed system is determined using level sensors. They are mostly used to gauge fuel levels, but are also found in hydraulics, medical equipment, compressors, and reservoir levels as well as in the sea and on land.

h. **Image Sensors:** Digital cameras, medical imaging, night vision, and biometric equipment all use image sensors. They are employed in the automobile industry as well and are crucial to the advancement of autonomous vehicles.
i. **Accelerometer Sensors:** Accelerometer sensors are used to identify acceleration, tilting, and vibration in an object. Applications for consumer electronics such as cell phones and pedometers include antitheft devices and fleet monitoring for vehicles, aircraft, and the aviation industry.
j. **Gyroscope Sensors:** Gyroscope sensors are used in conjunction with accelerometers to detect angular velocity, which is the rate at which an object rotates around an axis. Robotics, consumer electronics, game controllers, and automobile navigation systems are among their primary uses.

5 Key features of IoT smart buildings

IoT smart buildings have a number of key features in common.

Real-time data analytics: The DNA of a smart building is comprised of real-time data collection and analytics. Facilities managers and other decision-makers can use this information to choose the best course of action, as they can quickly comprehend and unravel user behavior, spot patterns, and foresee hazards due to the raw data collected by sensors and other devices throughout the facility.

Wireless Communication: The development of wireless communications has been essential to the smarter construction of buildings. Real-time communication and data exchange are made possible by this technology in an approach that is both simple to use and reasonably priced.

User Interface: People can better understand complex or information-heavy data with the use of simple user interfaces. A central dashboard with data gathering, analytics, reporting, and management features that make information meaningful and usable is part of a smart building solution. IoT smart building software can go one step further and incorporate personalization based on each user's unique behavior, further enhancing their experience.

These days, cloud technology makes sure that these software programs are always available from any location at any time on any connected device [23].

These real-time communications solutions' simplified user interfaces offer the following advantages:

- Improved user experience
- Faster, simpler collaboration and information exchange
- Improved awareness of the situation
- Improved diagnosis
- Enhanced service
- Faster responses

5.1 Plan for IoT integration into new or existing buildings

There are things to consider and recommended practices to follow when preparing to integrate IoT for better building energy management and other benefits. Using tried-and-true IoT integration techniques boosts gradual digital transformation into a fully smart building while ensuring excellent building performance.

Employ a modular approach: It is wise to start simple and eliminate complexity in order to properly redesign the building systems with IoT. It is important to identify the many construction requirements, plan out every aspect, and do a pilot project on the most important elements, such as lighting or HVAC. Remember that each subsystem ultimately needs to interface with other building systems and subsystems efficiently. Hence, it is advised to ensure end-to-end compatibility with other systems.

Define the key building requirements: Building owners or managers can create a strong IoT integration plan by outlining the major quantifiable goals that must be accomplished. The plan should be made at the project's start for it to succeed. Software, hardware, infrastructure, and safety and security measures could all be included in the criteria.

Adopt an all-inclusive approach: All stakeholders must be involved in creating a solid IoT integration master plan, starting with the operations, security, administration, and maintenance staff. Stakeholders can get involved by setting sustainability standards, operational objectives, and productivity targets.

Evaluate the return on investment (ROI): It is feasible to evaluate the return on investment and build a successful plan based on the specified parameters. One approach is to start with a trial project and then broaden it to include the entire Internet of Things.

Collaborate with reliable vendors or contractors: Based on aspects including knowledge, pricing, ease of access to services, and organizational objectives, choose a qualified vendor or contractor. Several IoT businesses have their own branding for intelligent buildings and knowledgeable teams of experts. These businesses can provide IoT services with the highest reliability, efficiency, and quality.

5.2 Critical factors to be considered during IoT design for a smart building

Planning ahead and adopting a proactive attitude are essential for a designer to achieve successful IoT integration into a smart building's energy management system.

a. Interoperability and integration with other systems: A holistic approach should be implemented to ensure that the IoT energy management system operates efficiently. The multiple devices chosen for IoT integration should include essential qualities like adaptability and scalability. Additionally, they must be able to operate as a single system and interface flawlessly with other systems. Future expansion provisions should be provided because of new developments in IoT technology in smart buildings.

b. **Privacy and Cybersecurity:** IoT increases the volume and rate of data flow between systems and devices through a wireless backbone network, increasing the danger of hacking, unintentional data spills, and data breaches. To address cybersecurity and data privacy concerns, the IoT Institute has developed IoT data protection frameworks for integrating smart building systems. Starting with the design phase, these privacy frameworks should be followed [24,25].

6 Key features of IoT smart cities

IoT-enabled smart cities have a variety of applications, from promoting a healthier environment and enhancing traffic to strengthening public safety and improving street lighting. The most well-liked features already in use in smart cities worldwide are summarized here.

Road Traffic: Smart cities ensure that residents can travel from point A to point B as safely and effectively as feasible. Municipalities use IoT development and smart traffic systems to achieve this.

To estimate the quantity, location, and speed of vehicles, smart traffic solutions use a variety of sensors as well as GPS information from drivers' smartphones. In order to avoid congestion, smart traffic lights connected to a cloud management platform allow monitoring green light timings and automatically changing the lights based on the scenario. Smart traffic management technologies can also forecast future traffic patterns and take action to avoid congestion by leveraging historical data to make these predictions.

Smart Parking: Smart parking systems can tell whether parking spaces are occupied or available by using GPS data from drivers' smartphones (or road-surface sensors buried in the ground on parking places) and can generate a real-time parking map. Instead of driving aimlessly around looking for parking spaces, users receive a signal when the closest parking spot becomes available and may utilize the map on their phone to find a spot more quickly and easily.

Public Transport: IoT sensor data can be used to identify trends in how people use transportation. Public transportation operators can utilize this information to improve the traveling experience and reach a higher standard of safety and punctuality. Smart public transportation solutions can incorporate different sources, such as ticket sales and traffic data, to perform a more complex analysis.

Utilities: IoT-enabled smart cities give residents more control over their residential utilities, allowing them to save money.

IoT enables various methods for smart utilities:

a. **Smart Meters and Billing:** Municipalities can affordably connect their residents to the IT systems of utility providers by using a network of smart meters. A public utility can receive accurate meter readings by using smart linked meters to relay data straight to it over a telecom network. Utility providers can precisely bill customers for the quantity of water, gas, and electricity they use by using smart meters.
b. **Revealing Consumption Patterns:** A smart meter network allows utilities to gain greater visibility into how their customers use energy and water. Utility

companies can use a network of smart meters to monitor demand in real time, redirect resources as needed, or encourage consumers to use less energy or water during times of scarcity.

c. **Remote Monitoring:** IoT smart city solutions can make utility management services available to citizens. With the use of these services, citizens may remotely monitor and manage their usage using their smart meters. For instance, a homeowner can use a cell phone to switch off the central heating in their home. In addition, utility companies can alert homeowners and dispatch experts to remedy any issues that arise (such as a water leak).

Street Lighting: Street lamp control and maintenance are more accessible and more affordable in IoT-based smart cities. Streetlights with sensors and a cloud management system are connected to adjust the lighting schedule to the lighting zone.

In order to optimize the lighting schedule, smart lighting solutions gather information on lighting conditions, traffic flow, and the movement of people and vehicles. They then combine this information with historical and contextual data (such as information on holidays, special occasions, public transportation schedules, time of day and year, etc.). Because of this, a smart lighting system "tells" a streetlight how bright or dark to make it, as well as whether to turn the lights on or off based on the weather [26–28].

Waste Management: The majority of waste collection operators empty containers on predetermined schedules. This is not an efficient approach because waste containers are used inefficiently and waste collection trucks consume unnecessary fuel. Smart city solutions powered by IoT assist in optimizing waste collection schedules by tracking waste levels and providing route optimization and operational analytics.

Each waste container is outfitted with a sensor that collects data on the level of waste in the container. When it approaches a certain threshold, the waste management solution receives a sensor record, processes it, and sends a notification to the mobile app of a truck driver. As a result, the truck driver empties a full container rather than half-full ones.

Environment: IoT-driven smart city solutions enable tracking parameters necessary for a healthy environment and keeping them at an ideal level. For instance, a city can set up a network of sensors throughout the water system and connect them to a cloud management platform to monitor water quality. Sensors gauge pH, the quantity of dissolved oxygen, and the concentration of dissolved ions. The cloud platform initiates a user-defined output when a leak occurs and the chemical composition of the water changes. For instance, if a nitrate (NO_3^-) level surpasses 1 mg/L, a water quality management solution notifies maintenance teams of contamination and instantly generates a case for field employees, who then begin resolving the issue.

Monitoring air quality is another application. A network of sensors is set up near busy roads and in close proximity to plants. Sensors measure the amount of CO, nitrogen, and sulfur oxides, and the data is then analyzed and visualized by a central cloud platform. Users of the platform can view the map of air quality and use this information to identify areas with particularly high levels of air pollution and develop recommendations for the public.

Public Safety: Real-time monitoring, analytics, and decision-making tools are provided by IoT-based smart city technology to improve public safety. Public safety systems can identify probable crime situations by evaluating data from social media feeds, sound sensors, and CCTV cameras placed across the city. This will allow the police to apprehend or locate suspected offenders.

For instance, a gunshot detection technology is used in more than 90 communities in the United States. The remedy involves networked microphones placed all over a metropolis. The cloud platform receives the data from the microphones and processes it to identify a gunshot after sound analysis. The platform calculates the gun's location and measures the time it takes for the sound to reach the microphone. Cloud software notifies the police via a mobile app when the gunshot and its location are identified [29–31].

6.1 Iterative approach in implementing smart city solutions

The applications for smart cities are extremely varied. Their method of implementation is what unites them. Municipalities should begin with the foundation, or a fundamental smart city platform, whether they intend to automate waste collection or enhance street lighting. Future smart city services can be added to an existing architecture without having to rebuild it if a municipality chooses to do so completely.

Following are the six steps for implementing an effective and scalable IoT architecture for a smart city.

Step 1: Basic IoT-based smart city platform

The first step in implementing a scalable smart city is to create a basic architecture. This architecture will act as a foundation for any future improvements and allow the addition of new services without compromising functionality. Four elements make up an IoT solution for smart cities at its core:

a. **The Network of Smart Things:** Smart items with sensors and actuators are used in smart cities, just as in any IoT system. The sensors' immediate objectives are data collection and transmission to a central cloud management platform. Devices can act with the help of actuators, which also change the lights, stop water from flowing into a leaky pipe, etc.
b. **Gateways:** Any IoT system consists of two components: a cloud component and a "physical" component made up of IoT devices and network nodes. Data cannot just "flow" from one component to another. Field gateways and doors are necessary. Field gateways make data collection and compression easier by cleaning and filtering data before sending it to the cloud. Between field gateways and the cloud component of a smart city solution, the cloud gateway enables safe data transmission.
c. **Data Lake:** A data lake's primary function is to store data. Data lakes maintain information in its unprocessed form. The large data warehouse receives the extracted data when it is required for insightful analysis.
d. **Big Data Warehouse:** This is the single data repository that makes up a huge data warehouse. It solely includes structured data, unlike data lakes. Following

the definition of the data's value, it is extracted, converted, and fed into a large data warehouse. Additionally, it saves the commands that control applications send to the actuators of linked devices and contextual information about connected things, such as the date that sensors were installed.

Step 2: Monitoring and Basic Analytics: Data analytics makes it possible to keep an eye on the surroundings of devices and establish rules for control applications to follow to complete a particular activity. Cities, for instance, can design regulations for the electronic valves to close or open depending on the recognized moisture level by evaluating the data from soil moisture sensors installed across a smart park. Users can see the data gathered using sensors on a single platform dashboard to see how each park zone is currently doing.

Step 3: Deep Analytics: When city administrations process IoT-generated data, they can go beyond monitoring and simple analytics to find patterns and hidden relationships in sensor data. Machine learning (ML) and statistical analysis are two sophisticated methods used in data analytics. Machine learning algorithms examine previous sensor data kept in the big data warehouse to find trends and develop forecast models on them. Control applications, which instruct the actuators of IoT devices with commands, use the models. How it works can be described as follows.

A smart traffic light may adjust signal timings to the traffic situation, unlike a typical traffic light that is programmed to display a specific signal for a set amount of time. In order to identify traffic patterns and modify signal timings, machine learning algorithms are applied to previous sensor data. This increases the average vehicle speed and helps to prevent traffic jams.

Step 4: Smart Control: By providing commands to their actuators, control applications improve the automation of smart city items. In essence, they instruct actuators on how to do a specific duty. Applications for control that use ML and rules both exist. While models produced by ML algorithms are used in ML-based control applications, rules for rule-based control applications are manually written.

Step 5: Instant interaction with citizens via user applications: There should always be a mechanism for users to affect how smart city applications behave, in addition to the potential for automated control (for example, in case of emergency). User applications perform this work. In order to monitor and control IoT devices and receive notifications and alarms, citizens can connect to the central smart city administration platform through user applications. For instance, a smart traffic management system can identify a traffic jam using GPS information from drivers' smart phones. The technology automatically notifies drivers in the region to choose an alternate route to avoid even more congestion.

A desktop user software sends a "congestion alert" to traffic control center staff. They give the actuators of the traffic lights instructions to change the signals in order to ease the congestion and reroute some of the traffic [32–34].

Step 6: Integrating Several Solutions: It takes time and effort to become "smart," and it cannot be done all at once. Municipalities should consider the services they might want to implement in the future while developing IoT-based smart city solutions. Increasing the number of functionalities is more significant than simply adding additional sensors. Let's use the example of a smart city traffic monitoring technology to demonstrate this functional scalability.

A city implements a traffic management system to monitor congestion in real time and control traffic signals to lessen traffic in congested regions. After some time, the city decides to link the smart air quality monitoring technology with the traffic management solution to ensure that city traffic doesn't negatively impact the environment. Integrating many solutions enables dynamic control of the city's traffic and air quality.

For this functionality, air quality sensors might be installed in traffic or streetlights next to roads. Sensors assess CO, NO, and NO_2 in the air and transmit data records to a centralized air quality management platform for processing. Control applications apply rules or use models to execute an output action, such as "change traffic lights," if the level of dangerous gases in the air is critical. Prior to that, it must be confirmed that changing traffic signals won't result in mishaps or obstructions in nearby locations. This is achievable because the air quality management system and the traffic management solution are integrated. Real-time analysis is done by the traffic management platform to determine whether changing the traffic signals is an option. Control apps deliver a command to the traffic light actuators, who carry out the command if changing the lights is approved [35].

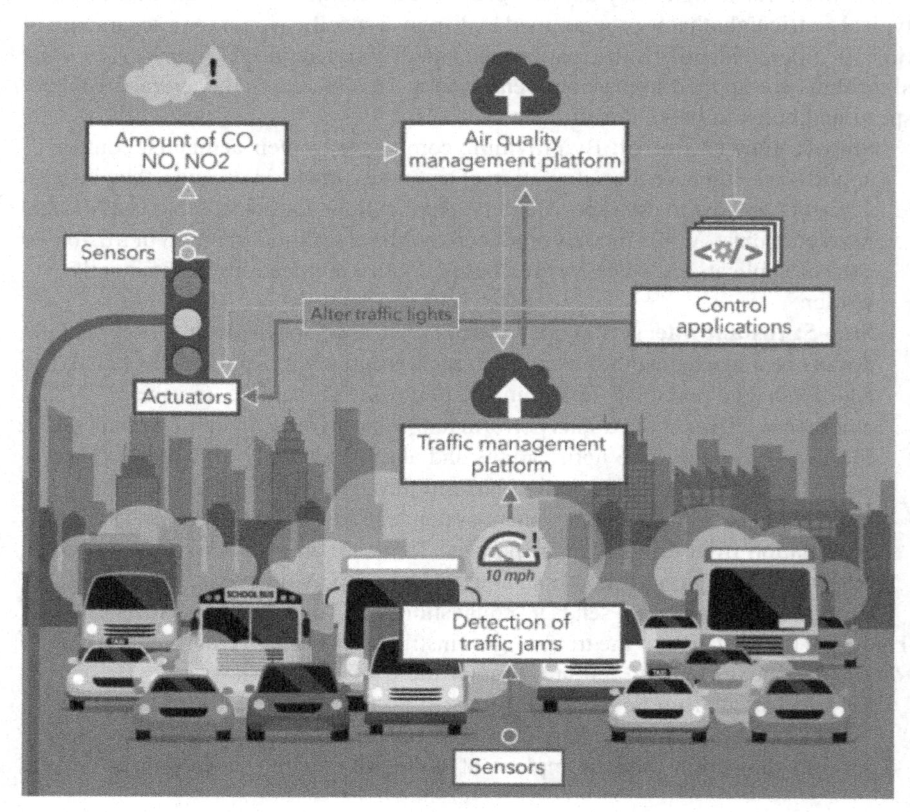

Using an iterative strategy enables governments to lower implementation costs, get a quicker payoff, and make the advantages of smart solutions more immediately apparent to citizens.

7 Key features of IoT smart homes

Home automation with the Internet of Things (IoT) is the ability to control domestic appliances with Internet-connected hardware. It can entail preprogramming intricate heating and lighting systems, alarms, and home security controls that are all connected via a central hub and managed remotely via a smartphone app.

However, there doesn't seem to be much discussion about why consumers haven't clearly bought into the IoT home "hype," or whether domestic life has improved as a result of it in this always-connected IoT home of mood-sensing music systems, smart lighting, intelligent heating and cooling, motorized blinds, and automated windows and doors.

The American Association of Home Builders first used the term "smart house" in 1984, but IoT smart homes, or IoT home automation, gained popularity in the early 2000s. Smart houses became a more viable option for customers as various lower priced smart home technologies appeared. (The first home automation system, the Echo IV, was developed in 1966. It weighed 800 lbs!)

In an IoT smart home, each individual device is often managed by a central hub that is connected to the Internet. Then, using a smartphone app, the main smart home hub is managed. Setting up an IoT smart home can be challenging, as a variety of attachments can only be used with particular products. A home that has been automated is referred to as "home automation" or a "smart home" or "smart house." The IoT home automation ecosystem allows the user to monitor and/or control various appliances, entertainment systems, lighting, and temperatures. Also, it can have components for home security such as alarm systems and access control. Once connected to the Internet, household appliances become an important part of the Internet of Things. A central hub or gateway in a home automation system is typically connected to controlled devices. Wall-mounted terminals, tablet or desktop computers, a smartphone app, or even an online interface that can be accessed from off-site through the Internet are all possible ways to access the system's control program [36–38].

7.1 Benefits of IoT in smart homes

IoT-based automation in smart homes includes a range of appliances for temperature control, security, and home entertainment. The advantages of IoT in smart homes include the following.

a. **Cloud-based IoT Software:** Be notified of events instantly: users may easily control their device from a distance because the sensors and equipment are connected to the cloud. The cloud-based program enables data management and archiving for cloud-based home automation. Users can submit commands or signals to the hub, subsequently sending a signal to the device. This causes the action to be completed. Once that is finished, the hub updates the action's status across the cloud network and notifies the user.

b. **Optimize Energy Consumption:** IoT devices use this data to optimize their performance in addition to monitoring usage. The devices offer thorough time- and energy-based consumption statistics. This allows for the efficient modification of smart home systems. Devices can be turned off manually or automatically when not in use to conserve energy.
c. **Convenience and Security:** Consider the advantages of a smart home the next time you worry about your pet's safety, or whether you forgot to turn off the gas or turn on the lights. IoT technology makes it easy for users to instantly access information via a cloud-based server. It makes automation and information delivery to the user simpler. IoT-based smart home devices have already made their mark in terms of security. IoT devices that collaborate with one another include security cameras, smart locks, and smoke sensors, to mention a few. Whether away at work or on vacation, users can access the home's security and be alerted immediately if there is an incident.
d. **Customized Preferences and Safety Settings:** IoT devices offer individual preferences and safety settings and save energy. A thermostat can be programmed to automatically change the temperature to a specific setting based on time, or components such as lights can be set to dim after a specific time of day. The processing and collecting of data in these devices is significantly aided by AI integration. An older audience who might forget to turn off electronics will find this to be especially helpful. Energy-saving capabilities and, in some situations, the ability to program an alarm are both possible. In general, home automation isn't only about expensive equipment. For optimum and personalized use, it's about making everyday tools smarter [39–41].

7.2 Applications of IoT in smart homes

The IoT can be applied in a number of areas in smart homes.

a. **Lighting:** The lighting in the home can now be automatically changed to suit the needs of the person. For instance, when people start watching a movie, the lights might be adjusted to dim automatically, to prevent watchers from being drawn away from the movie plot. It's possible that when entering the home, the lights will turn on without the need to press a button. The system may automatically turn down the lights when people leave their home, to conserve energy. All of the home's lights may be controlled using a smartphone, laptop, and other connected devices. As a result, users could program the app to turn on the light when their alarm goes off in the morning.

b. **Bathrooms:** IoT technology could improve the convenience and efficiency of time spent in the bathroom. Smart mirrors can connect to other devices like computers and smartphones, recognize the faces of family members in their line of sight, and display information relevant to them, such as news stories, weather predictions, or certain websites.

Special sensors can detect movement and turn off the water automatically if no one is using the restroom. Smart shower controls may also identify individuals, set their preferred water pressure and temperature preferences, and impose a time limit on shower sessions to conserve water. Users of automated jacuzzis may unwind and enjoy their bath without having to manually adjust their preferred temperature and air-jet regime, or choose their preferred music, as the app will take care of everything.

c. **Gardens:** For those who desire to grow their own fruits, vegetables, and herbs at home, sensors can be quite helpful. The software allows users to check the plant's temperature, moisture level, and sunlight exposure. The program can monitor the soil's current condition, assess whether it has enough moisture, and, if necessary, turn on a smart irrigation system. The sensor recognizes when the moisture level reaches the desired threshold and turns off the irrigation system, saving water. Significant advancements in gardening made possible by IoT technology will profoundly alter how plants are cultivated in the future.

d. **Kitchen:** By integrating AI technology, IoT devices can increase the safety and simplicity of cooking. Smart sensors can monitor the temperature, humidity, and levels of smoke and carbon monoxide in your kitchen to ensure everything is operating as it should. There are specific built-in programs that monitor if the user has enough food in the refrigerator (and order more if necessary), provide recipe recommendations, and evaluate the nutritional value of meals. For instance, smart spoons encourage better eating habits.

e. **Security Systems:** For those who often doublecheck their home's doors, windows, and other electrical devices to make sure they are all off before leaving the house, smart security systems can help with this using particular sensors. These devices can lock the door, close the shutters, turn off the electronics, and ensure the home is secure from human and animal intruders. Users may check on their homes and remotely control the temperature, humidity, and lighting using the app on their phones. The devices can also let users watch over their elderly family members and offer assistance if necessary.

f. **Safety Sensors:** Intelligent devices known as safety sensors are capable of spotting problems in the home. They can take immediate action to avert hazards and notify consumers of potential risks. Only a smartphone with Internet connectivity and sensors installed in the home are required. Regular air quality checks can be performed in the home using temperature, humidity, and gas controllers. Internet-based notifications are sent if the readings fall outside the recommended range. Safety sensors can be used to stop fires, explosions, water leaks, and gas leaks. Proximity and video sensors can detect an attempted break-in, immediately activate the alarm, and notify the police.

g. **Temperature Control**: The home's temperature can be set to the most comfortable level with automated temperature management. Users can set up smart thermostats to control the temperature based on their settings and preferences. These thermostats can determine what the people in the house are doing right now and change the temperature accordingly. For instance, users can use the app to automatically increase the temperature in their bath or shower. The temperature will drop to help them stay cool if they prefer to exercise at home, whether doing yoga, Pilates, or another type of workout.

h. **Doors and Windows:** Doors in the future won't need keys. Face recognition technology may be used by the smart door to unlock the house. Anyone entering the facility who is not accompanied by a resident must be identified as a resident. The doors might even be programmed to open as the user approaches their home and close when they leave. They might also start a chain reaction in other home appliances. The entry door may identify authorized people and open, followed by the light going on; additional doors in the home may then open; the TV and coffee maker may then be turned on.

   Smart windows can be configured to react to signals from various devices in addition to triggering events. As the user leaves the house, the system will check and close the windows if necessary, so users don't have to worry about remembering to do so. Windows may be set to open or close at predetermined intervals, and shutters may operate in accordance with the time of day, such as raised in the morning and lowered at night. Weather events, including rain, snow, storms, or strong winds, may activate these devices.

i. **Home Routine:** With the use of Al and ML technologies, the home's temperature, lighting, and security system can be controlled and regulated. The technology can send reminders about purchases needed via an online app, seek information on the Internet, present news updates, order food, schedule an appointment, and reserve a flight or hotel. The condition of the home automation system may also be checked from anywhere. Since the lights, security, and other Internet-connected devices are connected, the app can be used to check on them when the user is away from home [42–46].

8 Key features of IoT smart factories

A highly digitized factory floor, known as a "smart factory," continuously gathers and shares data using networked equipment, devices, and production methods. For businesses that concentrate on manufacturing and supply chain management, this can be a factory where physical production processes and operations are merged with digital technology, smart computing, and big data to produce a more opportunistic system. The smart factory is a component of Industry 4.0, the fourth Industrial Revolution, which emphasizes embedded sensors, networking, automation, machine learning, and real-time manufacturing and logistical data.

Businesses must reconsider their approaches to everything from automation initiatives to workforce development strategies, as factories change in response to the digital revolution. Manufacturers will require modernized tools along the way, such as powerful, adaptable enterprise resource planning systems as a data and transactional backbone, to help them adjust rapidly as they move toward a future of smart factories.

Smart factories are able to monitor every aspect of production, from individual shop floor workers to manufacturing equipment and the supply chain, by fusing the real and digital worlds [47,48].

The following conditions must be fulfilled in order for a factory to become smart:

a. **Smart Usage of Data:** Massive volumes of data are being gathered as production is running at full capacity. To fully utilize the potential of the Industrial Internet of Things (IIoT), it is crucial to gather, analyze, comprehend, and deploy this enormous amount of data effectively and safely.
b. **Ubiquitous, Reliable Connections:** The heartbeat of every smart factory is the reliable and secure connectivity of people, machines, and "things" in physical and virtual domains via wired and wireless technologies, allowing high-speed communication for complex automation.
c. **Autonomous Production:** People, machines, systems, and products must communicate and react independently in order to achieve autonomous production. Smartness, like artificial intelligence and machine learning, helps to reduce need for human intervention.
d. **Transparent and Flexible Supply Chains:** Supply chains may be made intelligent and resilient when real-time, end-to-end visibility is shared with customers and suppliers. Early warning signs of problems allow for their early prevention or correction.
e. **Sustainability:** Reliable data gathered by sensors can be used to meet rising energy demand while improving energy efficiency and lowering CO_2 emissions. If properly interpreted, they can result in energy savings by identifying ineffective manufacturing processes, improving production and logistics planning, and foreseeing repair requirements [49,50].

8.1 The levels of smart factories

In order to evaluate progress toward becoming a smart manufacturer, four levels can be used:

a. **Level 1: Basic Data Availability:** A plant or facility isn't "smart" at this point. Despite being accessible, the data cannot be quickly analyzed. When done incorrectly, data analysis can make your production process less efficient and take a lot of time.
b. **Level 2: Proactive Data Analysis:** The data can be accessed at this level in a more organized and intelligible format. The data will be centralized and organized, and displays and visualization tools will aid in processing it. All of this

makes it possible for proactive data analysis, albeit some work will still be required.

c. **Level 3: Active Data:** At this level, machine learning and artificial intelligence can help analyze the data, producing insight with less human oversight. The system is more automated than at level 2 and can anticipate significant problems or abnormalities in order to foresee probable breakdowns.
d. **Level 4: Action-Oriented Data:** This level expands on the active character of level 3 to develop solutions to problems and, occasionally, take action to solve them or enhance a process without human interaction. At this stage, data is gathered and examined for issues before being used to produce solutions that require as little human involvement as feasible.

8.2 Application of IoT in smart factory

The capabilities of the IoT can be applied in several ways to benefit smart factories.

a. **Intelligent Product Enhancements:** The Internet of Things (IoT) in manufacturing improves production quality similarly to other IoT applications. Before IoT, extensive market research and customer feedback were necessary for product development. Now, owners have access to a wealth of data and information. The IoT ensures more revenues by serving as a trustworthy source of data about any product.
b. **Dynamic Response to Market Demands:** The ability to meet market needs is influenced by a variety of variables, including customer expectations, population income, taste and preferences, and financial resources of the nation. Continuous research is necessary to meet demand, and the current supply could result in significant losses for the firm and future choices. The Internet of Things continuously stores and retrieves information with little assistance from humans. It manages supply chains since the data the IoT collects is largely correct.
c. **Improved Facility Service:** The IoT enhances workplace conditions and provides safety and security to any conventional facility. Safety managers use applications to communicate and retrieve real-time data on hazards and safety events. This enables businesses to keep an eye on events, improve communication, and boost output.
d. **Product Safety:** Despite a complex system of procedures insuring client safety, risks and dangers continue to enter the market. Serious incidents may arise for unknown reasons. The IoT uses sensitivity, control, and management approaches to monitor such situations and send out notifications in the event of possible hazards.
e. **Lower Costs, Optimized resource use, and waste reduction:** In several fields, the IoT takes the place of manual labor. It lessens the need for product background checks to be conducted by people. The majority of the time and money spent on maintenance checks and tests is manual labor. With sensors and security webcams, one can use the IoT to monitor the state of the business remotely. The

IoT provides strategies for managing and maximizing the use of resources like people and minerals. It offers workable solutions to challenging issues that are affordable.

f. **Quality Control:** The IoT suggests real-time monitoring of industry items and equipment. Instead of waiting for a machine to fail, manufacturers can provide remedies when particular machinery parts are likely to fail. The IoT helps systems by keeping track of the health of its mechanisms, machinery, and engines. The automation of some operations decreases the dependency on manual labor.

g. **Predictive Maintenance:** In the past, manufacturers have performed maintenance checks on equipment and engines using a time-based approach. However, routine checks are automated with the advent of IoT. In other words, the devices perform their own maintenance without outside assistance and notify consumers of hazards via mobile applications. IoT sensors keep an eye on things and analyze data stored in clouds in real time. Process automation has been made possible via the IoT, and such automation includes predictive maintenance. The device schedules a recurring self-maintenance check in this location to monitor its performance. It informs the authorities of any errors and harm, and they respond by taking appropriate action to address the problems. Due to the machine automatically performing periodic system checks at predetermined intervals, owners are not required to complete maintenance checks manually.

h. **Inventory Management:** Inventory management can be a smooth and effective operation using RFID and IoT. An RFID tag is attached to every piece of merchandise, and each tag produces its own unique identifier (UID). Today, most organizations rely heavily on the data that RFID tags collect to function. The systems keep track of RFID tag output and alert users when inventory is missing by sending notifications.

i. **Smart Packaging:** The Internet of Things is being applied to smart packaging, which employs technology to wrap goods and does more than just keep them safe. Users can communicate with the package and get their questions answered about the contents, the goods, or delivery. Sensors, QR codes, and other choices are all part of how IoT and packaging operate together. The basic objective is communicating with the consumer and gathering the required information.

j. **Smart Metering:** This decreases resource waste and enables the usage of resources more optimally. Water, fuel, and power consumption are monitored via smart meters. They track how these resources are being used and implement strategies to use them more effectively.

k. **Supply Chain Management:** The IoT devices track and keep an eye on supply chain data that is received in real time. Authorities can operate delivery systems, machinery, and equipment from a distance. In addition to providing enterprise resource planning (ERP) software, some IoT systems also do away with the requirement for manual process documentation.

l. **Workshop Monitoring:** Machine workshops are facilities where materials and tools are made. These workshops have a high energy use but low efficiency, as these tools are made using a complex energy flow. In order to improve the process, the IoT can be used to create a reliable monitoring system to collect and track the energy used by these workshops. It controls the production process, which reduces prices and energy use.

m. **Production Flow Monitoring:** Production flow is one of the crucial procedures in manufacturing. It becomes challenging to control and monitor the manufacturing flow manually. The IoT employs sensors to track the prediction by giving owners real-time data. These sensors provide information on the components of machines and trigger service calls when they detect a malfunction or damaged components.
n. **Digital Twins:** Digital twins are a technique that uses the cloud to make digital models that are exact copies or reproductions of real physical items. In order to test and deploy these models before publishing the real-world item, IoT scientists and IT officials first construct these digital models. Large structures, construction sites, and cities are now using this technology [51–53].

9 Conclusion

The exploration of the role of the Internet of Things (IoT) for sustainable enhancement in smart spaces presents numerous promising opportunities and challenges. This study has identified several key findings that underscore the significance of leveraging IoT technologies to promote sustainability and efficiency in various domains. First and foremost, IoT technology has proven to be a transformative force in the creation of smart spaces. By enabling the seamless integration of physical devices and systems with data-driven intelligence, the IoT has facilitated the development of interconnected, intelligent environments that can optimize resource consumption, enhance user experiences, and promote sustainability.

One of the primary areas where the IoT has showcased its potential is in energy management. Smart spaces equipped with IoT-enabled devices can monitor and control energy consumption, leading to reduced waste and enhanced energy efficiency. By implementing real-time data analytics and intelligent algorithms, energy consumption patterns can be optimized, leading to lower operational costs and a reduced environmental footprint. Moreover, IoT-driven solutions in smart spaces have also demonstrated significant benefits in waste management and resource allocation. With sensors and interconnected systems, waste collection and recycling processes can be optimized, leading to reduced landfill usage and increased recycling rates. This has a direct positive impact on environmental preservation and the promotion of circular economy practices.

Additionally, the integration of IoT in transportation systems has played a crucial role in improving urban mobility and reducing emissions. Smart transportation systems powered by real-time data and predictive analytics have enabled better traffic management, more efficient route planning, and increased adoption of public transportation alternatives, all of which contribute to reducing the carbon footprint of urban areas. However, this exploration has also revealed several challenges that must be addressed to fully realize the potential of the IoT in sustainable smart spaces. Security and privacy concerns remain critical issues that demand attention. As smart spaces gather vast amounts of data from diverse sources, protecting this data from breaches and unauthorized access is of paramount importance. Robust encryption, authentication mechanisms, and adherence to privacy regulations are necessary to build trust in IoT-enabled smart spaces.

Furthermore, the interoperability of IoT devices and systems is another obstacle that needs to be overcome. For the seamless functioning of smart spaces, a standardized framework that allows different devices and platforms to communicate effectively is essential. This will enable scalability, flexibility, and better integration of new technologies in existing infrastructures. The Internet of Things (IoT) is quickly changing the world by bringing connection and novel experiences into our daily lives. Forward-thinking CEOs are focusing on the spaces that they occupy as the advantages of IoT become easier. Building owners and their tenants are searching for ways to boost energy efficiency, maximize space use, and boost productivity through IoT efforts in offices, hospitals, schools, factories, and retail spaces all over the world. Additionally, as smart buildings and spaces spread and become more linked, they open the door for entire communities and smart cities built on a foundation of IoT-enabled insights.

In conclusion, the integration of IoT in smart spaces holds immense promise for sustainable enhancement across various sectors. From energy management to waste reduction and efficient transportation, IoT technology has demonstrated its potential to drive positive environmental and social changes. However, to fully capitalize on these opportunities, addressing security, privacy, and interoperability concerns is crucial. With concerted efforts from stakeholders, IoT can undoubtedly play a pivotal role in creating more sustainable and intelligent environments for future generations.

References

[1] S.E. Bibri, A. Alexandre, A. Sharifi, et al., Environmentally sustainable smart cities and their converging AI, IoT, and big data technologies and solutions: an integrated approach to an extensive literature review, Energy Inform. 6 (2023) 9, https://doi.org/10.1186/s42162-023-00259-2.

[2] L. Belli, et al., IoT-enabled smart sustainable cities: challenges and approaches, Smart Cities 3 (3) (2020) 1039–1071, https://doi.org/10.3390/smartcities3030052.

[3] S.E. Bibri, On the sustainability of smart and smarter cities in the era of big data: an interdisciplinary and transdisciplinary literature review, J. Big Data 6 (2019) 25.

[4] M. Chen, S. Gonzalez, V.C. Leung, V.C. Leung, Smart spaces and energy management in the Internet of Things era, Wirel. Commun. Mob. Comput. 2017 (2017).

[5] K. Ashton, That 'Internet of Things' thing, RFID J. 22 (7) (2009) 97–114.

[6] V.D.V. Athanasios, C. Alexandros, S.V. Georgios, G.S. Dimitrios, Smart spaces and smart homes applications and services for the future Internet of Things, Future Internet 9 (3) (2017) 42.

[7] J. Carretero, A. Fernández-Montes, M. González, P. Sánchez, Smart spaces orchestration technologies for the internet of things, Futur. Gener. Comput. Syst. 36 (2014) 254–269.

[8] C.D. Nugent, J.C. Augusto, M. Huch (Eds.), Designing Smart Homes: The Role of Artificial Intelligence, Springer, 2016.
[9] M. Satyanarayanan, Pervasive computing: vision and challenges, IEEE Pers. Commun. 8 (4) (2001) 10–17.
[10] V. Kostakos (Ed.), Internet of Things: Principles and Paradigms, CRC Press, 2018.
[11] Smarter Spaces: How Technology is Shaping the Physical World (PwC report): https://www.pwc.com/gx/en/services/consulting/technology/emerging-technology/smarter-spaces.html.
[12] L. Atzori, A. Iera, G. Morabito, The Internet of Things: a survey, Comput. Netw. 54 (15) (2010) 2787–2805.
[13] E. Borgia, The Internet of Things vision: key features, applications, and open issues, Comput. Commun. 54 (2014) 1–31.
[14] J. Gubbi, R. Buyya, S. Marusic, M. Palaniswami, Internet of Things (IoT): a vision, architectural elements, and future directions, Futur. Gener. Comput. Syst. 29 (7) (2013) 1645–1660.
[15] D.A. Hall (Ed.), Sensors: A Comprehensive Survey, vols. 1–7, VCH Publishers, New York, 1991.
[16] S. Roy (Ed.), Handbook of Industrial Automation, CRC Press, Boca Raton, FL, 2004.
[17] D.C. Patranabis, Sensors and Transducers, third ed., PHI Learning Pvt. Ltd, 2010.
[18] J.G. Webster (Ed.), Measurement, Instrumentation, and Sensors Handbook, CRC Press, 1998.
[19] J. Cooper, Y. Patt, Environmental Monitoring With Arduino: Building Simple Devices to Collect Data About the World Around Us, Maker Media, Inc., 2003.
[20] J.W. Gardner, V.K. Varadan, Micro Sensors, MEMS, and Smart Devices, Wiley-IEEE Press, 2001.
[21] M. Beinert, Sensing Technologies for Precision Irrigation, first ed., Academic Press, 2019.
[22] J.A. Shaw, A.P.F. Turner (Eds.), Sensors and Sensing in Biology and Engineering, CRC Press, 2017.
[23] Y. Gao, W. Li, J. Luo, An energy-efficient indoor air quality monitoring system based on wireless sensor networks, Energy Build. 102 (2015) 183–195.
[24] Y. Zhou, Y. Zhang, Smart home energy management systems: concept, configurations, and scheduling strategies, Renew. Sustain. Energy Rev. 56 (2016) 30–40.
[25] L. Yao, S. Tang, G. Yang, An IoT-enabled system for management and energy optimization in smart buildings, IEEE Internet Things J. 4 (6) (2017) 1936–1948.
[26] E. Park, A. del Pobil, S. Kwon, The role of Internet of Things (IoT) in smart cities: technology roadmap-oriented approaches, Sustainability 10 (5) (2018) 1388.
[27] S. Al-Nasrawi, C. Adams, A. El-Zaart, A conceptual multidimensional model for assessing smart sustainable cities, J. Inf. Syst. Technol. Manag. 12 (2015) 541–558.
[28] V. Albino, U. Berardi, R.M. Dangelico, Smart cities: definitions, dimensions, performance, and initiatives, J. Urban Technol. 22 (1) (2015) 3–21.
[29] M. Höjer, J. Wangel, Smart sustainable cities: definition and challenges, in: ICT Innovations for Sustainability, Springer International Publishing, 2015.
[30] A. Zanella, N. Bui, A. Castellani, L. Vangelista, M. Zorzi, Internet of Things for smart cities, IEEE Internet Things J. 1 (1) (2014) 22–32, https://doi.org/10.1109/JIOT.2014.2306328.
[31] B.D. Silva, M.P. Moreno, J.J. Rodrigues, A survey of smart cities: analysis of requirements, data privacy, security, and applications, IEEE Commun. Surv. Tutorials 21 (3) (2019) 2347–2376, https://doi.org/10.1109/COMST.2019.2890111.
[32] A. Caragliu, C. Del Bo, P. Nijkamp, Smart cities in Europe, J. Urban Technol. 18 (2) (2011) 65–82, https://doi.org/10.1080/10630732.2011.601117.

[33] L.G. Anthopoulos, P. Fitsilis, A survey of smart city initiatives, J. Clean. Prod. 88 (2015) 112–121, https://doi.org/10.1016/j.jclepro.2014.09.083.
[34] L. Atzori, A. Iera, G. Morabito, From "smart objects" to "social objects": the next evolutionary step of the internet of things, IEEE Commun. Mag. 55 (2) (2017) 78–84, https://doi.org/10.1109/MCOM.2017.1600596CM.
[35] N. Komninos, F. Branco, C. Kakderi, P. Tsarchopoulos, Smart cities and digital rights: the need for a new Viable Utopia, J. Urban Technol. 23 (1) (2016) 71–86, https://doi.org/10.1080/10630732.2015.1135455.
[36] N. Baccour, A. Hasnaoui, L.A. Saidane, A survey on IoT-based energy management systems for smart homes, Renew. Sustain. Energy Rev. 82 (2018) 1329–1335.
[37] P.H.J. Chong, S. Kumar, Smart homes: technologies, developments and challenges, Renew. Sustain. Energy Rev. 102 (2019) 138–147.
[38] M.S. Hossain, G. Muhammad, M. Abdullah-Al-Wadud, A. Alamri, Smart home energy management system using IoT technologies, IEEE Access 3 (2015) 1018–1028, https://doi.org/10.1109/ACCESS.2015.2450482.
[39] G. Saha, S. Misra, H.N. Saha, Internet of Things (IoT) in agriculture: a comprehensive survey and its future direction, Comput. Electron. Agric. 139 (2017) 8–32.
[40] M. Chan, D. Estève, C. Escriba, E. Campo, A review of smart homes—present state and future challenges, Comput. Methods Prog. Biomed. 91 (1) (2008) 55–81.
[41] R. Rana, T. Choudhury, S. Chatterjee, Smart home: an IoT-based approach to enhance security and provide assistance, IEEE Consum. Electron. Mag. 5 (3) (2016) 75–83, https://doi.org/10.1109/MCE.2016.2541358.
[42] N. Ahuja, N. Dey, Security issues in IoT based smart homes: a survey, J. Ambient. Intell. Humaniz. Comput. 10 (10) (2019) 3917–3938, https://doi.org/10.1007/s12652-019-01462-z.
[43] A. Biswas, S. Misra, Security and privacy issues in IoT-enabled smart homes: a survey, IEEE Internet Things J. 7 (3) (2020) 1932–1953, https://doi.org/10.1109/JIOT.2019.2962148.
[44] C. Antonopoulos, N. Bessis, I.N. Fovino, A survey on security and privacy issues in Internet of Things for smart homes, J. Netw. Comput. Appl. 66 (2016) 1–10.
[45] Y. Yao, J. Yao, X. Xu, Y. Zhong, A survey of Internet of Things (IoT) for smart home: past, present, and future trends, IEEE Internet Things J. 3 (6) (2016) 1, https://doi.org/10.1109/JIOT.2016.2525204.
[46] E.A. Nigus, T. Leppänen, K. Väänänen-Vainio-Mattila, Internet of Things (IoT) for smart home: a systematic review, J. Ambient Intell. Smart Environ. 8 (3) (2016) 321–340, https://doi.org/10.3233/AIS-160402.
[47] Y. Lu, L.D. Xu, X. Ruan, Internet of Things (IoT)-based smart manufacturing: a case study on the framework of chocolate factory, IEEE Trans. Ind. Inform. 13 (4) (2017) 2133–2142, https://doi.org/10.1109/TII.2017.2660061.
[48] M. Chen, S. Mao, Y. Liu, Big data: a survey, Mob. Netw. Appl. 19 (2) (2014) 171–209, https://doi.org/10.1007/s11036-013-0489-0.
[49] M. Sharma, P. Kumar, A. Deep, Internet of Things (IoT)-enabled smart manufacturing: a systematic literature review and research agenda, Robot. Comput. Integr. Manuf. 56 (2019) 1–13, https://doi.org/10.1016/j.rcim.2018.08.001.
[50] F. Tao, J. Cheng, Q. Qi, M. Zhang, H. Zhang, F. Sui, Digital twin-driven product design, manufacturing and service with big data, Int. J. Adv. Manuf. Technol. 94 (9–12) (2018) 3563–3576, https://doi.org/10.1007/s00170-017-1270-1.

[51] L. Li, X. Xu, D. Ma, Industrial Internet of Things (IIoT)-enabled smart manufacturing: a literature review, J. Manuf. Syst. 53 (2019) 261–279, https://doi.org/10.1016/j.jmsy.2019.06.010.
[52] Y. Ding, Z. Jiang, M. Fu, W. Zhu, Smart manufacturing systems for Industry 4.0: conceptual framework, scenarios, and future perspectives. Frontiers of, Mech. Eng. 14 (1) (2019) 137–150, https://doi.org/10.1007/s11465-019-0520-1.
[53] J. Wan, H. Cai, K. Zhou, H. Suo, A manufacturing big data solution for active preventive maintenance, J. Ind. Inf. Integr. 1 (2016) 4–11, https://doi.org/10.1016/j.jii.2016.06.001.

CHAPTER

9 Machine learning frameworks in IoT systems: A survey, case study, and future research directions

Zheyi Chen, Pu Tian, Cheng Qian, Weixian Liao, Adamu Hussaini, and Wei Yu
Department of Computer and Information Sciences, Towson University, Towson, MD, United States

1 Introduction

Internet of Things (IoT)-based smart systems, along with the development of cyber-physical systems (CPS), leverage massive smart sensing and actuating devices to empower intelligence, reliability, and efficiency in different application domains (smart city, smart transportation, smart home, etc.) [1–9]. As a result, the IoT represents a technological paradigm shift that aims to build an intelligent world, in which "smart things" can communicate with one another over the Internet. However, the inherent complexity of IoT-based systems, such as massive information from local devices, heterogeneous smart devices, and resource constraints from IoT sensors, gateways, and network resources bring challenges to the system operator that makes decisions by using traditional data collection and analytics methods. To this end, machine learning (especially deep learning) has been considered a viable method, showing great potential for various tasks, including computer vision, image processing, and speech recognition, among others [10–12].

Fig. 1 shows several machine learning architectures. To be specific, with the centralized machine learning framework, the server handles the training job, while the local client collects and transmits data samples. Recently published research has demonstrated that centralized machine learning techniques performed excellently in various IoT applications [13]. Nonetheless, centralized machine learning training uses expensive communication and enormous computing resources. Furthermore, transmitting sensitive data to a public cloud server could raise privacy issues [14]. On the one hand, federated learning, one of the representative distributed learning frameworks, is envisioned as a promising paradigm for training a machine learning model without compromising local privacy [15]. Federated learning aims to train a shared global learning model, in which the central parameter server and local clients

Smart Spaces. https://doi.org/10.1016/B978-0-443-13462-3.00015-7

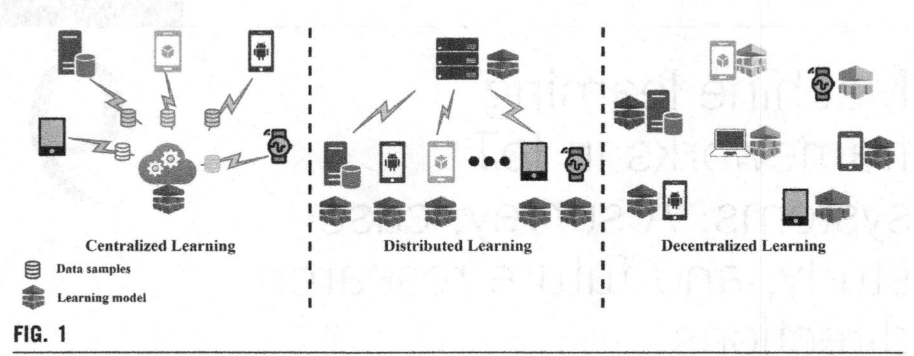

FIG. 1

Machine learning architectures.

conduct model aggregation and the training process, respectively. Local clients upload gradient updates to assist in model aggregation. However, in this stage, some open issues (e.g., communication efficiency, single-point failure, and heterogeneity) make it challenging to deploy the federated learning framework under IoT systems. Finally, the decentralized learning framework offers a more flexible, adaptable solution to train a machine learning model without needing a central parameter server. Instead, all local clients conduct model aggregation in a peer-to-peer fashion [16]. The absence of a centralized server makes it challenging to use traditional defensive mechanisms to identify or detect malicious behavior. Some research projects designed blockchain-based consensus methods that charge local clients extra maintenance costs so that device or node behaviors can be tracked [17].

There are some survey papers that examine machine learning technologies in IoT from various perspectives. For instance, Xiao et al. [18] studied machine learning-based IoT security techniques such as authentication, secure offloading, and node access control. In addition, several studies and tutorials examined the use of centralized machine learning in various IoT systems, such as smart transportation [19], smart cities [20], smart grid [21], and smart healthcare [22]. Li et al. [23] also reviewed the studies on privacy challenges, heterogeneity problems, and large-scale federated learning. For IoT systems, Nguyen et al. [24] concentrated on federated learning. Different from the existing survey works, we make the following contributions:

- *Framework:* We systematically investigate the development of machine learning frameworks under IoT-based systems, including centralized machine learning, distributed machine learning, and decentralized machine learning. Then, to show the scope of this paper, we propose a three-dimensional problem space, in which the X dimension presents three learning frameworks, the Y dimension aims to clarify data distribution, and the Z dimension demonstrates the two types of topics in research works in terms of performance and security.
- *Case Study:* Decentralized learning allows the model training process peer-to-peer; however, it raises security concerns for local clients. Thus, in the

case study, we investigate the security threat under the decentralized learning framework, showing that it is important to develop security mechanisms for dealing with such issues. This case study can be mapped to one subspace in our defined problem space (i.e., $< X_3, Y_1/Y_2, Z_2 >$). Finally, we discuss some unsolved issues and point them out as future research directions.

The remainder of this chapter is organized as follows. In Section 2, we propose a three-dimensional problem space and investigate various representative research efforts based on the proposed problem space. We carry out one case study in Section 3. In Section 4, we discuss and outline several future research directions. Finally, we conclude the paper in Section 5.

2 Machine learning and IoT systems

In this chapter, we review various machine learning (deep learning) frameworks in IoT-based systems. To show the scope of this study, we propose a three-dimensional problem space, including three orthogonal dimensions as shown in Fig. 2. Specifically, the X dimension presents the three learning frameworks, which contain centralized machine learning, distributed machine learning, and decentralized machine learning; the Y dimension aims to clarify data distribution in IoT-based systems, which includes independently and identically distributed (i.i.d.) and non-i.i.d.; the Z dimension demonstrates the two types of topics in research works, covering both learning performance and security.

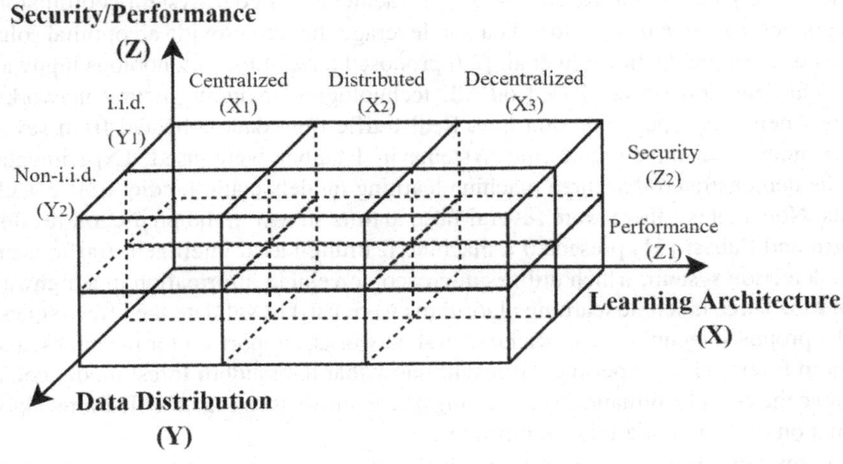

FIG. 2

Problem space.

2.1 Centralized machine learning

We now review the research efforts that apply various centralized ML technologies in IoT-based systems. ML technologies have been widely adopted as an effective and efficient way of aggregating massive local information in different application domains. For example, in smart transportation, traffic flow forecasting plays a vital role in optimizing traffic flow in urban areas, where machine learning technologies show significant capability in both long-term and short-term tasks [25–28]. Hou et al. [25] investigated the traffic flow issues for work zone events. To forecast traffic flow under urban work zone scenarios, four machine learning models—nonparametric regression, multilayer forward-propagation neural network, regression tree, and random forest—were investigated. In the experiments, historical traffic data was considered to forecast long-term tasks and real-time temporal and spatial traffic data was used to forecast short-term tasks. Experimental results showed that the random forest-based approach is capable of outperforming other learning models in long-term and short-term tasks. In addition, Alipio et al. [26] developed a road monitor system to provide real-time routing by analyzing vehicle traffic and flooded areas from local sensors. The proposed framework deploys a Bayesian network over a centralized cloud platform so that all collected real-time data can be aggregated, such as water level and real-time road images. Possible reroute areas would be sent back to users via mobile devices. Based on real-world road data, the proposed system was demonstrated to reduce delay time and prevent impassable roads. Similar to Hou et al. [25], Fusco et al. [27] studied the short-term traffic flow forecasting in urban areas and discussed the explicit and implicit models' features in the prediction of travel time.

Traffic accident detection is also vital for the smart transportation system, which aims to make transportation systems more adaptive and intelligent so that road and vehicle safety can be improved [29–32]. To achieve this goal, system coordinators can collect real-time traffic-flow data and leverage them to provide an optimal solution. For example, Ozbayoglu et al. [29] proposed a real-time autonomous highway accident-detection system based on ML technologies, including neural networks, nearest neighbor, and regression tree. Real traffic-flow data collected from seven independent real-time monitoring systems in Istanbul were used. Experimental results demonstrated that three machine learning models could predict traffic accidents. Nonetheless, there were several false alarms shown in the prediction results. Dogru and Subasi [31] presented a machine learning-based intelligent traffic accident detection system, which utilizes microscopic vehicle information in a highway scenario. Three machine learning algorithms were used to validate the effectiveness of the proposed techniques: artificial neural networks, support vector machines, and random forest. Their experimental results show that the random forest model could achieve the best performance in the testing phase, outperforming other learning algorithms on detection and false-alarm rates.

A smart city is a vital IoT system, which aims to use information and communication technologies to promote the efficiency of public resources and the quality of citizen welfare. For example, for parking lot management, convolutional neural networks, support vector machine (SVM), and other machine learning technologies show their great potential to recognize the status of parking information [33–35]. These systems could leverage the information provided by smart cameras to allocate vacant parking spaces in complex urban areas, such as exhibiting occlusions and challenging viewpoints. Amato et al. [33] proposed a deep convolutional neural network-based vision system for smart cameras to detect parking lot status. To validate the efficacy of the proposed vision system, the experiments were carried out under two datasets, in which PKLot was used for typical scenes, while CNRPark-EXT made in this work was used for some cases (e.g., occlusions, viewpoints, and various weather conditions).

The smart healthcare system is another IoT-based system that is expected to leverage new data mining and analytics techniques to provide more watchful, seamless, and professional services in healthcare applications [36–39]. Related to this direction, Hussain et al. [36] developed an intelligent wearable system based on a convolutional neural network, which can automatically monitor activity and food recognition. Various sensors embedded in wearable devices (e.g., motion sensors in a smartwatch, and piezoelectric sensors in a necklace) are responsible for recognizing physical activities, detecting food contents, and data transmission. Additionally, a new signal segmentation algorithm was proposed to benefit the annotation of data samples. Ismail et al. [38] developed a health monitoring application that introduces a convolutional neural network model to analyze human health factors. By feeding health-related parameters (e.g., body temperature, blood pressure) into a convolutional neural network learning model, the proposed method could provide different signal patterns, which can link health parameters so that the links between co-occurring diseases and symptoms can be explored. Likewise, Zahin et al. [40] proposed a semisupervised classifier using a deep generative model (VAE+CNN) to recognize the human activities in the smart healthcare system. The researchers employed the Actitracker dataset to evaluate the efficacy of their proposed framework. The results obtained from the experiment show that their model could outperform the conventional supervised learning with a small amount of labeled data.

On the other hand, some research efforts focus on studying security topics, including applying machine learning technologies to tackle security issues and improve the business of machine learning applications in IoT systems. For example, in Refs. [52–57], the authors adopted centralized machine learning models to tackle various security issues in IoT systems, such as malware detection, data integrity attacks, false data injection, malicious behavior mitigation, and others. Anand et al. [52] proposed a machine learning-based approach to protecting sensitive healthcare information from malware attacks. In their study, an eight-layer learning model was used to build a classifier, including a convolutional layer, ReLU, pooling layer, and dropout layer. Similar to Anand et al. [52], Zaza et al. [58] proposed a lightweight approach to detect malware under 5G-IoT scenarios, in which a convolution neural network-based machine learning model was developed to recognize malicious behaviors in the network.

IoT-based systems are expected to integrate massive smart devices and advanced network techniques to increase the efficiency of production tasks. Machine learning technologies could introduce intelligence and efficiency in various application domains. Nonetheless, from the adversary's perspective, the nature and characteristics of these learning technologies not only provide a promising solution for the traditional application, but also reveal an emerging interface for malicious behaviors [59]. To fill this gap, we notice that some research works aim to increase the robustness of machine learning- based applications in IoT systems. For example, on the attack side, Zhao et al. [60] investigated the vulnerability of reinforcement learning, in which the proposed attack model uses sequence-to-sequence models to predict the action space of the learning node. Additionally, the proposed attack model could perform effective malicious behaviors in white-box, gray-box, and black-box cases, which cause a trained node to misbehave with a given time duration. Bhagoji et al. [61] proposed an attack model, denoted as gradient estimation attacks, to generate adversarial examples for fooling learning systems, in which the adversary requires zero knowledge about the dataset and the model architecture. In their experiments, the proposed attack model could achieve acceptable performance on MNIST and CIFAR-10 datasets, which successfully compromise a real-world image classifier based on deep neural networks.

On the defensive side, Feinman et al. [62] developed two complementary approaches to detect adversarial samples for deep neural networks, including density estimation and Bayesian uncertainty estimation. Density estimation is a powerful tool to measure the distance between an adversarial sample and the correct data manifold. Bayesian uncertainty estimation leverages the statistical information provided by the dropout layer to distinguish adversarial samples from normal samples. Likewise, Zhang et al. [63] proposed a modified Markov decision process to investigate the malicious behavior in deep reinforcement learning. Based on the proposed theory, a robust policy regularizer was designed to improve the robustness of different reinforcement learning algorithms, including deep deterministic policy gradient, deep-Q networks, and proximal policy optimization.

Summary: Research works have widely studied how to apply centralized learning technologies in various IoT-based systems, which show their impressive capabilities to promote the intelligence and reliability of services. Building a centralized learning-based application generally requires integrating various components, including different sensors for data collection, smart devices for data processing, and a cloud platform for data analysis. We use Table 1 to classify research works

Table 1 Applying centralized learning model in IoT application domains.

$< X_1, Y_1/Y_2, Z_1$ (Smart Transportation) $>$	[25–31]
$< X_1, Y_1/Y_2, Z_1$ (Smart City) $>$	[33–35,41–43]
$< X_1, Y_1/Y_2, Z_1$ (Smart Home) $>$	[44–47]
$< X_1, Y_1/Y_2, Z_1$ (Smart Healthcare) $>$	[36–39]
$< X_1, Y_1/Y_2, Z_1$ (others) $>$	[32,48–51]

in various application domains, where these research works could be mapped to subspace $< X_1, Y_1/Y_2, Z_1 >$. Moreover, we discuss security topics for both applying machine learning technologies to solve security issues and increase the robustness of machine learning applications in IoT-based systems, where these research works could be mapped to subspace $< X_1, Y_1/Y_2, Z_2 >$.

2.2 Distributed machine learning

We now discuss distributed machine learning in IoT-based systems. Unlike centralized machine learning, distributed machine learning aims to avoid the massive amount of data being transmitted for training. For machine learning tasks in IoT systems, federated learning (FL) has been proposed as a flexible and workable alternative. Federated learning can build the global model collaboratively by coordinating distributed clients with the central server only using weight parameters [64]. While the FL architecture offers the benefit of data privacy and reduced network burdens of data transmission, the data distribution can affect performance. Earlier works [64–67] in FL focused on homogeneous datasets over client nodes. They considered the optimization of device performance capacity and network resource limitations, as shown in $< X_2, Y_1, Z_1 >$ in Fig. 2. Zheng et al. [65] considered the issue that slow client uploads would cause inconsistency for global model aggregation, and then proposed the use of Taylor expansion to compensate for the delay for SGD-based optimizers. Wang et al. [66] argued that the central server might not be able to collect all data and proposed a control algorithm to determine when the global aggregation would run after receiving local parameter uploads. Likewise, Nishio and Yonetani [67] proposed a client selection approach to mitigate the sluggish performance due to delayed local uploads.

Moreover, the heterogeneity in data distributions poses another challenge to deploying federated learning in IoT systems, which corresponds to the $< X_2, Y_2, Z_1 >$ subspace in Fig. 2. According to Zhao et al. [68], the non-i.i.d. datasets over clients could cause the performance to worsen up to 55%. Thus it is crucial to consider the impact of heterogeneous datasets distributed in IoT systems. To this end, research efforts were conducted on both the server and client sides to tackle the issue. For example, Zhao et al. [68] used a shared i.i.d. dataset between the server and clients to counteract the differences. Li et al. [69] introduced a proximal term during the server aggregation procedure. Likewise, Hsieh et al. [70] adopted the client clustering approach to control the impact of deviation of learning nodes. Nonetheless, the participation of client nodes could raise security threats, and the methods for training a uniform model for all clients may not achieve the optimal performance for each distributed pattern. To deal with the challenges, Tian et al. [71] proposed a cosine distance-based clustering approach, and the process was only implemented on the server side.

In addition to the performance problem (Z_1 dimension in Fig. 2), security (Z_2) is another major issue to consider when deploying a robust FL system in IoT systems. Although the FL training process does not need the transmission of training datasets over public networks or their storage in any specific location, the exchange of model parameters can expose sensitive client dataset information [72,73]. For example, Melis et al. [74] studied the fact that malicious clients could join the training process to infer key privacy features, such as locations of other peers, and launch further attacks. Moreover, Wang et al. [75] argued that even the server could be malicious in the FL framework rather than distributed nodes. In such a case, an adversary could collect all the sensitive information of participating nodes. Furthermore, instead of obtaining sensitive information, adversaries can further launch adversarial attacks by exploiting servers or distributed nodes [76]. The adversary can inject tampered information through a compromised server or controlled client.

More precisely, there are two common types of adversarial attacks: model poisoning and data poisoning. One intentionally uses malicious training data instances to tamper with the joint model through compromised clients to launch the data poisoning [77–79]. Typically, there are two ways to poison a dataset. The first one is to maliciously alter the features of the input instances, which will prevent the jointly trained model from identifying patterns and result in incorrect labels. For example, Zhang et al. [80] proposed a generative adversarial network (GAN) called PoisonGAN to generate adversarial training samples while the malicious node acts as FL participants. The PoisonGAN could eliminate unrealistic attack assumptions and improve attack efficacy. The other common form of data poisoning is label flipping. Instead of altering the input features, it changes the ground truth labels of the training datasets. Similar to the former approach, it could affect the performance of the trained model. Li et al. [79] proposed a label-flipping defense strategy based on KPCA and K-mean clustering to mitigate adversary attacks toward FL. Intuitively, the more clients an adversary controls, the more serious the impact will be. Moreover, due to the non-i.i.d. property of client datasets in FL (X_2, Y_2, Z_2 dimension in Fig. 2), it is difficult to distinguish the normal heterogeneous clients from malicious ones only based on the statistical data patterns themselves.

Rather than the data poisoning attack with corrupted training datasets, the model poisoning attack targets the model parameters [81]. The adversary can control clients or the central server to change the weight parameters before sending to other peers in the training process [82]. The impact is more serious than data poisoning, since the adversary does not have to compromise a lot of data to subvert the model. Additionally, detecting model poisoning has to consider the non-i.i.d. feature of input datasets [83]. In other words, the defense approach shall minimize the side effects on normal heterogeneous clients while combating poisoned models [84].

Summary: We discuss the heterogeneity challenges (in communication synchronization and statistical data patterns) and the security issues, and we then summarize research works. From those works, we can conclude that the robustness in federated learning has been significantly studied in recent research works, but there are still problems when considering both heterogeneity and security in the federated learning deployment [85].

2.3 Decentralized machine learning

We now discuss research efforts on decentralized machine learning. To make the learning process adapt to various IoT systems, decentralized machine learning is envisioned as a promising solution. Decentralized machine learning would conduct model aggregation in a peer-to-peer (P2P) manner, where a central parameter server is not required to manage the training process. By doing this, decentralized machine learning becomes more flexible for the real-world IoT systems [16,86–91]. For example, Lalitha et al. [86] developed a fully decentralized learning framework in a network (graph) where the model aggregation would be executed between two one-hop nodes. Pei et al. [87] proposed a decentralized, federated learning framework that aims to make local clients collaboratively train a graph learning model without a central parameter server. The model aggregation is conducted in a peer-to-peer network structure. Moreover, the Diffie-Hellman key exchange method was introduced into model aggregation to provide privacy protection. Hegedűs et al. [89] presented a comparison between conventional federated learning and gossip learning, showing that gossip learning could achieve an acceptable learning performance under various scenarios. These research efforts could be mapped to subspace $< X_3, Y_1, Z_1 >$ in the defined problem space. Likewise, Li et al. [91] introduced a transfer learning technology into a decentralized learning framework to tackle the degraded learning performance and slow convergence. The mutual knowledge transfer algorithm improved knowledge sharing in the P2P model aggregation.

When decentralized machine learning techniques are used in IoT systems, several unsolved issues exist, including communication efficiency, gradient divergence, learning fairness, and security threats, among others [92]. For example, the inherent nature of real-world IoT systems, including heterogeneous devices, unbalanced datasets, non-i.i.d., and data distribution, among others, could significantly affect the learning efficiency and waste the communication bandwidth. To fill this gap, Kong et al. [93] leveraged a distance metric called consensus distance to investigate the performance gap between centralized and decentralized learning. The consensus distance theoretically reveals the average divergence between each local client and the average learning model over local clusters to guide the decentralized learning process with low communication expenses. Dai et al. [94] developed a decentralized personalized federated learning framework, which adopts personalized sparse masks to train local models with a P2P communication link. The proposed framework uses the decentralized sparse training technique, consisting of weighted average, sparse local training, and mask update to solve the heterogeneity issues.

Also, Sun et al. [95] developed the decentralized learning framework using momentum to improve communication efficiency over local training groups, where the learning protocol provides flexible communication to share local updates. Additionally, the proposed framework was extended to its quantized version for nonconvex assumptions. Cao et al. [96] proposed a decentralized, federated learning framework for the heterogeneous computing scenario. The proposed framework uses version-sensitive probabilistic partial model aggregation for training lagging local clients through the asynchronous process. Li et al. [97] developed a decentralized, federated learning framework based on adaptive neighbor matching to promote learning efficiency under non-i.i.d. data distribution. Two metrics, based on losses and gradients, were proposed to measure the pair-wise similarity for local clients. A decentralized learning algorithm includes matching same-cluster neighbors and adopts a heuristic mechanism based on expectation maximization to search for more local clients with similar objectives. Chen et al. [98] proposed a decentralized, federated learning framework to deal with the gradient divergence problem using dynamic average consensus. The proposed framework adopts a symmetric doubly stochastic matrix to present all local clients in an undirected graph for model aggregation. Additionally, the proposed framework considers the local training iterations as discrete-time series. The average model was estimated using a first-order dynamic average consensus method without relying on a central parameter server.

Learning fairness is another critical issue in the noncentralized learning framework. Notably, in distributed and decentralized learning, the training process would be assigned on the local side to preserve privacy, where accuracy distributions over local clients might be skewed [70,99–101]. To address this problem, Chen et al. [100] proposed a P2P decentralized learning framework based on a fairness-aware aggregation scheme. The proposed framework adopts an adaptive aggregation scheme to make the local model fairer, allowing each client to aggregate the shared model with its local preferences. Experiments showed that the proposed framework could outperform two basic decentralized learning frameworks under both i.i.d. and non-i.i.d. data distribution. Hsieh et al. [70] empirically investigated the learning performance of decentralized deep neural network training over non-i.i.d. data distribution. The relationship between skewed data distribution and the accuracy loss of decentralized learning was investigated. Moreover, a system-level approach was proposed to evaluate the degree of skew and adaptively modify the communication frequency. Likewise, Li et al. [101] proposed a decentralized learning framework to train personalized models for user-customized computer-vision tasks. An attention mechanism was used to optimize the mixing weights for non-i.i.d. learning scenarios. The proposed approach would take a pretraining process to improve its generalization. It achieves an acceptable performance on unseen data samples.

In IoT systems, decentralized learning shows great potential to achieve task objectives with a flexible communication topology while preserving privacy [88,102–105]. For example, Bi et al. [104] designed a doctor recommendation system based on a decentralized, federated learning framework using electronic health records, which provides more privacy-sensitive service between doctors and patients. Specifically, the proposed approach adopts a graph attention network to present heterogeneous nodes' information, including patients, doctors, and medical services. Nair et al. [88] presented the application of decentralized, federated reinforcement learning in multirobot scenarios using two open source tools, Webots and Tartarus. To perform the decentralized, federated reinforcement learning objective, a mobile agent was adopted to aggregate and share the learned models in a decentralized manner, resulting in a better Q-table.

The aforementioned research efforts demonstrate that the decentralized learning framework could benefit various IoT systems by providing a more flexible communication topology and preserving sensitive user data. However, decentralized learning introduces potential security threats into the training process, in which the lack of a central parameter server makes it difficult to eliminate malicious behaviors from adversaries. For instance, a group of malicious nodes could collaboratively train a learning model with backdoors, which would be disseminated to other benign nodes. In this sense, research efforts regarding security issues, denoted as subspace $< X_3, Y_1/Y_2, Z_2 >$ in our defined problem space, are vital to the deployment of decentralized learning frameworks in IoT systems. In addition, Che et al. [106] proposed a decentralized learning framework based on the committee mechanism to increase the robustness of the training process. In the training process, a scoring system would sort each local update and use it to elect committee members for the next round. This research work can be mapped to subspace $< X_3, Y_1, Z_2 >$ in our defined problem space. Hu et al. [107] proposed a blockchain-based, decentralized FL framework, which introduces a consistent hashing algorithm into the system to increase communication efficiency. On the security side, the blockchain-based approach adopts the consensus mechanism to verify each transaction and trace malicious behaviors under non-i.i.d. scenarios. This research work could be mapped to subspace $< X_3, Y_2, Z_2 >$.

Summary: In this subsection, we discuss research works in decentralized learning, including developing decentralized learning frameworks, improving communication efficiency, increasing convergence rate, applying decentralized learning technologies in IoT systems, and mitigating security threats. All these research efforts have shown that decentralized learning is a promising way of leveraging on-device data samples while preserving user privacy. As an example of summarizing some research topics in decentralized learning, Table 2 classifies research works that can be mapped to subspace $< X_3, Y_1/Y_2, Z_1 >$. However, due to the nature of the decentralized learning framework, such as the lack of central parameter servers, it becomes difficult to design and deploy defensive mechanisms so that malicious behaviors can be dealt with. As a potential solution, some research efforts show that blockchain-based approaches could alleviate some security issues, but it would take extra resources to maintain the consensus mechanism [109].

Table 2 The development of decentralized learning framework.

$< X_1, Y_1, Z_1$ (Learning Framework) $>$	[16,86–91]
$< X_1, Y_1, Z_1$ (Communication Efficiency) $>$	[93–96,108]
$< X_1, Y_1, Z_1$ (Applications) $>$	[88,102–105]
$< X_1, Y_2, Z_1$ (Gradient Divergence) $>$	[97,98]
$< X_1, Y_2, Z_1$ (Learning Fairness) $>$	[70,99–101]

3 Case study

In this section, we carry out a case study to investigate the security threat in a decentralized learning framework. The nature of decentralized learning, such as no central parameter server, makes it difficult to deploy defensive mechanisms, so that decentralized learning might introduce new attack interfaces into local groups. This case study could be mapped to subspace $< X_3, Y_2, Z_2 >$.

Non-i.i.d. data distribution: We study the security threat of decentralized learning framework under non-i.i.d. data distribution. We deploy the decentralized learning framework over the public E-MNIST dataset, where all local clients would adopt the learning model. E-MNIST has 240,000 training samples and 40,000 testing samples for 10 balanced classes. To perform non-i.i.d. data distribution, each local client collects ζ ($\zeta = 2, 3$) classes from the E-MNIST dataset and the size of the local dataset is $\zeta \times 500$.

Benign clients and threat model: We assume benign clients would do local training and find a learning pair in a P2P manner. We set up several malicious nodes to spread backdoors in the decentralized training process. To enhance their ability, all learning models in a malicious group were pretrained to reach a global $accuracy = 96\%$. To this end, each malicious model has an acceptable learning performance on targeted backdoors and normal data samples, which could further increase the stealth of backdoors. In model aggregation, benign clients adopt a weight averaging-based algorithm to update the local model, and malicious clients reject updating the local model. This is because the purpose of malicious clients is to deliver backdoors to normal clients as much as possible, not to train a good local model. We use the poisoning data to generate backdoors by manipulating the label information of data samples. For example, we collect some images "5" ("0") to change the label as "7" ("8") [110]. The key parameters used in this case study are listed in Table 3.

Performance metrics: We aim to investigate backdoor dissemination in a decentralized learning approach. First, we shall check the performance of the malicious backdoor to confirm that it corrupts local clients. In this sense, we would flag a local client as "corrupted" if the malicious backdoor's accuracy reaches a threshold ($accuracy = 90\%$) on the local side.

Table 3 List of parameters.

Parameter	Value
Total node number	20
Local epoch	2
Non-i.i.d. coefficient ζ	2, 3
Minibatch size	100
Learning pair	1000

Table 4 Numerical results for backdoor ("5" to "7").

Numerical results in backdoor dissemination ("5" to "7")				
	$m=1$	$m=2$	$m=3$	$m=4$
200 iterations	0	0	1	1
400 iterations	0	1	1	2
600 iterations	1	1	4	5
800 iterations	3	4	9	12
1000 iterations	4	6	12	15

Note that m *represents the number of malicious clients.*

Table 5 Numerical results for backdoor ("0" to "8").

Numerical results in backdoor dissemination ("0" to "8")				
	$m=1$	$m=2$	$m=3$	$m=4$
200 iterations	0	0	1	1
400 iterations	0	1	2	2
600 iterations	1	2	4	7
800 iterations	2	5	9	13
1000 iterations	4	6	13	16

Note that m *represents the number of malicious clients.*

Evaluation Results: We conducted two experiments to study backdoor dissemination in a decentralized learning approach. In Tables 4 and 5, we record some numerical results showing the backdoor dissemination process. For instance, in Table 4, the malicious backdoor is successfully delivered to 4 benign clients with 1 malicious node ($m=1$), whereas the worst case ($m=4$) could convert 15 benign clients (original 16 benign clients) to corrupted models.

4 Discussion

We now discuss open and decentralized issues in federated learning and point out future research directions. Federated learning is a promising solution to leverage distributed local datasets without privacy concerns. However, some bottlenecks, such as communication efficiency, heterogeneity, and trustworthiness evaluation, make it difficult to deploy the federated learning framework in a real-world scenario. Federated learning aims to train a global learning model with local computing resources and datasets, in which local clients upload their local learning models to assist the model aggregation at the central server. In such a framework, the training process brings a high communication burden on the server side.

There are some research works that develop the asynchronous federated learning framework to alleviate the communication load [111]. Nonetheless, an increased number of clients would degrade communication efficiency, making it impossible to deploy a large-scale learning network [112]. Heterogeneity issues in federated learning could be translated into three aspects: non-i.i.d. data distribution, unbalanced datasets, and heterogeneous devices. In this stage, research efforts focus on improving the learning performance with non-i.i.d. data distribution. Moreover, unbalanced datasets and heterogeneous devices would degrade the performance of federated learning. For example, a global learning model would be skewed with unbalanced local datasets in federated learning due to weight averaging-based aggregation algorithms. Similarly, due to heterogeneous computing resources, it is difficult to ensure the synchronization of local clients in which some clients might contribute stale gradient updates to the central server.

To address these issues, it is important to develop transfer learning-based federated learning approaches, which leverage the existing learning models to benefit future training tasks. The learning efficiency could be significantly increased if a well-trained (learned) model is used to assist a new learning model. This research topic could be mapped to subspace $< X_2, Y_1/Y_2, Z_1 >$.

Decentralized learning is another vital learning framework aiming to leverage local datasets with a flexible communication topology. Nonetheless, a decentralized learning process needs to deal with the issues of reliability and robustness. As shown in the case study, in a 20-client group, one malicious node could successfully spread a target backdoor to 4 benign clients. In the worst scenario, an adversary could use 4 malicious nodes to corrupt 90% of benign clients. This is because decentralized learning (e.g., no central node) makes it hard to adopt traditional defense mechanisms. At this stage, some research works adopt a blockchain-based approach to trace node behaviors in the training process, but it would add extra cost to maintain the consensus mechanism [113]. In this sense, it is necessary to establish a trustworthiness evaluation to trace both short-term and long-term node behaviors. This research topic could be mapped to subspace $< X_3, Y_1/Y_2, Z_2 >$.

Meanwhile, decentralized machine learning is not limited only to model-based machine learning, but is also deployed to model-free machine learning techniques, such as reinforcement learning [12]. Due to the limited resources of local computing devices, decentralized architectures can be used for reinforcement learning to improve computational efficacy. Decentralized reinforcement learning aims to implement multiple agents on multiple sensors and provide more comprehensive feedback to improve efficiency and reduce convergence time. Recently, a decentralized algorithm called QMIX has been implemented into reinforcement learning to estimate global Q-values based on sensor observations [114]. The algorithm can find optimal policies to reduce convergence time based on sensor observations [115]. The adversary can also launch targeted attacks based on the channel state of distributed agents to disrupt a system with minimal resources [116]. This research topic could be mapped to subspace $< X_1 = X_2, Y_2, Z_1 = Z_2 >$.

5 Conclusion

In this chapter, we investigated the development of machine learning frameworks, including centralized machine learning, distributed machine learning, and decentralized machine learning. We proposed a three-dimensional problem space (Security/Performance, Data Distribution, and Learning Architecture) to show the scope of this survey chapter. Moreover, we conducted one case study to demonstrate research problems related to the security of the decentralized learning framework by showing that even a small number of compromised clients can pose a significant threat. Finally, we discussed potential research topics concerning communication efficiency, heterogeneity, and trustworthiness assessment as future research directions.

References

[1] X. Liu, C. Qian, W.G. Hatcher, H. Xu, W. Liao, W. Yu, Secure Internet of Things (IoT)-based smart-world critical infrastructures: survey, case study and research opportunities, IEEE Access 7 (2019) 79523–79544.

[2] J.A. Stankovic, Research directions for the internet of things, IEEE Internet Things J. 1 (1) (2014) 3–9.

[3] S. Mallapuram, N. Ngwum, F. Yuan, C. Lu, W. Yu, Smart city: the state of the art, datasets, and evaluation platforms, in: 2017 IEEE/ACIS 16th International Conference on Computer and Information Science (ICIS), 2017, pp. 447–452.

[4] J. Lin, W. Yu, N. Zhang, X. Yang, H. Zhang, W. Zhao, A survey on internet of things: architecture, enabling technologies, security and privacy, and applications, IEEE Internet Things J. 4 (5) (2017) 1125–1142.

[5] J. Song, C. Qian, X. Liu, H. Liang, C. Lu, W. Yu, Performance assessment of deep neural network on activity recognition in WiFi sensing, in: 2022 IEEE International Conferences on Internet of Things (iThings) and IEEE Green Computing & Communications (GreenCom) and IEEE Cyber, Physical & Social Computing (CPSCom) and IEEE Smart Data (SmartData) and IEEE Congress on Cybermatics (Cybermatics), 2022, pp. 510–517.

[6] C. Qian, X. Liu, C. Ripley, M. Qian, F. Liang, W. Yu, Digital twin—cyber replica of physical things: architecture, applications and future research directions, Future Internet 14 (2) (2022). [Online]. Available: https://www.mdpi.com/1999-5903/14/2/64.

[7] H.H. Addeen, Y. Xiao, J. Li, M. Guizani, A survey of cyber-physical attacks and detection methods in smart water distribution systems, IEEE Access 9 (2021) 99905–99921.

[8] F. Liang, C. Qian, W.G. Hatcher, W. Yu, Search engine for the internet of things: lessons from web search, vision, and opportunities, IEEE Access 7 (2019) 104673–104691.

[9] W.G. Hatcher, C. Qian, W. Gao, F. Liang, K. Hua, W. Yu, Towards efficient and intelligent internet of things search engine, IEEE Access 9 (2021) 15778–15795.

[10] W.G. Hatcher, W. Yu, A survey of deep learning: platforms, applications and emerging research trends, IEEE Access 6 (2018) 24411–24432.

[11] F. Liang, W.G. Hatcher, W. Liao, W. Gao, W. Yu, Machine learning for security and the internet of things: the good, the bad, and the ugly, IEEE Access 7 (2019) 158126–158147.

[12] X. Liu, H. Xu, W. Liao, W. Yu, Reinforcement learning for cyber-physical systems, in: 2019 IEEE International Conference on Industrial Internet (ICII), 2019, pp. 318–327.
[13] M.S. Mahdavinejad, M. Rezvan, M. Barekatain, P. Adibi, P. Barnaghi, A.P. Sheth, Machine learning for internet of things data analysis: a survey, Digit. Commun. Netw. 4 (3) (2018) 161–175.
[14] H. Tabrizchi, M. Kuchaki Rafsanjani, A survey on security challenges in cloud computing: issues, threats, and solutions, J. Supercomput. 76 (12) (2020) 9493–9532.
[15] H.B. McMahan, E. Moore, D. Ramage, S. Hampson, et al., Communication-Efficient Learning of Deep Networks From Decentralized Data, arXiv preprint arXiv:1602.05629, 2016.
[16] A. Lalitha, S. Shekhar, T. Javidi, F. Koushanfar, Fully decentralized federated learning, in: Third Workshop on Bayesian Deep Learning (NeurIPS), 2018.
[17] Y. Xu, G. Wang, J. Yang, J. Ren, Y. Zhang, C. Zhang, Towards secure network computing services for lightweight clients using blockchain, Wirel. Commun. Mob. Comput. 2018 (2018).
[18] L. Xiao, X. Wan, X. Lu, Y. Zhang, D. Wu, IoT security techniques based on machine learning: how do IoT devices use ai to enhance security? IEEE Signal Process. Mag. 35 (5) (2018) 41–49.
[19] F. Zantalis, G. Koulouras, S. Karabetsos, D. Kandris, A review of machine learning and IoT in smart transportation, Future Internet 11 (4) (2019) 94.
[20] Z. Ullah, F. Al-Turjman, L. Mostarda, R. Gagliardi, Applications of artificial intelligence and machine learning in smart cities, Comput. Commun. 154 (2020) 313–323.
[21] E. Hossain, I. Khan, F. Un-Noor, S.S. Sikander, M.S.H. Sunny, Application of big data and machine learning in smart grid, and associated security concerns: a review, IEEE Access 7 (2019) 13960–13988.
[22] W. Li, Y. Chai, F. Khan, S.R.U. Jan, S. Verma, V.G. Menon, X. Li, et al., A comprehensive survey on machine learning-based big data analytics for IoT-enabled smart healthcare system, Mob. Netw. Appl. 26 (1) (2021) 234–252.
[23] T. Li, A.K. Sahu, A. Talwalkar, V. Smith, Federated learning: challenges, methods, and future directions, IEEE Signal Process. Mag. 37 (3) (2020) 50–60.
[24] D.C. Nguyen, M. Ding, P.N. Pathirana, A. Seneviratne, J. Li, H.V. Poor, Federated learning for internet of things: a comprehensive survey, IEEE Commun. Surv. Tutor. 23 (3) (2021) 1622–1658.
[25] Y. Hou, P. Edara, C. Sun, Traffic flow forecasting for urban work zones, IEEE Trans. Intell. Transp. Syst. 16 (4) (2014) 1761–1770.
[26] M.I. Alipio, J.R.R. Bayanay, A.O. Casantusan, A.A. Dequeros, Vehicle traffic and flood monitoring with reroute system using Bayesian networks analysis, in: 2017 IEEE 6th Global Conference on Consumer Electronics (GCCE), IEEE, 2017, pp. 1–5.
[27] G. Fusco, C. Colombaroni, L. Comelli, N. Isaenko, Short-term traffic predictions on large urban traffic networks: applications of network-based machine learning models and dynamic traffic assignment models, in: 2015 International Conference on Models and Technologies for Intelligent Transportation Systems (MT-ITS), IEEE, 2015, pp. 93–101.
[28] M. Munoz-Organero, R. Ruiz-Blaquez, L. Sánchez-Fernández, Automatic detection of traffic lights, street crossings and urban roundabouts combining outlier detection and deep learning classification techniques based on GPS traces while driving, Comput. Environ. Urban. Syst. 68 (2018) 1–8.

[29] M. Ozbayoglu, G. Kucukayan, E. Dogdu, A real-time autonomous highway accident detection model based on big data processing and computational intelligence, in: 2016 IEEE International Conference on Big Data (Big Data), IEEE, 2016, pp. 1807–1813.

[30] S. Sadeky, A. Al-Hamadiy, B. Michaelisy, U. Sayed, Real-time automatic traffic accident recognition using HFG, in: 2010 20th International Conference on Pattern Recognition, IEEE, 2010, pp. 3348–3351.

[31] N. Dogru, A. Subasi, Traffic accident detection using random forest classifier, in: 2018 15th Learning and Technology Conference (L&T), IEEE, 2018, pp. 40–45.

[32] A. Ghosh, T. Chatterjee, S. Samanta, J. Aich, S. Roy, Distracted driving: a novel approach towards accident prevention, Adv. Comput. Sci. Technol. 10 (8) (2017) 2693–2705.

[33] G. Amato, F. Carrara, F. Falchi, C. Gennaro, C. Meghini, C. Vairo, Deep learning for decentralized parking lot occupancy detection, Expert Syst. Appl. 72 (2017) 327–334.

[34] Q. Wu, C. Huang, S.-y. Wang, W.-c. Chiu, T. Chen, Robust parking space detection considering inter-space correlation, in: 2007 IEEE International Conference on Multimedia and Expo, IEEE, 2007, pp. 659–662.

[35] P.R. De Almeida, L.S. Oliveira, A.S. Britto Jr., E.J. Silva Jr., A.L. Koerich, PKLot—a robust dataset for parking lot classification, Expert Syst. Appl. 42 (11) (2015) 4937–4949.

[36] G. Hussain, M.K. Maheshwari, M.L. Memon, M.S. Jabbar, K. Javed, A CNN based automated activity and food recognition using wearable sensor for preventive healthcare, Electronics 8 (12) (2019) 1425.

[37] Y. Yang, Medical multimedia big data analysis modeling based on DBN algorithm, IEEE Access 8 (2020) 16350–16361.

[38] W.N. Ismail, M.M. Hassan, H.A. Alsalamah, G. Fortino, CNN-based health model for regular health factors analysis in internet-of-medical things environment, IEEE Access 8 (2020) 52541–52549.

[39] M. Alhussein, G. Muhammad, M.S. Hossain, S.U. Amin, Cognitive IoT-cloud integration for smart healthcare: case study for epileptic seizure detection and monitoring, Mob. Netw. Appl. 23 (6) (2018) 1624–1635.

[40] A. Zahin, L.T. Tan, R.Q. Hu, Sensor-based human activity recognition for smart healthcare: a semi-supervised machine learning, in: S. Han, L. Ye, W. Meng (Eds.), Artificial Intelligence for Communications and Networks, Springer International Publishing, Cham, 2019, pp. 450–472.

[41] I. Kök, M.U. Şimşek, S. Özdemir, A deep learning model for air quality prediction in smart cities, in: 2017 IEEE International Conference on Big Data (Big Data), IEEE, 2017, pp. 1983–1990.

[42] M. Mohammadi, A. Al-Fuqaha, M. Guizani, J.-S. Oh, Semisupervised deep reinforcement learning in support of IoT and smart city services, IEEE Internet Things J. 5 (2) (2017) 624–635.

[43] J. Chin, V. Callaghan, I. Lam, Understanding and personalising smart city services using machine learning, the Internet-of-Things and Big Data, in: 2017 IEEE 26th International Symposium on Industrial Electronics (ISIE), IEEE, 2017, pp. 2050–2055.

[44] H. Berlink, A.H. Costa, Batch reinforcement learning for smart home energy management, in: Twenty-Fourth International Joint Conference on Artificial Intelligence, 2015.

[45] A. Shojaei-Hashemi, P. Nasiopoulos, J.J. Little, M.T. Pourazad, Video-based human fall detection in smart homes using deep learning, in: 2018 IEEE International Symposium on Circuits and Systems (ISCAS), IEEE, 2018, pp. 1–5.
[46] M.M. Hassan, M.Z. Uddin, A. Mohamed, A. Almogren, A robust human activity recognition system using smartphone sensors and deep learning, Futur. Gener. Comput. Syst. 81 (2018) 307–313.
[47] A. Wang, G. Chen, C. Shang, M. Zhang, L. Liu, Human activity recognition in a smart home environment with stacked denoising autoencoders, in: International Conference on Web-Age Information Management, Springer, 2016, pp. 29–40.
[48] R. Moorthy, V. Upadhya, V.V. Holla, S.S. Shetty, V. Tantry, CNN based smart surveillance system: a smart IoT application post Covid-19 era, in: 2020 Fourth International Conference on I-SMAC (IoT in Social, Mobile, Analytics and Cloud) (I-SMAC), IEEE, 2020, pp. 72–77.
[49] S.-W. Chen, X.-W. Gu, J.-J. Wang, H.-S. Zhu, AIoT used for COVID-19 pandemic prevention and control, Contrast Media Mol. Imaging 2021 (2021).
[50] E.D. Nugroho, A.G. Putrada, A. Rakhmatsyah, Predictive control on lettuce NFT-based hydroponic IoT using deep neural network, in: 2021 International Symposium on Electronics and Smart Devices (ISESD), IEEE, 2021, pp. 1–6.
[51] S. Karunịawati, A.G. Putrada, A. Rakhmatsyah, Optimization of grow lights control in IoT-based aeroponic systems with sensor fusion and random forest classification, in: 2021 International Symposium on Electronics and Smart Devices (ISESD), IEEE, 2021, pp. 1–6.
[52] A. Anand, S. Rani, D. Anand, H.M. Aljahdali, D. Kerr, An efficient CNN-based deep learning model to detect malware attacks (CNN-DMA) in 5G-IoT healthcare applications, Sensors 21 (19) (2021) 6346.
[53] D. An, Q. Yang, W. Liu, Y. Zhang, Defending against data integrity attacks in smart grid: a deep reinforcement learning-based approach, IEEE Access 7 (2019) 110835–110845.
[54] H. Alkahtani, T.H. Aldhyani, Botnet attack detection by using CNN-LSTM model for Internet of Things applications, Secur. Commun. Netw. 2021 (2021).
[55] X. Niu, J. Li, J. Sun, K. Tomsovic, Dynamic detection of false data injection attack in smart grid using deep learning, in: 2019 IEEE Power & Energy Society Innovative Smart Grid Technologies Conference (ISGT), IEEE, 2019, pp. 1–6.
[56] S.O.M. Kamel, S.A. Elhamayed, Mitigating the impact of IoT routing attacks on power consumption in IoT healthcare environment using convolutional neural network, Int. J. Comput. Netw. Inf. Secur. 12 (4) (2020) 11–29.
[57] M.N. Hasan, R.N. Toma, A.-A. Nahid, M.M. Islam, J.-M. Kim, Electricity theft detection in smart grid systems: a CNN-LSTM based approach, Energies 12 (17) (2019) 3310.
[58] A.M. Zaza, S.K. Kharroub, K. Abualsaud, Lightweight IoT malware detection solution using CNN classification, in: 2020 IEEE 3rd 5G World Forum (5GWF), IEEE, 2020, pp. 212–217.
[59] A. Kurakin, I. Goodfellow, S. Bengio, Adversarial Machine Learning at Scale, arXiv preprint arXiv:1611.01236, 2016.
[60] Y. Zhao, I. Shumailov, H. Cui, X. Gao, R. Mullins, R. Anderson, Blackbox attacks on reinforcement learning agents using approximated temporal information, in: 2020 50th Annual IEEE/IFIP International Conference on Dependable Systems and Networks Workshops (DSN-W), IEEE, 2020, pp. 16–24.

[61] A.N. Bhagoji, W. He, B. Li, D. Song, Exploring the Space of Black-Box Attacks on Deep Neural Networks, arXiv preprint arXiv:1712.09491, 2017.
[62] R. Feinman, R.R. Curtin, S. Shintre, A.B. Gardner, Detecting Adversarial Samples From Artifacts, arXiv preprint arXiv:1703.00410, 2017.
[63] H. Zhang, H. Chen, C. Xiao, B. Li, M. Liu, D. Boning, C.-J. Hsieh, Robust deep reinforcement learning against adversarial perturbations on state observations, Adv. Neural Inf. Process. Syst. 33 (2020) 21024–21037.
[64] B. McMahan, et al., Communication-efficient learning of deep networks from decentralized data, in: Proceedings of the 20th AISTATS, 2016.
[65] S. Zheng, Q. Meng, T. Wang, W. Chen, N. Yu, Z. Ma, T.-Y. Liu, Asynchronous Stochastic Gradient Descent With Delay Compensation for Distributed Deep Learning, arXiv preprint arXiv:1609.08326, 2016.
[66] S. Wang, T. Tuor, T. Salonidis, K.K. Leung, C. Makaya, T. He, K. Chan, Adaptive federated learning in resource constrained edge computing systems, IEEE J. Sel. Areas Commun. 37 (6) (2019) 1205–1221.
[67] T. Nishio, R. Yonetani, Client selection for federated learning with heterogeneous resources in mobile edge, in: ICC 2019-2019 IEEE International Conference on Communications (ICC), IEEE, 2019.
[68] Y. Zhao, M. Li, L. Lai, Federated learning with non-IID data, CoRR (2018). abs/1806.00582, [Online]. Available: http://arxiv.org/abs/1806.00582.
[69] T. Li, A.K. Sahu, M. Zaheer, M. Sanjabi, A. Talwalkar, V. Smith, Federated Optimization in Heterogeneous Networks, arXiv preprint arXiv:1812.06127, vol. 3, 2018, p. 3.
[70] K. Hsieh, A. Phanishayee, O. Mutlu, P. Gibbons, The non-IID data quagmire of decentralized machine learning, in: International Conference on Machine Learning, PMLR, 2020, pp. 4387–4398.
[71] P. Tian, W. Liao, W. Yu, E. Blasch, WSCC: a weight-similarity-based client clustering approach for non-IID federated learning, IEEE Internet Things J. 9 (20) (2022) 20243–20256, https://doi.org/10.1109/JIOT.2022.3175149.
[72] P. Kairouz, H. McMahan, B. Avent, A. Bellet, M. Bennis, A. Bhagoji, K. Bonawitz, Z. Charles, G. Cormode, R. Cummings, et al., Advances and Open Problems in Federated Learning, arxiv, arXiv preprint arXiv:1912.04977, 2019.
[73] A.N. Bhagoji, S. Chakraborty, P. Mittal, S. Calo, Analyzing Federated Learning Through an Adversarial Lens, arXiv preprint arXiv:1811.12470, 2018.
[74] L. Melis, C. Song, E. De Cristofaro, V. Shmatikov, Exploiting unintended feature leakage in collaborative learning, in: 2019 IEEE Symposium on Security and Privacy (SP), IEEE, 2019, pp. 691–706.
[75] Z. Wang, M. Song, Z. Zhang, Y. Song, Q. Wang, H. Qi, Beyond inferring class representatives: user-level privacy leakage from federated learning, CoRR (2018). abs/1812.00535, arXiv preprint arXiv:1812.00535.
[76] N. Bouacida, P. Mohapatra, Vulnerabilities in federated learning, IEEE Access 9 (2021) 63229–63249.
[77] V. Tolpegin, S. Truex, M.E. Gursoy, L. Liu, Data poisoning attacks against federated learning systems, in: European Symposium on Research in Computer Security, Springer, 2020, pp. 480–501.
[78] M. Goldblum, D. Tsipras, C. Xie, X. Chen, A. Schwarzschild, D. Song, A. Madry, B. Li, T. Goldstein, Dataset security for machine learning: data poisoning, backdoor attacks, and defenses, IEEE Trans. Pattern Anal. Mach. Intell. (2022).

[79] D. Li, W.E. Wong, W. Wang, Y. Yao, M. Chau, Detection and mitigation of label-flipping attacks in federated learning systems with KPCA and K-means, in: 2021 8th International Conference on Dependable Systems and Their Applications (DSA), IEEE, 2021, pp. 551–559.
[80] J. Zhang, B. Chen, X. Cheng, H.T.T. Binh, S. Yu, PoisonGAN: generative poisoning attacks against federated learning in edge computing systems, IEEE Internet Things J. 8 (5) (2021) 3310–3322.
[81] M. Fang, X. Cao, J. Jia, N. Gong, Local model poisoning attacks to {Byzantine-Robust} federated learning, in: 29th USENIX Security Symposium (USENIX Security 20), 2020, pp. 1605–1622.
[82] X. Chen, C. Liu, B. Li, K. Lu, D. Song, Targeted Backdoor Attacks on Deep Learning Systems Using Data Poisoning, arXiv preprint arXiv:1712.05526, 2017.
[83] G. Sun, Y. Cong, J. Dong, Q. Wang, L. Lyu, J. Liu, Data poisoning attacks on federated machine learning, IEEE Internet Things J. 9 (13) (2022) 11365–11375, https://doi.org/10.1109/JIOT.2021.3128646.
[84] Y. Wang, P. Mianjy, R. Arora, Robust learning for data poisoning attacks, in: International Conference on Machine Learning, PMLR, 2021, pp. 10859–10869.
[85] Q. Li, Z. Wen, Z. Wu, S. Hu, N. Wang, Y. Li, X. Liu, B. He, A survey on federated learning systems: vision, hype and reality for data privacy and protection, IEEE Trans. Knowl. Data Eng. 35 (4) (2023) 3347–3366, https://doi.org/10.1109/TKDE.2021.3124599.
[86] A. Lalitha, O.C. Kilinc, T. Javidi, F. Koushanfar, Peer-to-Peer Federated Learning on Graphs, arXiv preprint arXiv:1901.11173, 2019.
[87] Y. Pei, R. Mao, Y. Liu, C. Chen, S. Xu, F. Qiang, B.E. Tech, Decentralized federated graph neural networks, in: International Workshop on Federated and Transfer Learning for Data Sparsity and Confidentiality in Conjunction With IJCAI, 2021.
[88] J.S. Nair, D.D. Kulkarni, A. Joshi, S. Suresh, On Decentralizing Federated Reinforcement Learning in Multi-Robot Scenarios, arXiv preprint arXiv:2207.09372, 2022.
[89] I. Hegedűs, G. Danner, M. Jelasity, Gossip learning as a decentralized alternative to federated learning, in: IFIP International Conference on Distributed Applications and Interoperable Systems, Springer, 2019, pp. 74–90.
[90] C. Hu, J. Jiang, Z. Wang, Decentralized Federated Learning: A Segmented Gossip Approach, arXiv preprint arXiv:1908.07782, 2019.
[91] C. Li, G. Li, P.K. Varshney, Decentralized federated learning via mutual knowledge transfer, IEEE Internet Things J. 9 (2) (2021) 1136–1147.
[92] P. Kairouz, H.B. McMahan, B. Avent, A. Bellet, M. Bennis, A.N. Bhagoji, K. Bonawitz, Z. Charles, G. Cormode, R. Cummings, et al., Advances and open problems in federated learning, Found. Trends Mach. Learn. 14 (1–2) (2021) 1–210.
[93] L. Kong, T. Lin, A. Koloskova, M. Jaggi, S. Stich, Consensus control for decentralized deep learning, in: International Conference on Machine Learning, PMLR, 2021, pp. 5686–5696.
[94] R. Dai, L. Shen, F. He, X. Tian, D. Tao, DisPFL: Towards Communication-Efficient Personalized Federated Learning Via Decentralized Sparse Training, arXiv preprint arXiv:2206.00187, 2022.
[95] T. Sun, D. Li, B. Wang, Decentralized Federated Averaging, arXiv preprint arXiv:2104.11375, 2021.
[96] J. Cao, Z. Lian, W. Liu, Z. Zhu, C. Ji, HADFL: heterogeneity-aware decentralized federated learning framework, in: 2021 58th ACM/IEEE Design Automation Conference (DAC), IEEE, 2021, pp. 1–6.

[97] Z. Li, J. Lu, S. Luo, D. Zhu, Y. Shao, Y. Li, Z. Zhang, C. Wu, Mining Latent Relationships Among Clients: Peer-to-Peer Federated Learning With Adaptive Neighbor Matching, arXiv preprint arXiv:2203.12285, 2022.

[98] Z. Chen, D. Li, J. Zhu, S. Zhang, DACFL: Dynamic Average Consensus Based Federated Learning in Decentralized Topology, arXiv preprint arXiv:2111.05505, 2021.

[99] T. Li, M. Sanjabi, A. Beirami, V. Smith, Fair Resource Allocation in Federated Learning, arXiv preprint arXiv:1905.10497, 2019.

[100] Z. Chen, W. Liao, P. Tian, Q. Wang, W. Yu, A fairness-aware peer-to-peer decentralized learning framework with heterogeneous devices, Future Internet 14 (5) (2022) 138.

[101] S. Li, T. Zhou, X. Tian, D. Tao, Learning to collaborate in decentralized learning of personalized models, in: Proceedings of the IEEE/CVF Conference on Computer Vision and Pattern Recognition, 2022, pp. 9766–9775.

[102] A.G. Roy, S. Siddiqui, S. Pölsterl, N. Navab, C. Wachinger, Braintorrent: A Peer-to-Peer Environment for Decentralized Federated Learning, arXiv preprint arXiv:1905.06731, 2019.

[103] S. Lu, Y. Zhang, Y. Wang, C. Mack, Learn Electronic Health Records by Fully Decentralized Federated Learning, arXiv preprint arXiv:1912.01792, 2019.

[104] L. Bi, Y. Wang, F. Zhang, Z. Liu, Y. Cai, E. Zhao, FD-GATDR: A Federated-Decentralized-Learning Graph Attention Network for Doctor Recommendation Using EHR, arXiv preprint arXiv:2207.05750, 2022.

[105] Q. Chen, Z. Wang, W. Zhang, X. Lin, PPT: A Privacy-Preserving Global Model Training Protocol for Federated Learning in P2P Networks, arXiv preprint arXiv:2105.14408, 2021.

[106] C. Che, X. Li, C. Chen, X. He, Z. Zheng, A Decentralized Federated Learning Framework Via Committee Mechanism With Convergence Guarantee, arXiv preprint arXiv:2108.00365, 2021.

[107] Y. Hu, Y. Zhou, J. Xiao, C. Wu, GFL: A Decentralized Federated Learning Framework Based on Blockchain, arXiv preprint arXiv:2010.10996, 2020.

[108] S. Wu, D. Huang, H. Wang, Network gradient descent algorithm for decentralized federated learning, J. Bus. Econ. Stat. 41 (3) (2023) 806–818, https://doi.org/10.1080/07350015.2022.2074426.

[109] W. Gao, W.G. Hatcher, W. Yu, A survey of blockchain: techniques, applications, and challenges, in: 2018 27th International Conference on Computer Communication and Networks (ICCCN), 2018, pp. 1–11.

[110] Z. Chen, P. Tian, W. Liao, W. Yu, Zero knowledge clustering based adversarial mitigation in heterogeneous federated learning, IEEE Trans. Netw. Sci. Eng. 8 (2) (2021) 1070–1083, https://doi.org/10.1109/TNSE.2020.3002796.

[111] C.-H. Hu, Z. Chen, E.G. Larsson, Device scheduling and update aggregation policies for asynchronous federated learning, in: 2021 IEEE 22nd International Workshop on Signal Processing Advances in Wireless Communications (SPAWC), IEEE, 2021, pp. 281–285.

[112] T. Chen, X. Jin, Y. Sun, W. Yin, VAFL: A Method of Vertical Asynchronous Federated Learning, arXiv preprint arXiv:2007.06081, 2020.

[113] M. Qi, Z. Wang, F. Wu, R. Hanson, S. Chen, Y. Xiang, L. Zhu, A blockchain-enabled federated learning model for privacy preservation: system design, in: Australasian Conference on Information Security and Privacy, Springer, 2021, pp. 473–489.

[114] Y. Yang, X. Ma, L. Chenghao, Z. Zheng, Q. Zhang, G. Huang, J. Yang, Q. Zhao, Believe what you see: implicit constraint approach for offline multi-agent reinforcement learning, Adv. Neural Inf. Process. Syst. 34 (2021) 10299–10312.
[115] S. Wang, M. Chen, Z. Yang, C. Yin, W. Saad, S. Cui, H.V. Poor, Distributed reinforcement learning for age of information minimization in real-time IoT systems, IEEE J. Sel. Top. Signal Process. 16 (3) (2022) 501–515.
[116] P. Dai, W. Yu, H. Wang, G. Wen, Y. Lv, Distributed reinforcement learning for cyber-physical system with multiple remote state estimation under DoS attacker, IEEE Trans. Netw. Sci. Eng. 7 (4) (2020) 3212–3222.

CHAPTER

Augmented reality content and relations of power in smart spaces

10

Michal Rzeszewski[a] and Leighton Evans[b]
[a]*Adam Mickiewicz University in Poznań, Poland,* [b]*Swansea University, Swansea, Wales, United Kingdom*

1 Introduction

The experience of place is never unmediated. Each place we encounter as an embodied human being embodies its own associations drawn from a mixture of culture and memory [1]. Over time, computing and computational devices have become an important agent in this mix. Ubiquitous computational devices such as smartphones act as familiars that contribute to the framing of perception of place, as actors in the field of reality itself [1]. The field of computation has never stood still, and we now stand on the precipice of a new kind of computing. Augmented reality (AR) is a technology that allows users to overlay computer-generated content onto the real-world environment. AR typically involves the use of a device, such as a smartphone, tablet, or AR headset, that captures the user's view of the real world and enhances it with additional information or digital objects. AR can be used in a wide range of applications, from entertainment and gaming to education, training, and professional fields such as medicine and engineering.

The newness of AR comes from the embodiment in place that emerges from use of the medium. AR is intended as a lens, a way of *seeing through* onto the world, where that lens exerts a mediating force based on the overlay of information. As such, this is not a focal medium or an embodied medium alone—AR is a hermeneutic medium that will transform the meaning of place through providing the view of place in the first instance. The role of AR in perception means that the meanings ascribed to place in AR will become part of the perceptual experience of place, and in themselves will become a salient aspect of the meaning of place. Therefore a number of questions about the power of AR become obvious: who ascribes meaning, who "writes" AR, and how is meaning attached to objects in the real world. These questions are essential as augmentation will not happen by itself. The effect of AR on understandings of place and the power dynamics of AR in that process are aspects of the coming medium that require close attention.

Smart Spaces. https://doi.org/10.1016/B978-0-443-13462-3.00008-X

The chapter's main aim is to interrogate how AR content can change the relations of power within a place and transform the perception of place from being a material background for social interactions into a living agent that can have its own agency. We posit, therefore, that AR can be seen as a physical manifestation of the agency of place. This position can have consequences for the practical development of smart spaces and for theoretical consideration of the human-technology interaction in urban space in its material and digital dimensions. In this chapter, we try to combine the quality of smart space as a smart environment that actively, through affordances of smart technology, influences people's plans and intentions [2] with the rich theoretical thought on human and more-than-human spatial interactions. To do this, we investigate AR's role through sociotechnical imaginaries [3].

Apart from this introduction, the chapter structure includes six sections. We begin with a theoretical introduction to the agency of place, and then we briefly discuss our use of the concept of sociotechnical imaginary. In the following sections, we describe our methodology and results. Finally, we discuss the relations of power in place in the context of AR-based agency of place and then examine the most important consequences that AR and its influence over place can have for the development of urban smart spaces.

2 Agency of place

The concept of "agency of place" refers to the idea that places can shape human actions and social processes. In other words, places are not passive backdrops for human activities, but rather have their own agency and influence on human behavior. Agency of place is based on the understanding that places are not just physical locations, but are also imbued with social, cultural, historical, and ecological meanings and values. For example, a particular neighborhood may have a sense of community and history that encourages its residents to feel a sense of belonging and to engage in certain activities that are specific to that place. This sense of belonging is an aspect of the notion of the "geographical self" [4]. The geographical self conceptualizes the human as being situated in place and being embodied in an all-encompassing landscape [4]. The self is, in the geographical sense view, not the product of disembodied, immaterial processes but instead is emergent from our relationship with place. Our sense of self is deeply entangled with place through our embodiment, activity, and interactions with the places where we enact our existence. For Jeff Malpas [5], this entanglement with place is the foundation of consciousness itself. Contra Descartes, the thinking creature experiences the world and always finds itself in a complex but unitary place that encompasses the individual and all other things in that place. Therefore the lived body is essential to this encountering—and so are the actions of the lived body, including our use of things to experience place as a thinking thing. For Casey [4] there is "no place without self and no self without place." Place and self are not independent conditions, but are mutually essential conditions for being itself [6].

Place is therefore a critical coconstituent of our sense of being and being as a thinking being in the world. Access to place as an embodied being though is not unmediated. As an embodied being, we are embodied in place—a tautology. Our manner of embodiment, or mode of embodiment, is a function of what we are doing in that place. This doing is engagement with other things or beings in place. A now highly developed line of scholarship has developed around how engagements with media lead to mediated understandings of place. Studies of locative media have typically considered the impact of this phenomenon on phenomenological understandings of space, place [7–10] and culture [11,12], and everyday life [13–15]. Research has shown that the embodiment of space with mobile media [16] can augment the urban environment [17], craft new environmental experiences [18], and turn ordinary life "into a game" [19]. Likewise, location-based applications can reshape mobilities [20–23].

Research in this area has largely focused on the use of smartphones and smartphone-based applications. In the 2020s, AR is emerging as a new kind of mobile, place-based media that will also have profound effects on embodied being-in-the-world. The possible ability of AR to augment and change the appreciation and understanding of place through the overlay of a digital layer onto physical space challenges any notion that place is simply a material background for social interaction. The possible affordances of AR will allow for spaces to be interrogated, labeled, redesignated, and assigned new meanings through a digital interface, without a physical change to the underlying space. New senses of place and meaningful interactions with place could occur only through the mediation of AR, without an alteration to the spatial characteristics of the space *sans* AR. The potential of AR is, in terms of the agency of place, to mediate space in a manner that space is untouched, yet the sense of place will be a function of a *space-AR-human* technological relation where the AR mediates both the experience and appreciation of place. AR would be a significant medium in the everyday experience of place because it allows for significant interventions and inscriptions that will become a focal part of the understanding of that place for the embodied self. For example, the agency of place is often discussed in the context of urban planning and architecture, where the design of physical spaces is seen as having a significant impact on social and cultural dynamics. By understanding the agency of place, planners and designers can create spaces that better meet the needs and desires of the people who use them. AR will create a new layer of spatial creation, where multiple meanings can be designed and disseminated to users of space that will have an influence on the appreciation of space as place that comes from being an embodied agent that uses media to understand space.

3 Sociotechnical imaginaries

AR development is still at an early stage for a medium, with low commercial and consumer uptake. As such, the positing of AR as a significant mediator in the agency of place belongs in the realm of the sociotechnical imaginary. The sociotechnical imaginary is a concept that describes the shared beliefs and values that people hold about the relationships between technology and society. It is a broader and more complex idea than the technological imaginary because it considers the social, cultural, and political contexts that shape the ways in which technology is designed, implemented, and used. The sociotechnical imaginary recognizes that technological change is not just a technical process but is deeply embedded in social relations and cultural practices. It is the product of the interplay between social norms, economic interests, political power, and technological capabilities. Sociotechnical imaginaries shape the way in which people think about and use technology. They inform the design of technologies and the policies that regulate them. For example, the sociotechnical imaginary of the 20th century emphasized the benefits of mass production and consumption, leading to the development of technologies that could produce goods cheaply and quickly, such as the assembly line and consumer appliances. In contrast, the sociotechnical imaginary of the 21st century has shifted toward sustainability, social justice, and human well-being. This has led to the development of technologies that are more environmentally friendly, socially responsible, and designed to enhance human well-being. The concept of the sociotechnical imaginary has been recently employed as a vehicle for an explanation of various complicated assemblages of sociological, cultural, and technological phenomena. Originally defined by Jasanoff and Kim [3] as "collectively imagined forms of social life and social order reflected in the design and fulfillment of nation-specific and/or technological projects," it has been redefined since and acquired several meanings. The main appeal of this concept is that through it, we can capture how human imagination, technological objects, and social norms are increasingly inseparable in practice [24]. Visions of the desirable future that are produced and held collectively within a society are filtered through the lenses of social norms and social orders and shaped to fit the attainable advances in science and technology [24]. At the same time, those visions can have a measurable, direct or indirect, impact on the technologies, shaping their development through managing resources such as public grants or producing goals, expectations, and directing legislation. This inclusion of the interplay between material and mental processes is crucial for the understanding of AR development. It is hard to think about any imaginary related to AR that is not involving in its discourse the questions of what is real and what is virtual.

Moreover, within the concept of sociotechnical imaginary, science and technology are no longer regarded as the sole domain of facts and figures but are also influenced by storytelling, imaging, and imagining [25]. Even more essentially, we may no longer consider the dyadic relations of mental and material phenomena as important [26], which opens the possibilities of investigating a new constellation of interactions between material and digital dimensions of space, place, and everyday spatialities [27]. While initially focused on state and corporate actors, sociotechnical imaginaries have also been used in recent years to analyze contested views of social groups at different spatial scales [28]. Here, we adopt an approach that considers the imaginary being formed, developed, and expressed in media outlets by a community of AR developers and users working within the realm of urban issues. Doing this, we also acknowledge that the AR imaginary we explore here is closely connected to a larger sociotechnical imaginary of a smart city [26].

4 Methodology

Following the assumption that sociotechnical imaginaries can be seen as one of the dominant forces that will guide the use of AR in future urban spaces, we have aimed to discover the current dominant narrations on this theme. To achieve this goal, we have conducted a thematic analysis of the media content related to AR. To gather source material, we systematically searched for relevant texts, images, and videos on Reddit, Twitter, YouTube, and using Google search engines. We have included a diverse set of marketing materials, promotional videos, blog posts, and articles. To be taken into consideration material needed to fulfill a set of criteria: (1) the main topic needs to be AR; (2) context needs to be urban space; (3) date of creation needs to be 2020 or later; (4) it should be authored by a person/entity directly involved with development of AR technology or software or a person/entity describing its direct use of AR technology and software, including interviews with such persons; (5) it should not consist of usability testing or software comparison. This means that we have filtered out most of the review materials, including academic papers. With this approach, we wanted to uncover narratives within the AR developer's community without influencing our analysis with metareviews and narratives developed by real-life AR users. Focusing on urban space also means that we are not analyzing a great number of materials that discuss AR in an industrial setting, which is now, as of the writing of this chapter, its most popular practical application [29]. A time filter was placed to exclude past narratives, as we felt that it was necessary to do this when analyzing the dynamic landscape of AR technology. The constructed database consisted of 154 media objects. The next step was conducting inductive, open coding without precoding assumptions. We have searched for content that described the current and future influence of AR on urban space. We were able to distinguish 48 unique codes that fit this search. Afterward, we aggregated 30 codes into two broader themes: smart information in place and subversion of place meaning (Table 1).

5 Results—Dominant themes of urban AR

We could distinguish two dominant narratives in the gathered materials about AR technology and use. These themes encompass a wide array of thoughts and perspectives on the role of AR in urban space, both in its current stage of development and in the future. The first theme, which we have named **Smart information in place**, is focused on the AR as a component of a much larger imaginary of smart and digital cities [26,30]. This narrative is predominantly present in marketing materials and technical descriptions of AR platforms and software applications. In those cases, AR is perceived as one of the technologies, others being virtual/mixed reality and the Internet of Things, that can transform the way people interact with space. While smart spaces are rarely explicitly named, the narrative involves the ability of space to constantly adapt to its users' needs and communicate through digital layers of AR content, which fits the most basic definitions of smart space [31]. The underlying

Table 1 Individual codes and main themes.

Theme 1: Smart information in place		Theme 2: Subversion of place meaning	
AR glasses	Tourism	AR art and performance	Superimposition
AR headsets	Interactive space	AR games	Virtual protest
AR navigation	Location-based services	AR graffiti	Visual changes
Automation	Marketing	AR markers	
BIM	On-demand information	Digital object	
Commodification of space	Smart city	Directing attention	
Digital twin	Smart infrastructure	Intrusion into public space	
HUD displays	Smart navigation	Pokémon Go	
Information overlay	Virtual space	Privacy	

idea is a creation of a kind of hybrid digital/material environment in which digital layers of information can be accessed on-site through augmented reality. As Magic Leap describe their idea of smart city overlays:

> *The creation of a city environment that is filled with information-rich digital data that can be digitally augmented [32]*
>
> *(...)platforms evolve and developers discover new ways to integrate and use actionable digital information within the real world. [33]*

Much emphasis is placed on the ability of AR information layers to be dynamic and location based. Contrary to other kinds of communication technologies, the content of AR is seen as being customized both by the user's needs and its location in geographical space. CGI describes this in their material about the Kiruna case in Sweden:

> *As the user's location changes, so does the view of the data within AR. Greater insights are available as artificial information about the environment is stored and retrieved as an additional information layer atop a real-world view. [34]*

While this vision is currently limited to a couple of examples, one strand of narration already illustrates existing and impactful applications in urban space—the construction of new buildings and city infrastructures. Here, technologies such as building information modeling (BIM) and digital twins are being closely connected to AR (and XR in general) use. According to XYZ (cytat), its Engineering-Grade augmented reality technology "(.,..) could not exist without BIM and one that represents a noticeable paradigm shift for construction." In the construction, there is a kind of reverse AR use, where digital layers come before an actual physical object rather than being overlaid upon existing material structure. AR is being perceived as something that will inevitably be linked to BIM processes, as a communication layer that could streamline procedures and allow a more comprehensive understanding among stakeholders, including those without domain knowledge. In the latter case, AR can present design in a way that is easily understood without the necessity of having expert spatial knowledge. As Magic Leap promotes its solutions:

> *Architects and construction companies are already (....) eliminating costly construction change orders by presenting clients with immersive designs and reviewing plans on-site and at scale. [33]*

Changing the perception of space and place through additional layers of AR information is also something that permeates AR use in the automotive industry and end-user applications. The most popular take is to supplement driver vision with heads-up displays (HUDs), which is seen as beneficial to the awareness of the driver, enhancing the surroundings with vital information based on the precise location of the car. AR can bring detailed navigational cues that are more cognitively accessible than voiced GPS instructions:

> *a complex intersection might involve multiple right turns. By overlaying the instructions onto the road itself, AR-HUD enables you to instantly see which turn to take. [35]*

Smart information through AR also has a strong presence as an imaginary related to tourism and heritage. In this setting, it is perceived as a tool to bring the possibility of interaction into a given location, as well as the ability to present additional information related to a given place, enhancing the experience of the visitors:

> *By bringing forward information, showing areas and artefacts in context, and bringing liveliness and dynamism to historical settings, it can add to the visitor experience and increase understanding. [36]*

But at the same time, AR in heritage sites can also be seen as something that can supplement *a place* with stimuli and feelings that make it more real, as Manuel Charr describes in a blog post:

> *stimuli that make the setting feel more real, such as the sound of swords clashing on the battlefield or the rush of wind howling through a mountain village. [36]*

The potential of AR for building place meanings forms the primary basis for the second theme—Subversion of place meaning. Contrary to the first one, it is much more diverse, encompassing a wide array of imaginations, that are nevertheless connected with a main theme of seeing AR content as a layer that can change the meanings associated with place. This change can be either by adding a digital object or replacing most of the surroundings, bringing the experience closer to virtual reality. No matter the technical proposition, however, the intention to somehow change the place through AR is common to the narrations we have grouped in this theme.

AR (virtual) graffiti is the most widely digital object associated with this purpose. Artwork often pushes the boundaries of technology applications, and AR is no exception, as it is seen as especially fitting for various forms of street art. This is because AR is, by its very nature, ephemeral, dynamic, and yet closely connected to place. Adding digital objects can bring additional meanings to place or replace existing perceptions in locations where physical interventions are not practical, are too demanding, or are even forbidden:

> *The ability to merge the analog with the digital makes the augmented murals so interesting. Layers complete one another, bringing life to parts of a city that are otherwise invisible. [37]*

The ability of AR to easily subvert the meanings of place is mainly because visual cues are one of the most important factors determining our spatial perception [38]. The other factors that play a role in this replacement process are the technology's attractiveness and immersion's effect. Immersion, while mainly associated with virtual reality [39], can still be an important part of the AR experience, where it is based on the connection between augmented reality environments and the real context of the material world [40]. The importance of location's role is mainly present in the materials we gathered related to various forms of AR used as a protest tool in urban public spaces. Public spaces, apart from their numerous functions [41], provide an essential spatial context for political discourse [42] and make places meaningful for people [43]. For the expression and democratization of social and political processes [44], it's crucial to be present in a particular place. And AR can make the place "particular" by subverting its meaning. AR activism is often mentioned in the context of large-scale protests such as #OccupyWallStreet [45], where AR apps added the ability to place virtual protesters in spaces where the physical manifestations were restricted. More recent mentions of applications were connected to the Black Lives Matter (BLM) movement. During the protest, various Instagram and Snapchat AR filters were created ad hoc for people to put their mark on public space without vandalizing it. And what is even more interesting, after the protests subsided, AR was used to combat the inevitable return of places that were the scene of the protests to their former, "normal states." As the creator of the Amp'Up Seattle app put it:

> *But you won't be catching Pikachus or other fantasy critters. Instead, you'll be viewing a different perspective of Seattle, one inspired by the recent protests for racial justice. [46]*

In the Pedestal Project the goal was even more directly related to subversion of place meaning. AR was used to give people the power to adorn empty monument pedestals with important figures from recent black history:

> *The Pedestal Project uses technology to see a world where statues finally portray worthy idols for future generations. [47]*

While AR protests have their own limitations, they are perceived as something that will have an impact on social activism since they have the power to overcome the limitation of material spaces [48].

One strand of narration that goes beyond simply adding digital AR content to place and that postulates the possibility of wholly replacing spatial experience can be found within materials from automotive brands. The first variant of this idea is aimed at including passengers in an AR experience of travel to combat boredom and provide enterprise, as exemplified by Audi Holoride:

> *Now it seems as if the Audi e-tron is no longer driving past the city's grey building facades. Instead it is moving through a colorful fantasy world populated by blue and white little chickens, so it seems. [49]*

The second variant of the AR-based replacement of the real sensory experience of car travel is considering future solutions that involve autonomous vehicles. In this setting AR is no longer necessary for the driver, since the role is no longer present. Instead, AR content is aimed at providing environment that could facilitate a kind of "mobile third space" that is replacing the insides of the vehicle:

> *Huawei predicts that autonomous cars will become the "mobile third space", where we'll be able to enjoy all kinds of leisure and entertainment activities as we'll no longer need to focus our attention on the act of driving [35]*

In both of these cases there is an underlying assumption that the material world around us is somewhat imperfect, and it needs either additional stimuli or, even better, it should be displaced by computer-generated reality to be entertaining and bearable. While not directly, this narration is present in a lot of other marketing materials, such as the slogan for Snapchat Spectacles that proclaims, "*Create the world you want to see*" [50].

6 Discussion—Relations of power and AR-based agency of place

The sociotechnical imaginary of AR, as expressed in the documentation discussed here, is nothing short of revolutionary, akin to a modernist sociotechnical imaginary that sees technology as a transformative aspect not only in the world but also in being itself. Indeed, the notion of AR as a bringing-forth of information is indicative of a position where information is the key aspect of contemporary life. This critique of modern technology is resonant with Martin Heidegger's critique of modern technology. Heidegger argued that "Bringing-forth brings out of concealment into unconcealment" [51]. From the imaginaries of AR, what is brought forth is information or data—without a necessary questioning of what that data or information is, where it emerges from, or who constructed it. Heidegger argued that modern technology does not bring forth the world itself, but instead reveals all entities in the world as resources to be used. Technology is an ordering: ordering us to see other entities as resource and ordering the world as a resource to be used in the furtherment of intentionality and efficiency. The sociotechnical imaginaries in these extracts—smart information in place which transforms the understanding of place for the geographical self and subversion of place meaning in another transformation—can be read as instantiations of this revealing of the world as AR developers and companies see the technology now. The agency of place in this view is limited and stymied by AR. AR acts not as an augmentation of place, but as a reducer of the experience of place down to seeing the world as information and data for the achievement of efficiencies in late capitalism.

On the other hand, AR, in the view of the narrations we presented, can also be seen as transcending the traditional constellations of power relations that shape places and spaces of modern cities. Spatial relations of physical accessibility and material symbols present in place result from those relations that are manifestations of, often conflicting, interests of various actors—city officials, urban designers, city dwellers, private corporations, and urban activists. The discourse on the symbolical and actual ownership of urban space shows how important this issue is for modern cities, where contrasting views on who should manage public space and how it should be managed form a central part of high tensions in an ongoing urban debate [52,53]. Digital technologies, such as AR and other spatial media [54], influence this discussion by introducing complicated assemblages of dynamic digital content and code produced by social actors that can be only loosely geographically connected to a given location. This process is transductive and ontogenetic, constantly changing the conditions under which space and place are made and remade [55,56]. As Graham and Zook [57] argue, technological artifacts, content, and code help produce and mediate place, exerting power from social and software actors. The consequence of this phenomenon is that place can no longer be as easily controlled and appropriated through purely material means. This also may suggest that the balance of power to create a place could be skewed toward nonhuman actors, of which AR may be the most visible and influential. And the narration of the subversion of meaning seems to, in part, follow this sentiment, imagining commercial and noncommercial uses for AR in increasingly smart spaces.

AR can introduce shifts in the balance of place-making power in various dimensions that warrant a more critical examination. Firstly, there is an inevitable imbalance between **AR users and AR nonusers**. Someone with access to AR technology, who knows that AR content is present or has been informed about this by location-based service, can interact with place differently. AR can give access to additional layers of information and ways of communicating within a place and about a place. In contrast, a nonuser does not have access to additional services, affordances, and digital objects, which creates an imbalance in power. Secondly, there is a volatile issue of **the ownership of space and place**. The current legal status of AR content is vague in most legislations, and they are ill-equipped to deal with specific problems that arise when AR content is anchored in private and public spaces [58]. A glimpse of this problem was already visible in the numerous examples of digital objects of the AR game Pokémon Go that need to be placed in physical locations [59] and that were located on private land or in places deemed as inappropriate [60]. It is even more essential to consider the possible implications of AR content not only trespassing material dimensions of space but also invading symbolic meanings of place, changing the perception of what is sacred and profane. AR-based interventions can instill a

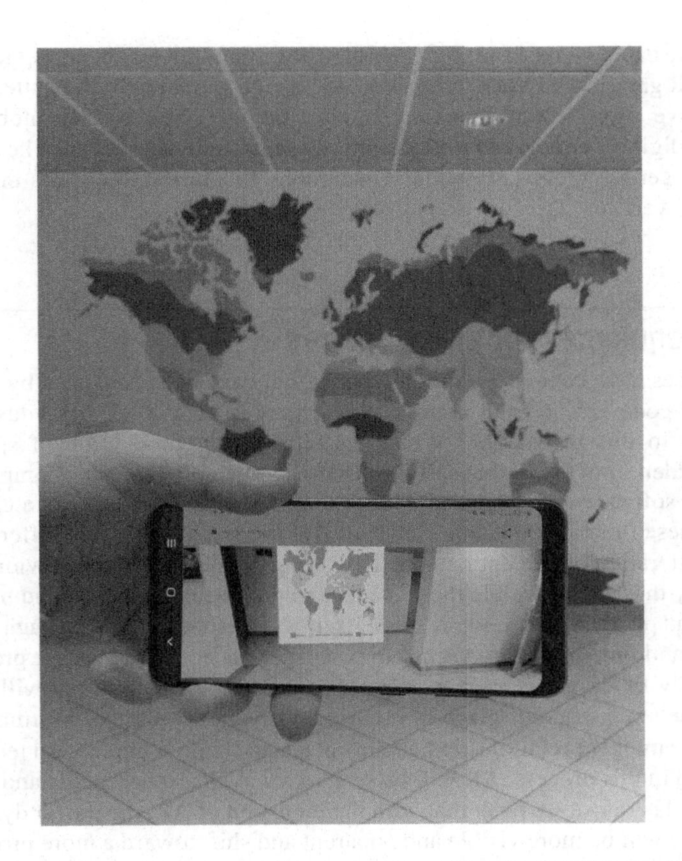

FIG. 1

Example showing the custom-made AR application (developed as a part of the Authors' project) that replaces wall maps with up-to-date information about climate change impacts.

debate on the nature of a given space and the content that should be displayed (Fig. 1). Another dimension of power that can be sensitive to the introduction of AR content is the issue of **machine vs human perception of space and place**. It is possible to imagine that urban design will include guidelines for adopting spaces to accommodate nonhuman users, such as autonomous cars or drones, using AR elements such as specific street signs and navigational cues. The question of power is already being discussed as various ways to deceive machine perception are being developed [61]. Finally, there is an issue related to the fact that AR is strongly dependent on physical location, that there may be mechanisms introducing a kind of **forced presence and forced behavior**. If AR content is necessary to obtain certain goods and services, traveling to a specific location would also be necessary. Similarly, when present in a smart space that can be interacted with using AR software, this interaction often requires some kind of specific activity, such as pointing a smartphone camera and scanning an AR anchor. It has already been suggested that playing AR games such as Pokémon Go can change human mobility patterns, often in a positive way [23,62]. However, it must be noted that forcing mobility and requiring digital literacy and possession of AR-capable devices has the potential to exclude certain groups of people. Once more, this introduces shifts in the relations of power.

7 Consequences for smart spaces

Smart spaces have been thought of as spaces that have been produced by the functioning of code [55,56] and computational processes. AR clearly adds another dimension to this production. While the ontogenetic production of space was largely hidden from the spatial self in smart spaces (through opaque computational processes, software operations, and algorithmic functioning) there is a clear visibility of these functions in the presence of AR. Through the visible differentiation of users, alteration of perception, and the forcing of presence and behavior, AR is a technology that makes visible the systems and processes of control and mediation of space and place. As such, smart space itself will become visible through the presence, use, and functioning of AR in a manner that has not been the case previously. AR not only functions as an interface for data and information, but will actively mediate the experience of place in real time through contextual information that affects the emergent relationship between place and self. While digital technology is involved in this process, AR will be seen in a way that other media and technology is not. Therefore we posit that with the inclusion of AR, the power dynamic of smart space will be more visible and apparent and shift toward a more pronounced agency of place. What is also significant in this narration is that the power created in place through the AR-based agency of place is based on the prevalent sociotechnical imaginaries. This disconnects this power from being solely dependent on the presence of technology. Within AR imaginaries, any space is potentially smart, which changes human behavior, perception, and placemaking processes. This effect can be seen even when the technology is not present. In essence, within reach of sociotechnical imaginaries, nonsmart space "disappears" and humans are no longer the sole agent creating the place. All these factors may be considered in the design of smart spaces. Using AR may be an excellent way to communicate additional technologically mediated interaction possibilities, with a resulting shift in the power relations.

Acknowledgments

This work was supported by the National Science Centre, Poland under grant 2019/33/B/HS4/00057.

References

[1] M. Pesce, Augmented Reality: Unboxing Tech's Next Big Thing, John Wiley & Sons, 2020.
[2] A. Ricci, M. Piunti, L. Tummolini, C. Castelfranchi, The mirror world: preparing for mixed-reality living, IEEE Pervasive Comput. 14 (2015) 60–63, https://doi.org/10.1109/MPRV.2015.44.
[3] S. Jasanoff, S.-H. Kim, Containing the atom: sociotechnical imaginaries and nuclear power in the United States and South Korea, Minerva 47 (2009) 119–146.
[4] E.S. Casey, Between geography and philosophy: what does it mean to be in the place-world? Ann. Assoc. Am. Geogr. 91 (2001). http://www.tandfonline.com/doi/pdf/10.1111/0004-5608.00266. (Accessed 10 October 2016).
[5] J. Malpas, Place and Experience, a Philosophical Topography, Cambridge University Press, 1999.
[6] S.C. Larsen, J.T. Johnson, The agency of place: toward a more-than-human geographical self, GeoHumanities 2 (2016) 149–166, https://doi.org/10.1080/2373566X.2016.1157003.
[7] L. Evans, Being-towards the social: mood and orientation to location-based social media, computational things and applications, New Media Soc. 17 (2015) 845–860.
[8] L. Evans, M. Saker, Location-Based Social Media, Springer International Publishing, Cham, 2017, https://doi.org/10.1007/978-3-319-49472-2.
[9] J. Farman, Location-based media, in: Dialogues on Mobile Communication, Routledge, 2016, pp. 161–177.
[10] J. Hamilton, OurPlace: the convergence of locative media and online participatory culture, in: Proceedings of the 21st Annual Conference of the Australian Computer-Human Interaction Special Interest Group: Design: Open 24/7, 2009, pp. 393–396.
[11] M. Ward, A. Galloway, Locative media as socialising and spatialising practices: learning from archaeology, Leonardo Electron, Alm. 14 (2006) (Locative Media Special).
[12] C. Speed, Developing a sense of place with locative media: an "Underview Effect", Leonardo 43 (2010) 169–174.
[13] L. Hjorth, S. Pink, H. Horst, Being at home with privacy: privacy and mundane intimacy through same-sex locative media practices, Int. J. Commun. 12 (2018) 1209–1227.
[14] D. Özkul, Location as a sense of place: everyday life, mobile, and spatial practices in urban spaces, in: Mobility and Locative Media, Routledge, 2014, pp. 121–136.
[15] M. Saker, L. Evans, Everyday life and locative play: an exploration of Foursquare and playful engagements with space and place, Media Cult. Soc. 38 (2016) 1169–1183.
[16] J. Farman, Mobile Interface Theory: Embodied Space and Locative Media, Routledge, 2020.
[17] A. Townsend, Augmenting public space and authoring public art: the role of locative media, ArtNodes (8) (2008) (article online) https://doi.org/10.7238/a.v0i8.771.
[18] J. Southern, Comobility: How Proximity and Distance Travel Together in Locative Media, 2012.
[19] J. Frith, Turning life into a game: Foursquare, gamification, and personal mobility, Mob. Media Commun. 1 (2013) 248–262.
[20] M. De Lange, From always-on to always-there: locative media and playful identities, Digit. Form. 57 (2009).
[21] A. Lemos, Post—mass media functions, locative media, and informational territories: new ways of thinking about territory, place, and mobility in contemporary society, Space Cult. 13 (2010) 403–420.

[22] C. McGarrigle, The construction of locative situations: locative media and the Situationist International, recuperation or redux? Digit. Creat. 21 (2010) 55–62.

[23] M. Saker, L. Evans, Intergenerational Locative Play Augmenting Family, Emerald Publishing, 2021. http://www.vlebooks.com/vleweb/product/openreader?id=none&isbn=9781839091391. (Accessed 28 July 2021).

[24] S. Jasanoff, Future imperfect: science, technology, and the imaginations of modernity, in: Dreamscapes of Modernity: Sociotechnical Imaginaries and the Fabrication of Power, The University of Chicago Press, Chicago/London, 2015, pp. 1–33.

[25] D. McNeill, Global firms and smart technologies: IBM and the reduction of cities, Trans. Inst. Br. Geogr. 40 (2015) 562–574.

[26] J. Sadowski, R. Bendor, Selling smartness: corporate narratives and the smart city as a sociotechnical imaginary, Sci. Technol. Hum. Values 44 (2019) 540–563, https://doi.org/10.1177/0162243918806061.

[27] J. Kotus, M. Rzeszewski, A. Olejniczak, Material and digital dimensions of urban public spaces through the lens of social distancing, Cities 130 (2022) 103856, https://doi.org/10.1016/j.cities.2022.103856.

[28] D.J. Hess, B.K. Sovacool, Sociotechnical matters: reviewing and integrating science and technology studies with energy social science, Energy Res. Soc. Sci. 65 (2020) 101462, https://doi.org/10.1016/j.erss.2020.101462.

[29] T. Alsop, Augmented reality (AR)-statistics & facts, 2020. https://www.statista.com/topics/3286/augmented-reality-ar. (Accessed 6 October 2020).

[30] I. Zubizarreta, A. Seravalli, S. Arrizabalaga, Smart city concept: what it is and what it should be, J. Urban Plan. Dev. 142 (2016) 04015005, https://doi.org/10.1061/(ASCE)UP.1943-5444.0000282.

[31] D. Cook, S.K. Das, Smart Environments: Technology, Protocols, and Applications, John Wiley & Sons, 2004.

[32] S. Ochanji, Magic Leap Plans to Build Digital Overlays to Create Smart Cities, Virtual Reality Times, 2019. https://virtualrealitytimes.com/2019/07/10/magic-leap-plans-to-build-digital-overlays-to-create-smart-cities/. (Accessed 28 February 2023).

[33] Augmented Reality 101: What Is AR and How Does It Work? 2023. https://www.magicleap.com/news/augmented-reality-101-what-is-ar-and-how-does-it-work. (Accessed 28 February 2023).

[34] CGI, Using Augmented Reality and Precision Data to Enable the Future Smart City, 2018. https://www.cgi.com/sites/default/files/2018-06/cgi-case-study-kiruna-sweden_0.pdf. (Accessed 28 February 2023).

[35] AR-HUD: A Heads Up on the Road Ahead, Huawei BLOG, 2022. https://blog.huawei.com/2022/01/04/ar-hud-road-ahead-transportation/. (Accessed 28 February 2023).

[36] M. Charr, What Can AR Do to Bring Heritage Sites to Life?, MuseumNext, 2020. https://www.museumnext.com/article/what-can-ar-do-to-bring-heritage-sites-to-life/. (Accessed 28 February 2023).

[37] B. Habelsberger, Augmented Reality and Street Art: Social and Cultural Engagement in the City, Artivive, 2018. https://artivive.com/augmented-reality-and-street-art/. (Accessed 28 February 2023).

[38] M. Eimer, Multisensory integration: how visual experience shapes spatial perception, Curr. Biol. 14 (2004) R115–R117, https://doi.org/10.1016/j.cub.2004.01.018.

[39] M. Slater, Immersion and the illusion of presence in virtual reality, Br. J. Psychol. 109 (2018) 431–433, https://doi.org/10.1111/bjop.12305.

[40] M.J. Kim, A framework for context immersion in mobile augmented reality, Autom. Constr. 33 (2013) 79–85, https://doi.org/10.1016/j.autcon.2012.10.020.

[41] E. Duivenvoorden, T. Hartmann, M. Brinkhuijsen, T. Hesselmans, Managing public space—a blind spot of urban planning and design, Cities 109 (2021) 103032.

[42] A. Rosenthal, Spectacle, fear, and protest: a guide to the history of urban public space in Latin America, Soc. Sci. Hist. 24 (2000) 33–73.

[43] P. Gustafson, Meanings of place: everyday experience and theoretical conceptualizations, J. Environ. Psychol. 21 (2001) 5–16, https://doi.org/10.1006/jevp.2000.0185.

[44] D. Mitchell, The Right to the City: Social Justice and the Fight for Public Space, Guilford Press, 2003.

[45] J.S. Juris, Reflections on #Occupy Everywhere: social media, public space, and emerging logics of aggregation, Am. Ethnol. 39 (2012) 259–279, https://doi.org/10.1111/j.1548-1425.2012.01362.x.

[46] M. Vansynghel, Augmented Reality App Reveals Seattle Protest Art in Surprising Places, Crosscut, 2020. https://crosscut.com/culture/2020/08/augmented-reality-app-reveals-seattle-protest-art-surprising-places. (Accessed 28 February 2023).

[47] A. Rasilla, Damien McDuffie Sees Augmented Reality as the Future for Preserving Oakland Stories, The Oaklandside, 2021. https://oaklandside.org/2021/08/18/damien-mcduffie-augmented-reality-app-black-panthers-preserving-oakland-stories/. (Accessed 28 February 2023).

[48] R.M.L. Silva, E. Principe Cruz, D.K. Rosner, D. Kelly, A. Monroy-Hernández, F. Liu, Understanding AR activism: an interview study with creators of augmented reality experiences for social change, in: CHI Conference on Human Factors in Computing Systems, ACM, New Orleans, LA, 2022, pp. 1–15, https://doi.org/10.1145/3491102.3517605.

[49] Holoride: Virtual Reality Meets the Real World, Audi.Com, n.d. https://www.audi.com/en/innovation/development/holoride-virtual-reality-meets-the-real-world.html. (Accessed 28 February 2023).

[50] Spectacles od Snap Inc., Spectacles nowej generacji, n.d. https://www.spectacles.com/pl/. (Accessed 28 February 2023).

[51] M. Heidegger, The question concerning technology, N. Y. 214 (1977).

[52] M. Carmona, Contemporary public space: critique and classification, part one: critique, J. Urban Des. 15 (2010) 123–148, https://doi.org/10.1080/13574800903435651.

[53] J. Németh, Controlling the commons: how public is public space? Urban Aff. Rev. 48 (2012) 811–835, https://doi.org/10.1177/1078087412446445.

[54] R. Kitchin, Thinking critically about and researching algorithms, Inf. Commun. Soc. 20 (2017) 14–29, https://doi.org/10.1080/1369118X.2016.1154087.

[55] R. Kitchin, M. Dodge, Code/Space: Software and Everyday Life, MIT Press, Cambridge, MA, 2011.

[56] M. Dodge, R. Kitchin, Code and the transduction of space, in: Machine Learning and the City: Applications in Architecture and Urban Design, John Wiley & Sons Ltd, 2022, pp. 309–339.

[57] M. Graham, M. Zook, Augmented reality in urban places, in: Machine Learning and the City: Applications in Architecture and Urban Design, John Wiley & Sons Ltd, 2022.

[58] E.F. Judge, T.E. Brown, A right not to be mapped? Augmented reality, real property, and zoning, Laws 7 (2018) 23, https://doi.org/10.3390/laws7020023.

[59] A.M. Clark, M.T. Clark, Pokémon Go and research: qualitative, mixed methods research, and the supercomplexity of interventions, Int. J. Qual. Methods 15 (2016). 1609406916667765.

[60] H. Gong, R. Hassink, G. Maus, What does Pokémon Go teach us about geography? Geogr. Helv. 72 (2017) 227–230, https://doi.org/10.5194/gh-72-227-2017.
[61] C. Sitawarin, A.N. Bhagoji, A. Mosenia, M. Chiang, P. Mittal, DARTS: Deceiving Autonomous Cars With Toxic Signs, 2018, https://doi.org/10.48550/arXiv.1802.06430.
[62] L. Hjorth, I. Richardson, Pokémon GO: mobile media play, place-making, and the digital wayfarer, Mob. Media Commun. 5 (2017) 3–14, https://doi.org/10.1177/2050157916680015.

CHAPTER

Ensuring a harmonious state of smart space when there is a conflict of interest of its elements

11

Valeria Shvedenko[a] and Vladimir N. Shvedenko[b]

[a]*T-INNOVATIK, Limited Liability Company, St. Petersburg, Russia,* [b]*FSBUN All-Russian Institute of Scientific and Technical Information of the Russian Academy of Sciences (VINITI RAS), Moscow, Russia*

1 Introduction and motivation

The modern developments of science and technology have made it possible to create smart spaces, which are understood as a controlled environment where optimal conditions for life and work are maintained with the help of sensor systems, actuator systems, and control systems.

Their distinctive feature is the creation of a controlled intelligent complex based on the included remote control devices—smart space elements, the presence of stable wireless communication between them, the functions implemented to collect, process, and distribute information about the state of the smart space elements, risk prediction related to possible changes in the specified parameters, and the causes of their occurrence using intelligent data processing methods.

The concept of a smart space is applicable to objects of any complexity, from a "smart house" to "smart territories" [1–5].

Smart space elements are located in the same spatial and temporal coordinates, and are part of one or more functional slices (layers) of a controlled smart complex.

Currently, the scientific direction of organization and management of smart spaces is being widely developed and applied in energy, transport, housing, and communal complexes, security systems, monitoring of environmental problems, and efficient resource management [6–16], and others.

The scope of smart space applications will continue to expand and cover more and more spheres of human activity. These tasks are becoming increasingly important due to the current and ongoing global energy crisis, the lack of storage and processing capacity, as well as the new opportunities offered by Industry 4.0, improved geoinformation technology, and artificial intelligence.

Smart Spaces. https://doi.org/10.1016/B978-0-443-13462-3.00005-4

However, the implementation and operation of smart spaces is a costly and time-consuming undertaking. Therefore there is a need for further improvement of theoretical research and development of practical recommendations to the level of reference solutions aimed at ensuring effective and harmonious interaction of the elements of smart spaces.

This chapter considers the smart space paradigm from the perspective of a system of systems and the direction of its development.

We propose to evaluate the effectiveness of smart space functioning according to the criterion of maintaining the level of its harmonious state. For this purpose, the notions of a "harmonious state of smart spaces" and "harmonious development of smart spaces" have been introduced. From these positions, the interaction of smart space elements is considered.

For the first time, the notion of a digital ecological boundary of the system is given, and the indicators of influence, growth, and containment of changes in the state of the properties of objects included in smart spaces are highlighted.

From the position of the theory of systems, we present a scheme of mutual influence of changes in the values of indicators of one smart space element on other smart space elements, which allows the influence of various factors of the external and internal environment of smart spaces to be taken into account, to maintain a harmonious state.

An original scheme for organizing the information and control superstructure of smart spaces in the form of a polystructure is presented.

The principles of metamodel formation of topology and topology of information space of the smart space polystructure system are described.

The conclusions of the work, prospects, and future direction of the authors' research are given.

2 Background, definitions, and designations

From the perspective of a system of systems (SoS), a smart space can be represented as a multilayered information structure, bound to single space-time coordinates, each layer of which is an independent complex system (see Fig. 1).

The smart space paradigm implies that a smart space has a set of the following capabilities:

- Includes new elements, details the hierarchy of its constituent objects and the relationships between objects and object properties, making it open to change;
- Provides data exchange between the elements of the system or systems that it comprises and collects primary data in space-time coordinates;
- Ensures the coordination of information flows of the elements of the system or its constituent systems, to achieve the targets for which it was created. In the process of adjusting smart space targets, the direction, intensity, and content of information flows of the system elements or constituent systems may change;

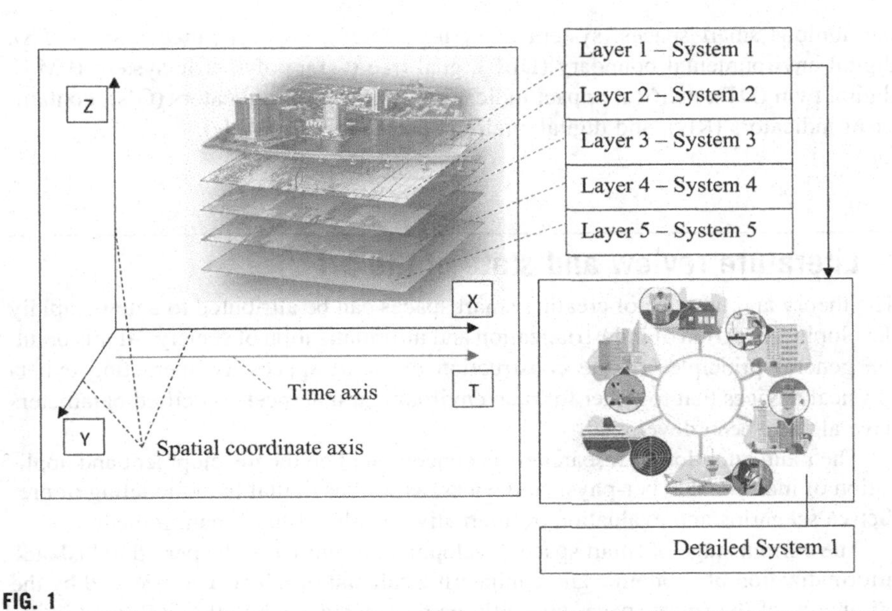

FIG. 1

Representation of the smart space as multilayered information structure.

- Uses methods of artificial intelligence in data processing to ensure the selection of optimal characteristics of the interaction of its elements, in order to achieve a given or more effective result;
- Serves as wide a variety of interests of smart space subjects as possible;
- Focuses on the simultaneous solution of different classes of tasks within the framework of the unified information space being formed.

Each layer of a smart space, according to Fig. 1, is an independent complex system consisting of a set of interacting objects, processes, and functions implemented in accordance with the goals set for the system. Each layer has its own set of sensors; cyber-physical devices; primary, intermediate, and final data storage; and methods of collection, evaluation, interpretation, processing, and management of the results obtained.

All smart space layers are located in a single system of spatial and temporal coordinates, which enables their interaction, connection to each other's resources, and evaluation of mutual influence on the values of state indicators of system properties, stages of performed processes, and efficiency of functional blocks.

This goal is achievable if we consider a smart space from the system of systems (SoS) perspective, which is subject to appropriate constraints and requirements according to the paradigm presented earlier.

We offer this presentation of the theoretical foundations of the design and management of smart spaces in the context of the following definitions and concepts:

harmonious smart spaces, system of systems (SoS), polystructured system (PS), digital environmental boundary (DEB), goal tree (GT), polymetric system (PMS), digital twin (DT), metrics, impact indicators (IIs), growth indicators (GIs), containment indicators (RIs), and digital environmental indicators (EIs).

3 Literature review and state of the art

The theory and practice of creating smart spaces can be attributed to a new, rapidly developing direction of industrialization and informatization of society. At this point, the general principles of the construction of smart spaces as interacting cyber-physical devices that together form an environment that meets specified parameters have already been developed.

The main attention of researchers is concentrated on the development and application of individual cyber-physical devices, as well as digital twin modeling of predictive scenarios and evaluation of alternative results of their management.

The current stage of smart space development is similar to the period of "island" informatization of economic and engineering calculations. This is evidenced by the originality of the smart spaces currently being created, each within a limited information field. However, reference solutions have not yet been created. There are no universal methods and industry recommendations, for example, as implemented in the design of buildings and structures, mechanisms, and machines, etc.

As in the period of the formation of the information society, methods are currently being applied involving creating "islands" of intelligent data processing and solving coordination and control of a narrow range of tasks. Consequently, the idea of a smart space is at the stage of accumulation and systematization of experience, analysis of mistakes made, development of clear algorithms, and transfer of the results of theoretical research into the plane of solving reference engineering problems.

Analysis of monographs and articles on the subject reveals a clustering of scientific directions both by geography of research conducted and by types of problems involved, with models developed for their smoothing and control over the current situation.

The main groups of publications and implementation of solutions refer to the problems of comfort assessment and management [17–19], safety [20,21], resource conservation [22], resource allocation [23–25], navigation of connected multiple objects (e.g., tourism infrastructure, educational space, etc.) [26], control [27], and system elements or interacting of two or more systems.

To effectively build smart spaces requires:

- first, the development of reference industry solutions for individual systems, for example, systems for monitoring and prevention of natural and man-made disasters, systems for integrated management of housing and transport

communications of residential spaces, systems for monitoring the environmental safety of industrial complexes, etc.;
- second, organization of effective information and communication interaction between the systems that are components of smart spaces, which should be considered from the SoS point of view;
- third, ensuring harmonious target solutions for the SoS elements that form the smart space.

4 Problem definition

This chapter presents smart spaces in terms of SoS as a set of self-sufficient, developing complex systems interacting with each other, and located in a single information space within common space-time coordinates.

If the interacting elements of a smart space implement the principles, objectives, and goals of SoS, then we can talk about its stable operation.

However, achieving a state of consistency in the interests of each individual complex system that is part of the SoS, and even more so the maintenance of such a state over a sufficiently long period of time, is a difficult and sometimes insoluble task. The factors of the external environment that affect the state of the characteristics of the SoS elements are dynamically changing. The priorities in goals to be achieved change. There is a shortage of required resources. It becomes necessary to use new achievements from scientific and technological progress. The tools to assess the accuracy and weighting of decisions are always improving. The subjects of smart spaces are changing. New alternative opportunities for the establishment of more profitable alliances of systems, etc., are appearing. Therefore it is proposed to use the notion of a harmonious state of smart spaces.

By *harmonious state of smart spaces* we will understand the provision of target settings of SoS, which are within the acceptable range of regulation and ensure the dynamic equilibrium of the whole system, as well as its constituent elements.

Harmonious development of a smart space is a process of development in accordance with the specified targets that does not worsen the values of the indicators of functioning of its constituent systems and that has the possibility of achieving positive synergistic effects from their interaction.

The reaction time to restore the harmony of a smart space is related to the inertial characteristics of its individual elements.

Changes in the target settings of the SoS will be associated only with sharp fluctuations in the resource availability of the SoS, and as a consequence the imbalance of the target settings of its elements. These circumstances may arise due to natural, man-made, economic, social, military-political, and other events.

Maintaining a harmonious state of a smart space is similar to maintaining dynamic equilibrium in mechanical systems.

5 Proposed solution

Each system that is part of a smart space affects the other systems that interact with it. The influence of one system on another system is always mutual, but not equal. This influence can be measured through the values of changes in the values of the indicators of the properties of the system objects, indicators of the execution of the system process stages, as well as the values of deviations in the results of the functions performed.

Target orientations of the complex systems that form the smart space layers can be in resonance with each other (see Fig. 2), partially coincide (see Fig. 3), be neutral

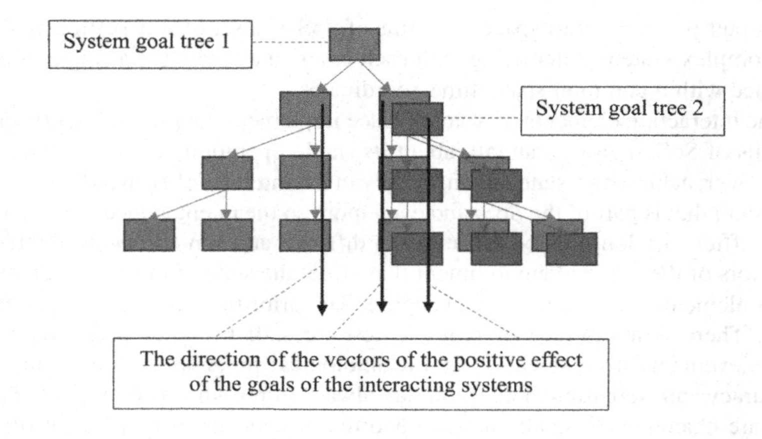

FIG. 2

Resonance (amplification from the convergence) of the goals of interacting systems.

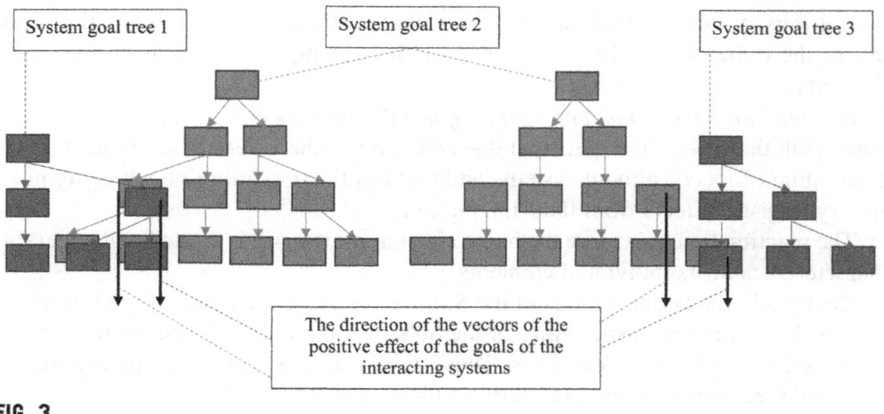

FIG. 3

Partial coincidence of the target benchmarks of the interacting systems.

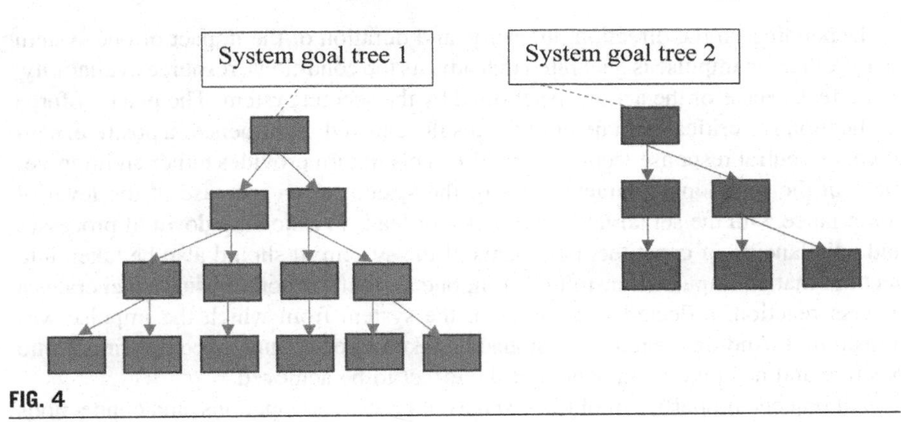

FIG. 4

Nonintersecting goal systems.

in relation to each other (see Fig. 4), or be in a state of conflict of interest due to the lack of shared resources or because of the multidirectional vectors of positive achieved effects by different systems (see Fig. 5).

Based on the established targets, the systems whose functioning and reactions are considered in the information field of unified space-time coordinates interact with each other, weakening or strengthening each other's characteristics. Some of them are in antagonistic contradiction. Some are neutral to each other. A number of systems, combining into more powerful structures, exhibit an additional synergistic effect, increasing resilience and reducing the necessary energy capacity of each of them.

Functioning independently of each other, but having points of intersection of target benchmarks, these systems influence the characteristics of the smart space, increasing or decreasing its controllability, as well as the possible accuracy of making necessary adjustments to comply with the regulatory requirements of the smart space.

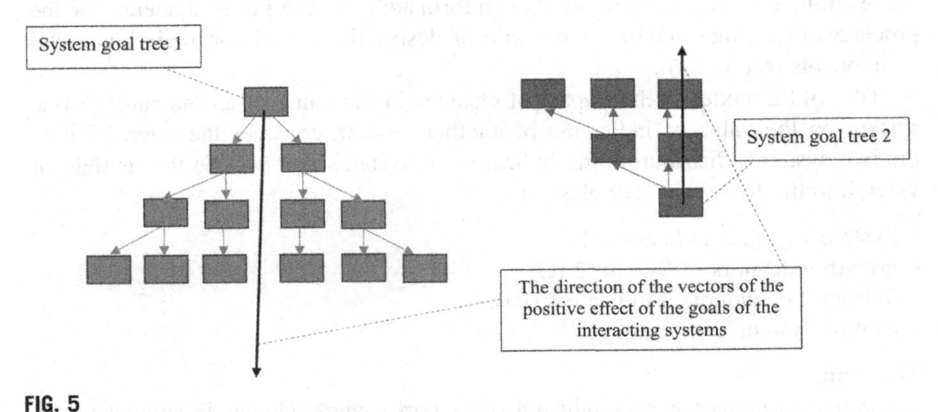

FIG. 5

Differently directed vectors of positive reached effects of different systems.

Depending on the direction, intensity, and duration of the impact of one system on the other, an impulse is generated that adjusts the conditions, resource availability, and effectiveness of the actions performed by the second system. The place of force application is a critical parameter of the result achieved. It influences a positive, negative, or neutral response from the impulse. This in turn provides either an improvement in the functional characteristics of the system or an increase in the level of compliance with the set target benchmarks, or leads to a slowing down of processes and unbalancing of constituent elements of the system. It should also be taken into account that an impulse transmitted from one system to another usually generates a reverse reaction, reflected in changes in the system from which the impulse was transmitted. And this reaction must also be taken into account, decomposing it into positive and negative components of the effect to be achieved.

All objects, properties of objects, stages of processes, functions, and connections between them can be considered as points, which can be influenced by other systems.

Depending on the type of impact, only some of them are in the "sensitive zone" (the zone of positive and negative reactions).

It is important to identify these critical points (objects, properties of objects, stages of processes, functions, decision-making centers) and establish which list of indicators of the system controls the emergence of reactions to external perturbing factors, as well as to determine the source of this impact and the level of its intensity.

Some influences stimulate the process of growth and are the reason for achieving positive effects. Others, on the contrary, worsen the situation or slow down the positive dynamics of the process. Others remain neutral to perturbing influences and only absorb energy and resources of the first system without bringing a proper result.

It is important to know the points of effective impact on the system, to understand the degree of sufficiency or redundancy of efforts applied to them, as well as to assess the risks of impact, which could destroy connections or worsen the values of indicators of vital activity of the system or other systems.

If more than two systems are interacting, each of them influences the other, and the resulting effects from their interaction form additional impulses that enhance the processes of change, having synergistic or destructive effects on their functional components (see Fig. 6).

To build a model of the impact of changes in the value of an indicator of one system on the value of indicators of another system, consider the interaction of the two systems, distributing the indicators of systems that will be in our field of view, into the following four classes:

- System 1 impact indicators (II);
- growth indicators of System 2 (GI);
- System 2 containment indicators (RI);
- neutral System 2 indicators (NI).

That said:

An impact indicator is an indicator of System 1 whose change in value causes a pulse that has an impact on the change in values of System 2 indicators. Typically,

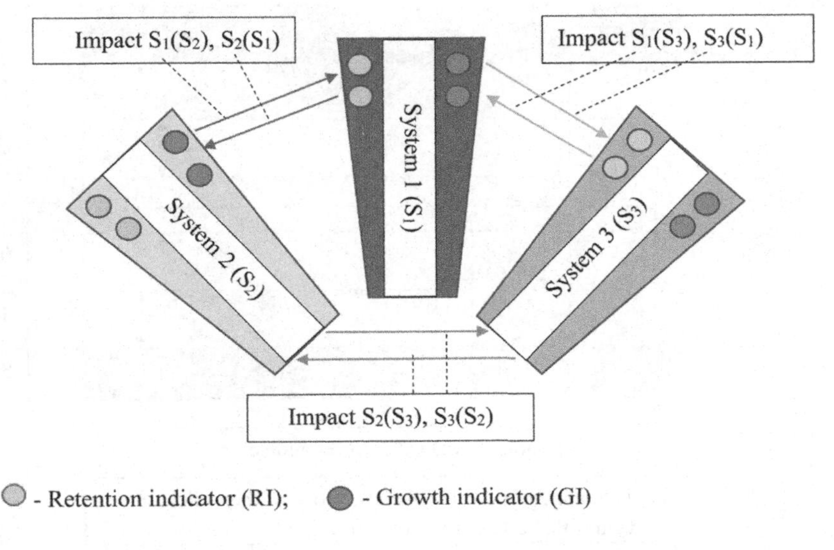

FIG. 6

Diagram of the influence of systems.

the impact indicator of System 1 corresponds to one or more indicators of growth and/or containment of System 2.

A neutral indicator is an indicator that is not involved in the process of assessing the interaction of changes in the parameters of one system on the parameters of the other system.

A growth indicator is an indicator of system efficiency increase, which value change depends on the presence of an external impulse (change in values) of another indicator (increase in power, minimization of resources, or achievement of new effects). The source of the transmitted impulse can be a change in the value of the indicator of another system, or a change in the value of the indicator of the system to which it refers.

A constraint indicator is an indicator of restraint of development, unbalance, deterioration, or critical destruction of the functionality of the system. It can also be called an indicator of digital environmental safety of the system. The normative value of this indicator is set according to the standards of safety of the system functioning and human life (if humans are a participant in this system).

It is important to monitor the range in which its value is located to be able to make the necessary corrective actions to bring it to the specified standards. Ranges of possible variants of deviations of values of the containment indicator are shown in Fig. 7.

The "Growth Indicator" or "Containment Indicator" status is assigned to the system indicator in case the impact of changes in the values of the indicators of one system on the change in the values of the indicator (indicators) of the other system is considered. The set (number and list) of growth and containment indicators may be different

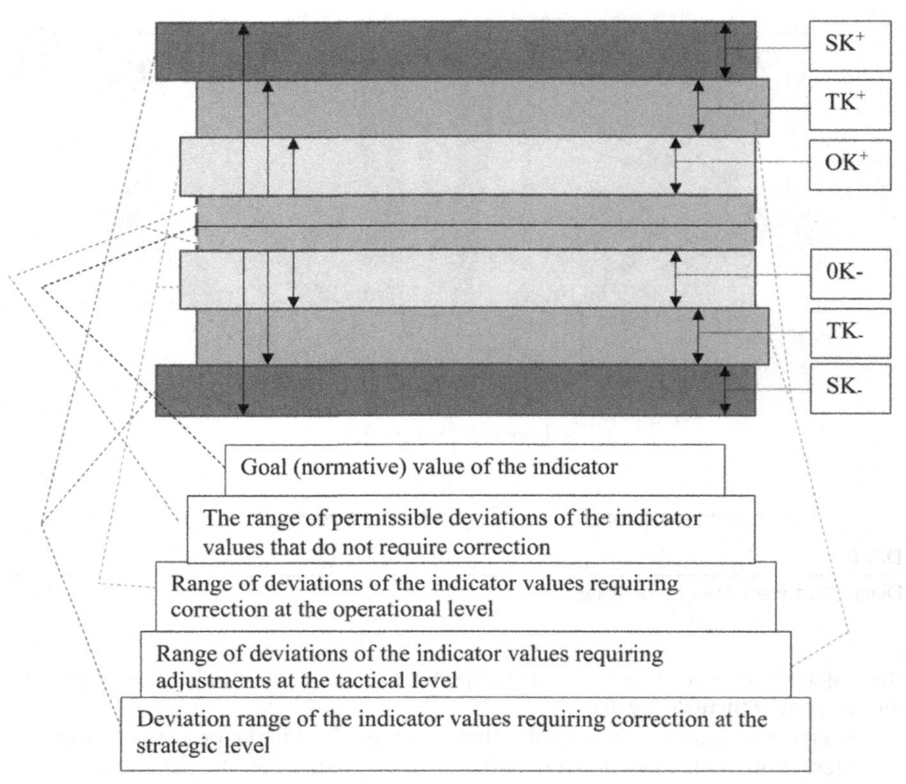

FIG. 7

Deviation ranges of the deterrence indicator values and the level of response to their correction.

for two different systems interacting with each other. The intensity of the impact of changes in the value of one indicator (the first) of the system on changes in the value of the indicator (the second) of the system may differ (see Table 1).

Types of dependencies of changes in the values of System 1 to System 2 are shown in Figs. 8 and 9.

To establish the target (normative) value of the growth indicator of System 1, it is necessary to analyze the resulting changes in the values of other indicators, which have mutual or unilateral influence on the functionality (improvement or deterioration of their characteristics) of the two systems interacting with each other (see Figs. 10 and 11).

Such analysis is carried out for each growth and containment indicator of the first and second systems. In addition, changes of values of indicators in the system to which the growth or deterrence indicator belongs are taken into account. The influence matrices are constructed on the basis of the data obtained; thus (if such statistics

Table 1 Classification of the intensity of the influence of the value of the indicator of the first system on the change in the value of the second system.

The intensity of the effect of the value of the indicator of the first system on the change in the value of the second system	Moves the value of the indicator to a critical value. Requires an urgent response
	Requires prompt intervention in the function block and adjustment of its settings
	Is not significant, may not be taken into account when maintaining the current level of intensity of exposure
	Acts as "white noise," a distraction, but has no effect on the change in the value of the indicator
The intensity of the effect of the value of the indicator of the first system on the change in the value of the second system	It is a constant value
	It has a tendency to increase
	It has a tendency to fade
	Provokes avalanche-like processes of system growth
	Provokes avalanche-like processes of destruction of the system
The intensity of the effect of the value of the indicator of the first system on the change in the value of the second system	Affects the change in values of a single indicator of the system
	Affects changes in the values of several indicators of the system

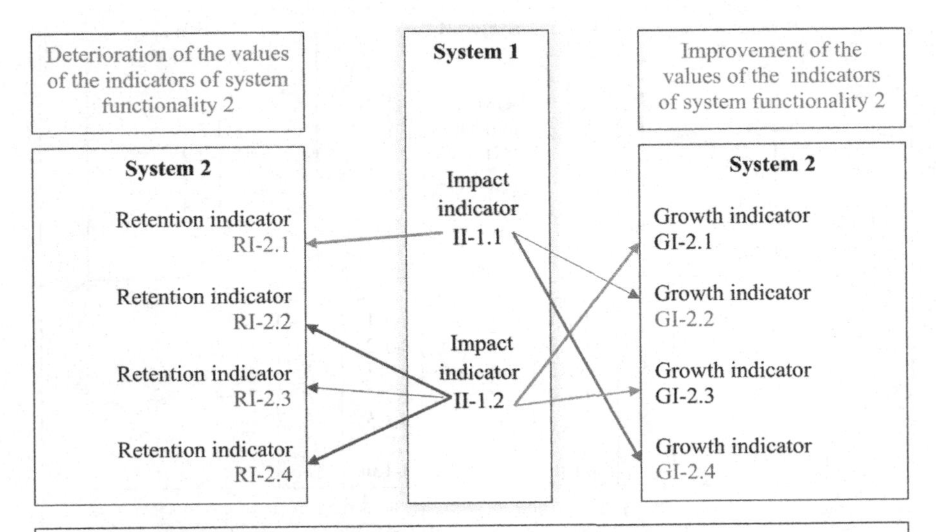

FIG. 8

Diagram of links between System 1 impact indicators and System 2 containment and growth indicators.

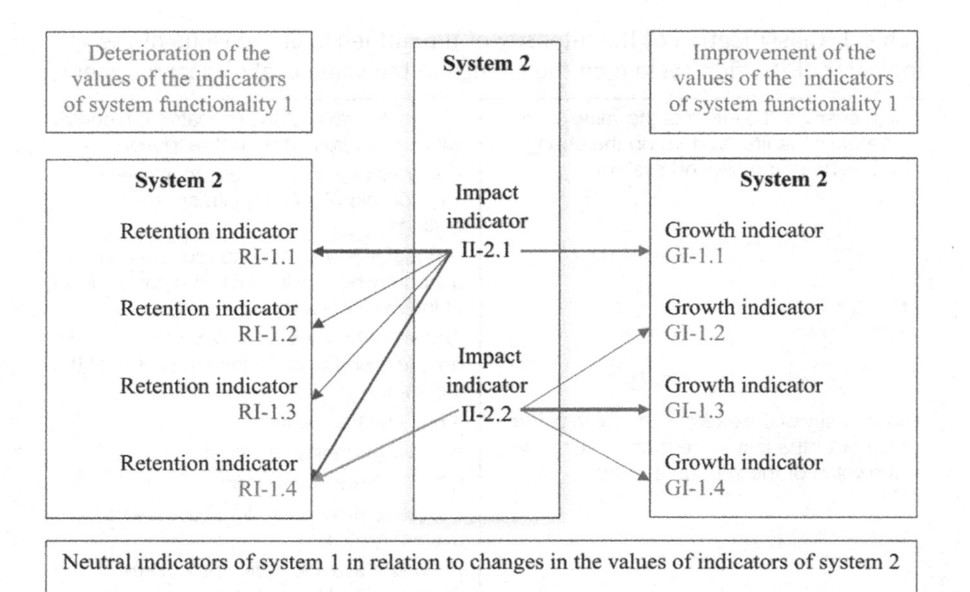

FIG. 9

Diagram of links between System 2 impact indicators and System 1 containment and growth indicators.

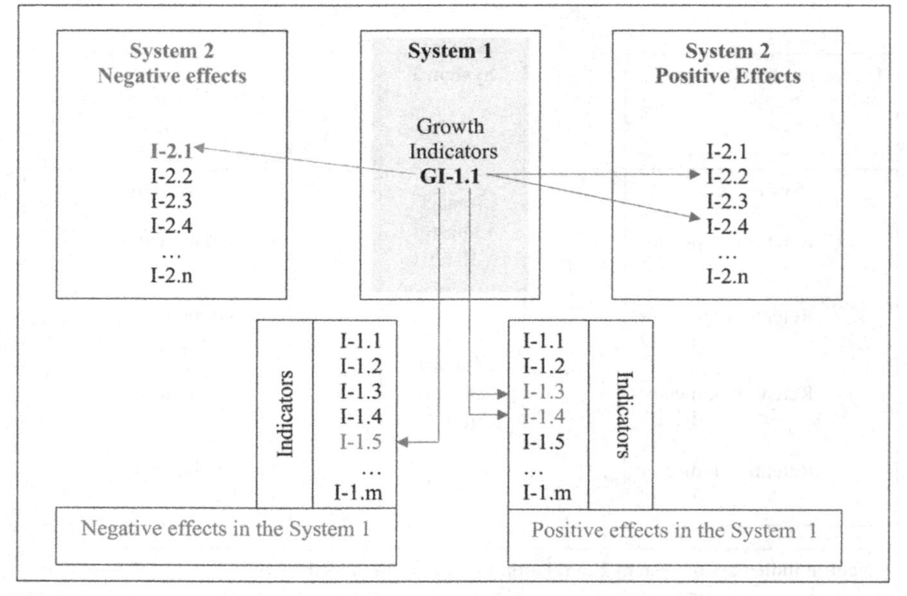

FIG. 10

Diagram of relations of the growth indicator of System 1 on the achieved positive and negative effects in Systems 1 and 2.

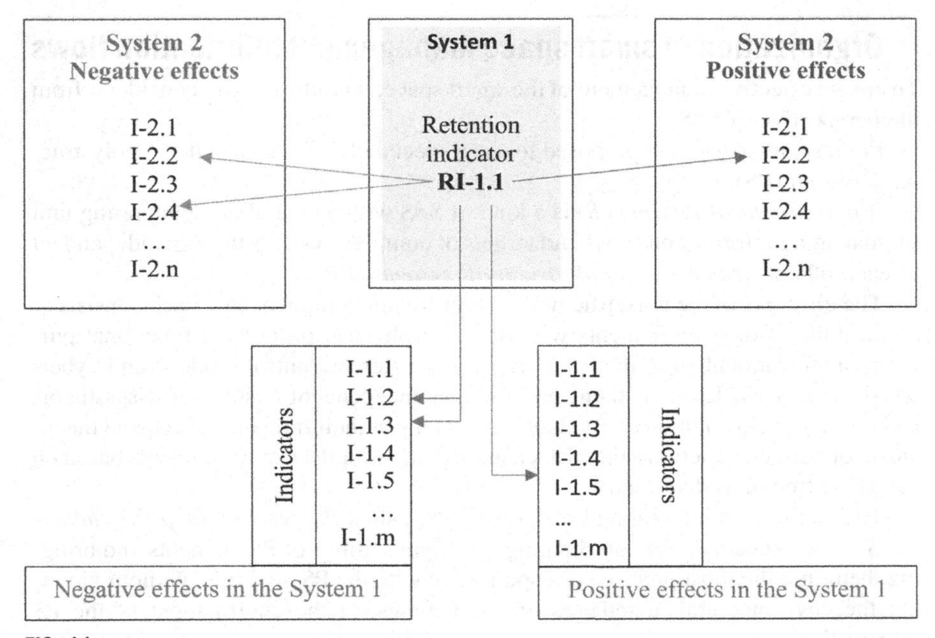

FIG. 11

Diagram of relations of System 1 containment indicator on achieved positive and negative effects in Systems 1 and 2.

are accumulated or there are predictive expert judgments) each indicator is credited with its corresponding intensity of influence on value changes.

If the data on the impact intensity is poorly studied, a point system of estimates can be used (0—no impact, 1—weak impact, which can be neglected; 2—small impact, no adjustment of the system functionality; 3—medium impact, which requires adjustments to the system functionality; 4—strong impact, which requires significant adjustments to the system functionality; 5—critical impact requiring immediate response).

It is important to synchronize the data to establish dependencies. This is achieved through timing (calendar of events and synchronization of data received from the stages of the processes) and IoT devices included in the change monitoring system.

Measurement lags can be changed, achieving the appropriate accuracy (calibration) of the measurements of the achieved effects, both positive and negative. System management should be considered in terms of resource availability of reactions to operational, tactical and strategic deviations from the specified target benchmarks of the system development indicators and indicators of system containment. The fewer reactions to deviations, the more effective the management system is.

The implementation of controlling the values of impact (II), growth (GI), and containment (RI) indicators to assess the harmonious state of interaction of the systems included in the smart space can be carried out by the polymetric system (PS).

6 Organization of smart space management information flows

To ensure effective management of the smart space, it is advisable to consider it from the perspective of SoS.

For a smart space, it is proposed to use a special class of SoS, called a polystructural system (PS).

A polystructural system (PS) is a kind of SoS with a centralized integrating unit for managing information flows and targets of complex systems that are independent of each other (*elements of a polystructural system—*EPS_i).

The elements of the polystructural system are not homogeneous. Each consists of a multitude of diverse elements with different physical properties, functional purpose, organizational structures and processes, different control models, and cyber-physical devices. Each of them has its own independent system of assessments and measurements of indicator values, algorithms of their transformation, and mechanism of correcting actions aimed at changing values of the achieved results based on the given tree of system goals.

The centralized integrating block of the PS, called *the body of the polystructure (BPS)*, is a mechanism for coordinating the target settings of PS elements and bringing them into the most precise correspondence with the PS goal tree, without violating the environmental boundaries of its elements, in the environment of the PS information space.

The digital environmental frontier of a PS element is understood as the state of the values of its indicators such that they are within the limits of acceptable deviations from the established normative values of the system indicators in accordance with its goal tree.

The body of the polystructure (BPS) is connected by PS elements through informational two-way gateways. Any interaction of two or more PS elements is carried out through the integrating center of PS (the body of the polystructure). It records changes, validates input data into terms, concepts, indicators and units of change of PS. The body of the polystructure accumulates and systematizes data on the state of its elements, compliance and direction of the values of the achieved indicators of PS elements; predictive modeling of scenarios of effective interaction of PS elements is conducted in order to develop solutions according to the PS goals tree.

Fig. 12 presents a PS information flow control structural diagram, which considers the interaction between the PS integrating unit (BPS) and one of its elements (EPS_i), where:

- GT_BPS—the goal tree of the BPS polystructure body;
- GT_EPS_i—goal tree EPS_i;
- PMS_BPS—polymetric BPS;
- MS_EPS_i—EPS_i metric system;
- DT_BPS—digital twin of BPS;
- DT_ EPS_i—digital twin of EPS_i;
- $V_{BPS\_EPS_i}$—data conversion validator from MS_EPS_i to PMS_BPS and vice versa.

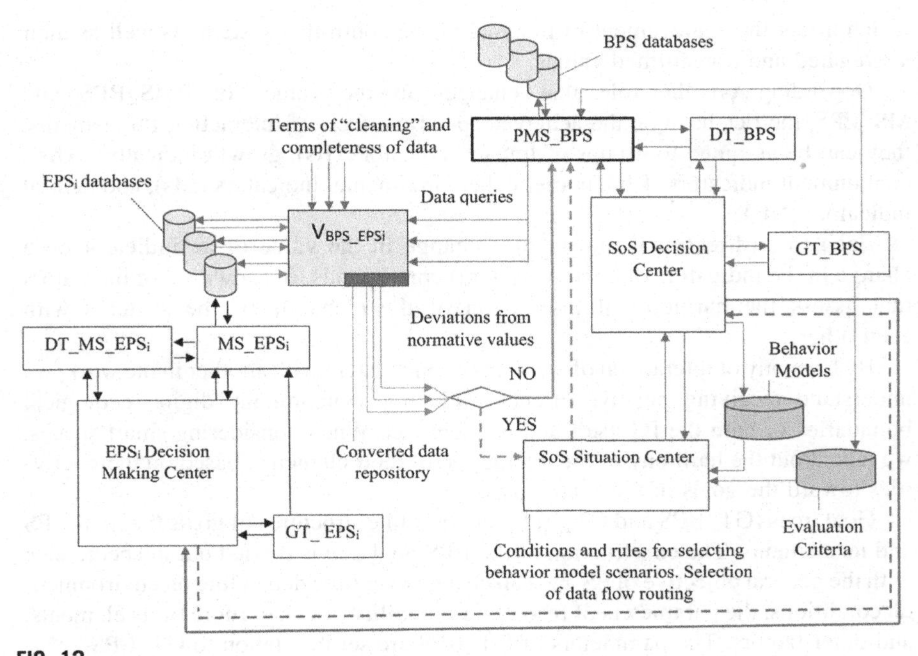

FIG. 12

PS information flow control mechanism structure diagram.

The Polymetric Structure Body System (PMS_BPS) is a mechanism built into the PS that performs the function of shaping and coordinating data flows both in the BPS and between the BPS and EPS_i. PMS_BPS tasks include:

- assigning "cleanup" conditions and completeness to a set of input data streams based on specified business logic;
- defining the rules for translating input data into a single measurement system and their further processing (conversion, aggregation) to obtain metrics, the main indicators used in decision-making at the operational, tactical, and strategic levels of PS management;
- monitoring of deviations for specified time intervals of actual values of metrics from their target values, which are normative values of the objectives tree PS;
- calculation of the size of deviations of actual metric values to determine the type of control action;
- establishing rules for transmitting signals on the emergence of abnormal situations;
- routing of data on existing or predicted deviations of metrics values to the relevant decision-making centers.

Metrics are formed from the number of indicators that characterize the state of a property or group of properties of the system objects, which can be regulated and which affect the achievement of the goal of the controlled system, as well as their aggregated and transformed values.

Depending on the role that changes in the values of PMS_BPS and MS_EPS_i metrics have on the achieved specific effects of interacting PS elements, they can be assigned to groups of impact indicators (IIs), growth indicators (GIs), containment indicators (RIs), numerical environmental indicators (EIs), and neutral indicators (NIs).

Each role (directional impact of a change in the value of an indicator on a change in the indicators of another system) corresponds to its own set of indicators that assess the harmony of interaction of the elements of the systems with each other.

By harmony of interaction of elements of one system with another in the work, we understand receiving positive effects while not violating the digital ecological boundaries of both the PS itself and its elements. When considering smart spaces, we talk about the harmony of the interaction of their elements, based on their activities toward the goals of the smart space.

Goal trees (GT_BPS and GT_EPS_i) are tree-like structures of goals facing the PS and its elements. The construction of the BPS goal tree is carried out in accordance with the general objective of PS and can change over time depending on environmental conditions, the parameters of resource availability of one or more of its elements, and other factors. The parameters of GT_BPS are set in relation to GT_EPS_i.

The hierarchical structure of GT_EPS_i contains the branches of the goal tree corresponding to the functions of the EPS_i objects and its leaves are the normative values of the indicators characterizing the state of individual properties of the objects changing during the execution of one or more processes. These normative values of indicators further act as a basis for calculation of deviations of the actual values obtained.

Several leaves in the structure of the goals tree can correspond to each object of the controlled system. That is, the state of each object or object property can be measured and evaluated by one or more indicators. The GT_EPS_i goal tree can be periodically restructured depending on changes in the target vector of PS development and its individual elements.

When adjusting the alignment of BPS and EPS_i target vectors, the normative values of GT_BPS leaves are loaded into PMS_BPS, and serve as the basis for calculating the absolute and relative deviations of the actual values of indicators of the state of objects and parameters of the stages of processes analyzed in BPS for decision-making to ensure the effectiveness of PS functioning and development. Then the corrective adjustment of the GT_EPS_i target tree leaf values is performed (see Fig. 13).

Digital Twin (DT)—“a complete virtual description of a physical product with micro- and macro-level accuracy”. A digital twin allows simulation of a given scenario in a virtual environment, to investigate the emerging processes and the consequences of the processes, to assess the level of harmony of the interaction of PS elements on each other.

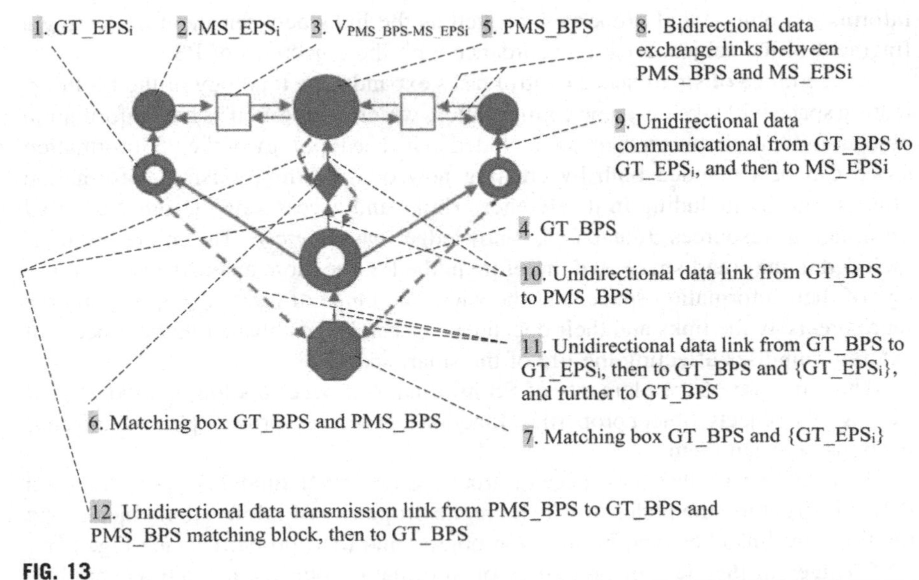

FIG. 13

Principal diagram of information interaction GT_BPS—GT_EPS$_i$, PMS_BPS—MS_EPS$_i$.

The digital twin should act as a predictor of proactive management. This function must become dominant in the digital twin to maintain a harmonious state of the system.

The proposed diagram of information support for the maintenance of smart space functionality allows it to be implemented as a reference model in a computer system.

7 Topology of smart space information space

The smart space information space topology corresponds to the digital environmental boundaries of the PS information space topology and is defined by its constituent:

- subject area objects, their metamodels, functional and structural relationships between them;
- a set of structured and weakly structured data describing the properties of the objects of the subject area and the links between them;
- methods of data processing and achieving the objectives of the management of objects in the subject area;
- other elements that perform the creation, storage, modification, processing, transfer, and deletion of data within a given information space.

The metamodel of information space topology in relation to each considered subject area PS is a set of information objects included in the aggregates (fragments of information object) and properties, as well as the links between them, reflecting a fragment of the subject area in accordance with the objectives of PS.

As the range of smart space control tasks expands, the topology of the PS information space will take on a new configuration, which is modified as new information objects and links between them are included in it. The topology of the PS information space can be expanded both by creating new or upgrading existing information objects, and by including in it references (links and access keys) to other external information resources (databases, knowledge bases, etc.). The more systems included in the smart space, and therefore in the PS, the more extensive is the topology of their information space, and the wider the range of tasks it supports and the more precisely the links and their dependencies can be established in the process of external impulses affecting the life of the smart space.

The structure of the elements of PS information space topology includes: goal trees, goals, objects, object properties, functions, processes, process phases, and connections between them.

The topology of the smart space information space metamodel is represented as a network structure, the nodes of which are the properties of information objects, and the links are links between information objects and their properties (see Fig. 14).

Changes in the state of properties of information objects, as well as stability, intensity, and direction of relations of properties of information objects, allow the intensity and direction of the influence of some indicators of the system on others to be estimated, and thus their values to be managed.

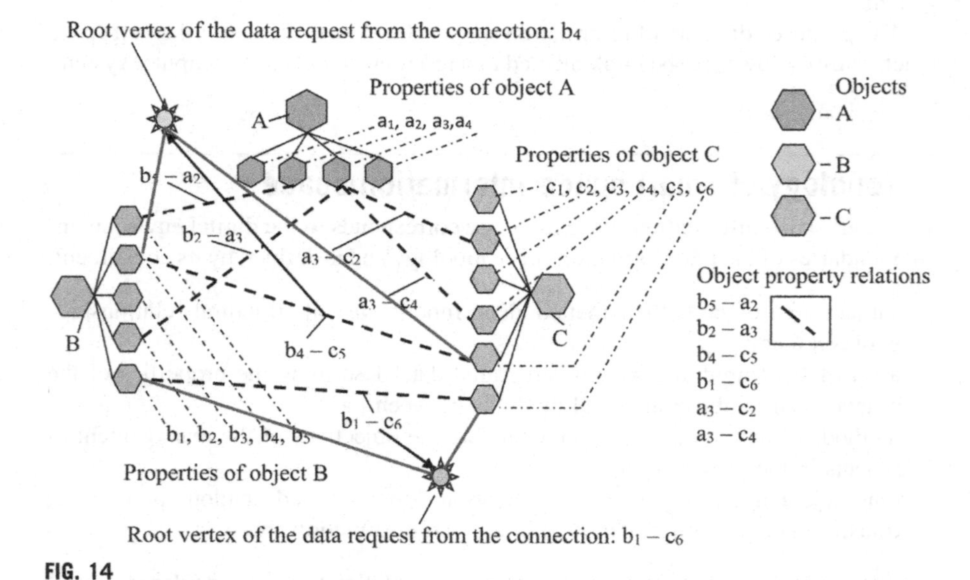

FIG. 14

Example of a scheme for retrieving data about the state of object properties.

Assigning as the root nodes of the query the indicators of influence of an element of one system included in PS on its other systems, it is possible to trace the chain of passing changes and to develop recommendations, with the help of DT, concerning the choice of normative target benchmarks of interacting systems, corresponding to the requirements of a harmonious state of the smart space.

8 Conclusions

The development of smart spaces will be associated with the gradual build-up of a certain territory by cyber-physical systems to automate certain activities affecting humans.

The interactions of smart space elements, which can be both antagonistic and friendly interactions, are considered from a system perspective. It is shown that in order to create a control system for smart space elements, it is necessary to use methods of SoS theory and practice.

A polystructural system (PS), which can act as an integrator of individual elements of smart spaces, is described. Its structure is presented. The process of smart space integration is carried out through the setting of targets of each system in such a way as to smooth out possible contradictions when there are severe constraints on shared resources.

The combination of target settings that leads to minimal damage to interacting systems is called its harmonious state. In this regard, an important place is given to digital twins of smart space elements, which provide prediction of the results of the active influence of systems on each other.

For the practical implementation of the proposed solution, a model of information support for the smart space integrator was developed.

9 Perspective and future work

To some extent, existing social networks such as Facebook, Telegram, etc. have already created a semblance of a smart space on a global scale. The trend toward smart assistants and smart human life support systems will undoubtedly evolve and improve. There are already trends toward integrating various information systems for faster results in achieving goals, including the exchange of data with other systems.

In the future it is expected that the theoretical foundations of intersystem interaction as applied to polystructural systems will be developed. The theory and practice of digital twins should be developed in the direction of their participation in the management of both elements within smart spaces and various smart spaces with each other in unified spatial and temporal coordinates.

For practical implementation it is supposed to create appropriate services for the formation of the topology and metamodels of the topology of the information space smart space.

References

[1] A. Asvadi, A. Mitriakov, C. Lohr, P. Papadakis, Digital twin driven smart home: a feasibility study, in: 9th International Conference On Smart Living and Public Health (ICOST 2022), Paris, France, 2022 (hal-03691144) https://hal.archives-ouvertes.fr/hal-03691144.

[2] I. Roussaki, K. Doolin, A. Skarmeta, G. Routis, J.A. Lopez-Morales, E. Claffey, et al., Building an interoperable space for smart agriculture, Digit. Commun. Netw. 9 (1) (2023) 183–193. Special Issue.

[3] K.B. Clark, Smart device-driven corticolimbic plasticity in cognitive-emotional restructuring of space-related neuropsychiatric disease and injury, Life 12 (2) (2022) 236.

[4] A. Cocchia, Smart and digital city: a systematic literature review, in: R.P. Dameri, C. Rosenthal-Sabroux (Eds.), Smart City, Springer, Cham, 2014, pp. 13–43.

[5] S.-L. Shaw, D. Sui, Understanding the new human dynamics in smart spaces and places: toward a splatial framework, Ann. Am. Assoc. Geogr. 112 (2019) 505–521.

[6] F.P. Appio, M. Lima, S. Paroutis, Understanding Smart Cities: innovation ecosystems, technological advancements, and societal challenges, Technol. Forecast. Soc. Chang. 142 (2019) 1–14.

[7] S. Bansal, A. Gupta, IoT-enabled intelligent traffic management system, in: IoT Based Smart Applications, Springer, Cham, 2023, pp. 89–111.

[8] A. Caramizaru, A. Uihlein, Energy Communities: An Overview of Energy and Social Innovation, European Commission, 2020.

[9] M. Clark, M.W. Newman, P. Dutta, ARticulate: one-shot interactions with intelligent assistants in unfamiliar smart spaces using augmented reality, Proc. ACM Interact. Mob. Wearable Ubiquitous Technol. 6 (1) (2022) 1–24.

[10] S. Diefenbach, L. Christoforakos, D. Ullrich, A. Butz, Invisible but understandable: in search of the sweet spot between technology invisibility and transparency in smart spaces and beyond, Multimodal Technol. Interact. 6 (10) (2022) 95.

[11] G. Dodig-Crnkovic, S. Ljungblad, M. Obaid, 4th space as smart information ecology with design requirements of sustainability, ethics and inclusion, Proceedings 81 (1) (2022) 124. MDPI.

[12] J. Käll, Posthuman Property and Law: Commodification and Control Through Information, Smart Spaces and Artificial Intelligence, first ed., Taylor & Francis, 2022.

[13] M.J. Kim, H.J. Jun, Intelligence sensors and sensing spaces for smart home and environment, Sensors 22 (8) (2022) 2898.

[14] Y. Koucheryavy, S. Balandin, S. Andreev (Eds.), Internet of Things, smart spaces, and next generation networks and systems, 21st International Conference, NEW2AN 2021, and 14th Conference, ruSMART 2021, St. Petersburg, Russia, August 26–27, 2021, Proceedings, Vol. 13158, Springer Nature, 2022.

[15] J.W. Lee, A. Helal, Modeling and reasoning of contexts in smart spaces based on stochastic analysis of sensor data, Appl. Sci. 12 (5) (2022) 2452.

[16] L. Mora, R. Bolici, M. Deakin, The first two decades of smart-city research: a bibliometric analysis, J. Urban Technol. 24 (1) (2017) 3–27.
[17] J.R. Gil-Garcia, T.A. Pardo, T. Nam, A comprehensive view of the 21st century city: smartness as technologies and innovation in urban contexts, in: Smarter as the New Urban Agenda, Springer, Cham, 2016, pp. 1–19.
[18] M.M. Goh, K. Irvine, M.I. Ung, To bus or not to bus: structural equation modelling of ridership perceptions among university students as a planning tool to increase use of public transit in Phnom Penh, Cambodia, J. Archit. Plan. Res. Stud. 19 (2) (2022) 105–124.
[19] N. Streitz, Reconciling humans and technology: the role of ambient intelligence, in: European Conference on Ambient Intelligence, Springer, Cham, 2017, pp. 1–16.
[20] E.G. Carayannis, D.F. Campbell, Conclusion: smart quintuple helix innovation systems, in: Smart Quintuple Helix Innovation Systems, Springer, Cham, 2019, pp. 51–54.
[21] Frost & Sullivan, Smart Cities, 2019. https://ww2.frost.com/wp-content/uploads/2019/01/SmartCities.pdf.
[22] S. Taweesaengsakulthai, S. Laochankham, P. Kamnuansilpa, S. Wongthanavasu, Thailand smart cities: what is the path to success? Asian Polit. Policy 11 (2019) 144–156.
[23] A. Caragliu, C.F. Del Bo, Smart innovative cities: the impact of Smart City policies on urban innovation, Technol. Forecast. Soc. Chang. 142 (2019) 373–383.
[24] N. Komninos, C. Kakdery, L. Mora, A. Panori, E. Sefertzi, Towards high impact smart cities: a universal architecture based on connected intelligence spaces, J. Knowl. Econ. 13 (2022) 1169–1197.
[25] T. Yigitcanlar, M. Foth, M. Kamruzzaman, Towards post-anthropocentric cities: reconceptualizing smart cities to evade urban ecocide, J. Urban Technol. 26 (2019) 147–152.
[26] C. Kuehnl, D. Jozic, C. Homburg, Effective customer journey design: consumers' conception, measurement, and consequences, J. Acad. Mark. Sci. 47 (2019) 551–568.
[27] A. Chio, D. Jiang, P. Gupta, G. Bouloukakis, R. Yus, S. Mehrotra, N. Venkatasubramanian, Smartspec: customizable smart space datasets via event-driven simulations, in: 2022 IEEE International Conference on Pervasive Computing and Communications (PerCom), 2022, March, pp. 152–162.

CHAPTER

Assessment and monitoring of human emotional state and behavior in a smart space environment

12

Vladimir N. Shvedenko

FSBUN All-Russian Institute of Scientific and Technical Information of the Russian Academy of Sciences (VINITI RAS), Moscow, Russia

1 Introduction

The purpose of smart spaces is to provide a comfortable human condition. In these spaces there is a constant information exchange between subjects and objects of these systems. In addition to explicit mutual influence, this information flow affects unconscious areas of subjects' psyches, changing their emotional state, or putting them into a state of stress. The influence can be directed to achieve certain goals, or can be spontaneous, observed in particular in cyberbullying in social networks [1]. Regardless of the goals of such exposure, its effects are destructive to people's mental and physiological state, and significantly affect their ability to make rational decisions.

Some studies suggest that computers should analyze and respond to human emotions [2–4].

An important issue in the assessment of emotional state is the identification and establishment of a measurement system for human emotions. For example, to assess a person's emotional state, it has been proposed to measure their head movement, the content of the text they write, and a number of other indicators that can be found in various works [5–11].

It is believed that human-computer interaction would be more natural if computers could perceive and respond to human nonverbal communication such as emotions [12]. Facial expression analysis is most commonly used to assess a person's emotional state; see references [13,14]. A detailed review of human emotion recognition can be found in Ref. [15].

Nowadays one can find quite a large number of techniques and various psychological practices aimed at maintaining the stable emotional state of a person;

Smart Spaces. https://doi.org/10.1016/B978-0-443-13462-3.00020-0

however, in relation to smart spaces, it is necessary to develop a special system of continuous control of a person's emotional state. In this chapter we propose to consider mechanisms of protection of subjects of smart spaces, both decision-makers with regard to its functioning, and participants and consumers of information resources in this environment, so as to prevent the manifestation of destructive factors of information flows on their emotional state that may cause stressful situations.

To determine the necessary methods of protection against external influences on the emotional state of a person, we analyzed the most vulnerable groups of subjects to influence, the existing threats, and mechanisms for determining the impact on the emotional state of a person.

2 Definitions and notations

We will consider a socio-computer system as a set of software and hardware united through local and software computer networks, and a set of users.

Bots are examples of such hardware and software. By the term socioactive bot we will understand an automated software object that has certain algorithms of action and is used to influence the subjects of a socio-computer system to achieve its goals [16].

One of the attributes of a socio-computer system is the presence of empathy. In a socio-computer system it is possible, using certain external manifestations, to detect the emotions experienced by the subject.

A feature of socio-computer systems is the ability to control and manage the emotional state of subjects. Therefore we introduce the concept of computer empathy as the ability of the object of a socio-computer system, by detecting various external manifestations of the subject, to determine the subject's emotional states.

3 Literature review

In the following we will use well-known psychological concepts: an emotional state, emotion, emotional peace, joy, surprise, suffering, fear, anger [17].

To determine a subject's current emotional state, markers are used: facial expression, motor skills, electroencephalogram, breathing, alpha rhythm, beta rhythm, delta rhythm, and gamma rhythm [18].

In socio-computer systems, a new class of tools has emerged that can be used to determine a person's emotional states.

According to Kragel [19], a neural network was developed that was able to determine a person's emotional response from an image with high-enough accuracy—more than 95% of correct answers from 25,000 tests.

4 Problem

In the classical variant of man-machine system, only mutual influence of the subject and the software-hardware part is observed. Emotional states are not considered at all. However, a person's emotional state can have a significant impact on their performance and management decisions.

Some groups of subjects of socio-computer systems are more vulnerable to external influence on emotional state, including:

- subjects of socio-computer systems with insufficient personal maturity (children or adolescents);
- people with functional disorders of the central nervous system, unable to adequately assess the incoming information;
- individuals prone to conformist thinking.

The emotional state of these groups changes most rapidly, and they are more likely to become victims of external influences. At the same time, the complexity of determining exposure is due to the large number of ways to influence a person's emotional state. Influence can be through a conversation, a certain color palette, special sound accompaniment, or through other channels of information transfer. Consequently, the impact should be determined by studying changes in a person's emotional state.

Modern socio-computer systems are an accessible tool used by all social groups. In today's online community, cyberbullying is becoming an increasingly common phenomenon, negatively affecting both individual subjects of socio-computer systems and the systems themselves, distorting and clogging the information exchange between users. And harmful online challenges such as "death groups" and "blue whale" cause significant damage to the mental health of users who have been affected by these phenomena.

Emotional states of users of socio-computer systems can be considered as an integral element of socio-computer interaction of the information system (object), and a set of users (subjects), which interact with each other to achieve common or different goals, and in the process of interaction, mutually influence each other.

5 Proposed solution

To evaluate a person's emotional state, we propose a system of coordinates. As the beginning of the coordinate system, we take a person in the state of emotional rest. Then we introduce two axes: positive and negative emotional background. The whole range of emotions is quite complex, but the following discrete emotional states are introduced to mark the axes, between which their refined characteristics can be found [20].

To determine the emotional states, a system of evaluation of their external manifestations and technical means for their registration is proposed. For creation of the system of measurements, a classification of external manifestations of emotional states (Table 1) was carried out.

The offered classification allows to control the emotional state on several markers simultaneously.

Having the data on markers of emotional states, which it is possible to detect by different devices, the possibility of objective detection of the current emotional state of a person in a socio-computer system has appeared.

Table 2 shows devices for detecting the previously mentioned characters of external manifestations of emotional states. Computer sensors are physical devices,

Table 1 External manifestations of emotional states.

Emotional state	Nature of external manifestations	External manifestations
Emotional rest	Respiration	Ratio of inhalation time to exhalation time is 0.43
	Circulation	Pulse rate is 60–70 beats per minute
Concentration at rest	Respiration	Ratio of inspiratory time to exhalation time equal to 0.3, tends to decrease as concentration increases
	Circulation	Pulse rate is equal to 60–70 beats per minute
	Mimicry	Eyebrows slightly raised or lowered, while eyelids slightly dilated or narrowed
	Brain electrical activity	Gamma rhythm is decreased throughout the cortex
Surprise	Respiration	Ratio of inhalation time to exhalation time is 0.71
	Circulation	Pulse rate is 70–80 beats per minute
	Mimicry	The raised eyebrows form wrinkles on the forehead, the eyes are dilated, and the ajar mouth has a rounded shape
Fear	Respiration	The ratio of inhalation time to exhalation time is 0.75
	Circulation	The pulse rate is more than 90 beats per minute. In addition, blood pressure rises 15–30 mmHg above normal
	Mimicry	The eyebrows are slightly raised but straight, their inner corners are shifted and horizontal wrinkles run across the forehead, the eyes are dilated, with the lower eyelid tense and the upper slightly raised, the mouth may be open and its corners pulled back, pulling and straightening the lips over the teeth (the latter is just indicative of the intensity of the emotion…); when only the aforementioned eyebrow position is present, it is controlled fear
	Brain electrical activity	Suppression of alpha rhythm (8–13 Hz), significant amplification of beta rhythm (18–30 Hz), desynchronization in the band alpha-2 (10–12 Hz) and beta-1 (12–18 Hz)
	Body temperature	Skin temperature decreases
Anger	Circulation	Slight increase in pulse and blood pressure
	Hearing	Increased perception of sounds in the right ear due to more intensive work of the left hemisphere in relation to the right
	Mimicry	Forehead muscles moved inward and downward, arranging a threatening or

Table 1 External manifestations of emotional states—cont'd

Emotional state	Nature of external manifestations	External manifestations
		frowning expression, nostrils dilated and wings of the nose raised, lips either tightly compressed or pulled back, assuming a rectangular shape and exposing clenched teeth, face reddened
	Printed speech	Use of short words, significant increase in speed of typing and sending messages, occurrence primarily of syntactic errors in speech, increased force of keyboard presses when typing
	Body temperature	Slight elevation in consequence of redistribution of blood flow in favor of muscles and brain
	Brain electrical activity	Strengthening of alpha rhythm (8–13 Hz), significant strengthening of beta rhythm (18–30 Hz)
Joy	Typed speech	Repetition of vowel letters in writing, significant increase in typing and texting speed, occurrence of primarily grammatical errors in speech.
	Mimicry	The lips are curved and the corners of the lips are pulled back; fine wrinkles appear around the eyes; facial expressions become more active
	Motor skills	Increased muscle activity, unconscious small movements, increased motor skills
	Body temperature	Body temperature is elevated
	Electrical activity of the brain	Increased alpha rhythm (8–13 Hz)
Sadness (suffering)	Motility	
	Body temperature	Motions sluggish, slowed, no involuntary movements
	Oral speech	Skin is cooled due to outflow of blood from the muscular cover
	Mimicry	Verbal speech is slow, quiet. Absence of emotional fluctuations during conversation.

allowing a computer (or other complex technical device) to register various manifestations of emotional state markers.

The emotional state of a subject of a socio-computer system can be detected by sensors that are located in devices used while in the system. Such devices as a fitness bracelet or a smart watch allow detecting changes in pulse or blood pressure, a

Table 2 Computer sensors for detecting markers of subject's emotional state.

Nature of external manifestations	Device
Respiration	Microphone, video camera
Blood circulation	Sphygmograph, tonometer
Mimicry	Video camera
Body temperature	Thermometer
Hearing	No device at this time, only indirect detection possible (subjective audiometry)
Typewritten speech	Keypad with force-tracking function
Oral speech	Microphone
Motor skills	Video camera, mouse, fitness bracelet
Brain electrical activity	Neurodetector (neurointerface)

camera records changes in a person's facial expressions and motor skills, a keyboard allows determining the degree of concentration through the frequency of spelling mistakes and the speed of writing. The aggregate of these parameters allows the current emotional state of the person to be determined with respect to the state of emotional rest.

Having determined the external influence of the socio-computer system on a subject's emotional state, the subject should be warned about possible negative influence, or immediately limited from exposure to the source of external influence, if the subject belongs to vulnerable groups. Likewise, measures should be taken to return the subject to a state of emotional rest.

Thus, by analyzing a person's emotions over a long period of time, it is possible to identify a predisposition to certain emotional states and, upon a sharp atypical change, to assume an external influence from the socio-computer system.

6 Analysis (qualitative/quantitative)

When working in modern network environments such as computer games, online forums, or social networks, the occurrence of situations that cause a person emotional stress, which in turn affects the somatic state, is possible.

In order to prevent the occurrence of stress in time, it is necessary to offer a method of detecting a prestress situation and a way of preventing it.

In order for a person to escape from a prestress situation to a state of emotional calm, it is suggested to use music chosen according to the individual characteristics of a person's personality.

The fastest and easiest way is to listen to musical compositions through speakers or other sound output device from a computer. Moreover, information systems such as Apple Music and others are capable of selecting a music collection suitable for each individual.

Perception of music is one of the main elements of musical psychology, because the perception of music is accompanied by a very strong emotional experience. Therefore an individually chosen musical collection can be an effective tool for controlling and managing stressful situations.

In order to determine the dependence of a person's emotional state on listening to music, a study of music's impact on the emotional state of a person was conducted.

During the research an anonymous online survey of a group of social network users was conducted, including questions about the age of the survey participants, their musical preferences, including frequency of listening to music and preferred styles of music, as well as their testing on the PSM-25 scale of psychological stress to measure stressful feelings based on somatic, behavioral, and emotional traits.

The PSM-25 scale was chosen because it focuses not only on the inner feelings of those questioned, but also on their somatic readings, such as headaches, feeling cold, etc., which are easily recorded by the individual.

The question about the participant's age was necessary to account for age-related changes in music perception. It is known that older people react less emotionally to the harmonious and disharmonious aspects of a piece of music. Also, with age, manifestations of stress become more distinct and obvious, and the dependence of the emotional state on external stimuli becomes weaker.

The questionnaire also asked about other ways the interviewees cope with stress besides listening to music.

When analyzing the data obtained, the questions included in the stress level scale were processed:

$$R = \sum x(i),$$

where R, result (25–200); i, question number (1–25); x, score for the answer (1–8).

Then, according to the rules of interpretation of this test, the test results were compared with the control values, and for clarity, a graph of the level of stress of the subjects was drawn (Fig. 1).

This figure shows overlapping scales for frequency of music listening (the rarer the listening, the higher the value of the scale) and stress level.

According to research, as people age, their stress tolerance decreases, and consequently their stress level increases, according to the graph in Fig. 2.

The age of the survey participants and their use of other (not related to listening to music) ways of coping with stress are additional variables.

Listening to music can be considered as a technique to prevent a stressful situation, but when high levels of emotional tension are reached, listening to music is no longer able to lead a person away from stress to a state of emotional peace. That is why when a person works with computer systems, it is important to trace the moment of occurrence of slight stress, to inform the user about it and offer him/her the choice of listening to a musical composition.

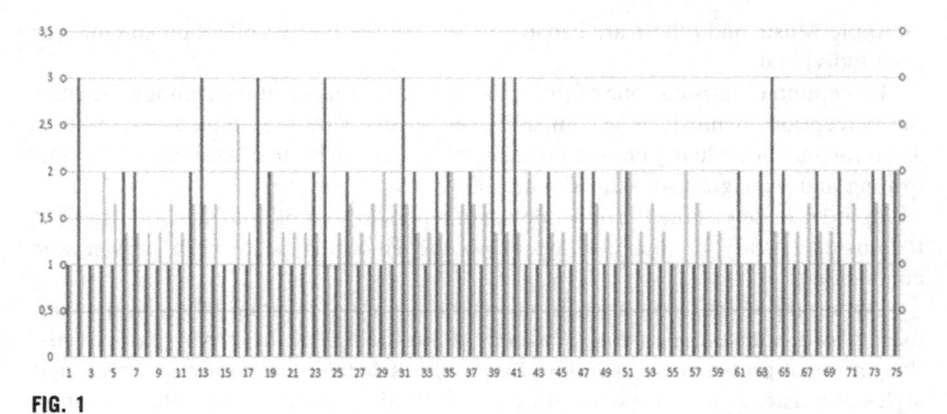

FIG. 1

Graphs of stress level and frequency of listening to music (*blue*—stress level, *orange*—frequency of listening to music).

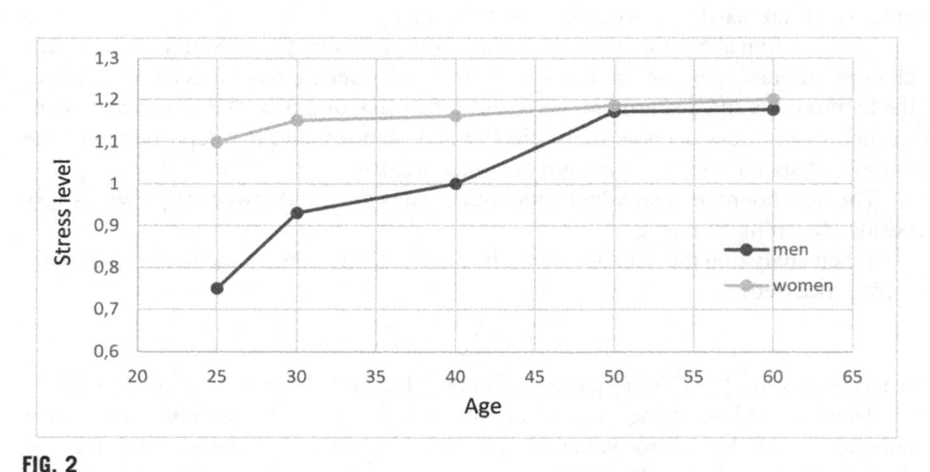

FIG. 2

Stress level at rest.

The moment of the appearance of light stress is tracked by such signs as frequent keystrokes, grammatical and lexical errors when typing letters, or sharp mouse movements, which can be determined by the user's computer.

7 Application

For distribution of the given mechanism of prevention of external influences on the emotional state of the person among a considerable number of users, an automated system for detection and prevention of such influences was developed. The work of the system of counteraction of external influences on the emotional state of the person includes the following stages.

Training of the system. Includes training of mechanisms for determination of facial expressions and motor skills, detection of pulse and other physiological indicators, allowing the emotional state to be detected.

Detection of emotional rest. Implies determination of the physiological indicators of the person at the moment of emotional rest, calculation of errors, and deviations from the normal state of various indicators during the transition to a different emotional state.

System adjustment. Modern information systems can perform the function of behavior and self-correction. Typical human emotional state indicators, which depend on temperament and character, can change under the influence of various factors. This stage assumes that the system will take into account both long-term user characteristics and short-term characteristics to increase its degree of objectivity.

Detection of external influences. With the help of various sensors, the system detects changes in the pulse, blood pressure, facial expressions, and other external manifestations of the person's emotions. If the combination of these indicators shows a change in the emotional state, and this change is atypical for this situation, the system detects it as an external influence. It stores the new external emotional state, self-corrects if necessary, and sends a command to neutralize the external influence.

Counteracting the external influence. At this point, the most effective method of counteracting the external influence on the emotional state is to inform the subject of the negative influence, and limit it from the source. Playing a musical composition also contributes to returning the user to a state of emotional rest.

8 Conclusions

In the course of this work groups of users of socio-computer systems that are most vulnerable to external influences on the emotional state of a person were identified. Threats from the information environment, such as the Internet, have been identified, and a mechanism for identifying and counteracting external influences has been proposed. An algorithm for detecting and correcting a person's emotional state has been proposed.

9 Perspective and future work

As a person in a smart space system is typically surrounded by computer systems, there is a threat of their impact on the person's emotional state. The realization of threats to affect a person's emotional state over the long run could be as powerful as the effects of biological or chemical weapons. At the same time, the financial cost of such action will be minuscule compared to the implementation of other types of threats to human health. The threat is to be expected both directly and indirectly through computer bots or other, not yet invented, psychotronic influences. From this point of view, there is a need to create special devices, like fitness bracelets, which will protect people from the threats discussed in this article.

We should also consider typical solutions to protect the emotions of people depending on the characteristics of their ethnic, national, social, and other groups.

References

[1] L. Wartberg, R. Thomasius, K. Paschke, The relevance of emotion regulation, procrastination, and perceived stress for problematic social media use in a representative sample of children and adolescents, Comput. Hum. Behav. (2021), https://doi.org/10.1016/j.chb.2021.106788.

[2] L. Cai, J. Dong, M. Wei, Multi-modal emotion recognition from speech and facial expression based on deep learning, in: Chinese Automation Congress (CAC), Shanghai, China, 2020, pp. 5726–5729, https://doi.org/10.1109/CAC51589.2020.9327178.

[3] R.W. Picard, Affective Computing, 1997, https://doi.org/10.7551/mitpress/1140.001.0001.

[4] M. Spezialetti, G. Placidi, S. Rossi, Emotion recognition for human-robot interaction: recent advances and future perspectives, Front. Robot. AI 7 (2020), https://doi.org/10.3389/frobt.2020.532279.

[5] M. Dubey, P.L. Singh, Automatic emotion recognition using facial expression: a review, Int. Res. J. Eng. Technol. 3 (2016).

[6] D. Joshi, M.B. Zalte, Speech emotion recognition: a review, IOSR J. Electron. Commun. Eng. 4 (2013) 34–37, https://doi.org/10.9790/2834-0443437.

[7] H. Liu, Y. Zhang, Y. Li, X. Kong, Review on emotion recognition based on electroencephalography, Front. Comput. Neurosci. 15 (2021) 758212, https://doi.org/10.3389/fncom.2021.758212.

[8] A. Seyeditabari, N. Tabari, W. Zadrozny, Emotion detection in text: a review, ArXiv (2018), https://doi.org/10.48550/arXiv.1806.00674.

[9] L. Stark, J. Hoey, The ethics of emotion in artificial intelligence systems, in: Proceedings of the 2021 ACM Conference on Fairness, Accountability, and Transparency, 2021, https://doi.org/10.1145/3442188.3445939.

[10] G.K. Verma, U.S. Tiwary, Multimodal fusion framework: a multiresolution approach for emotion classification and recognition from physiological signals, NeuroImage 102 (2014) 162–172, https://doi.org/10.1016/j.neuroimage.2013.11.007.

[11] J. Wright, Suspect AI: Vibraimage, emotion recognition technology, and algorithmic opacity, ArXiv (2020), https://doi.org/10.2139/ssrn.3682874.

[12] C. Busso, Z. Deng, S. Yıldırım, M. Bulut, C.M. Lee, E. Kazemzadeh, S. Lee, U. Neumann, S.S. Narayanan, Analysis of emotion recognition using facial expressions, speech and multimodal information, in: International Conference on Multimodal Interaction, 2004, https://doi.org/10.1145/1027933.1027968.

[13] R. Kalaiselvi, P. Kavitha, K.L. Shunmuganathan, Automatic emotion recognition in video, in: 2014 International Conference on Green Computing Communication and Electrical Engineering (ICGCCEE), 2014, pp. 1–5, https://doi.org/10.1109/ICGCCEE.2014.6921398.

[14] A. Saxena, A. Khanna, D. Gupta, Emotion recognition and detection methods: a comprehensive survey, AI & Soc. 2 (2020) 53–79, https://doi.org/10.33969/ais.2020.21005.

[15] J.M. Garcia-Garcia, V.M. Penichet, M.D. Lozano, Emotion detection: a technology review, in: Proceedings of the XVIII International Conference on Human Computer Interaction, 2017, https://doi.org/10.1145/3123818.3123852.

[16] C. Gaser, R. Dahnke, CAT—A Computational Anatomy Toolbox for the Analysis of Structural MRI Data, 2016, pp. 336–348.

[17] A.K. Seth, Interoceptive inference, emotion, and the embodied self, Trends Cogn. Sci. 17 (2013) 565–573.

[18] T. Naselaris, K.N. Kay, S. Nishimoto, J.L. Gallant, Encoding and decoding in fMRI, NeuroImage 56 (2011) 400–410.

[19] P.A. Kragel, Emotion schemas are embedded in the human visual system, Int. J. Comput. Integr. Manuf. 32 (2018) 1–12.

[20] L.F. Barrett, E. Bliss-Moreau, Affect as a psychological primitive, Adv. Exp. Soc. Psychol. 41 (2009) 167–218.

CHAPTER

Digital twins for smart city 13

Małgorzata Pańkowska and Mariusz Żytniewski
Faculty of Informatics and Communication, Department of Informatics, University of Economics in Katowice, Katowice, Poland

1 Introduction and motivation

In general, a city involves multiple stakeholders having various interests and information processing requirements. They use the city information system to process data about the physical urban objects. Therefore information integration from different distributed sources is a common requirement in many city development projects [1]. Although the term digital twin (DT) was developed in the context of a product life-cycle management almost 20 years ago, this concept and terms for establishing a DT context are also applicable to the smart city domain. In general, the DT concept involves a physical entity being represented by their digital counterpart. Hence, the DT is always connected to its real, physical counterpart via sensors and impacts on the actuators. The sensors observe the state of the physical entity as well as the state of the physical environments wherein the entity is located.

Although several architecture models have been developed for the construction and implementation of DTs in different application domains, in this study the authors present the DT development approach based on business process modeling notation. In a smart city supported by digital twins, the feedback between the physical objects and humans enables enhanced data-driven decision-making and helps city managers evaluate community policies and initiatives through what-if scenarios. Therefore city decision-makers are able to anticipate citizens' behaviors and understand how cities should be supported by new information communication technology (ICT) under various economic, environmental, and social conditions.

The authors have proposed the following research methodology:

- Presentation of DT definitions, background knowledge, and technology context.
- Literature survey on smart city, digital twin, and new generation of information technology.
- Identification of knowledge gaps, problem formulation, and constraints recognition.
- Defining the objectives of a proposed solution.

Smart Spaces. https://doi.org/10.1016/B978-0-443-13462-3.00014-5

- Design and development of business processes.
- Demonstration to expound the practical application of a typical scenario.
- Discussion on positive effects of the proposed solution.

Therefore the authors start by using a systematic literature survey regarding digital twin definitions, interpretations, and applications, as well as construction frameworks. In the next section, the authors propose a conceptual model and approach to enable DT planning, designing, and implementation. A use case exemplifies the applicability of the proposed approach. This study's objective is to contribute to the standardization of DT concepts, while applying digital twins for smart city business processes.

2 Background, definitions, and notations

According to the United Nations organization's prediction, the world is expected to have 43 megacities including more than 10 million citizens each by 2030 [2]. Such formations create many challenges. Urbanization and the rapid adoption of ICT by cities are generating complex interdependencies between humans, infrastructure, and technologies, which is creating uncertainties, poor management decisions, and difficulties in foreseeing future states and planning the development of cities. Therefore cities need to identify business objects and manage complex interdependent processes. The challenge is to provide in a virtual world a realistic representation of the city and its components, while the physical city is represented with its buildings, roads, bridges, supporting infrastructure, and its citizens. A smart city is one certain solution that integrates the real world and the digital world through big data and cloud computing, as well as through intelligent objects, sensors, and services for overseeing people and things. According to Lehtola et al. [3], cities need city planning and urban development to ensure good living conditions, primary and secondary education institutions, healthcare and emergency services, and infrastructure to be properly maintained. The European Commission defines a Smart City as "a place where traditional networks and services are made more efficient with the use of digital and telecommunication technologies for the benefits of its inhabitants and business" [4].

Ouafiq et al. [5] have identified six basic components for making a smart city, i.e., an intelligent governance system to exchange information among citizens, municipal services, and emergency services, intelligent economy of products, smart mobility, intelligent environment reducing energy consumption, smart residents having access to education, and intelligent lifestyle with access to health and culture services. Lately, the concept of metaverse has arrived as a result of a combination of various technologies in different city contexts. That concept is to promote the integration of real world and digital world, and provide technically advanced support for the construction of a smart city [6]. However, according to Allam et al. [7] the metaverse represents a "parallel and virtual world," which presents ways of living and working as an alternative to smart cities in the future. Both the metaverse and smart city use artificial intelligence, big data, and Internet of Things, having the potential to redefine city design activities. For years, building information models (BIMs) have offered several ideas and decision-making capabilities throughout the life cycle of the built environment [8]. The BIM is to integrate the information about all the stakeholders involved in a project, providing information on spatial geometry, topological relationships, semantic attributes, and temporal dimensions of building components [9]. Boje et al. [10] have noticed that while BIM includes procedures, technologies, and data schemas for semantic representation of building components and systems, the idea of digital twin covers more holistic sociotechnical and process-oriented characteristics of complex artifacts.

According to Farsi et al. [11] an urban digital twin is a collection of digital urban resources covering digital models of the physical environment, real-time data, static datasets, urban analyses, simulations, and visualizations. Urban digital twins are mainly used as a tool to help in the operations and maintenance stage [12]. A literature survey reveals many various examples of urban digital twins. For instance, the digital twin of Zurich is a 3D model representation of the city designed to simplify digitizing space and performing data processing through the creation of a spatial data infrastructure [13]. The digital twin of Herrenberg is a city information model (CIM) composed of different urban data to ensure public participation in democratic city governance [14]. The digital twin of Rotterdam simulates a 3D city and supports the management of urban development [15]. The digital twin of Helsinki supports citizen decision-making and open innovation development to reach sustainability goals [16].

So far, the digital twin has a variety of definitions. For instance, Ferre-Bigorra et al. [12] argue that the DT is "a virtual representation of a physical system, with its associated environment and processes," hence it is updated through the exchange of information between physical and virtual systems. Allam et al. [7] emphasize that the digital twin enables "what-if" scenario modeling at various institutions in the discipline of urban planning. Weber et al. [17] argue that the DT has the ability to collaborate with other digital twins and permits a holistic intelligence and self-control to be achieved. Stark and Damerau [18] highlight that a DT as a digital counterpart should contain features, conditions of work, and behavior of its selected object. Qian et al. [19] add that the design of a DT for Internet of Things (IoT) systems should include a specification of nonfunctional requirements, i.e., reliability, safety, latency, scalability, and security. Dalibor et al. [20] define the DT as a new paradigm in simulation in various use contexts, i.e., healthcare, maritime and shipping, design customization, manufacturing, product development, city management, and aerospace. NASA defined DT as "an integrated multi-physics, multi-scale, probabilistic simulation of a vehicle or system that uses the best available physical models, sensor updates, fleet history, etc., to mirror the life of its flying twin" [21]. ISO 23247-1:2021 presents the digital twin as a "manufacturing fit for purpose digital representation of an observable manufacturing element with synchronization between the element and its digital representation" [22]. Digital twins integrate various technologies, such as machine learning, deep learning, and IoT. Grieves [23] has argued that a DT model consists of three fundamental components, i.e., a physical product in real space, virtual objects in the virtual space, and the connections of information that link the virtual and real products together. Digital twins present some fundamental characteristics, i.e., integrating different data of physical objects, coexisting in the entire life cycle with physical objects, coevolving with them and accumulating knowledge, and optimizing physical objects' behavior [24]. Therefore the main steps in the DT design process should be as follows: analysis of requirements, physical system decomposition, services allocation and process designing, implementation and quality assurance, and verification and validation [25].

3 Literature review and state of the art

The objective of this literature survey is to present a comprehensive view of the DT technology and its implementation challenges and constraints in the most relevant domains. This study research question (RQ) is formulated as follows: What is the state of the art of DT technology implementing smart city applications? To answer the research question, prominent academic repositories, i.e., Association of Information Systems eLibrary (AIS eLib), IEEEXplore, Research Gate, Sage Journals, Science Direct, and Scopus, were used. The criteria applied for filtration were set equally. As the digital twin technology has evolved continuously since 2003, when the DT concept was first introduced, this search has included the years from 2003 to 2023 and excluded non-English articles. This study examines the RQ from the perspective of a digital twin for smart cities. For this literature survey, the authors utilized the PRISMA approach, which includes four steps: identification, selection, eligibility, and inclusion.

The identification step requires establishing the search parameters and defining the databases. This includes the searching keywords, and the time interval of publications chosen to conduct the survey. The research revealed 40 papers found in AIS eLib, 72 papers in IEEEXplore, 110 papers in Research Gate, 18 papers in Sage Journals, 705 papers in Science Direct, and 351 papers in Scopus—hence, a total of 1296 publications from 2003 to 2023. The search string "Digital Twin" AND "Smart City" was used in the databases.

In the selection step, publications identified from databases were screened to exclude articles that were duplicate or poorly connected with the search string. In this stage, 1113 publications were excluded, leaving 183 papers for the next step.

In the eligibility step, the titles, abstracts, and keywords of the articles were read, and it was assessed whether they should be included in the review. By the end of this step, 97 documents were excluded, leaving 86 papers for the last step. In the inclusion step, the articles were read and checked as to whether they were valuable for final inclusion in this study. Next, the articles were categorized into groups according to their contributions. Finally, 22 articles were included, shown in Table 1.

Table 1 Digital twin's application in smart cities.

Reference	Research findings
Transportation systems	
[26]	Extraction of city mobility trends as crowd context and prediction of long-term and long-distance movement at a metropolitan level.
[27]	Authors propose blockchain-based digital twin as a service (DTaaS) to support intelligent transportation system (ITS).
[28]	DT is used to model the transportation system, optimize it, conduct simulations based on various policy scenarios, and monitor the quality of transport data.
[29]	Authors propose a method for producing 3D city models and city entities' visualization coupled with data from IoT devices, heatmaps, traffic flows, bus routes, and cycling paths.
[30]	This study covers solutions for monitoring the state of the subway infrastructure. Authors discuss the digital twins in complex cyber-physical systems.
[31]	Authors propose a digital twin application for smart transportation and for smart energy grid. The DT system is combined with artificial intelligence, federated learning, edge computing, and automatic control.
[32]	Author provides a smart city architecture including digital twins, real-time streamed data ingestion, simulation, and analytics for proactive management of public transport systems.
Environmental protection, safety, and sustainability	
[33]	Authors introduce digital twin framework based on artificial intelligence and utilize data model on environmental pollution. The proposed framework is feasible for city policy evaluation and autonomous intervention.
[34]	Authors propose computational decision support by digital twins. They provide metric-driven framework for sustainability planning at various city levels.
[35]	Authors consider low-cost sensor application for air quality monitoring networks and air pollution modeling and measurement.
[36]	Author uses DT of a city by modeling different flows of information and 3D virtual representations to investigate safety and security during mass events, e.g., concerts, protests.
Infrastructure maintenance	
[37]	Authors propose a framework for digital twin-based management system for a solar heating system, efficient energy usage in real time, and integration of intelligent management systems into physical objects.
[38]	This study covers a model of digital twin of energy Internet of Things. Authors discuss projects on digital monitoring and diagnosis of power grid network, and energy IoT construction.
[39]	Digital twins are developed to improve the management of water distribution system (WDS) in Valencia Metropolitan Area in Spain.
[40]	Smart city digital twins are included in the model for disaster management and analysis on mitigation.

Continued

Table 1 Digital twin's application in smart cities—cont'd

Reference	Research findings
[41]	This study includes a simulation environment for analyses of the behavior of interconnected heat, power, and transport networks, as well as digital twins for the energy systems. The simulations support analyses of costs and pollution emissions.
[42]	Authors consider the linkage among the lighting system, the surveillance system, and digital twin lighting system to control energy consumption and electricity costs.
Buildings and construction	
[43]	This study provides a digital twin for urban regeneration business at the village level. The DT system supports decision-making on local regeneration, redevelopment, and reconstruction.
[44]	Authors present a digital twin for road and bridge management, construction monitoring and maintenance in West Java Province, Indonesia. Further, the DT is to support technical planning, budgeting, and procurement.
[45]	Authors focus on digital twin solution for the urban facility management (UFM) process, as well as for equipping UFM managers with a decision-making system.
[46]	Authors present an extension of the digital twin solution for the urban facility management (UFM) process, as well as for supporting stakeholders by enabling the scheduling of operations and minimizing the operation costs in a distributed geographical system.
[47]	Authors propose the deployment of a digital twin box (DTB) to roads construction. The DTB is supplemented by information on GPS location and measurements of temperature and humidity.

The literature survey allows the formulated research question to be answered. It is seen that DTs are used in several fields, including environmental protection and sustainability, urban management, electric power management, and transportation. They are highly desirable for construction, as they are helpful in what-if analyses. They support failure simulation, economic solution finding, reducing losses, and delivering the best possible outcome for construction. In social engineering, development and implementation of digital twins are still a challenge. Although urban managers want to uncover citizens' behaviors and attitudes, yet it appears difficult to evaluate and predict human behaviors. Therefore the digital twin must continuously learn from its physical counterpart. This allows for optimization of the diagnosis, maintenance, and predictions.

4 Identified research gap

The concepts of IoT, smart city, and digital twin are inseparably connected with the use of technology in the daily life of inhabitants. In its design, the IoT comprises four main areas: People, Data, Process, and Things. People use devices as part of the processes in which they participate, generating data that allow the processes to be better adapted to their expectations. The idea of a smart city assumes the use of IoT technology for supporting the activities of inhabitants and managing infrastructure. The third concept, the digital twin, provides a possibility for a better understanding of the processes in which the users participate, because it enables system performance analysis and behavior simulation. Such an approach to the presented concepts requires addressing the aspect of technology design and use with respect to the concept of digital transformation. Digital transformation, similarly to the process of constant improvement of an organization's operation, requires the understanding of the actions taking place in cities. Consequently, the concept of a smart city can be, due to people, processes, and data, supported by solutions well-known from organizational theory with respect to business process modeling. Modeling the processes taking place in a smart city leads to improvement of the process of monitoring city operation as well as improving the processes of collecting information, adjusting the services to user needs, and responding to threat situations. As the concepts indicated earlier are related, it is necessary to search for methods for designing sociotechnical systems. Currently used notations, such as BPMN and ARIS, can be used to design processes taking place in cities. However, a problem arises when modern technologies in the form of digital twin theory and IoT have to be addressed.

The limitations of BPMN notation, as pointed out by Ardito et al. [48], include a lack of the required elements in BPMN notation in designing IoT solutions, lack of semantic data sharing between processes, and problems with the interoperability of the applied technical solutions and a wide range of receivers. Extension of BPMN notation and the use of its principles in the process of modeling smart city processes provide, according to the concepts of digital twins and IoT, new possibilities connected with data collection and simulation of the course of processes built at the interface of humans and the technology surrounding them.

From the perspective of the theory of digital twins, the key question is what technology will be used to build solutions simulating the behavior of devices and people. The solution adopted in the present chapter is agent technology, with the JADE platform at the implementation level [49], which makes it possible to avoid the problems [48] connected with network services modeling. The use of a multiagent platform and the elements of its architecture offers a range of advantages. Firstly, software building is unified in terms of the classes and mechanisms used by the multiagent platform. This allows existing artifacts to be used in the process of system modeling. Object communication is supervised and takes place in accordance with the standards used by agent technologies (here FIPA [50]). Thanks to that, existing communication protocols can be used in the process of building systems with distributed architecture.

Thus, when combining the theories of IoT, smart city, and digital twin, it is necessary to indicate how the system will be modeled in terms of the four elements: People, Data, Process, and Things. As far as the first three are concerned, BPMN notation will be used in this chapter. As for the fourth element, this chapter will indicate the linkage of BPMN notation with new artifacts from the areas of digital twins and IoT. At the implementation level, the multiagent JADE platform will be used.

As Chang et al. [51] indicated, research into IoT modeling within BPMN notation can refer to three areas: BPEL modeling for IoT, XPDL modeling for IoT, and BPMN modeling for IoT. The first approach involves the use of BPMN systems and passing the execution of a BPMN process to the mechanism in control of process execution. Since the chapter assumed the use of digital twin theory, this approach will not be used. The second approach refers to transforming BPMN processes to possible XML-based model transformations. This approach to BPMN development has also not been applied in the chapter due to the use of a multiagent platform, where the diagram being built written in the form of XML is translated into agent instances present in the system. The theories and studies discussed further in the chapter refer to the third approach, which is connected with the development of BPMN theory for IoT in the area of digital twin and agent technologies.

Ardito et al. [48] addressed the integration of IoT theory and the BPMN metamodel, indicating the extension developed by them and proposing graphical elements used in this approach. A limitation of the authors' work is the lack of reference to multiagent systems and the fact that IoT integration is focused on network services. Marah and Challenger [52] addressed the issues of integrating IoT and digital twin within the JADE platform and presented a general concept for integrating these approaches without addressing the aspect of integrating the metamodels available in the literature or modeling the operation of such a model. Bocciarelli et al. [53] proposed their own metamodel of an IoT system and its integration with a BPMN model concentrating on the aspect of testing model transformation correctness based on the standard proposed by Carrez et al. [54]. Meyer et al. [55] proposed elements of IoT notation with respect to specific tasks performed in IoT systems, indicating the significance of the Sensing task and Actuation task. Moreover, they indicated the linkage between IoT Device class and Line class defined in the BPMN metamodel. These notions were adopted in the proposed metamodel as the main artifacts referring to the actions performed by IoT devices. Another publication in the area of BPMN for IoT was written by Sungur et al. [56]. They proposed BPMN notation's usage for handling WSN devices. The solution provided by them induced a limited range of artifacts used in the modeled BPMN process, mainly in terms of using new Lines. However, in the approach proposed by those authors, communication between Lines can take place based on standard messages, messages handled by IoT (e.g., based on network services), or in accordance with the FIPA standard. In a publication presented by Yousfi et al. [57], the authors focused on integrating the theory of ubiquitous processing and proposed new artifacts, which are related to tasks and events (e.g., processing of image, sound, text), and data representing Smart Objects. In the proposed approach, actions of this type are referred to as a DT Task and require a more in-depth writing through an agent action plan.

5 Proposal of IoTDT-BPMN metamodel

The research conducted identified a research gap related to the integration of IoT, smart city, digital twin, and Smart Agents with respect to modeling systems connected with the previously mentioned concepts. The approach proposed further in the chapter involves integration of the elements of the IoT metamodel [54,55], JADE platform metamodel, and BPMN metamodel, as well as an original extension of these metamodels and proposed artifacts for building extended IoTDT-BPMN diagrams. Fig. 1 depicts elements of the proposed IoTDT-BPMN metamodel.

The elements marked in *yellow (white in print version)* represent a simplified set of elements used in BPMN notation. The elements marked in *blue (dark gray in print version)* indicate connection with the IoT metamodel proposed by Carrez et al. [54], Bocciarelli et al. [53], and Meyer et al. [55]. In particular, the metamodel adopted Sensing tasks, used to designate tasks related to the reading of device parameters and Actuation tasks connected with controlling IoT devices. "IoT Task" requires specification of the operation executed on the device, access interface, and the structure of the message to be used in a given operation. This is particularly important when using communication protocols in IoT systems, e.g., MQTT [58]. The elements marked in *orange (dark gray in print version)* originate from the metamodel of JADE system architecture and refer to agent behaviors that they can execute and that must be programmed. This concept has been extended in the proposed diagram to include elements of an agent action plan consisting of a set of behaviors and, where the role of digital twin is fulfilled, IoT tasks. For graphic modeling of an agent action plan, an additional diagram of agent tasks was used and designed to develop the agent's action plan. Additionally, the JADE platform provides communication interfaces connected with communication acts executed by agents. For that reason, assuming that agents process messages in an asynchronous way, an artifact related to the FIPA message queue was introduced and linked with the class of messages in the BPMN diagram. This allows FIPA messages to be presented in a BPMN diagram. The elements marked in *green (light gray in print version)* are the elements binding the metamodel in terms of the proposed IoTDT-BPMN concept and are connected with previous research [59] into the integration of autopoietic systems within BPMN notation.

6 Example of modeling IoTDT-BPMN processes

As part of modeling of the processes based on IoT device operation within the concepts of smart city and digital twin, a software prototype was prepared that enabled modeling of extended BPMN diagrams in accordance with the proposed IoTDT-BPMN model. In order to demonstrate its capabilities, an example diagram has been developed that describes the operation of a user assistant collecting data on an inhabitant's geolocation. The task of the system is to determine the user's current position and, based on that, make a decision about what actions should be taken by the building's control system. Fig. 2 presents a diagram made in IoTDT-BPMN notation.

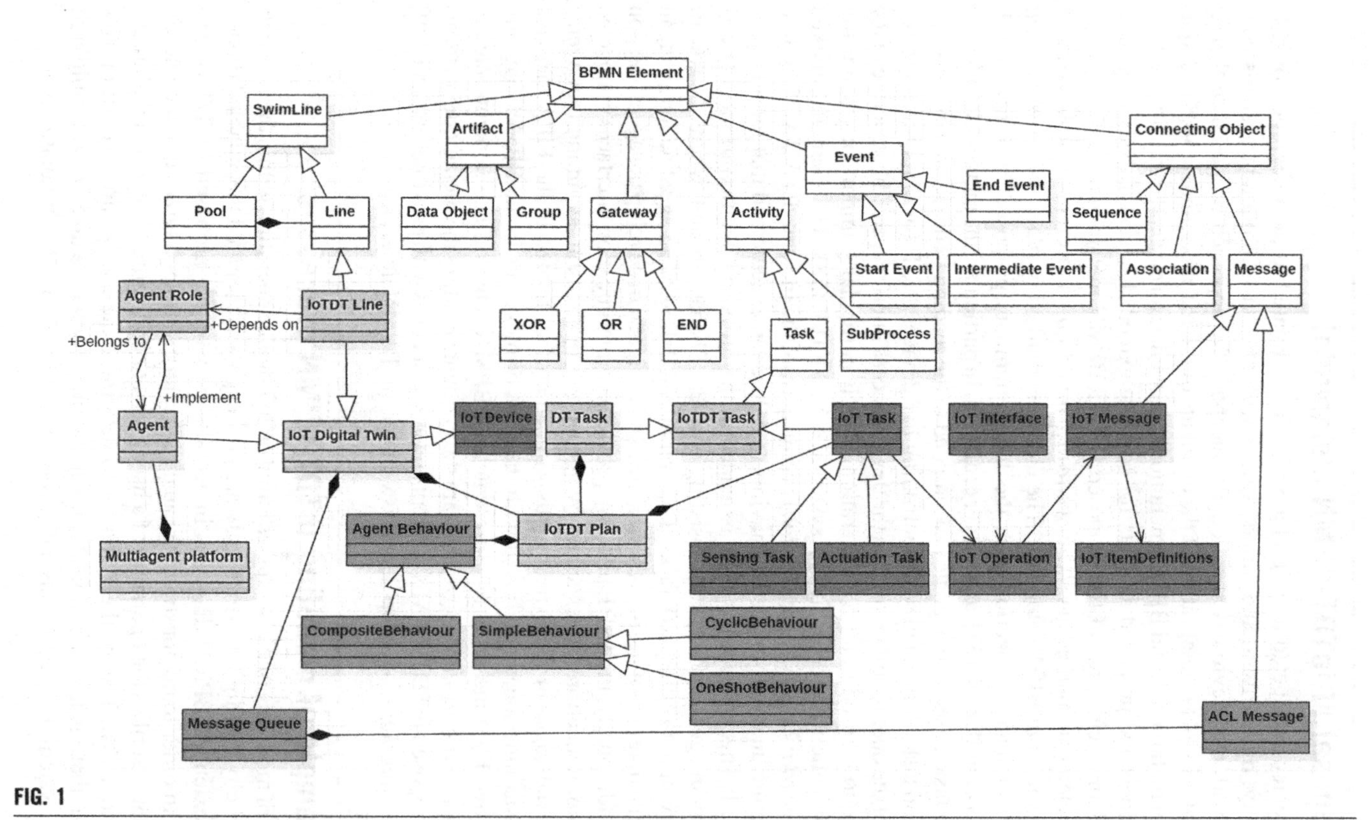

FIG. 1

Proposal of the IoTDT-BPMN metamodel.

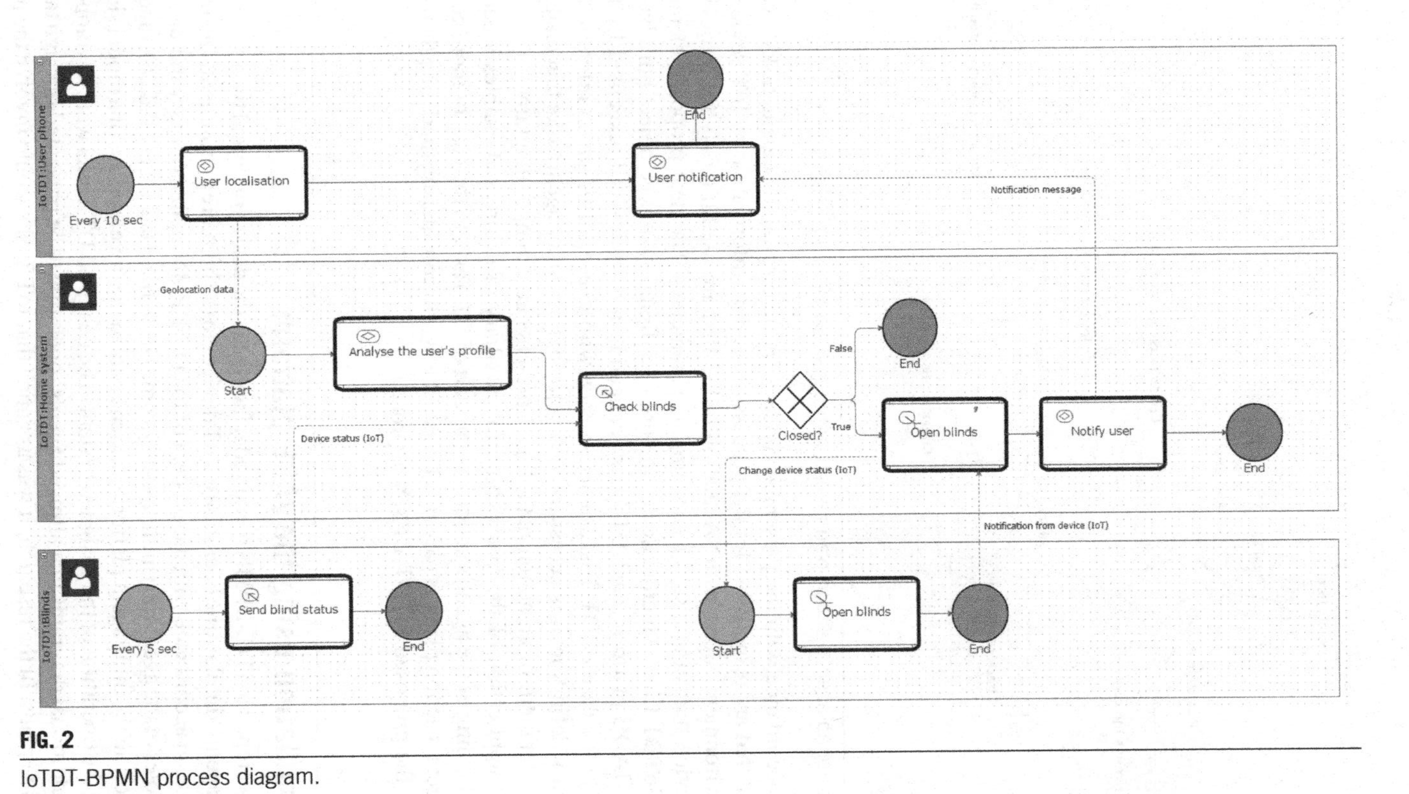

FIG. 2

IoTDT-BPMN process diagram.

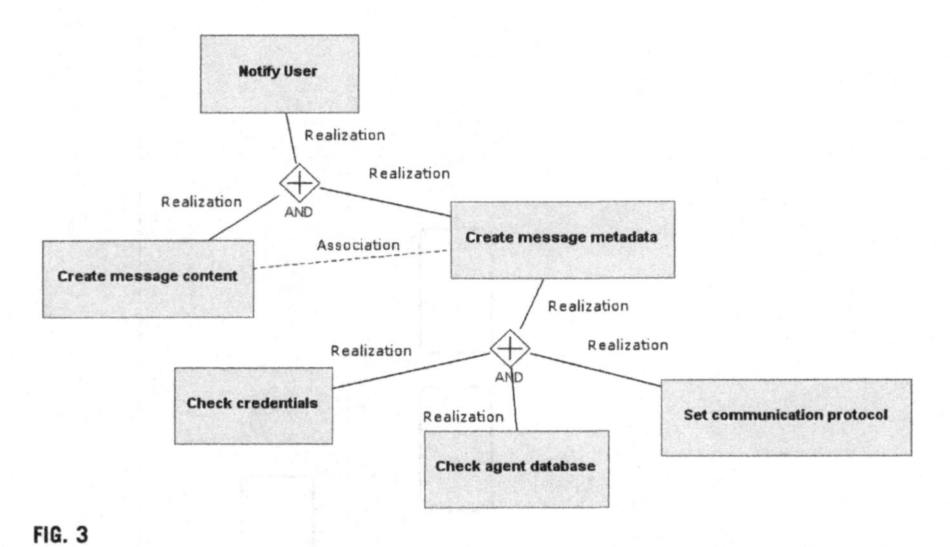

FIG. 3

Diagram of an agent's action plan as part of the task DTTask:NotifyUser.

In accordance with the presented concept, in the diagram, "Sensing tasks" were defined that are responsible for the process of sending information by IoT device to the environment, along with "Actuation tasks" that are used not only for retrieving IoT device status but also for changing its parameters. The next proposed element is the IoTDT Line, which indicates the use of a digital twin type object built based on the JADE platform, and DT Task, indicating a complex task executed by the agent. In the case of DT Task, in accordance with concept [9], which refers to the agent task definition, it is necessary to develop a diagram of agent tasks that enables the development of an action plan. Fig. 3 presents an example diagram.

The application of the task diagram is auxiliary in nature and enables a better understanding of the agent's action plan as part of DT Task. In the case of more complex agent's action plans as part of DT Task, it is worth considering modeling them as part of the proposed IoTDT-BPMN approach.

7 Simulation using the JADE platform

The developed IoTDT-BPMN metamodel facilitates the design of digital twin systems where agents simulate the operation of IoT devices in the human environment. Designing Lines in accordance with the concept of IoTDT-BPMN enables an easy indication of the roles of the agents in a multiagent system. The defined tasks were linked to agent behaviors in the multiagent platform, which enabled the development of a model simulating the operation of the designed system. The designed DT Tasks were developed using additional diagrams that allowed agents' operation to be presented through task execution plans. In the research, it was assumed that IoT Tasks would involve the use

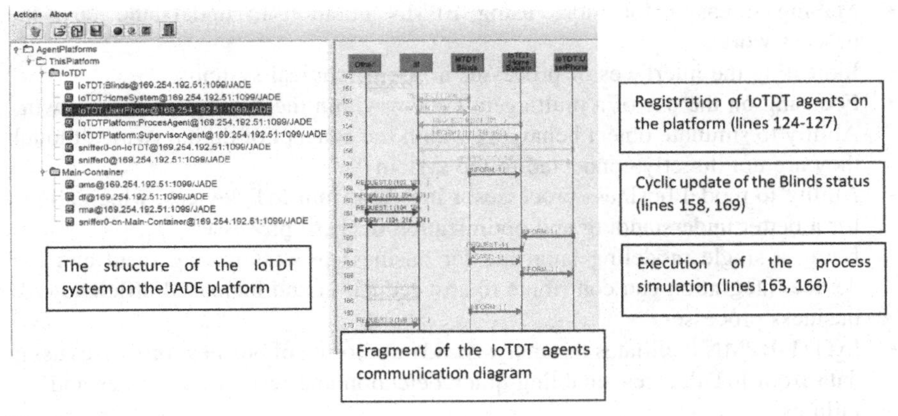

FIG. 4

Communication scheme of agents in the JADE platform.

of simple communication protocols such as MQTT or REST. Fig. 4 presents a finished project of a digital twin system developed using the JADE platform.

The use of a multiagent platform enables the introduction of a third communication layer. Apart from communication defined through a BPMN diagram and referring to, for example, transferring data to the user's phone, and apart from communication based on network services, e.g., REST, MQTT, within IoT devices, the use of a multiagent platform enables communication between simulated system elements as part of the developed agents and the FIPA standard. This facilitates the process of controlling and monitoring the operation of such a system. Such a prototype enables simulation of the behavior of devices as part of the concept of digital twin and can be used for:

- Building a user interface that supports the control of IoT systems' operation.
- Analysis of the operation of the actual IoT devices.
- Collection of information on a user operation profile performed by a multiagent.
- Identifying bottlenecks and carrying out process simulations.

8 Discussion

The limitations of the solution discussed in this chapter include: a focus on a single solution in the form of the JADE platform, the need to be familiar with BPMN notation in the process of modeling the process flow, and the paradigm of multiagent systems in designing agent behaviors. The advantages of the proposed IoTDT-BPMN metamodel include:

- Combining the concepts of IoT, digital twin, and multiagent systems within a metamodel that can be used in designing processes connected with the functioning of a smart city.

- Making it easier for those using BPMN notation to understand how the process works.
- Indicating the interfaces of processes in sociotechnical systems.
- Focusing on the use of a multiagent system within the concept of digital twin.
- Ability to simulate object behaviors within the concept of digital twin, although they are not directly connected to the system.
- Ability to model business processes using data from IoT devices, which allows for a better understanding and optimization of these processes.
- Using a single modeling language for business process management and IoT device integration can contribute to cost reduction and increased efficiency of business processes.
- IoTDT-BPMN facilitates the analysis and monitoring of business processes using data from IoT devices, enabling quick detection and response to issues and failures.
- This metamodel can be applied in various fields such as industry, logistics, healthcare, and energy, making it a very versatile tool for modeling business processes using IoT devices.

9 Conclusions

Digital transformation is an essential element of urban governance, enabling the efficient coordination and cooperation of multiple stakeholders involved in the city environment. Lately, the concept of the smart city has gained tremendous importance and public institutions have started thinking about innovative ways to develop solutions tackling the complexity of the various phenomena. The literature survey allows for the conclusion that digital twins are applied to support a city's functioning, mainly transportation, infrastructure maintenance, buildings and construction, as well as environment protection, safety, and sustainability. Application of digital twin in public governance is still a challenge. In this study, the authors present a BPMN-based approach for digital twin process modeling. An agent oriented programming (AOP) approach was employed for this purpose, along with utilization of the JADE multiagent platform. The implementation process of the presented solution was enhanced through the proactivity of JADE agents and utilization of its internal agent communication mechanism.

This chapter demonstrates that a digital twin can be built using already existing data, open source technologies, IoT, and data via API in a real-time representation to support decision-making, cost management, and risk control in the city environment.

10 Outlook and future work

To further advance the field of IoTDT-BPMN, several areas of future research can be identified. One area of research may concern the architecture and design of agents. The proposed approach assumes the existence of IoTDT agent plans, which are not part of the JADE metamodel. The second area of research is the automation of creating IoTDT agents based on the presented diagrams. The third research area encompasses the architecture of the developed system and a contextual perspective on the proposed metamodel. In the case study included in this chapter, the authors focused on a simple process concerning data on an inhabitant's geolocation.

In future work, the authors propose to further develop the application of digital twin to support public administration, urban planning, or security systems.

References

[1] M. Knezevic, A. Donaubauer, M. Moshrefzadeh, T.H. Kolbe, Managing urban digital twins with an extended catalog service, ISPRS Ann. Photogramm. Remote Sens. Spatial Inf. Sci. 10 (4/W3-2022) (2022) 119–126.

[2] Around 2.5 billion more people will be living in cities by 2050, projects new UN report, 2023. https://www.un.org/en/desa/around-25-billion-more-people-will-be-living-cities-2050-projects-new-un-report. (Accessed 23 February 2023).

[3] V.E. Lehtola, M. Koeva, S. Oude Elberink, P. Raposo, J.-P. Virtanen, F. Vahdatikhaki, S. Borsci, Digital twin of a city: review of technology serving city needs, Int. J. Appl. Earth Obs. Geoinf. 114 (2022) 102915.

[4] Smart Cities, 2015. https://wayback.archive-it.org/12090/20160322185320/https://ec.europa.eu/digital-single-market/en/smart-cities. (Accessed 20 February 2023).

[5] E.M. Ouafiq, M. Raif, A. Chehri, R. Saadane, Data architecture and big data analytics in smart city, in: 26th International Conference on Knowledge-Based and Intelligent Information and Engineering Systems, Procedia Computer Science, vol. 207, 2022, pp. 4123–4131.

[6] Z. Lv, W.-L. Shang, M. Guizani, Impact of digital twins and metaverse on cities: history, current situation, and application perspectives, Appl. Sci. (Switzerland) 12 (24) (2022) 12820.

[7] Z. Allam, A. Sharifi, S.E. Bibri, D.S. Jones, J. Krogstie, The metaverse as a virtual form of smart cities: opportunities and challenges for environmental, economic, and social sustainability in urban futures, Smart Cities 5 (3) (2022) 771–801.

[8] A.A. Dialite, L. Ng, J. Barton, M. Rigby, K. Williams, S. Barr, S. Zlatanova, Liveable city digital twin: a pilot project for the city of Liverpool (NSW, Australia), ISPRS Ann. Photogramm. Remote Sens. Spatial Inf. Sci. 10 (4/W2-2022) (2022) 45–52.

[9] H. Xia, Z. Liu, M. Efremochkina, X. Liu, C. Lin, Study on city digital twin technologies for sustainable smart city design: a review and bibliometric analysis of geographic information system and building information modeling integration, Sustain. Cities Soc. 84 (2022) 104009.

[10] C. Boje, A. Guerriero, S. Kubicki, Y. Regui, Towards a semantic construction digital twin: directions for future research, Autom. Constr. 114 (2020) 103179.

[11] M. Farsi, A. Daneshkhah, A. Hosseinian-Far, H. Jahankhani, Digital Twin Technologies and Smart Cities, Springer Nature Switzerland, 2020.

[12] J. Ferre-Bigorra, M. Casals, M. Gangolells, The adoption of urban digital twins, Cities 131 (2022) 10905.

[13] G. Schrotter, C. Hürzeler, The digital twin of the city of Zurich for urban planning, PFG J. Photogramm. Remote Sens. Geoinf. Sci. 88 (1) (2020) 99–112.

[14] P. Najafi, M. Mohammadi, P. van Wesemael, P.M. Le Blanc, A user-centred virtual city information model for inclusive community design: state-of-art, Cities 134 (2023) 104203.

[15] Rotterdam digital, Gemeente Rotterdam, 2023. https://www.rotterdam.nl/rotterdam-digitaa. (Accessed 23 February 2023).

[16] R. D'Hauwers, N. Walravens, P. Ballon, From an inside-in towards an outside-out urban digital twin: business models and implementation challenges, ISPRS Ann. Photogramm. Remote Sens. Spatial Inf. Sci. 8 (4/W1-2021) (2021) 25–32.

[17] C. Weber, J. Konigsberger, L. Kassner, B. Mitschang, M2DDM—a maturity model for data-driven manufacturing, Procedia CIRP 63 (2017) 173–178.

[18] R. Stark, T. Damerau, Digital Twin, CIRP Encyclopedia of Production Engineering, Springer, Berlin, Heidelberg, 2019, pp. 1–8.

[19] C. Qian, X. Liu, C. Ripley, M. Qian, F. Liang, W. Yu, Digital twin—cyber replica of physical things: architecture, applications and future research directions, Future Internet 14 (2) (2022) 64.

[20] M. Dalibor, N. Jansen, B. Rumpe, D. Schmalzing, L. Wachtmeister, M. Wimmer, A. Wortmann, A cross-domain systematic mapping study on software engineering for digital twins, J. Syst. Softw. 193 (2022) 111361.

[21] M. Shafto, M. Conroy, R. Doyle, E. Glaessgen, C. Kemp, J. LeMoigne, L. Wang, Draft Modeling, Simulation, Information Technology and Processing Roadmap, Technology Area 11, National Aeronautics and Space Administration, 2010. https://www.nasa.gov/pdf/501321main_TA11-MSITP-DRAFT-Nov2010-A1.pdf. (Accessed 25 February 2023).

[22] ISO 23247-1:2021 Automation Systems and Integration—Digital Twin Framework for Manufacturing—Part 1: Overview and General Principles, 2021. https://www.iso.org/standard/75066.html. (Accessed 25 February 2023).

[23] M. Grieves, Digital Twin: Manufacturing Excellence through Virtual Factory Replication, White Paper, Florida Institute of Technology, Melbourne, FL, 2014.

[24] T. Deng, K. Zhang, M. Shen, A systematic review of a digital twin city: a new pattern of urban governance toward smart cities, J. Manag. Sci. Eng. 6 (2) (2021) 125–134.

[25] C. Human, A.H. Basson, K. Kruger, A design framework for a system of digital twins and services, Comput. Ind. 144 (2023) 103796.

[26] Z. Fan, X. Yang, W. Yuan, R. Jiang, Q. Chen, X. Song, R. Shibasaki, Online trajectory prediction for metropolitan scale mobility digital twin, in: GIS: Proceedings of the ACM International Symposium on Advances on Geographic Information Systems, 2020.

[27] S. Liao, J. Wu, A.K. Bashir, W. Yang, J. Li, U. Tariq, Digital twin consensus for blockchain-enabled intelligent transportation systems in smart cities, IEEE Trans. Intell. Transp. Syst. 23 (11) (2022) 22619–22629.

[28] Sutani, F. Ramadhan, F. Hidayat, D.N. Hakim, U. Amaliah, W.S. Kuntoro, Study case of smart city: the usage of digital twin in transportation system, in: 2022 International Conference on Information Technology Systems and Innovation, ICITSI 2022—Proceedings, 2022, pp. 113–117.

[29] L. Adreani, P. Bellini, C. Colombo, M. Fanfani, P. Nesi, G. Pantaleo, R. Pisanu, Digital twin framework for smart city solutions, in: DMSVIVA 2022—Proceedings of the 28th International DMS Conference on Visualization and Visual Languages, 2022.

[30] A. Vodyaho, E. Stankova, N. Zhukova, A. Subbotin, M. Chervontsev, Use of digital twins and digital threads for subway infrastructure monitoring, in: 22nd International Conference on Computational Science and Its Applications, ICCSA 2022, Lecture Notes in Computer Science, 2022.

[31] Q. Lu, H. Jiang, S. Chen, Y. Gu, T. Gao, J. Zhang, Applications of digital twin system in a smart city system with multi-energy, in: Proceedings 2021 IEEE 1st International Conference on Digital Twins and Parallel Intelligence, DTPI2021, 2021, pp. 58–61.
[32] S. Van Den Berghe, A processing architecture for real-time predictive smart city digital twins, in: Proceedings—2021 IEEE International Conference on Big Data, Big Data 2021, 2021, pp. 2867–2874.
[33] E. Almirall, D. Callegaro, P. Bruins, M. Santamaria, P. Martinez, U. Cortes, Deep air—a smart city AI synthetic data digital twin solving the scalability data problems, Front. Artif. Intell. Appl. 356 (2022) 83–86.
[34] C.R. Corrado, S.M. DeLong, E.G. Holt, A. Tolk, Combining green metrics and digital twins for sustainability planning and governance of smart buildings and cities, Sustainability (Switzerland) 12 (20) (2022) 12988.
[35] G. Tancev, F.G. Toro, Towards a digital twin for air quality monitoring networks in smart cities, in: ISC2 2022—8th IEEE International Smart Cities Conference, 2022.
[36] F.J. Villanueva, C. Bolanos, A. Rubio, R. Cantarero, J. Fernandez-Bermejo, J. Dorado, Crowded event management in smart cities using a digital twin approach, in: ISC2 2022—8th IEEE International Smart Cities Conference, 2022.
[37] R.G. Silva, A. Araujo, Framework for the development of a digital twin for solar water heating systems, in: 2022 International Conference on Control, Automation, and Diagnosis, ICCAD, 2022.
[38] X. He, Q. Ai, J. Wang, F. Tao, B. Pan, R. Qiu, B. Yang, Situation awareness of energy Internet of Thing in smart city based on digital twin: from digitalization to informatization, IEEE Internet Things J. (2022) 1.
[39] P. Conejos Fuertes, F. Martinez Alzamora, M. Hervas Carot, J.C. Alonso Campos, Building and exploiting a Digital Twin for the management of drinking water distribution networks, Urban Water J. 17 (8) (2020) 704–713.
[40] D.N. Ford, C.M. Wolf, Smart Cities with Digital Twin systems for disaster management, J. Manag. Eng. 36 (4) (2020) 04020027.
[41] E. O'Dwyer, I. Pan, S. Acha, S. Gibbons, N. Shah, Modelling and evaluation of multi-vector energy networks in smart cities, in: International Conference on Smart Infrastructure and Construction 2019, ICSIS 2019: Driving Data-Informed Decision-Making, 2019, pp. 161–168.
[42] Y. Tan, P. Chen, W. Shou, A.-M. Sadick, Digital Twin-driven approach to improving energy efficiency of indoor lighting based on computer vision and dynamic BIM, Energy Build. 270 (2022) 112271.
[43] Y. Cho, J. Kim, A study on setting the direction of digital twin implementation for urban regeneration business, Int. J. Adv. Appl. Sci. 9 (4) (2022) 147–154.
[44] F. Hidayat, S.H. Supangkat, K. Hanafi, Digital twin of road and bridge construction monitoring and maintenance, in: ISC2 2022—8th IEEE International Smart Cities Conference, 2022.
[45] M. Mendula, A. Bujari, L. Foschini, P. Bellavista, A data-driven digital twin for urban activity monitoring, in: 2022 IEEE Symposium on Computers and Communications (ISCC), 2022, pp. 1–6.
[46] A. Bujari, A. Calvio, L. Foschini, A. Sabbioni, A. Corradi, A digital twin decision support system for the urban facility management process, Sensors 21 (24) (2021) 8460.
[47] O.E. Marai, T. Taleb, J. Song, Roads infrastructure digital twin: a step toward smarter cities realization, IEEE Netw. 35 (2) (2021) 136–143. 9267778.

[48] C. Ardito, D. Caivano, L. Colizzi, L. Verardi, BPMN extensions and semantic annotation in public administration service design, in: R. Bernhaupt, C. Ardito, S. Sauer (Eds.), Human-Centered Software Engineering. HCSE 2020, Lecture Notes in Computer Science, vol. 12481, Springer, Cham, 2020. https://link.springer.com/chapter/10.1007/978-3-030-64266-2_7.
[49] J.P. Müller, K. Fischer, Application impact of multiagent systems and technologies: a survey, in: O. Shehory, A. Sturm (Eds.), Agent-Oriented Software Engineering, Springer-Verlag, 2014, pp. 27–53.
[50] S. Poslad, Specifying protocols for multi-agent systems interaction, ACM Trans. Auton. Adapt. Syst. 2 (4) (2007), https://doi.org/10.1145/1293731.1293735. 15–es.
[51] C. Chang, S.N. Srirama, R. Buyya, Mobile cloud business process management system for the Internet of Things: a survey, ACM Comput. Surv. 49 (4) (2017) 70, https://doi.org/10.1145/3012000.
[52] H. Marah, M. Challenger, Intelligent agents and multi agent systems for modeling smart digital twins, in: 10th International Workshop on Engineering Multi-Agent Systems 9-10 May 2022, Auckland, New Zealand, 2022. https://emas.in.tu-clausthal.de/2022/papers/paper8.pdf.
[53] P. Bocciarelli, A. D'Ambrogio, T. Panetti, A model based framework for IoT-aware business process management, Future Internet 15 (2023) 50, https://doi.org/10.3390/fi15020050.
[54] F. Carrez, M. Bauer, M. Boussard, N. Bui, F. Carrez, C. Jardak, J.D. Loof, C. Magerkurth, S. Meissner, A. Nettstrater, A. Olivereau, M. Thoma, J.W. Walewski, J. Stefa, A. Salinas, Internet of Things—Architecture IoT-A Deliverable D1.5—Final Architectural Reference Model for the IoT v3.0, 2013.
[55] S. Meyer, A. Ruppen, C. Magerkurth, Internet of things-aware process modeling: integrating IoT devices as business process resources, in: C. Salinesi, M.C. Norrie, Ó. Pastor (Eds.), Advanced Information Systems Engineering. CAiSE 2013, Lecture Notes in Computer Science, vol. 7908, Springer, Berlin, Heidelberg, 2013, https://doi.org/10.1007/978-3-642-38709-8_6.
[56] T. Sungur, P. Spiess, N. Oertel, O. Kopp, Extending BPMN for wireless sensor networks, in: Proceedings of the 15th IEEE Conference on Business Informatics (CBI '13), IEEE, Vienna, 2013, pp. 109–116.
[57] A. Yousfi, C. Bauer, R. Saidi, A.K. Dey, uBPMN: a BPMN extension for modeling ubiquitous business processes, Inf. Softw. Technol. 74 (2016) 55–68.
[58] C. Hillar Gastón, MQTT Essentials: A Lightweight Iot Protocol: The Preferred Iot Publish-Subscribe Lightweight Messaging Protocol, Packt Publishing, Birmingham, 2017. http://www.myilibrary.com?id=1006335.
[59] M. Żytniewski, Use of a business process oriented autopoietic knowledge management support system in the process of auditing an organisation's personal data protection, in: E. Ziemba (Ed.), Information Technology for Management. Ongoing Research and Development. ISM AITM 2017, Lecture Notes in Business Information Processing, vol. 311, Springer, Cham, 2018, https://doi.org/10.1007/978-3-319-77721-4_4.

CHAPTER

From smart city to smart urban spaces: Prerequisites for the formation of smart urban spaces based on the participation of residents in the largest cities of Russia

14

Mikhail Vilenskii
Saint-Petersburg State University of Architecture and Civil Engineering, St. Petersburg, Russia

1 Introduction

Participation is one of the basic principles of the formation of smart cities [1]. The nature of complicity is linked to the formation of a democratic urban planning process. According to Fainstein [2], democracy, diversity, and equality are the three guiding principles of urban justice. Thus the medieval principle that "urban air makes a person free" [3] remains largely true today. Progressive urbanization and the changing urban environment create new social forms of interaction between people on the one hand and between people and the city on the other hand. According to a UN report, the share of the urban population in the total world population is projected to grow from 56% in 2021 to 68% in 2050.[a] These processes contribute to the expansion of urban planning conflicts and the expansion of the range of these conflicts [4–6], which in turn lead to the unification of communities that resist external influences.

We see on www.change.org thousands of petitions, gaining many votes, from hundreds to tens of thousands protesting against various forms of urbanization, and these voices are heard in all languages. The growth of urbanization leads to a change in the human habitat, both spatial and social. Spatial dimension is an integral part of social justice [7], since urban spaces determine the possibilities of residents

[a]www.unhabitat.org/wcr.

Smart Spaces. https://doi.org/10.1016/B978-0-443-13462-3.00004-2

both in obtaining urban resources and the ability to influence the management of the city's territorial resources.

The development of information and communication technologies (ICTs) that form new social connections, including in the process of finding solutions, becomes the basis for the emergence of local communities in an urbanized environment and contributes to the formation of boundaries of sociocultural spatial units that unite residents in the struggle for what Soja E. defined as sociospatial justice [8]. These boundaries are formed both in real and virtual space through confrontation and dialogue with the authorities and through awareness of common interests. In the absence of other effective tools for creating local communities in megacities with a high degree of disunity of residents, communities are formed on the basis of ICT tools. The tools of protest that are being formed change the ways of communication, cooperation, and activity [9].

New communication technologies used within the framework of ICT by government and business to solve their issues, and on the other hand, the use of social networks and ICT to empower citizens, joint projects, online activism, and civic engagement, stimulate the process of forming innovative tools for electronic participation of residents in conflict-free urban planning within both digital and physical space.

A 2018 study of smart cities commissioned by Intel, conducted by Juniper Research,[b] defined the following concept: a "smart" city is "an urban ecosystem that focuses on the use of digital technologies, the sharing of knowledge, increasing mobility, public safety, public health and economic productivity." The proposed methodology was based on the assessment of four directions of urban development: mobility, health, public safety, and city policy and technologies aimed at increasing economic productivity and democratizing services. Many researchers distinguish three types of innovations necessary to bring the city to smart standards: technological, organizational, and policy innovations [10], but at the heart of any socio-technological system is a person. Giffinger considers the following parameters of a smart city: smart economy, smart people, smart management, smart mobility, smart environment, and smart life. Defining a smart city through human activity and needs, the city as a social space develops in these areas on the basis of "a smart combination of endowments and activities of self-decisive, independent and aware citizens" [11]. A person is a measure of the measurement of this space, both digitally and physically.

Collectively, a smart city is an effectively organized and functioning space for a community of residents (representing a constantly learning social society with developed civic ideas); it is a "smart" (comfortable, safe, and environmentally neutral or reproducing) living environment, it has "smart" (effective, open, accessible to the population) management, it has a "smart" economy (high-performance, flexible, green), and finally it has "smart" mobility (logistics of population and goods).

[b]Juniper Research report. "Smart cities-what is in it for citizens?" 2018. URL: https://newsroom.intel.com/wp-content/uploads/sites/11/2018/03/smart-cities-whats-in-it-for-citizens.pdf.

All this refers us to the idea of ideal cities, which were written about during the Renaissance. Then Andrea Palladio [12] formulated his principle in the treatise *The Four Books of Architecture*: "a city is nothing but a certain big house, and vice versa, a house is a certain small city," which is true today, only with a new idea of technology and society, as well as taking into account scaling modern urban space. A city is a social organism, but what is important, it is a complex organism that consists of separate elements connected by complex algorithms. For comparison, the city is much closer to the coral reef with its biodiversity and a set of factors in which the reef as a system is associated with a variety of conditions for effective existence (temperature, salinity of water, sunlight, turbulence) [13] than to an anthill or a bee hive in which everything is based on the behavior of one species, but the vital activity of which also depends on many external factors. Prof. Bouvier, in his famous work *The Psychic Life of Insects*, (By E.L. Bouvier, 1918) described social processes among insects [14]. The urban community is diverse, and environmental, cultural, social, and economic views both divide and unite people. And large megacities proportionally increase this diversity; different approaches can be at the level of any type of space, including recreational and park areas [15]. As for simpler questions, if you ask residents of different condominiums what you want to build opposite, a store, a home, or a bar, we can get at least four different answers, depending on many factors.

Considering a smart city as a system striving, in addition to technological development, to resolve social conflicts and effectively transform existing socially differentiated spaces, we turn to the desire of its inhabitants for sociospatial justice. Through the prism of spatial justice, we can consider such different planning concepts as "New Urbanism" [16,17] or "Fifteen-minute City" [18].

Professional communities evaluate and determine how effective various spatial and technological solutions are, and at this time the very idea of spatial justice becomes a tool of politicians in different countries. The most famous example of this was the election of the mayor of Paris, A. Hidalgo, in 2020, at which she presented the concept of "La Ville du 1/4 d'Heure."[c] The 15-min City (FMC or 15mC) is an urban planning concept in which most of the necessities and services, such as work, shopping, education, healthcare, and leisure, can be reached in 15 min on foot or by bike from anywhere in the city.

In this regard, it is legitimate to ask how theoretical ideas about city planning take into account the interests and ideas of residents, and how they are practically implemented, taking into account public interests. According to Bengs [19], the public interest is not the real interest of each individual, but the potential interest of any person in the community. Urban planning in megacities is based on citywide tasks, thereby creating conditions for a conflict of citywide and local public interests and strengthening the positions of investors, developers and city authorities. Thus, the most important condition for achieving spatial justice will be an understanding of taking into account the interests of local communities, for which it is necessary to form an idea

[c]https://www.paris.fr/dossiers/paris-ville-du-quart-d-heure-ou-le-pari-de-la-proximite-37.

of their spatial boundaries. Within the framework of any positioned "smart city", the primary factor of its spatial planning development as a whole and its individual elements should be the sociocultural spatial elements of the urban environment that form local communities of residents—"smart urban spaces." Such spaces can be formed at different mego-, macro, meso-, and microlevels of the urban environment. At the same time, the mega-level is a city as a whole that unites all locally formed smart spaces.

In the presented work, the participation of residents in urban planning in the largest cities of Russia is considered, which, based on the use of ICT tools, leads to the formation of geographically united urban communities. According to the UN report, in 2022 Russia ranked 77th in terms of urbanization with an index of 74.4; 35,100,000 inhabitants lived in cities with a population of over 1 million people, i.e., 24% of all residents.

Big city data, the analysis of which makes it possible to identify local communities and related spatial elements, are becoming increasingly available for research and can be used in decision-making in the field of urban spatial development. According to Arribas-Hotel [20], these data can increase the level of "intelligence" necessary for understanding and describing the phenomena of social communication (in the ideas of Jane Jacobs) in modern smart cities [21].

Within the framework of the study, it was determined that the allocated sociocultural spatial elements as separate territorial units [22], combined with each other, become the basis for the formation of a smart city, where each of the units is a separate smart space formed on the basis of electronic democracy tools [22,23].

The most important basis for the development of smart spaces will be its own digital ecosystem.

Smart spaces are based on a broader sustainable development policy that protects the role of citizens and expands their rights and opportunities within the framework of local territorial elements or objects [24]. The formation of a smart city can go both from the general to the private, and from the private to the general, i.e., from individual smart spaces to the entire citywide system. In such a concept, the emphasis will be on "smart communities" rather than "smart cities," on ensuring and empowering citizens and supporting their individual and collective aspirations for well-being [25]. Such smart communities are formed by activating and using local institutions, participants, and resources on the platform of interaction using ICT [26].

The transformation of a city into a smart city is often a step-by-step procedure, rather than a large-scale change, both for technical reasons and because skeptical citizens need to be convinced of the benefits of new digital solutions [27].

Therefore an effective electronic democracy is necessary for an effectively functioning "smart" city. A smart city is a smart community that forms and develops smart spaces, whose residents, municipal administrations, developers, and investors, as well as urban planners and designers, need new approaches to organizing joint processes of transformation of these territories that adapt to various changing circumstances.

In Russia, the forms of management of urban territories have gone from post-Soviet, based on centralized management, to extreme forms of liberal markets, where regulation and intervention were not welcome and before the formation of a system of legal regulation, although in a rather curtailed form.

Centralized management was perceived by the population as permissible under certain conditions. The "Soviet form of city management" originally from the USSR is a centralized directive form of urban regulation and planning based on a planned economy. Such a system, in the absence of ownership rights to land and to any types of real estate, did not provide for any forms of participation of residents in making managerial or other decisions in the field of urban planning. The development of a market economy and the formation of a civil society made a centralized approach to urban transformation less and less appropriate. After 1991, new socioeconomic forms of development [28] led to the emergence of new forms of urban regulation, one of the tools of which were elements of participatory planning. However, only with the development of digital technologies has participatory planning moved from an exclusively state institution to a public institution.

New forms, digital formats of residents' associations, create prerequisites for new approaches to urban development. The participation of residents in the management of the city on the basis of electronic forms of interaction is one of the principles of the functioning and formation of a smart city and the formation of smart local spaces in it. In this study, the prerequisites for the development of such systems in Russia are considered.

2 Materials and methods

This study examines the development of smart urban spaces formed by communities of residents. It is based on the processing and interpretation of data on the participation of residents in participatory urban planning using ICT tools, which fundamentally distinguishes it from works where a smart city is more technology-oriented rather than citizen-oriented [29]. "Smart technologies" are a tool that can be turned in any direction. For example, according to the McKinsey Global Institute,[d] smart city technologies within the framework of the "security" direction will help reduce the number of emergencies by 20%–35%, reduce the number of crimes in general by 30%–40%, and reduce the number of accidents by 15%–20%. On the other hand, the use of video surveillance systems to control the mandatory wearing of the hijab by women in Iran, within the framework of the smart city video camera system, demonstrates the reverse side of the use of technology.[e]

"Smart city" is a system of "smart spaces" with "smart urban communities." In the study, smart urban spaces are considered within the framework of the spatial planning development of the city as a whole and its individual elements, as independently allocated sociocultural spatial elements of the urban environment.

In Russia, within the processes of interaction and confrontation between urban communities and city authorities, prerequisites for the formation of smart urban spaces are being formed.

[d]https://www.mckinsey.com/industries/public-and-social-sector/our-insights/smart-city-solutions-what-drives-citizen-adoption-around-the-globe#/.
[e]https://www.tasnimnews.com/fa/news/1402/01/19/2876943/.

This study examines the official and unofficial forms of participation of residents in urban planning.

The official forms of participatory planning are considered to be: public hearings and public discussions and their implementation in the largest cities of Russia on the basis of state and municipal ICT tools. Based on the statistics collected and analyzed for the first time, as well as documents, the actual results of public hearings are considered to be based on the example of 6 of the 16 largest cities with a population of over a million inhabitants, which made it possible to identify conflicts arising at the stage of urban planning, both at the city level and local urban communities. The information systems of municipalities used in the framework of the participation of residents in urban planning from the point of view of their effectiveness to achieve spatial justice are considered and systematized.

Various forms of self-organization of residents within the use of ICT tools are: social networks, messengers, and online petitions (Fb, VK, Twitter, Instagram, Telegram, Roi and Change.org); an analysis of the use of these tools by residents of the 16 largest cities as a whole and individual local urban communities was carried out. The main conflicts and the degree of participation of residents of the largest cities in them are considered. Such forms are considered to be collective forms of interaction to influence decisions made by city administrations or regional and federal authorities. The "self-formed" local digital ecosystems created by residents, combining all types of ICT tools, are identified and considered.

Based on the analysis of the results of public hearings and public discussions, search queries of residents on the Internet (based on wordstat tools. Yandex, Google Ads, and Google trends), and social group analysis (in Fb, VK, Twitter, Instagram, Telegram, Roi and Change.org), the maximum number of residents involved in the process of complicity in urban planning has been determined.

Based on Linders' three-level scheme [30], G2C (government for citizen), C2G (citizen for government), and C2C (citizen for citizen), the forms of interaction between residents and city administrations within the framework of ICT tools are considered.

Taking into account the use of ICT tools within the procedures of public discussions and public hearings, the principles of urban planning regulation of the city at different spatial levels are proposed as a basis for the formation of "smart urban spaces."

3 Public hearings and public discussions as official forms of participatory urban planning in Russia: Evolution, legal status, ICT

In the Soviet Union, the predecessor of Russia, no forms of participatory urban planning or design were envisaged. After its collapse in 1991, the first forms of complicity of residents appeared in Russia within the framework of new economic and political conditions.

In modern Russian legislation, participatory urban planning is regulated through two tools: public discussions and public hearings. The main provisions are established by federal legislation (Federal Law No. 212-FZ dated 21.07.2014 "On the Basics of Public Control in the Russian Federation" with Amendments; Federal Law No. 131-FZ dated 06.10.2003 "On General Principles of the Organization of Local Self-Government in the Russian Federation," Article 28. Public Hearings, Public Discussions; Federal Law No. 21.07.2014 212-FZ; "Town-Planning Code of the Russian Federation" dated 29.12.2004, Article 5.1. Public opinions, public hearings; No. 7-FZ 10.01.2002 "On Environmental Protection, " Article 32. Conducting an environmental impact assessment).

Public hearings and public discussions in the Russian legal field are a procedure for identifying collective opinions or clearly expressed differences of opinion. They are not a form of exercising power by the population; rather, their purpose is to develop recommendations on socially significant issues or to obtain a public assessment of a legal act (Ruling of the Constitutional Court of the Russian Federation No. 931—O-O of July 15, 2010, Resolution of the Constitutional Court of the Russian Federation No. 10-P of March 28, 2017). Taking into account the opinion of residents in the framework of such procedures is purely advisory in nature. The final decision is made by the local self-government body or state authorities (when transferring powers to them), and such a decision may not coincide with the results of the hearings and discussions, or with the recommendations of the body conducting them. Such procedures are applied to the following types of activities: urban planning, environmental impact assessment (EIA), urban development strategies and the municipality charter, as well as the draft budget of the municipality.

Federal Law No. 212-FZ (2014) "On the Fundamentals of Public Control in the Russian Federation" defines the concept of "public discussion" as "public discussion of socially significant issues used for public control purposes, as well as draft decisions of state authorities, local self-government bodies, with the mandatory participation of authorized representatives in such discussion of persons of the specified bodies and organizations, representatives of citizens and public associations whose interests are affected by the relevant decision."

Public hearings are understood to mean a meeting of citizens organized by a subject of public control, state authorities, and local self-government bodies to discuss issues related to the activities of these bodies and organizations and of particular public importance.

The fundamental difference between public discussions and public hearings is the absence of meetings of participants. In this regard, the following ICT tools are provided by law for public discussions: placement of the project to be considered at public discussions and information materials for it on the official website of the authorized local government body on the Internet information and telecommunications network (official website) and (or) in the state or municipal information system that ensures public discussions using Internet networks, or on the regional portal of state and municipal services and the opening of the exposition or expositions of such a project. With regard to public hearings, the use of ICT tools is provided in the form of placing the project to be considered at public hearings and information materials for it on the official website.

It is important to note that in relation to various types of activities and documentation within the framework of these types, for which procedures for hearings and discussions are established, different regulatory documents are applied that establish the territorial and extraterritorial nature of hearings and discussions, as well as requirements for their participants.

In particular, the extraterritorial nature of the hearings applies to EIA (Environmental Impact Assessment) documents, and a resident of the country can take part in them, regardless of the place of residence. There is no unified state database of the hearings held; however, in the conditions of extraterritoriality, environmental activists posted an updated unified database on the hearings on the website. Among the features of participation in such hearings, there is no need to register participants, as a questionnaire with comments can be sent via simple mail or e-mail. Various types of documents pass through the EIA system, from purely technical ones to those that have an impact on the spatial development of the city.

Hearings and discussions on development strategies, charters, and budgets of municipalities are territorial and provide that residents, public organizations, and companies that live and are located within the boundaries of the municipality participate in them. Such hearings or discussions can be considered extraterritorial within the boundaries of the municipality. The forms of such hearings or discussions may not provide for the participation of residents, since they are held in the format of expert invited groups or only representatives of municipal administrations (for example, in Novosibirsk).

In the field of urban planning (town-planning activity), citizens' participation was not envisaged until 1991; in 1992, the first forms of participation appeared, the declaration of which was "participation in the discussion of town-planning projects before their approval" within the framework of the Law on the Basics of Town-planning Activity (the Law "On the Basics of Town-planning Activity in the Russian Federation" of July 14, 1992 No. 3295-I). The document disclosed this form, and it was assumed that it was detailed at the regional level and at the level of municipalities.

In 1998, the Town-Planning Code of the Russian Federation (No. 73-FZ of May 7, 1998) was adopted, in which (Article 18) the participation of citizens and their associations in the discussion and decision-making in the field of town-planning activities was considered. The document provided that "informing citizens about urban planning activities is carried out by the relevant executive authorities, local self-government bodies through the media, through public discussions, as well as the organization of expositions and exhibitions." The rights of residents were determined: "before approving all types of urban planning documentation, discuss, make proposals and participate in the preparation of decisions in the field of urban planning in any form: participation in meetings (gatherings) of citizens, participation in public hearings, discussions of urban planning documentation and other forms, require consideration of their proposals." The possibility of independent examination of documentation was determined. The details of all these tools were not specified, and the procedure for their application was the responsibility of municipalities and regions.

In 2004, the "Town-Planning Code of the Russian Federation" 190-FZ established a uniform procedure for the participation of residents in the discussion of decisions in the field of urban development. The only form of complicity is established—public hearings with the presence of residents, applied to specific urban planning documentation: master plans, rules of land use and development (Russian abbreviation PZS), obtaining a permit for a conditionally permitted type of use (within the PZS), deviation from the limit parameters (within the PZS), documentation on the planning of the territory planning projects and surveying projects. The mandatory nature of the hearings is established, and their territorial nature is determined (for each type of documentation), including in relation to local transformations. There are general federal requirements for holding hearings (detailed at the municipal level). For the first time, municipalities are allowed to post information about the hearings on the Internet.

In 2006, amendments were made to the Town-Planning Code regarding the placement of information on town-planning documents, including during the hearings on the Internet on the official website of the municipality (subject to the availability of the official website of the municipality).

The format of public discussions (without meetings of residents) appeared in connection with the adoption of Federal Law No. 455-FZ dated December 29, 2017, "On Amendments to the Urban Planning Code of the Russian Federation and Certain Legislative Acts of the Russian Federation." The format of public hearings without a meeting of participants was justified by increasing the level of informing residents, as well as reducing time and organizational costs, and the possibility of expanded participation of residents. The necessity of accounting and identification of participants was determined (in connection with the territorial binding of the hearings).[f] The document defined informing residents via the Internet as part of the procedure of public hearings (through the websites of municipalities), and for the procedure of discussions, it provided for informing (posting materials) and directly conducting the procedure itself on the basis of a state or municipal information system or on a regional portal of state and municipal services.

Hearings and discussions in the field of urban planning are currently defined by the Urban Planning Code of the Russian Federation (as amended in 2022) (Article 5.1.) and are also territorial.

They are conducted on the following issues: approval of master plans and making changes to them, approval and making changes to the rules of land use and development, obtaining a permit for a conditional type of use (within the framework of the application of the rules of land use and development), deviations from the limit parameters (height, parking coefficients, building density, and others within the framework of the application of the rules of land use and development), approval of documents on the planning of the territory (planning projects and surveying projects), as well as approval of the rules of landscaping.

[f]https://sozd.duma.gov.ru/bill/133118-7.

Legislatively, at the federal level, the boundaries are defined in which hearings and discussions are held for each type of documentation, and a list of participants is defined depending on each type of documentation. According to master plans, PZS (rules of land use and development), and draft rules for the improvement of territories, hearings or discussions are citywide in nature; according to the documentation of the planning of the territory, they are held within the boundaries of the allocated elements of the planning structure (blocks, microdistricts, districts), and according to deviations from the limit parameters and conditionally permitted uses within the territorial zone (established in the PZS).

The necessity and procedure for holding public hearings and public discussions in Russia are regulated at the federal level. Regions and municipalities, depending on the redistribution of powers in the field of urban planning, have the possibility of adjustment in terms of the choice of procedure—hearings or discussions, as well as the order and form of their conduct (in accordance with Federal Law 455 of 12/22/2020).

In Russia, federal legislation allows regions (subjects of the federation) to redistribute the powers of municipalities. In a significant part of the subjects, powers have been redistributed in favor of the regions, including those for determining the nature of holding hearings and discussions on urban planning issues. Such a reduction in the powers of municipalities over the past 10 years is due to the ongoing policy of state centralization, actively implemented at the regional level from 2003 to 2008, and the consolidation of local municipalities, implemented since 2014, until now.

Currently, direct elections of mayors have been preserved only in seven regional and several district centers of the Russian Federation. In 2008, 73% of all heads of Russian cities passed through direct municipal elections, but by February 2023 it was only 8%. It is also important to note that at present (for 2023), except for the federal cities of Moscow and St. Petersburg, which have the status of a subject of the Russian Federation, there are no direct elections of the head of the city in any city with a population of over a million, as they were canceled as part of the policy of state centralization carried out in the Russian Federation. By 2019 elections were maintained only in Novosibirsk, the third largest city of the Russian Federation, and these were canceled during the preparation of this study.

Since 2021, after epidemiological restrictions in Russia related to COVID-19 (on March 2, 2020, the first case of COVID-19 was detected in Moscow), hearings and public discussions related to urban development have resumed and continue in all major cities. Epidemiological restrictions have led to the need to improve and legislate electronic forms of participation of residents in urban development. During 2020 all the hearings that took place in different cities were exclusively online, and in many cities the hearings were not held at all during this period, since the online form was not legally fixed. To launch a widespread process of public hearings, which, according to federal legislation in the field of urban planning, necessarily accompany the adoption of a number of documents and urban planning decisions, it was necessary to create legal conditions at the level of federal legislation. In 2021 the Federal Law "On Amendments to Article 28 of the Federal Law on General Principles of Organization of Local Self-Government in the Russian Federation" dated 01.07.2021 No. 289-FZ was adopted. According to this document, the necessity of mandatory and early posting by local self-government bodies on the

Internet of materials on issues that are submitted for public hearing and (or) public discussion, as well as the results of hearings (discussions), including a reasoned justification of the decisions taken, was determined. It is possible for residents of the municipality to submit their comments and suggestions on the projects submitted for discussion, including through the use of the official website on the Internet information and telecommunications network. In addition, citizens were given the opportunity by this law to send their proposals within the framework of these procedures in electronic form, including through the use of a Single portal of state and Municipal Services.[g]

In most cities of the Russian Federation, even after the end of restrictions related to COVID-19, public hearings have been transferred to the format of public discussions (for example, St. Petersburg, where face-to-face public hearings were canceled in 2020) and retain the character of being held exclusively online.

In 2022 and 2023, the cases of applying the procedure of public discussions and public hearings, in particular (on draft general plans, rules of land use and development, planning and surveying of territories, and amendments to them by Federal Law No. 58-FZ of 14.03.2022), were reduced.[h]

In some cases, the procedures for hearings and discussions on urban planning documentation projects have been canceled altogether. Thus the Government of the Russian Federation and the supreme executive authority of the subject have the right to provide that public discussions and public hearings will not be held on the territory of the subject during 2022. As of April 4, 2022, several regions had adopted such acts, including Moscow (Moscow Government Decree No. 438-PP of 22.03.2022). Also, public discussions of urban planning issues were partially canceled in the Nizhny Novgorod region, the center of which is Nizhny Novgorod. The simplified procedure for preparing construction documentation was first introduced at the beginning until the end of 2022, and then was extended until the end of 2023.

According to the legislation, in those subjects where public discussions and public hearings on urban planning documentation are preserved, the terms of their holding are reduced to 1 month and the circle of their participants is limited. Shortening the deadlines leads to a reduction in the awareness of residents—the less time, the fewer participants.

Thus, at present, the applicability and effectiveness of participatory planning at the level of legislative regulation in the field of urban planning is sharply reduced.

4 The evolution of the use of ICT tools within the framework of official forms of participation of residents

We can trace the stages of development of official participation procedures (over time) and their connection with the use of ICT tools by the state (municipalities).

The first stage (1992–2004)—informing and participating in discussions (with a meeting of participants) of residents with authorities and developers, including

[g]https://minjust.gov.ru/ru/events/47960/.

[h]https://www.consultant.ru/document/cons_doc_LAW_51040/fc77c7117187684ab0cb02c7ee53952df0de55be/.

mandatory consideration of alternative solutions, proposals, and public examinations. The hearing is subject to an expanded composition of urban planning documentation. At the first stage, there are no procedural requirements, and complicity is extraterritorial in nature. All the details of the application have been transferred to the level of regions and municipalities. The use of ICT is not provided.

The second stage (2004–17)—informing and participating in public hearings (with a meeting of participants) of residents with authorities and developers is the only form of participation. There is centralized regulation of the issues of holding hearings, reduction of the range of documents discussed, and detailing at the level of municipalities and regions. This stage provides for the possibility of informing residents through ICT (websites of municipalities, if available).

The third stage (2018–20) is informing and participating in hearings and discussions (discussions are the main form during the period of epidemiological restrictions), with cancellation of hearings and discussions in a number of regions and municipalities, taking into account the restrictions associated with COVID-19, and the use of state and municipal information systems to participate in public discussions.

The fourth stage (from 2022 to the present) is informing and participating in public hearings and public discussions. Public discussions are becoming the main form of complicity in most municipalities. There is a reduction of issues on which hearings and discussions are held, with cancellation of participatory planning procedures in a number of regions. The transition in most municipalities is to the format of online participation through regional and municipal information systems, through the use of a single portal system of state and municipal services for the identification of residents.

Back in 1969, Arnstein [31] in her research proposed a ladder of citizen participation in urban planning. At the lower stages, the real participation of citizens is replaced by actions that prevent citizens from actually influencing the initial result with formal participation in the process—this is termed "manipulation" and "therapy," allowing holders of authority the opportunity to lobby their own preferences by influencing participants. The next levels are symbolic cooperation, which allows those who do not have authority to have access to information and be heard through consultations. However, at this stage they have no way to guarantee that their opinion will be taken into account. When participation is limited to these levels, it remains interrupted and therefore does not provide a change in the situation. The next level is the level of consultations that participants can give to the authorities, waiting for their reaction. And the last levels are those corresponding to the actual powers of citizens; they are higher on the scale and include increasing degrees of influence on decision-making through partnership. At the top of the ladder, citizens who do not have authority get the majority of seats in decision-making bodies.

Thus, according to this model, all official forms of participation—public hearings and discussions—can be attributed to the level of symbolic cooperation and to the level of consultations, while since 1992 there has been a gradual transition from the levels of consultations and partnership to the level of information, with elements of therapy and influence. As a result, the official model of participation in urban planning in Russia over the past 30 years has been a progressive movement down the Arnstein ladder.

In the field of urban planning, within the framework of the state's ICT application tools in the field of participatory planning, we can, in turn, see an ascending system. The following stages can be distinguished: "the possibility of informing," "the obligation to inform, " "the possibility of participation of residents," "the main form of participation."

If we imagine such a system consisting of two elements in the form of a graph, we will see a descending system of involvement of residents with an ascending system of integration of ICT tools into it. In such a situation, can residents influence the formation and development of the city within the framework of existing legal instruments? Is it possible to make "smart residents" influence the formation of their "smart city" and their "smart urban spaces" using the official tools already available?

If we imagine even the existing inefficient system of official participation through the participation of tens, hundreds, or thousands of citizens, depending on the scale of the spaces and the scale of the city, we can see that access to information and consultations become an effective tool for effectively influencing decisions made by the authorities, even through forms of symbolic cooperation. The higher the number of participants, the more effective they are, and the more important their opinion is for the authorities. It is the ICT tools used by residents that can become the impetus that will allow them to start moving up the Arnstein ladder on the way to smart urban spaces and to the smart city as a whole—"up the stairs leading down."

5 Actual participation of residents in official formats—Hearings and discussions

Fundamental to the formation of smart cities and smart urban spaces within them is the question of to what extent people themselves are willing to become participants in the process of developing their city.

Until 2020, when public hearings were the main instrument of participation of residents in urban planning, their conduct was mandatory, and the issues on which they were conducted concerned most of the urban planning documentation and urban planning issues; during this period, the all-Russian statistics of public hearings were occasionally published.

According to the Ministry of Justice of Russia, 92,600 public hearings were held in 2018 and 14,600 hearings in the first 2 months of 2019, on information and analytical materials on the state and main directions of development of local self-government in the Russian Federation (data for 2018 to early 2019).[i]

[i]https://minjust. ru/razvitie-federativnyh-otnosheniy-i-mestnogo-samoupravleniya/doklado-sostoyanii-i-osnovnyh.

After the amendments to the legislation, official information on the total number of hearings and discussions on the Russian Federation and on most of its subjects was not posted. It should be noted that in the statistical data collected by municipalities, only the hearings and discussions themselves appear, and not the participants and the results of the hearings.

Various studies in the field of urban planning in Russia show that the number of residents who participate in official procedures of hearings and discussions sharply decreases depending on the population in cities [32].[j] According to studies conducted solely on the basis of sociological surveys, the maximum number of participants live in small towns, followed by medium-sized, large, and largest cities. The participation range is from 30% to 10%. But is this really the case?

According to the data of the statistical office for 2021 (based on the results of the population census),[k] the population of Russia is 147,182,000 people; the number of those 18 years and older, i.e., those who have the right to vote in the framework of public hearings, is 119,467,000. The urban population is 75% and the number of residents of cities with the right to vote is 90 million people. There are only 16 cities in Russia with a population of over 1 million people, with a total of 35,508,000 inhabitants, i.e., 24% of the population. The largest cities are Moscow, with a population of 13,010,000 and St. Petersburg with a population of 5,602,000. The rest of the cities have a population of 1,600,000 to 1,000,000. The population over the age of 18 living in the largest cities is 27,740,000 as of 2021, i.e., 23% of all eligible voters.

Thus not only almost a quarter of the entire population of Russia lives in the largest cities, but it is also almost a quarter of all those who have the right to vote.

In this study, to assess the actual participation of residents in urban planning in the field of urban development activities within the framework of the existing legal instruments of hearings and discussions, six major cities (St. Petersburg, Novosibirsk, Yekaterinburg, Kazan, Nizhny Novgorod, and Chelyabinsk) with a population over 18 years of 9,640,000 people, which is 8%, were considered from the entire population of Russia having the right to vote.

Moscow, which is the largest city in Russia, was not considered due to the lack of presentation of complete and sufficient information about the hearings and discussions, and also taking into account the fact that, as a result of the subsequent adjustment of the legislation of the Federal Law No. 58-FZ dated March 14, 2022 "on amendments to certain legislative acts of the Russian Federation," conditions were provided under which the procedures for urban planning documentation projects are canceled altogether. The Government of the Russian Federation and the supreme executive authority of the subject of the Russian Federation had the right to provide that public discussions and public hearings would not be held on the territory of the subject during 2022. Currently, several regions have adopted such acts, including Moscow (Moscow Government Decree No. 438-PP of 22.03.2022).

[j]https://wciom.ru/analytical-reports/analiticheskii-doklad/sreda-kotoraya-nas-formiruet-kak-rossiyane-oczenivayut-kachestvo-gorodskoj-sredy-i-dinamiku-ee-izmeneniya.
[k]https://rosstat.gov.ru/vpn_popul.

Thus, considering the six largest cities, it is possible to project this situation onto the rest and present a cross-section for all large, urbanized territories. Also, an important criterion for choosing these cities was the fact that during the period under review, there were no new master plans or land use and development rules in the documentation under discussion, which are citywide documents and affect the entire population of the city at the same time (adjustments of such documents affect only individual urban areas).

The study was carried out on the basis of conclusions from the results of hearings posted in the information systems of cities (at the same time, there is no single public, as well as nonpublic, electronic system in Russia, where the participation of residents and the results of hearings and discussions would be collected and analyzed or systematized).

The time frame of the presented study is one year—2021. This year is the first, after epidemiological restrictions in Russia related to COVID-19, when public discussions were held in all the cities studied.

The study analyzed the following characteristics of the system for resident participation in public discussions: assessment of the participation of residents in discussions/hearings on each of the documents which they are given, and assessment of the quantitative participation of residents in discussions/hearings by cities.

The results are presented in Fig. 1. As a result of the analysis of the conclusions on all hearings and discussions, it was revealed that the percentage of hearings and discussions in which residents participated, within the framework of local city documents: planning projects and surveying projects, as well as during the procedures for obtaining permits for conditionally permitted use and deviations from the limit parameters, the percentage of hearings/discussions does not exceed 70%, and the average level of participation in cities did not exceed 40%, which means that on average at least 60% of hearings and discussions took place without participants. In total, the number of discussions by city ranged from 364 in Novosibirsk to 69 in Kazan, and in all cities for the year, 1027.

According to citywide documents, in which all residents of the city participate: when making changes to the PZS and the master plan, the percentage of participation reaches 100%.

The total number of participants was 13,001. The maximum number of participants, in Novosibirsk, was 5570, and the minimum, in Kazan, was 279. The total number of participants from residents who have the right to vote (over 18 years old) did not exceed the maximum level of 0.4%.

Indicators of participation of residents within the framework of local, documents, depend on a variety of conditions, which can include the frequency of hearings, their nature, which territories fall within the boundaries of hearings, and the population density in these territories. However, in general, the indicators give us an idea of the limited number of those involved.

Since residents can participate in an unlimited number of hearings, it was possible to assume multiple repeated participation. In order to clarify this issue, an analysis of the comparison of participants was conducted at the hearings in Kazan in order to establish how often the same resident participated in hearings on various issues.

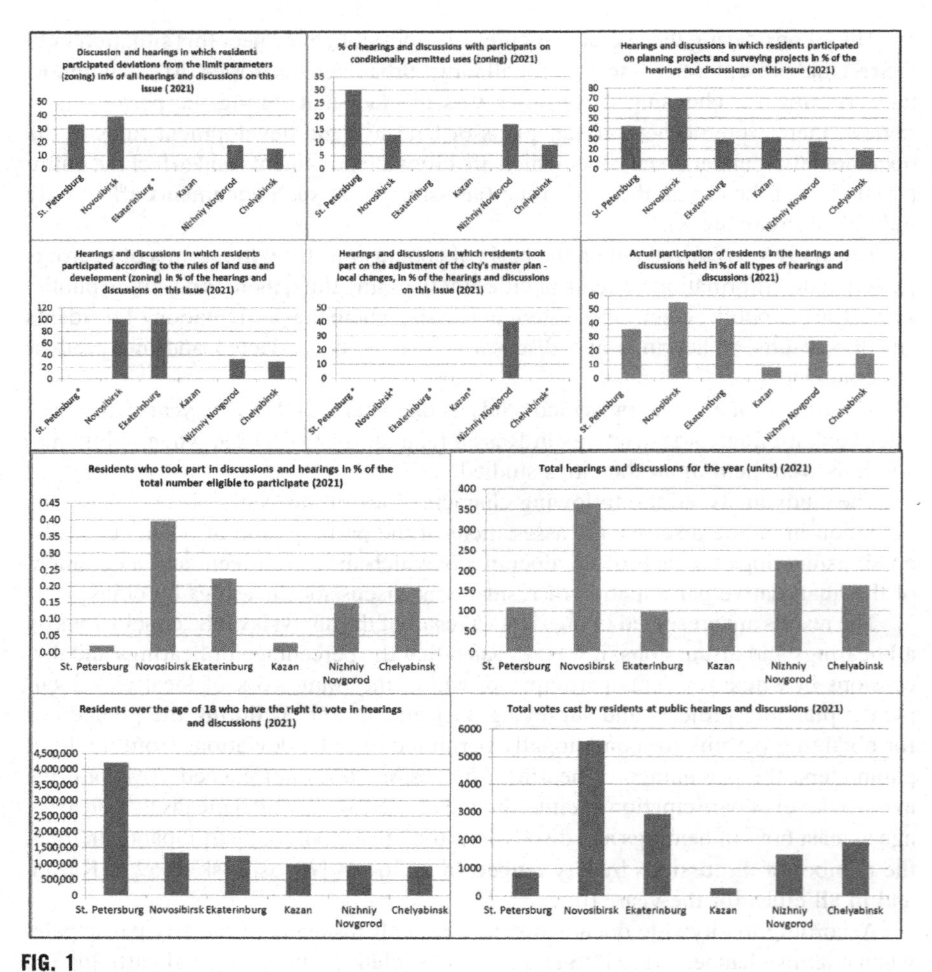

FIG. 1

Participation of residents in public discussions and public hearings (2021).

The results obtained showed that only 5% of surnames and initials were repeated. Thus it can be assumed that such an adjustment is permissible with a certain margin of error for other cities.

Is it really true that the actual number of participants in hearings and discussions is so low on average—only 0.4% of the possible number of voters, subject to possible multiple voting on various issues?

In order to answer this question, a control analysis was conducted in the same cities related to the main document of the territorial planning of the city—the master plan—in public hearings and discussions in which all residents of the city who have reached the age of 18 can participate. Since master plans were adopted in different cities, or their large-scale adjustments affecting the entire city were carried out in

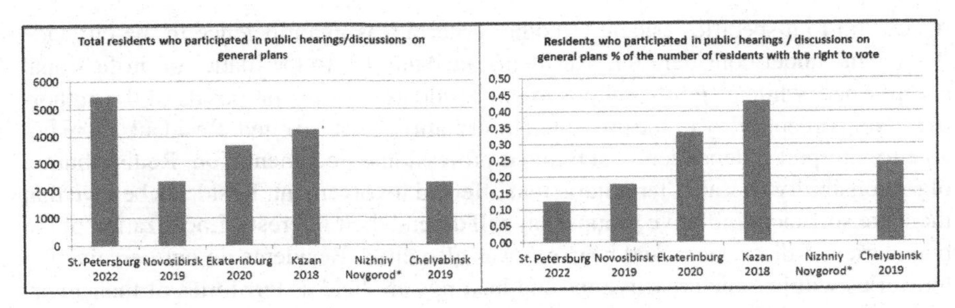

FIG. 2

Participation of residents in hearings/discussions on master plans (2018–22).

different years, the study examined the participation of residents in hearings/discussions in the period from 2018 to 2022. During this period, such events were held in five of the six cities considered.

In the results presented in Fig. 2, it can be seen that the total number of participants was 18,000 residents, and the percentage of participation from the total number of eligible voters did not exceed 0.43% (taking into account the population in cities for the period of the hearings and discussions). The number of participants ranged from 5465 in St. Petersburg to 2300 in Chelyabinsk.

An analysis of the discussions held in St. Petersburg in 2022 was also carried out, which showed that in 150 hearings the number of participants did not exceed 500. At the same time, in more than 50% of cases there was not a single participant.

Currently, for the 16 largest cities of the Russian Federation in 2023, in 14 of them, technologically public discussions are held online (on the main issues), and only in 2 are they held in person with the participation of residents (Chelyabinsk and Krasnoyarsk). Of the 14 cities, in 2 public discussions are practically not applied, taking into account the legislation on the temporary cancellation of this procedure (Moscow and Nizhny Novgorod).

What is the reason for such low participation in official forms of participatory planning, and are residents really not interested in participating in urban planning and shaping their own habitat? In order to answer this question, the opinions of residents posted on social networks were considered, as well as a survey of 50 residents of St. Petersburg and a survey of 30 professional architects-urban planners, about their participation in hearings and discussions as residents of the city outside of their direct professional activities.

As a result, we can identify the following set of criteria that determine the level of involvement of residents in the urban planning process. Awareness of residents about events and forms of participation in them is crucial. A significant part of the population, more than 90% of respondents, as well as those who express their opinions on social networks, did not know about the fact of holding hearings on certain issues, or about the time, place, and forms of discussions and hearings. At the same time, they say they are ready and would like to participate if such information was given to them. One of the most frequent responses was not having received an invitation to a hearing, in relation to a specific resident. Another criteria was incompetence in the citywide documents under consideration. Residents attribute this to the matter of professional competence, while a significant portion of residents trust the proposals of the authorities. Nonperceptibility of the presented information was also cited. Residents were not able to interpret independently submitted professional documentation. Redistribution of responsibility was another criteria that affected involvement. Residents believe that there are sufficient initiative groups that will defend their interests. Localization of the boundaries of discussions and hearings was also cited. Residents do not perceive the territories within which discussions and hearings are held as territories of their interests, because of their scale. Other criteria include the absence of accounts on the website of public services through which the hearings are held, and lack of time to study materials and participate in hearings and discussions.

One of the most frequently highlighted criteria was distrust of the effectiveness of the results of hearings and discussions. Residents do not perceive participatory planning as an instrument of effective influence on city authorities and developers. This criterion is important, taking into account the assessment of residents, both their experience and general ideas about the effectiveness of interaction with authorities. For example, the draft general plan of the city of Yekaterinburg (with a population of 1,238,000 residents over 18 years old) aroused the greatest interest of residents for the entire time of public discussions online.[1] Almost all proposals (567) were rejected, in 90% of cases—without explanation.

A survey of professional architects and planners showed that only 10% of them participated in hearings and discussions in the role of residents. The reasons for nonparticipation were identified as follows: lack of awareness within the framework of local projects, distrust of the effectiveness of official participatory planning tools, and conflicts of interest.

However, is the population really so passive about participating in urban planning? The answer to this question is provided by an overview of urban conflicts arising at local and citywide levels between residents and municipalities and developers. In every city, active forms of confrontation arise when the interests of residents diverge from the solutions offered to them. The confrontation spills out in the form of meetings and rallies, as well as active discussions of residents on social networks. However, the number of participants in such events is limited. An extremely small portion of conflict situations develop into court cases, since formally most of the solutions proposed by the authorities and developers are implemented within the framework of current legislation.

How do forms of participation of residents develop outside of the use of official tools of participation? Is it possible to increase the influence of participatory planning within the modern digital society, and can such tools as hearings and discussions become effective? This study shows how residents lay the foundations of smart urban spaces and what role ICT plays in this formation.

[1]https://66.ru/realty/news/227697/.

6 Information and communication technologies (ICT) in the system of formation of smart urban spaces of the largest city. ICT Tools—Between government, residents, and urban space

"Electronic participation" (*E*-participation) is a form of organization provided by modern information and communication for taking into account the opinion of citizens when authorities are making socially significant decisions, and taking actions to consider appeals addressed to authorities on various issues [33,34].

The 2030 Agenda for Sustainable Development, developed in 2015 by the UN General Assembly, calls on national governments to ensure the creation of a system of responsible, inclusive, representative, citizen-based decision-making at all levels of government (Goal number 16.7).[m] According to Holland [34], people and their interaction are key critical factors in the development of a smart city.

In different countries, including Russia, the authorities, primarily local and regional, are under increasing pressure from society to introduce e-participation technologies, to maximize the consideration of a wide range of interests and views, and to expand the participation of stakeholders in the decision-making process [35–37].

To assess the involvement and participation of residents in urban planning processes, various tools of information and communication technology (ICT) were considered within the relationship of residents with the authorities and urban space. With the development of technology, there are more opportunities for residents in all countries to act and influence urban structures. They organize themselves in the virtual world, but they think and act in the physical space. Technologies lead to the emergence of new common borders (zones of contact and, accordingly, interaction or confrontation) between citizens, the government, and urban spaces [38]. We are investigating whether various e-participation tools are related to the intensity of the use of e-participation technology and how to increase their effectiveness for all parties in the urban planning process, and in particular for local communities in major cities, to create sustainable and effectively developing "smart spatial elements of the urban planning structure."

The practice of using ICT within the framework of a "smart city" is changing both its development and management, as well as the civic consciousness of its residents. A city is defined as smart if it balances economic, social, and environmental development and if it is connected to democratic processes through government participation [39].

It should be noted that this definition lacks the most important characteristic of the development of the city—the formation of its spatial and functional structure, both as a whole and its local territories. The absence of this criterion does not allow us to consider a city at the territorial level, with a different scale of territories, concentrating on the citywide balance within the framework of several selected strategies.

[m]https://sdgs.un.org/2030agenda.

In many works of researchers, it is considered as ubiquitous urban sensing that big data and other technological options create conditions for citywide smart city models [40], but this approach contradicts how citywide systems and their tools are perceived by local communities and how much citywide solutions meet their interests. It is this contradiction between a global city and local territories within it that leads to the need to look at a smart city through its individual smart spaces that are connected to local communities of residents.

In this regard, the most important task is to create a smart multiscale city in which management, technological development, and social factors are correlated with the formation of the citywide space as a whole and its local elements. What forms of urban participatory planning arise in this process and how are they supported by digital practice? Rethinking a smart city through smart urban spaces of different scales based on ICT tools becomes the basis for the formation of local urban communities based on multiscale (from a city, district, quarter, park, street, to local territories at the microlevel) sociocultural spatial units uniting residents.

These boundaries, at different stages of the development of society and the city, are formed through a variety of different tools (social, urban planning, economic, and others). Currently, in Russia, ICT tools are becoming one of the foundations for the formation of urban and local civic communities based on participatory planning through interaction and confrontation with the authorities, which creates the basis for smart urban planning in the future and the creation of a "smart city" with "multiscale smart urban spaces."

7 Internet users in Russia

According to the Global Digital Report for 2023,[n] the use of digital technologies differs significantly in different countries. In 8 countries in the top of the rating, the prevalence of the Internet is at the level of more than 99, and in 55 countries, from 90%. In Russia, this figure is 88.2%, which puts her in 29th place in the world. In January 2023, there were 127.6 million Internet users and 106.0 million users of social networks in Russia, i.e. 87% of the total population used the Internet. According to Mediascope,[o] 71% of the population visits social networks, Telegram, or YouTube daily and 81% monthly. In terms of audience growth rates, the Global Digital Report indicates that COVID-19 accelerated the spread of social networks, so the growth rates in the period from 2020 to 2021 were almost two times higher than in the previous 12 months, and the growth continued until 2022. About 95% of users aged 16 to 64 years own smartphones, while 78% are owners of computers and laptops. At the same time, 91% of users use the Internet with mobile devices, and 80% use the Internet to search for information.[p]

[n]https://datareportal.com/reports/digital-2023-russian-federation.
[o]https://www.sostav.ru/publication/mediascope-57530.html.
[p]https://www.byyd.me/ru/blog/2022/03/digital-2022//.

8 Assessment of the number of participating residents in urban planning based on the analysis of internet search engines—From awareness to participation

How many residents actually participate and are interested in participating in urban planning? If we can estimate the number of actual participants within the framework of official tools by studying the reports on hearings and discussions, it is proposed to use ICT tools to make an enlarged assessment of the number of other participants.

Estimating the number of participating residents in the framework of urban planning of the largest cities, and the use of Internet resources in this process of interaction, the study suggested using search statistics through browsers. At this stage, we have determined that the main search queries will be the names of the official instruments of complicity: "public discussions" and "public hearings." The search for these expressions in the browser is a tool only for an enlarged assessment of participating residents in urban planning, both taking into account the presence of separate issues not related to urban planning to which similar legal forms of public participation apply, and taking into account residents who are not constantly involved in the process, since the main sources of information are the websites of municipalities, professional or public organizations, and social networks. Accordingly, residents already involved in the process use these resources without applying general requests. It can be assumed that the rest of the population is already interested to some extent in the process of participatory planning and receives information from specific dedicated electronic sources of information (specialized sites, social networks), which will be further considered in this study. Statistics https://gs.statcounter.com/ share of the search engine Yandex (Yandex LLC) is 48%; Google is 47%; and other search engines are less than 5%. Thus we can, taking into account the evaluation of Google Ads and Wordstat.yandex, determine the maximum number of interested residents.

Using the Google Ads resource[q] and taking into account search queries per month for February 2023 in Russia, in queries for the keywords "Public discussions" and "Public hearings," the indicators amounted to about 19,000 and 22,000 requests per month, respectively, and, respectively, for 2022, about 300,000 requests per year for each indicator. Another Google trends resource allows us to track the dynamics of popularity for these queries, which shows that, although public discussions are currently taking place in most regions of Russia instead of hearings, public hearings are still the most popular request.

The analysis carried out using the Wordstat.yandex (Table 1) showed that in 2021 there were 223,000 requests for the expression "public discussions," of which from mobile devices (smartphones and tablets) were only 50,000 requests. In 2022, this was 204,000 requests. Of these, from mobile devices (smartphones and tablets), only 42,000, or 20%. Requests on average per month were from 12,000–25,000. At the same time, in 2021 there were 326,000 requests for the expression "public hearings,"

[q]https://ads.google.com/intl/ru_ru/home/.

Table 1 Queries for the largest cities in Russia with a population of over 1 million people per month for February 2023 based on Wordstat. yandex.

		Queries by city					
		Public discussions			Public hearings		
		Stationary devices	Mobile devices	Requests from mobile devices in %	Stationary devices	Mobile devices	Requests from mobile devices in %
	Russia	19,730	5984	30	20,608	6167	30
1.	St. Petersburg	637	207	32	662	207	31
2.	Novosibirsk	363	90	25	373	95	25
3.	Ekaterinburg	210	64	30	228	79	35
4.	Kazan	161	27	17	156	23	15
5.	Nizhniy Novgorod	177	61	34	183	63	34
6.	Chelyabinsk	350	132	38	368	144	39
7.	Krasnoyarsk	250	68	27	263	70	27
8.	Samara	312	125	40	309	118	38
9.	Ufa	248	101	41	253	103	41
10.	Rostov-on-don	209	46	22	231	57	25
11.	Omsk	131	38	29	148	38	26
12.	Krasnodar	224	82	37	242	84	35
13.	Voronezh	177	57	32	200	72	36
14.	Perm	391	165	42	416	167	40
15.	Volgograd	133	48	36	129	50	39
	Total by city	3973	1311	33	4161	1370	33
	% of all Russia	20%	22%		20%	22%	
	Moscow	1817	564	31	1878	575	31
	Moscow in % of requests in Russia	9%	9%		9%	9%	

and in 2022, there were 284,000 requests. Of these, from mobile devices (smartphones and tablets) there were only 72,000, or 25%. Requests on average per month were from 16,000 to 38,000. Considering distribution by cities with a population of over 1 million, requests per month for February 2023 in Russia for "Public discussions" were 25,700 and for "Public hearings" were 26,675 requests.

Similar indicators, based on different search expressions, suggest the identification of these users who find out the nature of the events (public discussions or public hearings). We can assume that if we take each request as coming from a unique user, then their maximum number is only about 20,000 people in Russia.

Other queries seem irrelevant, because they do not allow singling out those who are looking for information purposefully. Therefore, given that the share of the search engine Yandex (Yandex LLC) is 48%, Google is 47%, and for other search engines less than 5%, we can identify the maximum total number of searches per month on all search engines as 40,000 to 50,000 due to the small variability of the query by month. There are no statistics on the distribution of the uniqueness of requests, but the data obtained give us the maximum number of users who are interested in this issue during the month in Russia. The actual number of unique requests will probably be less. People making inquiries in the field of urban planning regarding participation in hearings and discussions are looking for both specific information and analysis, and an overview of the situation, the reaction of the authorities and neighbors, but having found the source of information, they already turn to it.

Our survey of residents of St. Petersburg, conducted as part of public discussions on the Rules of land use and development in St. Petersburg in 2020, showed that residents rarely monitor (observe in the future) the current situation. To the question of how many times you have requested information about discussions via Internet search engines in general, before they were held, during, and after their conduct, out of 50 residents surveyed (the survey was conducted on the social network Vkontakte (VK) in one of the nonspecialized groups, the community of residents of the district), 20% had heard about the hearings and only 5% monitored the situation, making several requests at different stages of the hearings. On average, these 5% made requests two to three times. As for participation in the discussions, the number of participants was only 2% and only these 2% were members of specialized groups dedicated to urban planning in social networks.

Accordingly, if we proceed from the maximum number of requests of 50,000 per month in Russia, then about 5500 requests fall on the cities of millions, excluding Moscow, the largest city in Russia, which is home to 9,915,000 residents over the age of 18 who have the right to vote—8.4% of the total adult population. (In the whole country, this figure is 118,240,000 residents according to the Rosstat statistical bulletin for 2023.)[r]

The number of adults in all million person cities is 17,400,000 (excluding Moscow).

Thus, on average, within a month, no more than 0.02% of the adult population of the largest cities under consideration purposefully turn to Internet search engines for

[r]https://rosstat.gov.ru/storage/mediabank/Bul_chislen_nasel-pv_01-01-2023.pdf.

information on public hearings and public discussions in the field of urban planning. This indicator does not give an idea of all residents who are involved in the process of participatory planning (at the level of obtaining information), but together with the results of the analysis of social networks and other public electronic sources of information that we have considered in this study, it allows us to estimate their order of the total population.

Also, we can estimate that only 30% of requests for public hearings and public discussions in the field of urban planning requested through the Yandex search engine are made via mobile devices. As already discussed earlier, according to statistics, 95% of users aged 16 to 64 years own smartphones, while 78% are owners of computers and laptops.[s] At the same time, 91% of users use the Internet using mobile devices.[t]

These indicators reflect that the high degree of availability of mobile Internet is not directly correlated with the degree of interest of residents in participatory urban planning. Requests are not directly dependent on the level of technological development and access to the mobile Internet, primarily for residents of megacities.

9 Forms of interaction within the framework of ICT and their impact on the processes of citizens' participation

Urban development ideally pursues the creation of a comfortable, accessible, favorably sustainable living environment. However, such a representation is extremely conditional, with the presence of cardinal differences in views on these concepts on the part of governments, various groups of the population, and business. Ever since the struggle of Jane Jacobs against Robert Moses, the idea that urban planning is a confrontation between different groups with different interests has not changed fundamentally, although it has undergone a number of correlations in terms of the formation of institutions to reach compromises and develop optimization solutions (within the framework of participatory planning, the principles of new urbanism, etc.). The task remains the same, that is, the search for optimization solutions to prevent or resolve conflicts. Despite the fact that a large number of studies in the field of urban studies are devoted to the analysis of specific conflicts in specific cities and processes, the lack of a general and systematic approach to this issue is significant [41].

In the process of urban planning, various participants in urban planning activities interact, including planners, officials, politicians, entrepreneurs, and the general public. Their participation is based on personal, group, or institutional interests. Interaction involves participation in persuading other participants, power struggles, and negotiations. Consequently, the planning process is reduced not only to professional aspects, but also to the impact of all participants on both the process itself and its results [42].

[s]https://www.byyd.me.

[t]https://www.byyd.me/ru/blog/2022/03/digital-2022/.

The participation of the parties in the planning process can be divided into both *regulated by law*, when the norms of the law regulate the issues of public participation, and *spontaneous*, when these processes originate from within society, as a response to the actions or inaction of government or business, or as a desire to change their habitat.

The tools within technologies of electronic participation of residents in the field of urban planning, which are used by various participants in urban planning activities in Russia, can be classified, taking into account the transformation of the three-level scheme proposed by Linders [30], depending on who is the recipient and who is the supplier: G2C (government for citizen), C2G (citizen for government), and C2C (citizen for citizen), with the inclusion in this scheme of interaction between government and business G2B (government for business).

All tools can be divided into three blocks: the first is tools for obtaining information, the second is tools for communication and decision-making, and the third is tools for expressing one's views and interests, addressed both to decision-making institutions and to communities as a whole to attract supporters.

The G2C and G2B format assumes that the administration distributes information and data whose availability allows their processing and, thus, affects the participation of residents and businesses in the planning process. These are primarily the websites of administrations or planning departments, as well as state, regional, or municipal geoinformation systems. This can also include electronic platforms for voting, making proposals, and surveys, presented in the form of: separate systems, tools integrated into the structure of administration sites, or integrated into state or municipal GIS systems. All the structures considered are created taking into account the existence of mandatory, regulatory procedures, the placement of approved legal acts, informing the public, holding public hearings and public discussions, as well as for public opinion polls initiated by authorities of different levels (in isolated cases). Some of the state and municipal electronic platforms have their own mobile applications.

All these tools are part of a state institution and participation in the voting held on them, or other forms of feedback from a citizen or company, is associated with registration in local or federal identification and authentication systems. The main system is the federal state information system "Unified Identification and Authentication System in the infrastructure that provides information and technological interaction of information systems used to provide state and municipal services in electronic form" (ESIA). This system provides authorized access of participants in information interaction (citizens-applicants and officials of executive authorities) to information contained in state information systems and other information systems (Decree of the Government of the Russian Federation No. 977 of November 28, 2011 "On the Federal State Information System").

Registration in the ESIA is the basis for electronic registration on the Public Services website,[u] as well as other government websites or information systems with the

[u]https://esia.gosuslugi.ru.

possibility of requesting information, receiving information, and sending applications. To obtain an ESIA account, it is necessary to verify your identity using passport data and SNILS (individual insurance account number), including through the physical submission of an application and verification of identity.

According to the latest official statistics posted for 2021, the percentage of residents using the mechanism of obtaining state and municipal services in electronic form in urbanized regions is at least 70% (Chart 1).

Another tool of the G2C and G2B format is the official pages of the administration bodies in the field of urban planning in social networks and messengers. Their attribution to this format is determined by the one-sided format of interaction with residents and businesses: informing and, in some cases, surveys initiated by administrations.

One of the tools is also created by state or municipal authorities with the participation of external crowdsourcing companies, working on the principle of crowdsourcing with the involvement of a wide range of parties interested in urban planning and, first of all, active residents.

C2G-based tools are the main forms of electronic participation, as they involve collective forms of interaction initiated by residents to influence decisions made by city administrations or regional and federal authorities. On the one hand, they serve as tools that make it possible to discuss, collect ideas, vote, etc., and they can provide a joint solution to a certain problem. People can develop decision procedures, identify and prioritize requirements, participate in the search for solutions, etc. In the field of urban planning that we are investigating, such tools are platforms for online petitions. These are external platforms that are specialized, such as www.roi.ru and Change.org, or integrated as part of municipal or regional information portals with

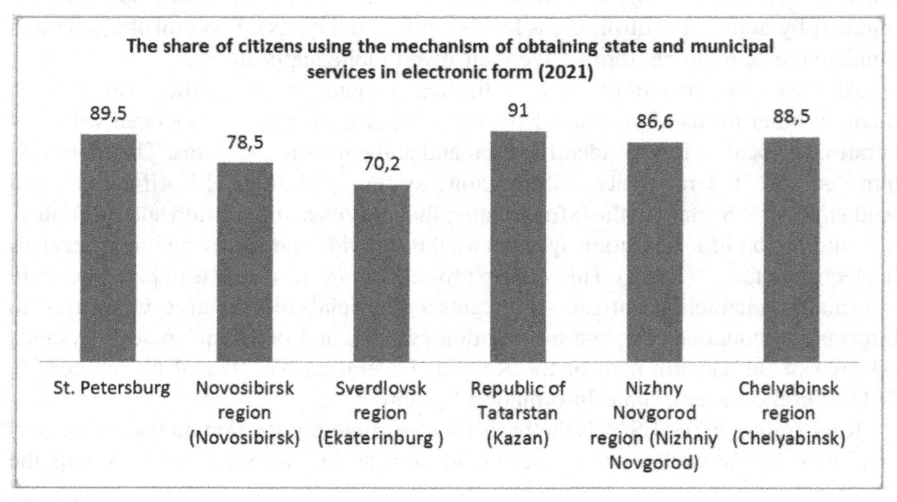

CHART 1

Share of citizens using the mechanism of receiving state and municipal services in electronic form (https://fedstat.ru/indicator/43568).

the identification of participants through the ESIA. Interactive GIS systems, in which elements of feedback with residents are integrated, can become such forms.

Tools based on C2C are the third type of electronic participation. These systems do not require the participation of the state, but instead give people the opportunity to organize in a virtual space to improve the life of the city [43]. In Russia, such tools are specialized websites and portals with active forums, as well as communities created in social networks and messengers.

10 G2C format tools

In 88% of cities (14 out of 16 cities including Moscow) with a population of over a million inhabitants, digital platforms are used to conduct public discussions in the field of urban planning (Table 2). In nine cities, these are local platforms or platforms based on the system of public services used by the municipality. In the remaining five are regional platforms that work for the entire subject of the Russian Federation, while two cities have the status of cities of federal significance (subjects of the Russian Federation, Moscow and St. Petersburg). Thus the digital platform for public discussions is the main form of holding hearings within the largest cities, within the framework of separate documents. The electronic form does not provide for meetings, including online discussion participants: residents or owners, city administration, designers, architects, and representatives of developers and investors.

Digital platforms for discussions in Russia can be divided into two types according to the nature of the registration of participants: using a system of public services

Table 2 Digital systems for the participation of residents in public discussions in urban planning.

		Not integrated not linked to the public services portal	Integrated with the possibility of registration through the portal of public services	Independent information system	Integrated into the administration's website	Integrated into GIS	Availability of a separate digital portal for surveys or acceptance of applications
1.	St. Petersburg	+			+		
2.	Novosibirsk		+				+
3.	Ekaterinburg		+		+		
4.	Kazan		+	+			
5.	Nizhniy Novgorod	+		+			
6.	Chelyabinsk						
7.	Krasnoyarsk						
8.	Samara		+		+		
9.	Ufa		+	+			
10.	Rostov-on-don		+				+
11.	Omsk		+				+
12.	Krasnodar		+	+			
13.	Voronezh		+				+
14.	Perm		+			+	
15.	Volgograd		+				+
	Total	2	13	4	3	1	5
	% of all cities	16%	84%	26%	25%	9%	41%

(84%), and platforms on which independent registration of residents is allowed (16%), taking into account the provision of personal data.

All platforms can be divided into two types: independent platforms that are not linked to the site of public services, or those integrated with sites for the provision of public services. It should be noted that in the personal account of residents on public services websites, there is no link to the ongoing public discussions.

Platforms for holding hearings that are not integrated into the platform for the provision of public services can also be divided into two types: separate information systems that are not associated with the websites of administrations and those integrated into the website of city administrations (for example, St. Petersburg).[v]

The study identified eight independent systems (when analyzing 15 cities with a population of over a million). Of these, it is possible to single out platforms for participation in discussions integrated into the urban geoinformation system, which includes the entire range of urban planning documentation, as well as other layers related to urban development (Perm).[w]

Separate platforms for holding hearings have different structures, taking into account the format of the discussions; they differ in the degree of information content and accessibility of information, as well as in the channels of interaction and the possibility of feedback between the initiators and participants of the discussions.[x,y,z]

Platforms for public discussions, due to an additional interface, can be used for holding various votes on urban issues, including on urban planning and transformation of the urban environment. In most cases, this is a separate tool.

Independent platforms are used for surveys; they are available in 90% of cities with a population of over a million, and they also include the possibility of residents submitting applications on environmental and landscaping issues and information urban content. These platforms in 80% of cases have mobile applications based on iOS and Android. The presence of a platform and applications is only the technical part of complicity, since the basis for their work is the nature of the content and the backlash to the statements of citizens. The results of the study show that the issues of urban development included depend on the policies implemented by administrations. All surveys and activities conducted on them, according to Arnstein's classification [31], can be divided into "manipulation and therapy" in terms of general declarative surveys and forms of interaction that do not bear actual consequences, before "informing." It is particularly necessary to highlight the almost complete disregard of the local level of surveys of concern to the local community at the level of neighborhoods or districts. The number of active users of applications and portals in Russian state and municipal systems is quite difficult to determine. However, by the number of downloads of applications (https://app.sensortower.com), it can be argued that the number of participants for a city

[v]https://kgainfo.spb.ru/38171/.
[w]https://isogd.gorodperm.ru.
[x]https://ag.mos.ru/debates?filters=all.
[y]https://dem.nso.ru/#/.
[z]https://dispute.kzn.ru.

with a population of more than a million does not exceed 50,000 (Yekaterinburg App ID: 1535796401, Voronezh App ID: com.osk_voronezh), i.e., no more than 4.5% of the adult population.

The total number of portal users together with mobile applications in relation to the total number of city residents does not exceed 8% (except in Moscow, where linking this portal to various city services increases the level of use). For the example portal of St. Petersburg "Our St. Petersburg",[aa] in 2023, the total number of users registered on the portal of messages (as of 09/01/2023) is 271,000.[ab]

Thus the digital tools at the disposal of city administrations—digital platforms and related mobile applications for the possibility of participatory planning used by residents and cities—are extremely limited. Cities practically ignore local communities and local problems. The policy for using digital tools is the most important factor limiting the effectiveness of the tool. Thus the level of digital integration of state and municipal systems does not achieve the goals of creating a smart city and smart urban spaces through the participation of residents.

11 Forms of interaction and feedback within the framework of holding hearings on the basis of ICT

The most important criterion for determining the effectiveness of public discussions based on ICT tools are the forms of interaction between the organizers (administration, designers, developers, and residents). As part of the study, all the forms used for the largest cities were analyzed (Table 3).

Since the ICT hearings are accompanied by mandatory nondigital formats of participation (for those who do not have access to the Internet), the main forms of interaction and feedback before the discussions, there are questions that residents and interested parties can ask by phone or at the exposition to representatives of the administration and designers. In practice, phone calls are used to solve technical issues of participating in discussions and obtaining information about previously posted information. There are usually no qualified specialists at the exhibition who are able to answer questions on projects; usually this is the place for interns or technical specialists. Interaction between the participants of discussions in the absence of face-to-face meetings is not provided.

Within the framework of digital platforms and websites, alternative feedback tools were used with residents. These tools can be divided into [1] informing on the submitted materials within the framework of which residents or other interested parties have the opportunity to receive additional information from the authorities and share information about discussions with other residents on social networks, and [2] forms of feedback from participants in discussions on the project as a whole

[aa]https://gorod.gov.spb.ru/.
[ab]https://www.gov.spb.ru/gov/otrasl/c_information/statistic/.

Table 3 Forms of interaction and feedback related to holding hearings based on ICT.

		Questions on the online platform	Reactions to discussion comments, to other speeches or answers to questions	Information about the platform - a direct link or banner on the administration's website	Possibility of feedback on the results of the hearings	Digital forms of explanations or comments before discussions	The opportunity to ask questions by phone or at the exhibition on the project	The ability to share a link on social networks
1.	St. Petersburg			+		+		+
2.	Novosibirsk		+		+		+	
3.	Ekaterinburg			+			+	
4.	Kazan	+		+				
5.	Nizhniy Novgorod	+					+	
6.	Chelyabinsk[a]							
7.	Krasnoyarsk[a]							
8.	Samara						+	
9.	Ufa			+				
10.	Rostov-on-don		+	+				+
11.	Omsk							
12.	Krasnodar			+			+	
13.	Voronezh			+			+	+
14.	Perm			+				
15.	Volgograd			+				
	Total	2	2	9	1	1	6	3
	% of all cities	15%	15%	69%	7%	7%	46%	23%

[a] There is no information system.

and suggestions or comments from other participants, as well as in general, based on the results of the discussion.

Information using ICT tools is used in only three cities (excluding Moscow). In St. Petersburg (as well as Moscow), these are individual video conferences that residents can sign up for through the portal, in automatic mode. The disadvantage of this form is the limited number of participants who can use it, no more than 10 participants per day. Video conferences use the Russian TrueConf Server video communication system. After signing up, the participant specifies the email address where the link with the invitation to the consultation will be sent; this does not require preinstallation on a computer or mobile device. In two cities, there is an opportunity to ask questions through the platform for holding hearings in offline format.

The feedback from residents is presented only through the placement of Like and Dislike buttons, in relation to the comments of other users (within the framework of their comments, opinions, and suggestions) and based on the results of the discussions in general. This format is presented only in two cities.

Another important ICT tool related to interaction with participants, as part of information platforms, is the ability to send a link to an event through social networks or messengers; this form is presented in three cities and allows information to be shared on the social networks Vkontakte (VK) and Telegram.

Taking into account the analysis of the applied forms of interaction, the systems in St. Petersburg, Novosibirsk, and Voronezh can be considered the most effective.

Thus, in the presence of possible forms for feedback or personal consultation of residents, they are rarely used by municipalities. At the same time, their role is extremely important in the effective use of public discussions. A significant degree of their integration within the existing information systems has enabled the online platform to be used as a tool for both G2C and C2G, and the expansion of tools for attracting discussion participants and integrating local discussions into it in the format of C2C forums. This is especially important when solving local issues concerning a limited number of participants. The expansion of such formats within the framework of online platforms can be effective in the formation of smart urban spaces and public perceptions for their consideration by designers, developers, and administrations.

12 Availability of data and materials for analysis

The study analyzed and evaluated the availability of materials for hearings and discussions. We did not aim to evaluate the informativeness and perception of the presented materials, but studied only the forms of transmission and provision of information that allow us to assess the effectiveness of their use related to the use of ICT tools.

Two criteria were identified: ease of use and geoinformativeness. As part of the "ease of use" criterion, the following characteristics were taken into account: the availability of source data—exposition materials (documents submitted for discussion) as part of an information system or the placement of these materials on the websites of municipal administrations (subjects), the placement of materials in preview format without the need to download them, the adaptability of websites and information systems (the possibility of effective photographing of the work on mobile devices), and the division of hearings by type of documents.

The geoinformativeness criterion is applied to the following characteristics: geolinking of the territory of the hearings, which allows determining the boundaries of the territory in which they are held; geolinking of the boundaries in which residents can participate in the hearings (may not coincide with the boundaries of the document development), for example, adjacent land plots or the boundaries of the territorial zone; linking of the materials placed into geoinformation systems to municipalities or subjects of the Russian Federation.

Regarding assessment of the ease of use of the exposition materials, the following conclusions can be drawn for cities using information systems: only in 41% of cases are the materials placed on the information platform for holding hearings; in all other cases, the materials are placed on the websites of city administrations. In almost all cases, the materials require downloading and cannot be viewed directly on the platform or on the website (two cities). In most cases, the materials are archives that require special preinstalled programs to be disclosed. Only 33% of sites and information platforms are adaptive, i.e., they are presented on mobile devices without distortion and taking into account scaling (adaptability was tested on the basis of two resources, Browserling and Adaptivator).[ac,ad] According to the results of the study, 33% of websites and online discussion platforms meet this requirement.

[ac]https://www.browserling.com/.
[ad]http://adaptivator.ru/.

Table 4 Accessibility and geoinformation of materials for public discussions.

		Ease of use					Geoinformation		
		Materials of the exhibition on the online platform	Materials of the exposition on the administration's website	Materials of the exhibition with a preview without downloading	Adaptability	Separation by type of documents	Geo-linking of the territory	Geo-linking of residence boundaries for possible participants in discussions	Linking to GIS
1.	St. Petersburg	+			+	+	+		+
2.	Novosibirsk	+				+			
3.	Ekaterinburg	+				+			
4.	Kazan	+			+	+	+	+	
5.	Nizhniy Novgorod		+	+	+				
6.	Chelyabinsk[a]		+						
7.	Krasnoyarsk[a]		+			+			
8.	Samara		+				+		
9.	Ufa		+						
10.	Rostov-on-don	+			+		+		
11.	Omsk		+			+			
12.	Krasnodar		+		+				
13.	Voronezh		+						
14.	Perm		+	+		+	+		+
15.	Volgograd		+						
	Total	5	10	2	5	7	5	1	2
	% of all cities	41 % *	66%	13%	33%	46%	41 % *	6% *	13% *

[a] Taking into account cities with online platforms.

Geoinformativeness of materials is the most important condition for the effective use of the discussion tool. After analyzing all the ICT-based systems for holding hearings, it was revealed that the link that allows identifying both the territory of the discussions and the localization of the territory within which residents and owners of land plots and real estate have the right to participate in the hearings has an extremely low level of use. Geolinking of the hearing territories is available as part of information systems and websites only in 41% of the cities studied (five cities) and only in one city (Kazan) are the boundaries of territories allocated in which residents and owners have the right to participate in local discussions. Of the five cities with geolinked territories, only two of them are linked to the urban GIS system, in which various layers of information are presented; these are Perm and St. Petersburg, and only in Perm are hearings held using the GIS portal. At the same time, all cities with a population of over a million inhabitants have operating publicly accessible municipal or state geoinformation systems, with a wide range of information provided [44].

Thus, after analyzing all cities according to the criteria of ease of use of materials and geoinformation, it can be concluded that the most effective system is found in three cities: St. Petersburg, Kazan, and Perm (Table 4).

13 Channels of participation of residents

Within the framework of the study, for all cities and proposed municipalities (and subjects of the Russian Federation, in case of redistribution of powers) considered, the channels of participation in public hearings and discussions (in relation to urban planning activities) are presented in the table (Table 5). The channels were examined in order to assess their inclusiveness and effectiveness in urban planning issues.

Table 5 Digital and nondigital voting channels.

		Using ICT			Without the use of ICT			Total channels
		Information system	Email	On the administration's website	During face-to-face hearings	Entry in the visitor log	Mail	
1.	St. Petersburg	+					+	2
2.	Novosibirsk	+	+				+	3
3.	Ekaterinburg	+				+	+	3
4.	Kazan	+	+			+	+	4
5.	Nizhniy Novgorod	+	+			+	+	4
6.	Chelyabinsk[a]		+		+	+	+	4
7.	Krasnoyarsk[a]				+	+	+	3
8.	Samara	+	+	+		+	+	5
9.	Ufa	+					+	2
10.	Rostov-on-don	+				+	+	3
11.	Omsk	+				+	+	3
12.	Krasnodar	+	+		+	+	+	5
13.	Voronezh	+	+				+	3
14.	Perm	+	+	+	+	+	+	6
15.	Volgograd	+			+	+	+	4
	Total	13	8	2	5	11	15	
	% of all cities	80%	53%	13%	33%	73%	100%	

[a] There are no information systems.

In the presence of digital forms of participation, nondigital forms of participation are mandatory, due to the requirements of federal legislation.

All types of channels were divided according to the principle of using or not using ICT tools. The channels using ICT tools included: voting through an information system (platform), by sending emails and by sending a package of documents for participation in hearings through the website of the administration of the municipality (subject of the Russian Federation).[ae] Channels without the use of ICT tools (traditional methods) included: mailings, journal entries at project expositions, statements at public hearings made in residents' speeches in cases when hearings are held.

In all cities that have information systems based on separate platforms or on a public service information system, these are the main channels of citizen participation. In 53% of all cities, the voting tool that allows taking into account the opinions and suggestions of residents is e-mail. In two cities, 13% of all such information can be provided by sending messages from the administration's website.

When voting in all formats using ICT, except for information systems, residents and owners must attach a full package of documents confirming their right to vote (including copies of a passport with registration data at the place of rèsidence, information about real estate objects, etc.) on a specific issue, taking into account the territorial nature of the hearings/discussions. Also, in most cases, it is necessary to download and fill out the consent forms for the processing of personal data separately.

The main tools among nondigital forms are the use of mail and the ability to leave an entry in the visitors' log at the project exhibition, or personally submit a letter to the administration. In all cases, a necessary condition is the submission of individual

[ae]http:/admkrgl.ru.

data that allows the identification of a resident with reference to the boundaries of the territory of discussion and the submission of materials confirming the rights of owners for owners of land and real estate. For cases when the formats of public hearings have been preserved, statements can be made by participants at such hearings. Taking into account the number of channels for submitting proposals and comments within the framework of public discussions without a meeting of residents, the leaders are the cities of Perm and Samara, which have five channels each.

It is not possible to determine the effectiveness of the use of channels for all cities, due to the lack of published statistics and the lack of reflection in the final protocols of most cities of channels for which applications from residents and owners were received.

For three cities where information systems are used (Kazan, Perm, and Samara), "conclusions based on the results of discussions" were selected, which indicated the channels through which applications were received. An analysis of 100 conclusions (with the participation of more than 50 residents) showed that 4.5% of residents' opinions were presented through a log entry from visits to the exposition, and 1.5% were sent by mail.

To assess the effectiveness of making entries in the journal at project expositions, it should be noted that the materials of the exposition are placed in the buildings of city administrations (as well as in district administrations for St. Petersburg, as the city is divided into 18 districts), or specially designated urban facilities. The exposition in most cities is open only on working days—3–4 h during working hours—and the scale and format of the documents presented at it is not regulated. Availability of printed handouts (brochures, leaflets, albums, or other materials) is absent both in regulatory documents and in practice. The timing of putting materials on display is also regulated at the local level and can range from 1 month to several days; there are even cases of only 1-day placement. All these conditions make such a form extremely inefficient. Mail messages are only considered received before the meeting of the commission of obedience/discussions.

From the analysis of channels and their use, it can be concluded that the electronic form of participation through a digital platform is the most inclusive and effective in the current legal and factual conditions and allows residents and owners to effectively participate in the participatory planning process at the local and citywide levels. Digital platforms may have different levels of efficiency, which we will consider further, but they are more efficient than other forms.

Thus, having considered all the 15 cities studied, it can be concluded that the most effective information support systems for participatory urban planning are provided in St. Petersburg, Perm, and Kazan.

14 Crowd platform

A separate tool of the G2C format is crowd platforms created with the participation of the state. Crowdsourcing is an effective tool for bringing stakeholders together in one place to create projects and exchange ideas and experience in order to solve problems facing business, government, and society. In 2020, separate attempts were made in Russia at the state level to use such forms of complicity outside the legal framework of the official system of urban planning regulation. They were implemented in the creation of an information crowd platform "100 cities"[af] with the participation of state institutions.

The platform was created for the involvement of residents, leadership training, and education. As part of the implementation of this platform, a "Standard for involving citizens in solving issues of urban environment development" was developed. The standard is focused exclusively on open and creative urban spaces and does not affect the most important aspects and documents of spatial and functional urban planning. The standard presented various ways and formats of participation and involvement of residents. The standard provides 20 formats for working with residents: focus groups, in-depth interviews, lectures and excursions, workshops, initiative budgeting, public-private partnership, and other options. At different stages, various applications of ICT tools were envisaged, from websites and Internet surveys to the creation of applications. However, the status of the standard has no legal force; these are recommendations only and its application is not mandatory and is not integrated into the existing legal field. The use of the tools listed in its composition does not bear any legal consequences. At the same time, its application in 3 years in different cities (primarily small ones) in the form of 127,000 votes cast (both for and against projects) in various projects within the framework of electronic surveys or other forms of complicity in the field of landscaping shows the degree of involvement of residents. However, after a year of active use, the project was actually frozen; after 2020 there are no active projects on the platform.

Thus, considering all the tools of the G2C format, it can be concluded that, despite the high level of technological achievements in the field of ICT, their effectiveness is quite low, and the format of their use does not allow taking into account the interests of local urban communities at different spatial and territorial levels. The availability of technical conditions is not a prerequisite for the formation of a smart city and smart urban spaces, creating only the effect of participation of residents in urban planning.

15 C2G-based tools

In Russia, in the absence of widespread access of local social groups to television, radio, and print media, residents are increasingly using online tools to call for social change, including attempting to influence urban planning processes. One of the widespread practices is the use of online petitions. Modern conditions, such as the coronavirus and the sociopolitical context, have transferred a significant part of the interactions between the population and the authorities to the virtual environment, which forms new reactive forms of communication between authorities and the population [45].

The interests and requests initiated by the local population strengthen the reactive nature of interaction in the context of urban planning and the creation of a

[af]https://100gorodov.ru/.

comfortable environment. In this regard, two digital platforms were considered: the Russian Public Initiative (www.roi.ru) and Change.org (www.change.org). The directions of social initiatives and the nature of interaction between the population and the authorities in the field of urban planning were highlighted.

ROI—an Internet resource for hosting public initiatives of citizens of the Russian Federation—is a state tool for collecting public initiatives created by Decree of the President of the Russian Federation dated 04.03.2013 No. 183 "On consideration of public Initiatives sent by Citizens of the Russian Federation using the Internet resource 'Russian Public Initiative.'"

Russian citizens authorized through the state-supported ESIA identification system can put forward various civic initiatives or vote for them. Initiatives are considered if, at the federal level and in the subjects of Russia with a population of more than 2 million, at least 100,000 votes were cast in support of the initiative, and at the regional and municipal level, at least 5% of the registered population.

Thus, for all cities of the Russian Federation with a population from 1 million to 2 million, more than 50,000 votes will be required to consider the petition of residents, and for Moscow and St. Petersburg where the population exceeds 2 million residents, 100,000 votes will be required.

There is no section on the ROI site of statistical and analytical reports on its work, and no information about the use of materials of the proposed initiatives and the nature of the use of user data. The total number of users is not known. There are no counters by city or by question type. The platform provides only voting; it is possible to forward the link to social networks (local Russian networks VKontakte and Classmates) and by e-mail. As of March 2017, out of 9800 initiatives published on the website, only 25 were considered, and 12 of them were adopted (including 3 municipal ones). By March 2023, the number of initiatives amounted to 21,000. Of these, according to the results of our study, only 400 initiatives were considered in all territories, including federal initiatives in the field of urban planning and development. There are only 36 initiatives with more than 1000 participants. Of these, 21 are geographically linked to cities with millions (Moscow—12, St. Petersburg—4, Yekaterinburg—4, Voronezh—1).

The largest initiatives are: "The Green Belt of Moscow"—100,000, "Against the construction of a temple on the central square in Yekaterinburg"—21,000, against the laying of a highway through a park in Moscow—11,000, all the rest—on average from 3000 to 1500. For all other initiatives, the average number of participants was 250 people.

Issues that are raised within the framework of petitions include: the planning and development of residential areas, the preservation of the historical environment, the preservation of natural components of the urban environment and the formation of new recreational elements of development, transport infrastructure, environmental conflicts of the placement of industrial and man-made facilities, and the preservation of public spaces as public urban spaces. All these problems are 90% concentrated within the framework of diverse but local territories, from the district to the quarter.

Thus we see only one initiative in the field of urban planning, which was considered at the level of cities with millions, which is the initiative "Green Belt of Moscow, " which gained 100,000 votes, which does not cover the local level, but is citywide. The decision on it was made, but it was not actually implemented.

Despite the fact that electronic petitions are not an official document but serve only as a feedback tool between the population and decision-making authorities, this form unites interested residents, although the platform itself does not allow for discussions, exchange of proposals, etc. Despite the fact that the portal itself and its legal form are deeply ineffective, since it does not resolve the issues for which it was created, the platform for online petitions as a tool of interaction creates conditions for the formation of positions of individual local communities and prerequisites for their unification within the framework of solving urban development and planning tasks, in fact becoming an element of electronic participation in the C2C format.

On Change.org—one of the world's largest platforms for online petitions—the number of participants from Russia amounted to 15 million people in 2020 (according to the head Change.org in Russia, Eastern Europe, and Central Asia), and all of them have signed or created a petition at least once.

Within the framework of this study, petitions from Russia presented on the website as part of the participation of residents in urban spatial planning, development, and urban infrastructure were considered (Table 6). The analysis was carried out by evaluating petitions by keywords (building, planning, heritage preservation, public spaces, layout, demolition, destruction, seizure, parks, neighborhoods), followed by

Table 6 Online petitions collected by residents of the largest cities of Russia on urban planning issues on Change.org

	Cities with a population of over 1 million inhabitants	Residents over 18 years of age	Petitions of over 1000 participants	Maximum petition, thousands	Average range of thousand participants
1.	Moscow	9,915,000	62	85	13.2
2.	St. Petersburg	4,208,000	28	96	17.3
3.	Novosibirsk	1,334,000	7	7	3.3
4.	Ekaterinburg	1,238,000	6	168	4[a]
5.	Kazan	984,000	16	60	8.9
6.	Nizhniy Novgorod	950,000	5	66	2[b]
7.	Chelyabinsk	926,000	5	3	2
8.	Krasnoyarsk	855,000	4	4	1.8
9.	Samara	904,000	4	4	1.9
10.	Ufa	892,000	4	3	1.5
11.	Rostov-on-don	890,000	4	7	3.1
12.	Omsk	903,000	4	5	6
13.	Krasnodar	899,000	3	15	5.7
14.	Voronezh	827,581	6	12	4.4
15.	Perm	825,000	2	3	1.5
16.	Volgograd	778,854	2	1	1
	Total	27,329,435	162		4.8

[a] *Excluding the petition for the construction of the temple.*
[b] *Without taking into account the petition against the development of the floodplain of the Volga River.*

context assessment. All appeals over a 5-year period were considered, from January 2018 to February 2023. The total number of petitions was 452. Petitions of a general nature without territorial reference were not taken into account. Of these, petitions were allocated by geographical reference to the largest cities; 320 petitions were allocated, i.e., 70.7%, respectively, and only 30% of petitions were attributed to cities with a population of less than 1 million. The number of residents over the age of 18 who live in the Russian Federation is 114,153,000; and 27,329,000 live in million-plus cities, i.e., 24% of all residents. Thus we can observe a situation in which the petition as a tool is used mainly by residents of the largest cities. Of all the petitions for the cities under study, petitions with more than 1000 signatories were singled out; there were 162 of them. Accordingly, 47.5% of petitions gained more than 1000 votes.

At least one petition with 1000 votes was submitted in all the million-plus cities considered. The maximum number of petitions that gathered more than 1000 votes was submitted from Moscow—62, while the petition "For the conservation of the Moose Island Nature Reserve" collected 83,000 votes, "To prevent the renovation of low-rise buildings in the Kuntsevo area" 85,000, and "Against the construction of a high-speed highway in the area of a radioactive waste burial ground within the borders of Moscow" 64,000. These petitions turned out to be some of the most massive. In addition to Moscow, cities with petitions with more than 50,000 participants were: St. Petersburg (3 petitions), including a petition for 96,000 participants for the preservation of the "300th Anniversary of St. Petersburg Park, " Yekaterinburg with the most massive petition of 168,000 votes against the construction of a temple in the park, Kazan with a petition against filling of the Kazanka River, and Nizhny Novgorod with a petition against construction in the floodplain of the Volga River. The cities with the largest number of petitions submitted that received more than 1000 votes were Moscow, St. Petersburg, and Kazan. And if in cities with a population of over 2 million, the average range of petitions under consideration reached 17,000, then in all others it ranged from 1000 to 8900. On average, the number of signatories of petitions for the studied cities was 4600 people. For the entire period in Moscow and St. Petersburg, petitions gained about 500,000 votes in each of the cities.

Two petitions gathered the largest number of votes in cities with a population of less than 2 million inhabitants, in Yekaterinburg and Nizhny Novgorod. The maximum number of votes in Russia was collected by the petition "Save the park at the Drama Theater—one of the few green parks in the center of Yekaterinburg" against the construction of a temple in Yekaterinburg (2019) on the site of the park, which gained 168,000 votes and caused popular protests. This movement received widespread news coverage at the national level, which led to discussion of the project in the highest executive authorities of the country. As a result of the influence of the urban community, the construction was canceled.[ag]

The second major peak number of votes for a city with a population of less than 2 million was collected by a petition to protect the Volga River floodplain in Nizhny

[ag]https://www.change.org/p/сохранить-парк-у-театра-драмы-один-из-немногих-зеленых-парков-в-центре-екатеринбурга?source_location=search.

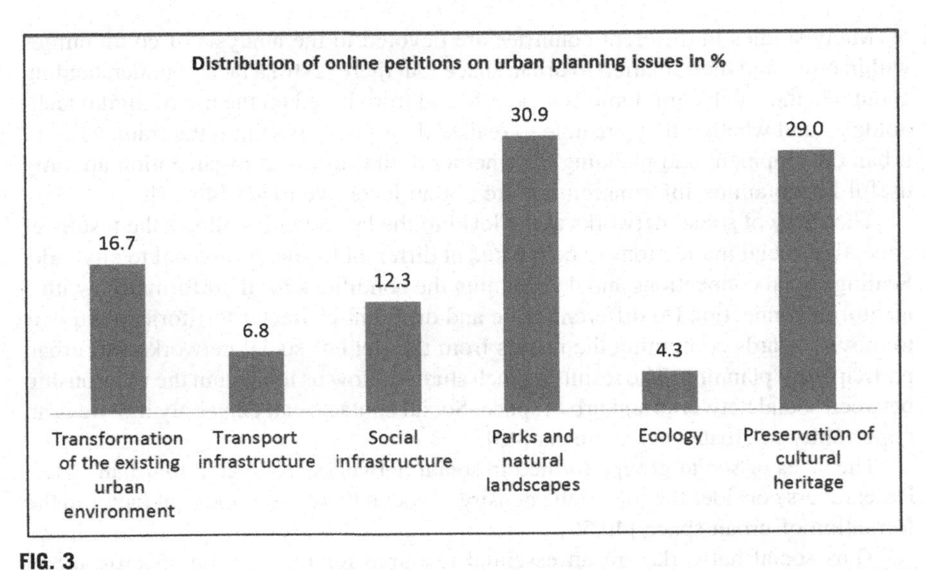

FIG. 3

Issues put to the vote in an online petition on Change.org.

Novgorod from development (2022), which gained 66,000 votes and also became the subject of discussions at the national level, but to a much lesser extent.

The issues raised as part of the petitions were distributed as follows (Fig. 3): 31% preservation of parks, squares, and urban natural spaces, 29% preservation of heritage and historical environment, 17% transformation of the existing built environment, and 12% social infrastructure.

It should be noted that only a small portion of the petitions included forwarding to the communities of residents in social networks, including links to videos on YouTube and in messages.

The allocated "victories," i.e., cases when positive results were achieved as a result of the petition, were no more than 10% for those who received more than 1000 votes (according to the links posted on the petition pages). However, the activity of users shows that the tool itself serves primarily to consolidate stakeholders, attract the attention of the authorities and the press, and is also used in real actions, such as organizing residents to participate in regulatory procedures for public hearings and public discussions, rallies, and other forms of civil influence on authorities.

16 Tools (C2C)

In this section, we consider how social networks in Russia are becoming an effective tool for participatory urban planning and an essential element in the formation of smart urban spaces and a smart city as a whole, based on localized urban communities striving to form a controlled and regulated urban space with their participation.

Many studies in different countries are devoted to the analysis of communities within cities and their relation to urban space, but there is still a lack of understanding about whether such communities can unite and form based on the use of digital technologies and whether they are able to realize their interests within the framework of urban development and planning, or whether digital forms of organization are only useful for obtaining information but are not an incentive to act [46,47].

The study of social networks at the level of the largest cities allows the results of collective social interactions to be tracked at different levels, from local to citywide. Scaling social connections and determining the conditions for their formation with a territorial connection (to different-scale and different-character territories) allow us to move towards combining the results from the field of social networks and urban participatory planning. The results of such studies allow us to explain the relationship between social networks and urban space. Social contacts can exist only if there is an opportunity to create such contacts [48].

The sizes of social groups formed in social networks are related to urban space. Researchers consider the role of the density of social ties as an important factor in the formation of urban space [49,50].

Thus social networks are an essential resource for research on specific urban aspects of development [9,20,51–53], including the impact on spatial-social systems, and their formation based on the participation of residents in urban planning [54].

A “smart city” is not only a digital city, but also, first of all, a social city based on the principles of spatial justice, starting at the local level, i.e., with communities of residents. The different nature of the use of social digital technologies and the impact of these technologies on the formation of the urban environment are studied in a wide range of works. The importance of protest forms in the digital space as stages of community formation at different levels is emphasized separately [9,51].

Social media channels are becoming a tool for informing and communicating among government bodies, the community of residents, and other interested participants, which makes social networks an essential element of the system of participation of residents in urban planning at different territorial levels [55].

Social networks in general in Russia have a high degree of penetration, have different sources, and determine the involvement of the population in different ways. According to a report by Mediascope[ah] at the end of 2022, 81% of residents visit social networks, Telegram, or YouTube at least once a month, and 71% of the population daily. In terms of population coverage among social networks, according to the company Mediascope, the maximum average daily coverage of the social network VKontakte has grown in a year to 43% of the population over 12 years old, whereas Instagram has 7% of the population, and Facebook has 1%.

[ah]https://mediascope.net/upload/iblock/5ab/8bh9sab0ioqdvufiv52lhw3ccruhq585/%D0%9D%D0%A0%D0%A4_SocialMedia_%D0%A1%D1%83%D0%B0%D0%BD%D0%BE%D0%B2%D0%B0_11.11.22.pdf.

In Russia, the main social networks that form information content and unite social communities in the field of urban participatory planning are the Russian social network VKontakte (VK), Instagram, Facebook, Twitter, and Telegram. As of February 2023, according to br-analytics, the penetration of Instagram was 9.9% (15,515,538 unique authors), Facebook was 1.73% (2,714,770 unique authors), Twitter was 0.21% (334,064 unique authors), and VK was 16.95% (26,568,765 unique authors).[ai]

In Russia, social networks developed in parallel with world practice, although access to them, taking into account the development of the Internet, was noticeably slower. The emergence of global social networks was briefly ahead of the emergence of Russian networks focused mainly on Russian-speaking users (in 1999, the Livejournal.com Russian Facebook service appeared, in 2004 Facebook, in 2005 YouTube, in 2006 Twitter as well as the Russian analogue of Facebook social network VKontakte (VK) and social network Odnoklassniki (OK.ru), and in 2010 Instagram).

Considering social networks and messengers as tools of participatory design when creating both smart cities and smart urban spaces, it is important to take into account the current nature and status of various networks and messengers in the country. This applies not only to usage statistics, but also to their legal status. Currently in Russia, Meta Platforms, which owns Instagram and Facebook, is considered an extremist organization, and its activities were banned in Russia by a court decision on 06/20/2022. Instagram and Facebook are banned, although private users can use them, while state and municipal authorities and organizations cannot use these networks. Twitter has been blocked in Russia since 2022 based on the request of the Prosecutor General's Office.

Access to social networks is currently possible only through a VPN. The Atlas VPN[aj] Implementation Index shows that the number of VPN downloads in Russia increased from 12.59 million in 2021 to 33.54 million in 2022, which represents an annualized increase of 167%. The VPN implementation rate was 22.98% until February 2022, when the VPN implementation rate was only 3.37%. However, a significant portion of VPN services in Russia does not work or operates with significant restrictions, falling under a ban in accordance with current Russian legislation. According to Mediascope, in 2022 the percentage of Facebook users decreased from 7% to 1%, and Instagram decreased from 30% to 7%.

Before the ban and blocking of Facebook and Twitter, despite a small share of population coverage, various communities (groups) developed and functioned in these social networks, positioning themselves in the field of urban participatory planning. The number of participants, as a share of the total number of Russian users, was extremely insignificant, primarily compared to VKontakte (VK). However, they were one of the tools of social organization of citizens interested in participatory planning. Also, these networks were used by official bodies involved in urban planning to post information, although the degree of informativeness was extremely

[ai]https://br-analytics.ru/statistics/am/vk/month/202302/total.

[aj]https://atlasvpn.com/vpn-adoption-index.

limited, as was the number of subscribers. Among 16 cities with a population of over 1 million, there are Twitter accounts (as of 2023 there were only 4 cities). Previously, some accounts were deleted; a similar situation is developing on Facebook (for example, the Ufa account). After the ban and recognition of these networks as extremist, all government and municipal bodies and organizations either deleted or stopped using their accounts.

As of March 13, 2022 (date of the last tweet) before the freezing of Twitter, the number of subscribers from the Department of Urban Planning Policy of Moscow was 6877, compared to 4463 subscribers[ak] in 2018. The channel of the Department of Urban Planning of the city of Samara[al] (currently deleted) had 3467 subscribers. In fact, every 330th resident of Samara had a subscription to the channel of the local urban planning authority with active feedback and the opportunity to receive a prompt response, without registration in state systems. In Chelyabinsk, the Department of Architecture had 489 subscribers and conducted surveys on the transformation of open spaces for specific territories. However, the informativeness of these sources on the use of official instruments of participation in hearings and discussions was minimal.

An analysis of the surviving accounts of the city authorities responsible for architectural and urban planning activities showed that there was no systematic placement of information on these issues or communication with residents on these issues.

Accounts created by residents and initiative groups, before the block, were used mainly as auxiliary to the main head groups of VK; their number rarely exceeded 100 participants, while the main groups had several thousand subscribers. For example, the Urban Protection St. Petersburg, @protect812, had 74 subscribers on Twitter and on VK 4000 subscribers; the account is currently frozen (the last tweet is dated August 24, 2022).

Despite the popularity of Instagram and TikTok resources, they are practically not used in the field of urban participatory planning by either residents or authorities. Exceptions are individual blogs on Instagram of local urban activists (concentrated in Moscow and St. Petersburg). The number of subscribers of local bloggers does not exceed 1000. There are also several large blogs of well-known opinion leaders who consider various aspects in their blogs, including the development of the urban environment and urban planning. The number of their subscribers varies from 500,000 to 1.5 million.

WhatsApp is mainly used by residents to organize private house-wide chats, as part of participation in the management of apartment buildings or as part of low-rise homeowner communities. The closeness of most groups does not allow us to fully assess the completeness of the participation of residents; however, a survey of residents of 10 apartment buildings in St. Petersburg allows us to conclude that the total number of chat participants does not exceed 40% of the number of apartments in houses. At the same time, the presence of such groups creates prerequisites for the formation of local communities of residents.

[ak]https://nvitter.,0111/DgpMos.
[al]https://t,vitter.com/depstroy/32.

YouTube is also used extremely rarely by urban communities; there are no separate videos on local or citywide planning and urban development issues, as they are posted outside the channels. Thus the previously considered citywide community "City Protection Petersburg" (protect812), dedicated to participatory planning in St. Petersburg, with 4000 subscribers on the VK network, has less than 100 subscribers on its YouTube channel, with 38 videos posted. At the same time, individual bloggers and civil society activists in different cities are posting videos on how to participate in public discussions through online platforms, compensating for the lack of such tools provided by municipalities and regional authorities.

State and municipal authorities responsible for urban planning also use this information resource extremely inefficiently.[am] Only the bodies responsible for urban planning in two cities—Moscow and Ufa—have channels. At the same time, 594 people have subscribed to the channel in Moscow, Russia's largest city (as of February 2023), and out of 148 uploaded videos, only 4 have more than 1000 views. It is important to note that within the existing channels of both residents and administrations, if the issues of discussions and hearings are considered, then it occurs mainly after the fact, after their completion and summing up.

One would assume that such use of the most popular video hosting platform only applies to Russia, but the Department of Urban Planning of New York, NYCPlanning, has only 1750 subscribers with 523 posted videos.[an]

Using this example, you can see the effectiveness of this ICT tool within different municipalities. All major cities are similar in their characteristics as biological organisms of the same order, but their layout, regulation, social structure, and the role of society in their planning may have significant differences.

Taking into account all of these factors, this study analyzed in detail the social activity in the field of urban participatory planning in VK and Telegram, which are the main sources of information and communication in this area.

17 Telegram and public hearings and discussions in the field of urban planning

Telegram is currently one of the main electronic communication platforms in Russia; for 2022, Telegram's Russian-speaking audience exceeded 55 million people (as follows from Mediascope data).[ao]

Telegram is the most popular messaging app, and private, personal, and public channels are created on its platform. Telegram in Russia is one of the most important sources of awareness of Internet users. According to the TGStat[ap] study (a study of the Russian-speaking audience of users based on the analysis of 70,000

[am]https://www.youtube.com/watch?app=desktop&v=ET0LCSlOmo8.
[an]https://www.youtube.com/@nycdepartmentofcityplanning/community.
[ao]https://www.vedomosti.ru/technology/articles/2022/03/20/914320-telegram-oboshel-whatsapp.
[ap]https://tgstat.ru/research-2021.

questionnaires), 32.5% of the users live in Moscow, 13.2% in St. Petersburg, and from 2.7% to 0.7% of users live in other million-plus cities. In total, 65% of Telegram users live in all cities with a population of more than 1 million. Direct extrapolation to the number of residents cannot be applied to analyze the actual users in the largest cities (otherwise the number of Telegram users would exceed the population of Moscow and St. Petersburg), but this shows the dynamics of distribution, which largely coincides with the number of residents.

The analysis of Telegram channels in the field of urban planning carried out during this study, within the framework of which information on public discussions and hearings is posted, showed the following results. In 40% of all million-resident cities studied, there are Telegram channels sanctifying public hearings and discussions in the field of urban planning (St. Petersburg, Yekaterinburg, Chelyabinsk, Samara, Omsk, and Voronezh), as presented in the table (Table 7).

Table 7 Telegram channels for public hearings and public discussions for the largest cities.

	Cities with a population of over 1 million inhabitants	Share in % of Telegram users by city according to the data gstat.ru	Subscribers unofficial Telegram channels	Subscribers official Telegram channels	Total subscribers
		2021	2023	2023	2023
1.	St. Petersburg	13.18	2106	210	2316
2.	Novosibirsk	1.86	–	–	–
3.	Ekaterinburg	2.73	195	234	429
4.	Kazan	1.76	–	–	–
5.	Nizhniy Novgorod	1.37	–	–	–
6.	Chelyabinsk	1.15	506	1038	1544
7.	Krasnoyarsk	1.03	–	–	–
8.	Samara	1.23	312	330	642
9.	Ufa	1	–	–	–
10.	Rostov-on-don	1.52			
11.	Omsk	0.7	56	–	56
12.	Krasnodar	1.96			
13.	Voronezh	1.04	–	644	644
14.	Perm	1.14	–	–	–
15.	Volgograd	0.75	–	–	–
	Total	32.42	3175	2456	5631

All channels can be divided into channels of public authorities and channels created by private users.

Private channels appear for the whole city as an alternative to insufficient and inefficient channels or, according to the authors, unreliable channels; as an alternative to municipal or regional sources of information about hearings and discussions; or as an initiative within the framework of a single event or a support group for residents and owners. They are mainly related to local projects affecting a particular project or territory. The term of their functioning is quite short—from the moment of the appearance of real prerequisites for a town-planning conflict or from the announcement of hearings or discussions. Most of these channels have a lifetime of 1–1.5 years. This study examines public private channels covering all hearings in the whole city.

Analysis of the citywide channels in 40% of all cities studied showed that they all appeared in the period from 2018 to 2020, i.e., in the prepandemic period. All channels work on the principle of collecting information from official sources, including using bots. Some channels provide their subscribers with additional information disclosure and analysis within the framework of the channel authors' vision. Since channels are information sources, they lack communication forms of interaction, such as commenting, as well as likes and dislikes or emojis.

The demand for this content in general turns out to be low, despite the large audience for Telegram in these cities. The number of subscribers ranges from 50 to 1000 people. At the same time, the channels have an extremely high share of traffic—more than 80% of the views from the number of subscribers for each posted news.

After the ban of Meta and Twitter social networks in Russia in 2022, the state regulator in the field of information technology, the Ministry of Digital Development, Communications and Mass Communications of the Russian Federation (the Ministry of Finance of Russia), recommended that all government agencies create accounts in Telegram and in the Russian social network VKontakte.[aq]

After that, VKontakte groups and official Telegram channels of various authorities at different levels began to appear actively, including individual departments responsible for various tasks, in particular those responsible for urban planning. An analysis of the allocated Telegram channels of the municipal authorities responsible for urban planning revealed that mostly official channels appeared only after the direct recommendation of the regulator in 2022. In just 33% of cases (in cities having 5 million-plus population), municipalities have organized such an independent Telegram channel. Of these, only two channels include information on public hearings and discussions, and none of them has the ability to post subscribers' likes or dislikes. (Like and Dislike). The channels do not provide users with anything other than links to official platforms and messages about holding hearings with a brief description of the situation (within the framework of mandatory officially posted

[aq]https://news.rambler.ru/internet/48255612/?utm_content=news_media&utm_medium=read_more&utm_source=copylink.

messages). Thus the Telegram channels of the authorities as a tool of participatory design in terms of urban planning are used in a very limited manner by administrations; in fact, this is a typical G2C format. At the same time, certain aspects of participatory design, such as discussions on the urban environment, including video files that contain recordings of speeches by invited lecturers, materials and discussions on architectural and urban planning competitions, information materials on the coordination of various documentation, and changes in legislation, presented in accordance with the format of the Telegram channel, are present in some of the channels (Voronezh, Chelyabinsk). These channels are characterized by the fact that they were created before the instructions of the regulator, as an initiative of local authorities. The number of subscribers on municipal channels exceeds private ones and averages from 200 to 1000, which corresponds to private channels.

An important factor revealed by the analysis of Telegram channels is that both private and official (state) channels are fixed in the same cities, and official channels appeared later than private ones, which indicates a competitive confrontation within this resource. Municipalities tend to intercept private initiative.

In general, it can be concluded that the effectiveness of the use of Telegram channels in the field of participatory city planning is still extremely low. At the same time, this resource could be one of the most advanced in the framework of the inclusion of the population in urban planning.

18 Social network Vkontakte (VK)

VKontakte (VK) is a Russian social network that allows users to send messages to each other, create their own pages and communities, exchange images and audio/video recordings, transfer money, and play browser games. It also positions itself as a platform for promoting business and solving everyday tasks with the help of mini-applications.[ar] It was created by the founder of Telegram, Pavel Durov, in 2006. According to Brand Analytics, a company engaged in monitoring and analyzing social media in Russia, as of February 2023, the network has 26,568,000 unique authors, i.e., 16.95% of the population.[as]

According to statistics, penetration in Moscow is 25% of the population, and in St. Petersburg it is 44%. Statistics are not kept for other major cities, but on average in the regions where they are located, penetration ranges from 12% to 29% of the population. The network ranks first in terms of the number of users in Russia against Facebook (1.3%) and Instagram (c. 9.9%), second only to YouTube.

Within the framework of the presented study, an analysis of groups in VK was carried out (as of February 2023). In each major city with a population of more than 1 million inhabitants, communities of residents were identified, united in groups in a

[ar]https://vk.com/about.

[as]https://br-analytics.ru/statistics/am/vk/month/202302/total.

social network and positioning themselves as part of participation in urban planning, including in the preservation and development of local territories and heritage sites.

All groups were considered in terms of size, issues, and format of the organization, and groups dedicated to public hearings and public discussions were singled out and considered separately. This analysis allows us to understand how such tools as discussions and hearings are integrated with social networks and how social networks create the conditions and the basis for new forms of participatory design and formation of sociospatial elements of the urban structure.

All the study groups devoted to the participation of residents can be divided into categories on a territorial basis into local and citywide. Citywide groups unite residents who advocate for various projects and discuss various aspects of urban spatial development, including forming a news agenda by attracting the press; organizing separate events, meetings, and rallies; as well as interacting with authorities. Citywide groups also include groups dedicated to tools for interaction with city authorities—hearings and discussions. The nature of citywide groups depends on the presence of public organizations or activists operating within these network communities. Citywide groups often come together within a separate local citywide facility or problem. It can be a landmark object for the city.

An example of this is the previously considered project for the construction of a temple in Yekaterinburg on the site of a park in the city center, which gathered 168,000 votes on change.org, while VKontakte in the group "For the protection of Parks," which organized the protest process, consisted of only 6800 people.[at] A similar example is the project in St. Petersburg for the preservation of historical buildings of the "Stable Department, " which gathered on change.org 60,000 signatures. As a result, the construction, which involved the demolition of part of the historical complex, was canceled, while the VKontakte group was only 6000 people.

Local groups are united by the problems of local communities within the framework of the spatial development of the city and the preservation of the environment and heritage sites. Using the example of 10 open local groups with more than 1000 participants in St. Petersburg, we conducted a comparison (by random reconciliation of 100 group members), which showed that no more than 35% of subscribers were present simultaneously in several local groups, so we can assume that the uniqueness coefficient is on average 65%. As part of the analysis of all groups of residents (regardless of the number of participants) that are associated with the spatial development of cities, several cities were considered: St. Petersburg, Yekaterinburg, Samara, Ufa and Perm, taking into account the population. The groups were identified based on the analysis of keywords (building, planning, hearings, discussions, demolition, heritage preservation, parks, open spaces) and subsequent reconciliation with the description of the group.

St. Petersburg is the second largest city and the city with the highest VK penetration—44%. The study revealed that the number of groups (with more than

[at]https://vk.com/parklandekb.

50 participants) meeting the criteria (as of January 2023) was 230 units. There are more than 220,000 subscribers in total. Of these, there are more than 60 groups associated with the preservation of individual parks and green areas. The rest are related to the preservation of the urban historical environment, the transformation of built-up and established territories, ecology, and transport. It is possible to distinguish groups—information centers that have become an association of a number of local groups: for example, the Green Coalition of St. Petersburg,[au] which unites 46 groups created to protect elements of the city's green infrastructure, and groups for the preservation of heritage and the existing urban environment. At the same time, the parent group itself unites only 6000 participants, and its member groups with a membership of 200–8000 people (not including large partner citywide associations) have more than 67,000 subscribers.

The largest dedicated group, "Beautiful Petersburg," which is the information center of one of the systems, consists of more than 60,000 subscribers.[av]

These societies are a system of not only related groups, but also groups that unite volunteers who carry out practical activities.

Almost a quarter of groups with more than 1000 participants create their own ecosystems, including websites, pages on Instagram.com and fb.me, Telegram channels, Twitter accounts, YouTube channels, and petitions promoted by them on www.change.org. Individual groups even develop their own applications; 40% of all groups are represented by several formats.

The analysis of the number of participants in relation to such ecosystems has shown that the maximum number of participants collect petitions for change.org; then there is the number of the VKontakte group, and a small number of participants in the remaining elements of the ecosystem (up to 100). An example of such an ecosystem is a VKontakte group organized to protect against the development of Murinsky Park in St. Petersburg,[aw] in which a petition is presented (https://www.change.org), Instagram page, Telegram channel, and Facebook page. The number of group members is 3800 people, and the petition gathered 5540 people. All other elements have less than 100 subscribers. This ratio can be traced for most groups. The least number of followers are on Twitter and Facebook.

The analysis of groups in the cities of Yekaterinburg, Samara, Ufa, and Perm showed that this situation is generally not typical for the largest cities of Russia. In Yekaterinburg, the fourth largest city in the country, the number of participants in all groups is 18,000, and the maximum number in one group does not exceed 6500; in Perm it is 7000, and in Samara 6000. At the same time, the principle of ecosystems is developing in all regions. As part of the development and functioning of individual elements of the ecosystem, we offer the example of a community in Samara (https://twitter.com/5prosekaS) on Twitter, which had 666 subscribers as of March 2022, and after Twitter was banned in Russia in 2022, moved to VKontakte

---

[au] https://vk.com/greencospb.

[av] https://vk.com/peterburg_krasiv?w=club38228859.

[aw] https://vk.com/kalininskiyzapark.

activity (https://vk.com/stopstroyka), where the number of the group was 2800 subscribers as of March 2023.

Separately, several large all-Russian communities in the field of participatory planning can be singled out; for example, "Urban Projects," which has more than 16,000 subscribers (currently in Russia, the authors of the project are well-known bloggers and activists, and Varlamov and M. Katz are recognized as foreign agents) https://vk.com/cityprojects. The same community on Twitter has 22,600 subscribers.

Thus it can be argued that social networks themselves are only a tool, but not prerequisites for the organization of united social communities. Only a "smart community," ready for initiatives, is able to organize itself into a spatial social community by attracting new participants associated with the real physical space. The number of people involved in the process of active participation is less than 1%–2% of the city's population. However, these groups are able to organize significant resources within the framework of local changes in the urban environment (densification of buildings, zoning changes, preservation of parks and the natural environment, preservation of local heritage sites), as well as large citywide projects affecting the interests of a significant part of the urban community as a whole. The localization of communities is determined through the perception of urban space (the principle of "friend–foe"). People perceive a spatial element as their own or someone else's, depending on a variety of characteristics (location of residence; time of residence in a given location—i.e., homeowners living on the territory are most interested, while tenants are much less interested in what is happening; sense of place—connection with the environment; how much residents use the local or district urban environment or infrastructure; a sense of urban identification of "I..."; and cultural level—the meaning of heritage, preservation of the memory of the place, etc.). Thus the scale of the transformation area and its sociocultural significance form the localization of interested residents, which affects who become subscribers of social groups created by interested communities, since all aspects of concern to communities, including natural and ecological recreational spaces, heritage preservation, and others, can have both local, quarterly, district, citywide, as well as regional and national level.

Urban planning conflict often becomes a source of unification of urban communities through Internet resources, and primarily through social networks. Only some significant cases of association in social networks within the framework of joint work with the city authorities on development projects were recorded. At the same time, established groups and associations are ready to offer their solutions to the city authorities and participate in projects. The forms of actual participation of residents in the transformations are different and are both of a volunteer nature within the framework of monitoring, various promotional processes, landscaping, and the nature of organized protection of their interests.

The only instruments of interaction with the population officially included in the legislation within the framework of urban development transformation of cities are hearings and discussions. In this regard, groups focused on informing the population and involving them in these procedures were separately considered.

Most of the previously reviewed groups include separate information and materials, including appeals prepared by the participants of the groups to the authorities regarding discussions and hearings. At the same time, among them there are groups devoted mainly to public hearings and public discussions. Considering this category of groups, we have included in the analysis groups created by the official bodies of municipalities responsible for urban planning and architecture. Thus all the groups devoted to hearings and discussions can be divided into two categories: official (groups of official authorities in the field of urban planning) and unofficial (groups not associated with authorities). As already discussed earlier, after the ban of Meta and Twitter social networks in Russia in 2022 (due to being recognized as extremist), the state regulator recommended that all government agencies create accounts in Telegram and in the Russian social network VKontakte. In the banned networks, all previously existing accounts of municipalities were deleted or frozen. Only 45% of the official groups were created before this requirement (the earliest is 2018), and the rest are dated 2022. For all time, groups of governing bodies in the field of urban planning were created for the 16 largest cities; similar groups were created in 12. Like any government communication, this tool can be considered as a means to implement the goals of real policy [56]. And if earlier the previous management methods were increasingly shifting towards consultations, which led to the introduction of new communication practices in different jurisdictions, then artificially created information elements that do not have real policy goals, and whose format is not regulated by law, perform an exclusively decorative function [57–59].

Thus groups created before 2021 reached 1000 people and ranged on average from 100 to 700; none of those created after 2021 reached the threshold of 200 participants. Only one-third of the groups included separate materials on the procedures of hearings and discussions, while none of them became a platform for discussions or proposals and they did not allow an opinion to be left on the planned transformation within the framework of the procedures under consideration.

Initiative (unofficial) groups formed by social activists to cover public hearings and public discussions were identified in each of the million-people cities studied. These groups can be divided into citywide and local by the nature of the issues covered. Citywide groups have been created and function as a tool for informing residents about hearings and discussions; they accumulate a wide range of information from various official and unofficial sources, and are addressed to various groups of stakeholders who are part of the circle of participants in urban development [60].

Currently existing citywide specialized groups have been registered on the VKontakte network since 2012, 6 years earlier than the first groups created by city administrations. In total, 10 groups were identified, one in each city; i.e., in 66% of cities there is such an initiative group. Citywide groups unite up to 500 participants, while there is a pattern that if an effective city administration group exists in the city that places a full range of information materials, then the smaller the number of initiative groups, and vice versa.

The groups themselves are often limited only to collecting and posting information and pay less attention to analysis. It is also important to note that not all aspects of the hearings and discussions fall into their areas of activity (everything related to local issues is related to conditionally permitted types or deviations from the limit parameters, and a number of projects on the layout of the territory are not covered by them, unless there are particularly significant changes); thereby there appears a specific selection determined by administrators or group members, so local changes affecting a limited number of participants do not actually get into the information space of the social network.

In addition to citywide groups, it is possible to distinguish groups created as part of the preparation of residents for public hearings and discussions within the framework of local city issues or documents. The number of participants in such groups at the time of discussion of the issue can reach 1500, but mostly it ranges only up to 400 people. However, after the hearings, these groups either transform into a local active community associated with a specific issue, or break up and are left by their participants.

A separate category of groups in the social network are groups formed by residents of districts, local territorial formations, and apartment buildings. These groups are formed within the framework of different principles and are united exclusively by localized interests that relate to all areas of the population's life: communication, entertainment, exchange of social and commercial information, local news, landscaping, transport development, construction, schools and kindergartens, ecology, solving local problems, and participating in the management of their homes.

All these groups on a territorial basis within the largest city can be divided into five levels: citywide, district (on a city scale, large administrative entities, or large planning structures that do not coincide with their borders are often such), local districts, less common quarterly groups within the boundaries of individual blocks, and groups of apartment buildings (the most extended type).

All groups, except for groups of multiapartment residential buildings, in most cases are open, whereas groups of multiapartment residential buildings are closed to access by external participants.

For an effective analysis of such practices in local neighborhoods and communities of condominium residents, it is necessary that the territory within which such groups are formed be singled out by residents as a separate sociospatial structure.

19 The peripheral district as an example of the formation of prerequisites for the creation of smart spaces and a smart city based on the use of participatory planning

For megacities, including Russian ones, such territories with an independently forming sociocultural environment are peripheral areas of multiapartment development that arise everywhere within the framework of intensive urbanization [61,62].

As part of the study, a separate local peripheral territory was considered. The considered city of Kudrovo is a new densely built-up and rapidly developing territory of a residential area within the boundaries of the St. Petersburg agglomeration, administratively not included in the borders of St. Petersburg, but closely bordering it. Kudrovo, being an independent city, does not have the status of a municipality: that is, it does not have elected authorities and its own powers, being part of the Zanevsky urban settlement. Most of the city's residents, except those employed in service systems, work in St. Petersburg.

Since 2010, when the modern development of this territory began, the number of residents has grown from 100 people to reach 80,000, of which 20,000 are residents living in rental housing, who are not taken into account by official statistics. It should be noted that, currently, taking into account the active construction of multiapartment housing, more than 100,000 residents are expected in the short term. Kudrovo is an intensively growing city with one of the highest levels of population density; within its borders there is the largest apartment building in Russia, the residential complex "Novy Okkerville, " in which 18,000 inhabitants live. The cost of housing in the city is 25%–30% lower compared to St. Petersburg.

Peripheral territories are emerging and developing urban spaces in which there is a high proportion of internal and external migrants, as the social system is formed primarily through digitalization [63–65].

Research on Kudrovo shows that identity is formed on the basis of a "hybrid neighborhood"—a digital human network that works 24 h a day, 7 days a week, and unites thousands of people as local residents [66].

Researchers identify a large number of initiatives within the community that have contributed to security, urban improvement, alternative economy, the formation of local identity, as well as leisure, educational, and other activities [67].

Within the framework of this study, it was considered how this community is formed through the use of the VK social network, and how it forms approaches participatory planning of the urban environment.

In total, 32 groups for each residential complex in the district were created in the social network by 2023. In the groups of 90% closed, 50% of all permanent residents of the district consisted of a total of 61,000 (according to the population census of 2021). In citywide open groups, a total of 18 groups were identified, with a total number of subscribers of more than 400,000. The largest group was 80,000 participants.[ax]

It is important to note that some of these groups were and are being created with the participation of developers and management companies created by them, selling apartments in residential complexes, and they are managed centrally, taking into account their use as a large advertising market. In these groups, the issues of participatory planning are found only at the level of local landscaping. In the remaining groups, separate discussions are held on the issues of landscaping, social

[ax]https://vk.com/kudrovodom.

infrastructure, transport infrastructure, but the issues of participation in official forms of participatory planning are practically not touched upon.

From all citywide groups, groups that positioned themselves as participating in urban planning, with a total number of subscribers of more than 15,000, were singled out separately. In these groups, discussions on the development of the city and its individual territories are regularly held, and materials are posted as part of public hearings.

It should be noted that, given the lack of an independent municipal status, all official forms are carried out within the municipality, which includes Kudrovo–Zanevsky urban settlement.

For a comparative assessment of the actual participation of residents in urban planning, public hearings held in 2021 on the General Plan of the entire Zanevsky settlement were considered (separately for Kudrovo, as taking into account the lack of the status of the municipality, the general plan is not being developed). According to the conclusion, 355 people took part in the hearings.[ay]

Thus it can be concluded that participatory planning with regard to official instruments, as well as interest in such planning, is significantly different. At the same time, at the level of local initiatives related to landscaping and open spaces, you can see the widest possible range of applications of participatory planning—workshops, discussions, gathering opinions, and initiatives.

At the level of open urban spaces, the design method involving residents, local communities, activists, representatives of administrative structures, local businesses, investors, representatives of the expert community, and other parties interested in the project works and the involvement of residents in Kudrovo occurs through social networks. Such results certainly suggest that the city is moving away from a smart city and smart urban spaces [68].

However, at the level of more complex issues related to the development of the city, transport and social infrastructure, reduction of building density, and establishment of requirements for the development of territories, the residents remain largely uninvolved. This is due to both the level of internal migration of residents from Kudrovo to St. Petersburg, their awareness of the events being held, as well as hypertrophied forms of development in which residents focus on interests within their condominiums and adjacent open city spaces. An important factor is that new internal and external migrants do not have the competencies to evaluate the solutions offered by the authorities and developers. At the same time, using the example of public hearings on the master plan held with the participation of residents, Kudrovo shows the highest indicator—0.7% of all residents compared to all cities with a population of over 1 million inhabitants considered in this study earlier.

The activity of residents of the urban planning area of the territory under consideration is not limited to using only official forms of complicity. For 5 years, 21 petitions have been posted on online petition platforms: for the preservation of open urban spaces, the development of social and transport infrastructure, reducing the

[ay]https://www.zanevkaorg.ru/sovet-deputatov/publichnyie-slushaniya/.

density of buildings, changing the number of floors of buildings, changing the administrative status—obtaining the status of a municipality, inclusion in the borders of St. Petersburg. The number of signatories has varied from 500 to 3000.

How can participatory planning based on digital technologies become an effective tool for creating smart spaces in a smart city such as Kudrovo? In the current status, we see that in the absence of effective systems for informing residents with regard to urban planning, there are prerequisites for the creation of alternative channels for obtaining information, which is actually one of the elements of Web 2.0. Residents become initiators of the creation of various alternative tools that allow disseminating information about the events.

The involvement of residents in the process of forming the environment at the level of local spaces creates prerequisites for the transition of influence to more complex elements of urban planning. This is facilitated by: the development of civic initiatives; the educational process, as most residents simply do not know and do not understand how certain documents submitted for discussion affect their lives; and joint interaction within the framework of local residential formations.

Smart technologies introduced into the urban environment, such as smart lighting, or a safe city (through the installation of surveillance cameras), or the placement of smart traffic control systems (https://vk.com/wall-51766355_2552652) do not change the environment dramatically. For more significant transformations, a social initiative is required that allows regulating urban development as a whole and within its individual elements, at different levels, from the quarter to the entire territory of the city, which is possible within such densely populated territories only through the use of ICT tools. To form such conditions of influence on city authorities and developers, there is a formed potential of the digital urban community, but its awareness and readiness for change requires the development of the perception of its territory as a separate locus, with the formation of ideas about its significance and value. The most important condition for such development can be educational programs in which residents receive the basics of ideas about the development of the city and the possibilities of their participation in its transformations.

20 Prospects and opportunities

These proposals are of a debatable nature and can potentially be considered only as a prototype of possible transformations, taking into account current trends in urban reality. The development of public institutions and the role of social networks in this, and the development of other forms based on ICT, suggest that in the near future their use in participatory urban planning will only increase. This trend allows for a new approach to participatory planning for the largest cities and their local territories, as the most important tool for the formation of a smart city based on social spatial justice. New approaches to urban planning, such as the 15-min city, show that trends such as localization of interests are becoming the most important condition for creating a comfortable and sustainable urban environment.

In Russia, participatory planning is currently at the lower levels of the Arnstein ladder [31]. Having analyzed participatory planning as institutions of the state, municipalities, and urban communities, within the framework of the future formation of smart cities and smart inclusive urban spaces in Russia, it can be argued that the current forms of participation and their principles not only do not meet the goals of a smart city consisting of smart spaces, but also directly contradict it. Profound changes are needed at all levels, from the state and municipal approach to city planning and development (including through the introduction of advanced smart technologies) with changes in current legislation, to the level of self-awareness of city residents. In this regard, it seems necessary to consider the city through the significance and influence on its inhabitants of a number of aspects: spatial, economic, social, ethnic, religious, and others for the formation of multilevel sociocultural spatial units, based on their perception by residents.

Such elements of urban space, formed at the mega-level—the city level, the meso-level—of individual districts; the macrolevel, of conditional quarters; and the microlevel of urban local spaces, regardless of the administrative and municipal division of urban territories, can be levels of participatory urban planning. At the same time, macro- and meso-levels should include not only the territory of the entire city, but individual territories of different scales, taking into account the significance of their perception and significance for the residents of the city as a whole.

Also, within such a system, it would be important, taking into account the opinion of residents, to determine the preliminary characteristics of the impact for all types of use on the transformation of the environment, including the degree and their impact on communities: considering the increase in density, the need to buy land and change the characteristics of the architectural and urban environment, and other transformations. This would allow, with the use of such specific uses or their characteristics, to attribute the territory to one of the levels of participatory planning. Thus each level would make it possible to involve various communities of residents and other stakeholders in discussing projects, both before their development and at the stage of evaluating the results obtained.

This model can be implemented on the basis of a citywide information system with the possibility of integrating all digital institutions at the level of urban communities into it. Such a system should be formed based on a GIS system with elements of social networks integrated into it, and the possibility of including in it as separate complementary elements digital ecosystems of local communities. It is proposed to create, taking into account localization, opportunities for discussions and voting of residents, where for each resident there will be a section where they can see at the initial stage all the options offered for participation, including nondigital forms of participation and also hearings offered for their participation at different city levels, including local ones in relation to the place of registration.

Such a system will create conditions for multifaceted participatory planning, with various forms of participation, and will attract the maximum number of urban communities to discuss decisions, increase the possible participation of all interested residents, and reduce the degree of conflict. Of course, such solutions will increase the project implementation time and will require significant investments in the creation of such an ICT tool. However, the effect of participation can create significant potential for the development of the city as a whole and its local parts.

The solutions proposed by the author are of a debatable nature and with the current legal framework in Russia cannot be implemented at the present time.

21 Discussion and conclusions

Rethinking the processes of participatory planning as a tool for the development of a smart city is becoming a definite challenge, reflecting the complexity and inconsistency of the rapid urbanization of the largest cities. With the problems presented at the beginning of the study, arising from the conditional use of participatory planning tools at different levels of urban development in the largest cities of Russia, each in its own way requires a revision of the methods and forms of interaction between municipalities and residents. The lack of effective tools and forms offered at the level of state and municipal authorities in the field of interaction between residents leads to their self-organization based on ICT and, first of all, social networks.

Most of the examples given indicate that in order to solve problematic situations, it is necessary to search for adequate approaches, choose appropriate methods, and find techniques for both interacting with residents and rethinking the structure of the city within the framework of this interaction. A smart city is not only technology, because all technologies are just tools, and smart spaces are not only a convenient and comfortable environment. First of all, these are urban communities that form a space around themselves, taking into account the possibilities of technological solutions based on social spatial justice. Without this, it is difficult to expect the emergence of prerequisites for sustainable development, as a city as a whole. So it is at any level of local territories and spaces. The vector of interaction with residents, based on ICT and taking into account their ideas and perceptions of space, becomes an indicator of the correctness of project decisions that meet the interests of both the city as a whole and local urban communities.

Thus the formed groups of residents in social networks can be defined as a self-developing system based on Web 2.0 that sets points of convergence, as well as points of divergence, between the city administration and residents. Its integrated use by city administrations at the stages of planning territorial transformations, prior to the development and adoption of decisions, can ensure rational joint planning of smart city activities, in general, and planning at the level of individual communities within the allocated spatial loci [69].

Currently, the number of residents participating in urban planning is no more than 5% of the adult population of the largest cities. This value is obtained from the analysis of various practices of participatory planning and analysis of residents' requests on the Internet. This indicator is critically low. The presented systematization of digital forms used for the interaction of municipalities and citizens regarding the development and planning of cities within the framework of legal regulatory procedures and urban initiatives, as well as the interaction of residents regarding independent initiatives, including urban planning conflicts, allows us to predict the possibilities of creating more perfect institutions and tools for their implementation.

The described research contributes to this transformation by presenting a methodology for this purpose. It is based on a detailed analysis of existing tools from an overview of established practices in Russia. The selected practices, despite the peculiarities of their application, can be considered as prerequisites for the formation of a smart city and smart urban local spaces that can change the situation in the city for the better. To implement them, digital tools and obvious prerequisites are associated with the self-organization of residents in the digital space of the city.

References

[1] H. Kopackova, J. Komarkova, Participatory technologies in smart cities: what citizens want and how to ask them, Telematics Inform. 47 (2020) 101325.

[2] S.S. Fainstein, The just city, Int. J. Urban Sci. 18 (1) (2014) 1–18.

[3] H. Mitteis, Über den Rechtsgrund des Satzes "Stadtluft macht frei", Münster u. a. (1952) 342–358.

[4] T.W. Ann, et al., Identifying risk factors of urban-rural conflict in urbanization: a case of China, Habitat Int. 44 (2014) 177–185.

[5] J. Skrede, S.K. Berg, Cultural heritage and sustainable development: the case of urban densification, The Historic Environment: Policy & Practice 10 (1) (2019) 83–102.

[6] H. Bekele, Urbanization and Urban Sprawl, Royal Institute of Technology, Stockholm, Sweden, 2005.

[7] E. Israel, A. Frenkel, Social justice and spatial inequality: towards a conceptual framework, Prog. Hum. Geogr. 42 (5) (2018) 647–665.

[8] E. Soja, et al., The city and spatial justice, Justice Spatiale/Spatial justice 1 (1) (2009) 1–5.

[9] R. Kelly Garrett, Protest in an information society: a review of literature on social movements and new ICTs, Inf. Commun. Soc. 9 (02) (2006) 202–224.

[10] Nam T., Pardo T. Smart city as urban innovation: focusing оп management, policy, and context. Estevez E., Janssen M. (Eds.), ICEGOV '11: Proceedings of the 5th International Conference оп Theory and Practice of Electronic Governance, Association for Computing Machinery, New York, NY, USA, 2011, pp. 185–194.

[11] R. Giffinger, N. Pichler-Milanović, Smart Cities: Ranking of European Medium-Sized Cities; Centre of Regional Science, Vienna University of Technology, Vienna, Austria, 2007.

[12] A. Palladio, I quatro libri dell Architectura, di Andrea Palladio, B. Carampelio, 1976.

[13] B. Rinkevich, Rebuilding coral reefs: does active reef restoration lead to sustainable reefs? Curr. Opin. Environ. Sustain. 7 (2014) 28–36.

[14] E.L. Bouvier, La Vie psychique des Insectes, Flammarion Ernest, 1918.

[15] M. Egerer, et al., Multicultural gardeners and park users benefit from and attach diverse values to urban nature spaces, Urban For. Urban Green. 46 (2019) 126445.

[16] A. Vanolo, Smartmentality: the smart city as disciplinary strategy, Urban Stud. 51 (5) (2014) 883–898.

[17] C. Ellis, The new urbanism: critiques and rebuttals, J. Urban Des. 7 (3) (2002) 261–291.

[18] C. Moreno, Z. Allam, D. Chabaud, C. Gall, F. Pratlong, Introducing the "15-Minute City": sustainability, resilience and place identity in future post-pandemic cities, Smart Cities 4 (1) (2021) 93–111, https://doi.org/10.3390/smartcities4010006.

[19] C. Bengs, Planning theory for the naïve? Eur. J. Spatial Dev. 3 (7) (2005) 1–12.

[20] D. Arribas-Bel, et al., Cyber cities: social media as a tool for understanding cities, Appl. Spat. Anal. Policy 8 (2015) 231–247.
[21] J. Jacobs, The Death and Life of Great American Cities, Random House, 1961.
[22] C. Grotherr, P. Vogel, M. Semmann, Multilevel Design for Smart Communities–The Case of Building a Local Online Neighborhood Social Community, 2020.
[23] T.M. Vinod Kumar, State of the art of e-democracy for smart cities, in: E-democracy for Smart Cities, Springer Singapore, 2017, pp. 1–47.
[24] R. Bull, M. Azennoud, Smart citizens for smart cities: participating in the future, Proc. Inst. Civ. Eng. Energy 169 (3) (2016) 93–101.
[25] M. Gurstein, Smart cities vs. smart communities: empowering citizens not market economics, J. Community Inform. 10 (3) (2014), https://doi.org/10.15353/joci.v10i3.3438.
[26] F. Elberzhager, et al., Towards a digital ecosystem for a smart city district: procedure, results, and lessons learned, Smart Cities 4 (2) (2021) 686–716.
[27] J. Bélissent, Smart City Leaders Need Better Governance Tools, Forrester for Vendor Strategy Professionals, Forrester, 2011.
[28] A. Shleifer, D. Treisman, A normal country: Russia after communism, J. Econ. Perspect. 19 (1) (2005) 151–174.
[29] A. Mamkaitis, M. Bezbradica, M. Helfert, Urban enterprise: a review of smart city frameworks from an enterprise architecture perspective, in: 2016 IEEE International Smart Cities Conference (ISC2), IEEE, 2016, pp. 1–5.
[30] D. Linders, From e-government to we-government: defining a typology for citizen coproduction in the age of social media, Gov. Inf. Q. 29 (4) (2012) 446–454.
[31] S. Arnstein, A ladder of citizen participation, J. Am. Inst. Plann. 35 (4) (1969) 216–224.
[32] M.S. Gunko, G.A. Pivovar, Participation/non-participation of the population in urban planning, Reg. Econ. Sociol. 2 (2018) 241–263.
[33] B.W. Wirtz, P. Daiser, B. Binkowska, E-participation: a strategic framework, Int. J. Public Adm. 41 (1) (2018) 1–12.
[34] R.G. Hollands, Will the real smart city please stand up? Intelligent, progressive or entrepreneurial? City 12 (3) (2008) 303–320.
[35] M.K. Feeney, E.W. Welch, Electronic participation technologies and perceived outcomes for local government managers, Public Manag. Rev. 14 (6) (2012) 815–833.
[36] G. Aichholzer, S. Strauß, Electronic participation in Europe, in: Electronic Democracy in Europe: Prospects and Challenges of E-Publics, E-Participation and E-Voting, Springer, 2016, pp. 55–132.
[37] A.V. Chugunov, Y. Kabanov, Evaluating e-participation institutional design. A pilot study of regional platforms in Russia, in: Electronic Participation: 10th IFIP WG 8.5 International Conference, ePart 2018, Krems, Austria, September 3–5, 2018, Proceedings 10, Springer International Publishing, 2018, pp. 13–25.
[38] G. Bugs, ICTs encouraging changes in the citizen's relationship with government and urban space: Brazilian examples, J. Community Inform. 10 (3) (2014).
[39] A. Caragliu, C. Del Bo, P. Nijkamp, Smart cities in Europe, J. Urban Technol. 18 (2) (2011) 65–82.
[40] E. Cosgrave, L. Doody, N. Walt, Delivering the Smart City-Governing Cities in the Digital Age, Arup, 2014.
[41] A. Sevilla-Buitrago, Debating contemporary urban conflicts: a survey of selected scholars, Cities 31 (2013) 454–468.
[42] G. Auerbach, Urban planning: politics vs. planning and politicians vs. planners, in: Horizons in Geography / היפגואגב םיקפוא, no. 79/80, 2012, JSTOR, 2023, pp. 49–69. http://www.jstor.org/stable/23718581. (Accessed 25 March 2023).

[43] H. Kopackova, J. Komarkova, O. Horak, Enhancing the diffusion of e-participation tools in smart cities, Cities 125 (2022) 103640. ISSN 0264-2751, https://doi.org/10.1016/j.cities.2022.103640.

[44] M. Vilenskii, O. Smirnova, Evaluation of web technologies in urban planning management in the largest cities of Russia, GeoJournal 87 (2) (2022) 1385–1397.

[45] Kranzeeva E.A., Golovatskiy E.V., Orlova A.V., Nyatina N.V., Burmakina A.L. Reactive Social and Political Interactions in Innovation Processes in the Russian Regions. Power and Administration in the East of Russia 2021. 2 (95). Pp. 86–102. DOI:10.22394/1818-4049-2021-95-2-86-102.

[46] F. Walsh, A. Pozdnoukhov, Spatial structure and dynamics of urban communities, in: Proceedings of the 2011 Workshop on Pervasive Urban Applications (PURBA) San Francisco, CA, USA, 2011 (Return to ref 27 in article).

[47] S. Gao, Y. Liu, Y. Wang, X. Ma, Discovering spatial interaction communities from mobile phone data, Trans. GIS 17 (2013) 463–481, https://doi.org/10.1111/tgis.12042.

[48] A. Noulas, S. Scellato, R. Lambiotte, M. Pontil, C. Mascolo, A tale of many cities: universal patterns in human urban mobility, PloS One 7 (2012) e37027, https://doi.org/10.1371/journal.pone.0037027.

[49] L.M. Bettencourt, J. Lobo, D. Helbing, C. Kühnert, G.B. West, Growth, innovation, scaling and the pace of life in cities, Proc. Natl. Acad. Sci. U. S. A. 104 (2007) 7301–7306.

[50] W. Pan, G. Ghoshal, C. Krumme, M. Cebrian, A. Pentland, Urban characteristics attributable to density-driven tie formation, Nat. Commun. 4 (2013), https://doi.org/10.1038/ncomms2961.

[51] L. Anselin, S. Williams, Digital neighborhoods, J. Urban. Int. Res. Placemak. Urban Sustain. 9 (4) (2016) 305–328.

[52] O. Roick, S. Heuser, Location based social networks—definition, current state of the art and research agenda, Trans. GIS 17 (5) (2013) 763–784.

[53] T. Shelton, A. Poorthuis, M. Zook, Social media and the city: rethinking urban socio-spatial inequality using user-generated geographic information, Landsc. Urban Plan. 142 (2015) 198–211.

[54] P. Verdegem, Social media for digital and social inclusion: challenges for information society 2.0 research; policies, tripleC—Cogn. Commun. Co-operation 9 (1) (2011) 28–38. Available from: http://urn.kb.se/resolve?urn=urn:nbn:se:uu:diva-142934.

[55] S.E. Bibri, Data-driven smart sustainable urbanism: the intertwined societal factors underlying its materialization, success, expansion, and evolution, GeoJournal (2019), https://doi.org/10.1007/sl0708-019-10061-x. URL: h ttps: / / www. resea rchga te. net/ pu b li cati on/ 3 3 465 5 362_ Data-D ri ven_Smart_Sustainable_ U rb an ism_ The In tertwin ed_Societal_Factors_ U nd erl ying_i ts_ Materialization_Success_Expansion_and_Evolution.

[56] M. Howlett, Government communication as a policy tool: a framework for analysis, Can. Polit. Sci. Rev. 3 (2) (2009) 23–37.

[57] M.S. Feldman, A.M. Khademian, The role of the public manager in inclusion: creating communities of participation, Governance 20 (2) (2007) 305–324.

[58] Y.M. Moiseev, Thresholds of uncertainty in the urban planning system: dissertation ··· Doctor of Architecture: 05.23.22 / Moiseev Yuri Mikhailovich; [Place of protection: Moscow. architecture in-t]. Moscow, 2017. 345 p.

[59] S.M. Zavattaro, A.J. Sementelli, A critical examination of social media adoption in government: introducing omnipresence, Gov. Inf. Q. 31 (2) (2014) 257–264.

[60] United Nations Centre for Human Settlements, The State of the World's Cities, 2001, UN-HABITAT, 2001, p. 27.

[61] R. Rudolph, I. Brade, Moscow: processes of restructuring in the post-soviet metropolitan periphery, Cities 22 (2) (2005) 135–150.
[62] M. Wojcik, Peripheral areas in geographical concepts and the context of Poland's regional diversity, Region and Regionalism 11 (2013) 255–265.
[63] K. Hampton, B. Wellman, Neighboring in Netville: how the internet supports community and social capital in a Wired Suburb, City Community 2 (4) (2003) 277–311, https://doi.org/10.1046/j.1535-6841.2003.00057.x.
[64] V. Voskresenskiy, I. Musabirov, D. Alexandrov, Private and public online groups in apartment buildings of St. Petersburg, in: 301–306 in WebSci '16: Proceedings of the 8th ACM Conference on Web Science, Association for Computing Machinery, New York, 2016.
[65] N. Komninos, M. Pallot, H. Schaffers, Special issue on smart cities and the future internet in Europe, J. Knowl. Econ. 4 (2013) 119–134.
[66] O. Gromasheva, Hybrid neighborhood in action: the example of Kudrovo, Russia, Lab. Russ. Rev. Soc. Res. 13 (2) (2021) 13–38, https://doi.org/10.25285/2078-1938-2021-13-2-13-38 (November 2021).
[67] L. Vitkova, M. Kolomeets, Approach to identification and analysis of information sources in social networks, in: Intelligent Distributed Computing XIII, Springer International Publishing, 2020, pp. 285–293.
[68] J. Clark, 6 smart cities as participatory planning, in: Uneven Innovation: The Work of Smart Cities, Columbia University Press, New York Chichester, West Sussex, 2020, pp. 156–180, https://doi.org/10.7312/clar18496-008.
[69] A. Rossi, The Architecture of the City, MIT Press, 1984.

CHAPTER

15 Smart and sustainable solutions for thriving tourism destinations

Tomáš Gajdošík, Zuzana Gajdošíková, and Matúš Marciš
Department of Tourism, Faculty of Economics, Matej Bel University, Banská Bystrica, Slovakia

1 Introduction

Smart technologies have been tightly knitted into the fabric of physical places and objects such that certain spaces can be called "smart." These spaces, responsive and adaptive to external changes and user behavior, have the potential to improve quality of life in many different domains, ranging from health and education, to work, entertainment, and tourism. Whereas there is a lively, ongoing discussion on smart cities as smart spaces, the research on smart tourism destinations lags behind.

Tourism is a highly significant economic sector globally and locally, and provides real prospects for enduring and inclusive economic growth. The sustainable development of tourism and the digital transformation of the sector are key issues [1]. The potential role of tourism with regard to smart spaces can be understood as how technology can be leveraged to enhance the tourist experience, the well-being of residents, and the competitiveness of businesses, and to create more sustainable and efficient spaces. Tourism can also play a role in the development of smart spaces by driving the adoption of new technologies and practices, since it is a major industry in many locations. This can lead to the development of more efficient and sustainable tourism practices, of benefit to both tourists and the local community.

The tourism sector, as well as the environment that affects it, is currently going through a turbulent period that is bringing significant changes in how tourism is perceived. We are witnessing the adoption of new ways of thinking and new patterns of consumer behavior, along with the establishment of new types of businesses and the creation of new products and services. To succeed in the tourism industry, it is necessary to quickly orient oneself to this new and hitherto unknown environment, understand its characteristics, and proactively respond to it. Tourism stakeholders will be successful, competitive, and integral parts of the tourism industry if they have a deep understanding of the environment in which they

Smart Spaces. https://doi.org/10.1016/B978-0-443-13462-3.00017-0

operate. To achieve this, it is crucial for tourism stakeholders to invest in technology and tools that allow them to collect and analyze relevant data on their target markets, as well as to monitor and measure the impact of their actions on the environment and the local community. Moreover, it is crucial to act in accordance with the principles of sustainable development to preserve resources for the development of tourism.

The pandemic, security breaches, and energy crisis have tested the adherence to sustainable principles in the tourism industry. However, researchers agree that the sector contributes only slightly to the goals of sustainable development [2,3]. To improve the contribution of tourism to sustainable development, it is appropriate to use a smart approach to tourism development. This approach is currently considered as a tool to respond to changing market conditions. The smart approach to tourism development is not the goal itself, but rather a means to achieve a sustainable, competitive, and resilient tourism sector. Despite its potential, the contribution of a smart approach to sustainable tourism development has not yet been sufficiently explored in theory or practice [4]. Several authors (e.g., Perles Ribes and Ivars Baidal [5]; Gelter et al. [6]) as well as international organizations and institutions such as the UN, OECD, UNWTO, and the European Commission call for a more thorough examination of the contribution of a smart approach to sustainable tourism development. The aim of this chapter is to examine the potential of the smart approach for sustainable development of tourism destinations.

The chapter fills a gap in the research by examining the link between smart and sustainable tourism development. Its main contribution lies in the analysis of methodological, managerial, and data problems of implementing sustainable principles in tourism destinations and the ways in which the smart approach could tackle them. The chapter establishes that the smart approach is suitable for destinations in a higher life-cycle phase, while the implementation of smart principles leads to higher destination competitiveness. Moreover, the analysis of implementation of smart technologies by destination management organizations (DMOs) critically assess their use. Based on the analysis, the chapter recommends the use of new data sources, focusing mainly on user-generated content, new devices, and electronic transactions. Consequently, requirements on intelligent information systems are proposed. To overcome the obstacles of data sharing and cooperation, the use of interactive game elements is welcomed, as well as cooperation with international organizations.

The concept of smart tourism destinations is derived from smart cities. Several cities are focusing on smart tourism development to meet the growing tourism demand and they recognize that travel is an important part of people's lives [7]. While smart cities prioritize the needs of local residents, smart tourism destinations also prioritize the needs of tourists. By leveraging technology and data, these destinations aim to create closer relationships between tourists, residents, businesses, local government, and tourist attractions to facilitate communication and education about the destination [8].

2 Developing smart solutions for sustainable tourism

In today's world, it is crucial to adapt to new challenges and trends in the tourism industry, while not forgetting the essence of tourism development. When it comes to explaining the principles of sustainable development, it becomes clear that complying with these principles is a challenging task for tourists, businesses, and tourism destinations alike. However, there is a positive potential in smart solutions, which, due to their capabilities, could contribute significantly to ensure the development of tourism according to sustainable principles.

The idea of smart development originated from the Sustainable Development Goals (SDGs) of the United Nations and later evolved into a strategy that complements sustainability in development efforts [9]. Sustainability is not adequately addressed in the theory and practice of smart tourism development [4]. As smart tourism development is significantly supported by information technologies, it is necessary to primarily focus on the impact of technology (hard smartness) on sustainable tourism development. Subsequently, it is also appropriate to focus on soft smartness, which supports innovation and knowledge.

Tourism has traditionally been viewed as a beneficial phenomenon that fulfills the needs of tourists, generates economic income, creates job opportunities, improves the standard of living of local residents, and develops destinations. However, this one-sided perspective has often emphasized the positive impacts of tourism and prioritized economic growth, leading to an increase in the number of tourists and the promotion of tourism as a means of maximizing economic benefits.

It is important to note that the sustainable development of tourism should not cause any environmental or social tension. When tourism exceeds the carrying capacity of a destination, negative effects can occur, outweighing the positives, resulting in slower growth or even a decrease in the number of tourists who are no longer satisfied with the destination. Therefore it is necessary to ensure that tourism development maximizes positive impacts and minimizes negative ones in a destination.

Sustainable tourism is defined by the UN World Tourism Organization as "tourism that takes full account of its current and future economic, social and environmental impacts, addressing the needs of tourists, the industry, the environment and host communities" [10]. Sustainable development recognizes the inseparability of the environment and human existence and seeks to meet basic human needs without irreversibly damaging natural resources [11,12]. Tourism, being reliant on natural resources, has limited growth opportunities [13]. Therefore tourism development must prioritize generating revenue, respecting environmental prerequisites, and improving the quality of life of local residents. Global tourism organizations (UNWTO, WTTC, OECD, etc.) emphasize the need for sustainable tourism through smart development. Smart tourism development is an efficient, inclusive, and sustainable approach that promotes innovation and addresses challenges in the tourism industry.

The concept of smart tourism development was created as a response to changes in the external environment that affect the tourism sector. It should be seen as a philosophy of tourism development, which creates opportunities for better satisfaction of tourist needs, creation of new business models that better link supply and demand, and more professional management of tourism destinations.

The concept of smart tourism development is applied at the level of tourism destinations with the goal of increasing competitiveness and improving the quality of life of all stakeholders, including residents and tourists [7]. This distinguishes it from the smart concepts applied in other industries. With the help of real-time data, decision-makers can make informed decisions and quickly adapt to changing circumstances.

A smart tourism destination can be defined as an "innovative tourist destination built on an infrastructure of state-of-the-art technology, which guarantees the sustainable development of tourist areas, facilitates the tourist interaction with and integration into his or her surroundings, increases the quality of the experience at the destination, and improves residents' quality of life" [14]. It is important to consider the life cycle phase of smart destination development, which can affect its competitive position in the market, as noted by Errichiello and Micera [15].

2.1 The impact of smart technologies on sustainable development in tourism destinations

The term "smart technology" does not imply that the technology itself is smart. Instead, it means that individuals and the sector can become smarter when using it, as it enables quick reactions, sensitive analysis, and anticipation of emerging situations [16]. These technologies can be classified into four groups: (1) technologies that provide information, (2) technologies related to data storage and management, (3) technologies that support experience, and (4) sensing and network technologies [17] (Fig. 1).

When examining the impact of information technologies on the sustainable development of tourism, we rely on the latest studies that have focused on both the demand and supply sides of tourism. According to Schmücker et al. [18], there are two main aspects in considering how information technology affects tourism demand. The first aspect is the impact of technology on the volume of tourists. In this case, technology acts as a double-edged sword in terms of sustainable development. On the one hand, it promotes increased participation in tourism, allows greater distances to be traveled, and extends the length of stays in destinations, thereby creating a larger carbon footprint. On the other hand, information technology can also serve as a substitute for traveling through digital communication and virtual environments.

The second way in which information technologies affect tourism demand is through changing tourist behavior, including their choice of transportation, accommodation, tourist attractions, and destinations. Information technologies can provide transparent information on the capacity of attractions in real time and enable

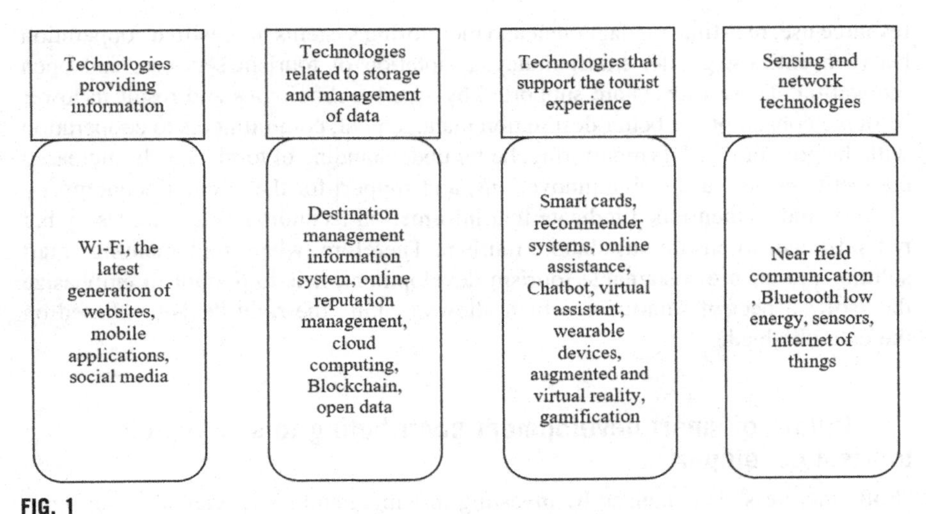

FIG. 1

Examples of smart technologies used in tourism destinations.

Source: Authors.

monitoring of the environment. They can also offer tourists intelligent recommendations and sustainable alternatives. Digitalization offers multiple opportunities to promote sustainable behavior, such as through reward systems, real-time benefits, transparent information, and awareness-raising campaigns. Similarly, Gössling [3,19] investigated the impact of information technology on sustainable tourism development from various perspectives, including sociology, transport, psychology, and management. His studies identified both positive and negative effects of information technology on the environmental, economic, and social pillars of sustainability.

From the perspective of tourism supply, businesses and organizations in tourist destinations have an influence on the sustainable development of tourism. Benckendorff et al. [20] followed up on the research of Ali and Frew [21] and they present the contribution of information technology to sustainable development from a supply point of view. They identified several technologies and specified their significant contribution to the environmental, economic, and social pillars of sustainability. Information technology can be used to support decisions necessary to ensure environmental protection. Its introduction helps to improve the awareness of stakeholders of the impact of tourism on the environment, local residents, and the economy of the destination. Perles Ribes and Ivars Baidal [5] share the view that the key to sustainability lies in data that can be collected and analyzed through technology to support decision-making. They apply this idea to tourism destinations and argue that a destination cannot be considered smart unless it is also sustainable. The authors propose a synergistic model of smart sustainability for tourism destinations, which includes key aspects such as long-term strategic development planning, more efficient resource use, real-time management and monitoring systems, intensified cooperation between entities, greater transparency, adaptation of tourism services, and open innovation. These aspects are supported by smart technologies and result in lower resource consumption, better destination management, commitments to cooperation with the possibility of its monitoring, better understanding of tourist needs, increased competitiveness, sustainable innovations, and support for the circular economy.

The studies mentioned indicate that information technologies are necessary but not sufficient to ensure sustainable tourism. Therefore, when implementing smart solutions to ensure sustainable tourism development, it is important to emphasize the "soft" aspect of smartness, which allows making the right decisions based on the data obtained.

2.2 Pillars of smart development contributing to sustainable tourism development

"Soft smartness" is enhanced by investing in human and social capital, open innovation, and participatory governance, which can contribute to collective competitiveness and promote social, economic, and environmental prosperity for all stakeholders, leading to sustainable competitive advantages [7]. The European Commission [22] also supports this concept and identifies five pillars of smart tourism development (Fig. 2), which include soft smartness in addition to tourism data.

Strategy and governance are interlinked with the ability to acquire and utilize insights from available data. To improve strategic planning and governance, it is crucial to use activities focused on data collection and analysis that can bring knowledge and help to overcome common issues in tourism [23]. This approach allows a wider focus on sustainable development, responsible resource use, progress in quality of life, and improvement of the well-being of stakeholders and the local community, rather than solely focus on economic growth. To achieve this, tourism destinations should involve both residents and tourists in policy formulation and decision-making

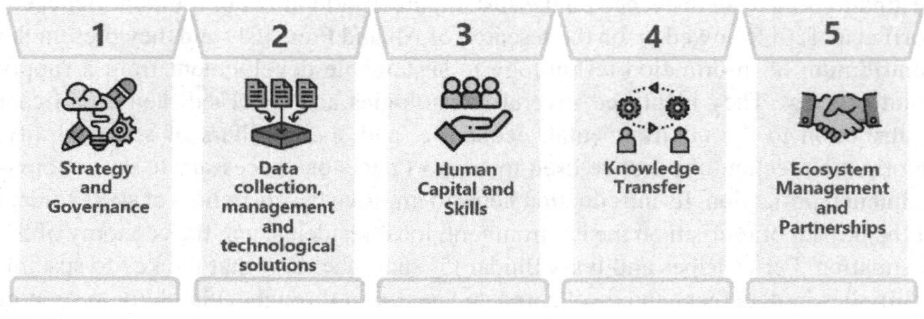

FIG. 2

Pillars of smart tourism development.

Source: European Commission, Mastering Data: A Toolkit for Tourism Destinations, 2023, 19 p. Available at https://smarttourismdestinations.eu/wp-content/uploads/2023/03/Smart-Tourism-Destinations_Toolkit.pdf.

in the tourism industry. Data are therefore seen as a tool for monitoring progress toward achieving goals and providing guidance for strategic decision-making. This new perspective on strategy and governance enables faster, more accurate, and reactive decision-making, leading to more sustainable, competitive, creative, and resilient tourism development in destinations.

Data is considered a new driving force in tourism, as its collection and analysis provide valuable information and optimize processes. An important prerequisite is having adequate information about tourists and their behavior. By understanding tourists' characteristic features and spatiotemporal behavior, businesses and tourism destinations can develop targeted marketing campaigns [24]. Data should also be integrated into overall strategic planning in addition to tactical marketing. Personal data, behavioral data, and location data should ensure that travel processes are efficient and that products are delivered to the right customers. For example, with advanced data analytics, tourist behavior can be analyzed in real time and services can be personalized. The digitalization of almost all processes provides new opportunities, while new sources of data are increasingly available to support the smart and sustainable development of tourism.

Human capital and skills are critical elements of a smart approach that supports sustainability. Adequate leadership is required to incorporate smart tourism development and data-driven decision-making into relevant development policies. Data analysis skills are becoming increasingly important for working in the tourism industry. Data analysts should support tourism managers who have limited involvement in data analysis, whether from an internal or external environment. To enable tourism managers to be proactive, they should be able to use or have access to predictive analytics (which provide scenarios of future development), as well as prescriptive analytics (which provide recommendations for decision-making) [25]. This often requires knowledge of data mining and working with algorithms based on artificial intelligence. However, tourism businesses and organizations often struggle with relatively small budgets and limited staff. Financial support for the development of smart tourism by national and international organizations can help overcome some of the budget problems mentioned here.

Knowledge transfer among stakeholders is an essential component of tourism destinations' competitiveness. Static reports, such as annual reports, are not suitable sources of relevant knowledge for the future, as the information they contain quickly becomes obsolete. To enable proactive and relevant decisions, data visualization through interactive dashboards is crucial. By combining real-time data, interactive dashboards facilitate a more efficient decision-making process. Data processing should be based on real-time processing and analysis, drawing from various sources. The use of open data stimulates knowledge transfer, creating opportunities for shared decision-making. As data sources differ, so do data owners. However, stakeholders are often unwilling to share data, fearing misuse of their data and a lack of trust or shared vision. Information technologies should provide better tools for cooperation and knowledge sharing. Open data should ensure the transparency of knowledge sharing as a first step.

The smart approach creates opportunities for better *partnerships* and promotes knowledge sharing. Human capital is crucial in smart development because the tourism product is centered on people. Smart destinations should become centers that attract people to discover, live, and do business in them [26]. Therefore tourism destinations should cooperate with local residents and tourists. Technology, such as smartphones, wearables, and social media connections, should be used to involve both locals and tourists in the creation of the tourism product. Providing training and supporting residents to develop their own technologies and crowd-sourced information can make these connected and creative residents more active and involved in creating the tourism product. In addition, today's tourists are more experienced, demanding, and active. They have higher expectations for tourism experiences [27]. By using technology and sharing data, tourists become cocreators of their experiences.

Based on the approaches mentioned, it can be concluded that the utilization of information technology and interconnected systems can optimize decision-making based on collected data, which ultimately enhances the tourist experience, provides business opportunities, and has a positive impact on the tourism destination. However, it is important to note that smart tourism development should not be considered as the ultimate goal, but rather as a means to achieve a better experience for tourists, improve the quality of life of local residents, increase the efficiency and competitiveness of businesses and tourism destinations, and promote sustainable tourism development in a dynamic environment.

3 Tourism destinations using smart solutions for sustainable tourism development

The principles of sustainable tourism development apply to all types of tourism destinations. In all destinations, the main goal of development should be the preservation of natural, cultural, and other attractions that motivate tourists and are the source of tourism growth, as well as the authenticity of local communities. Given the complexity of tourism destinations and the number of stakeholders, this task may seem daunting. However, smart technologies offer a solution to better implement these principles and have the potential to contribute significantly to sustainable tourism development.

In order to examine the potential of the smart approach for sustainable development of smart tourism destinations, multiple case study methodology is used (Fig. 3). This kind of methodology is used to fully and complexly examine the phenomenon of the smart approach in tourism destinations. The context of the research is geographically placed in Central and Eastern European countries, which are considered as digital challengers. This provides a useful knowledge for other destinations on the way to become smart and sustainable.

Firstly, the emphasis was put on the analysis of the importance of and barriers to implementing the sustainable principles in tourism destinations. A semistructured

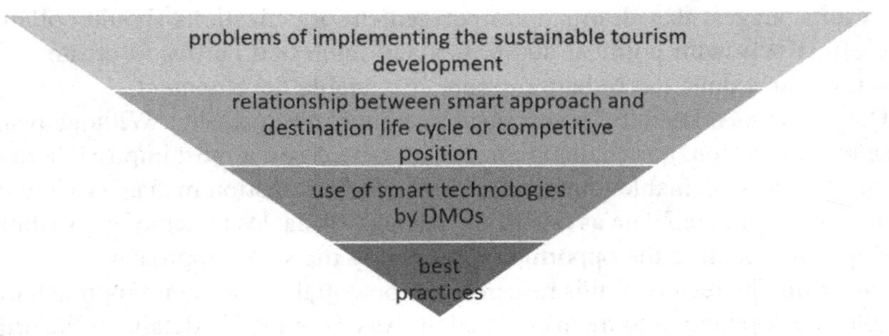

FIG. 3

Multiple case study methodology used.

Source: Authors.

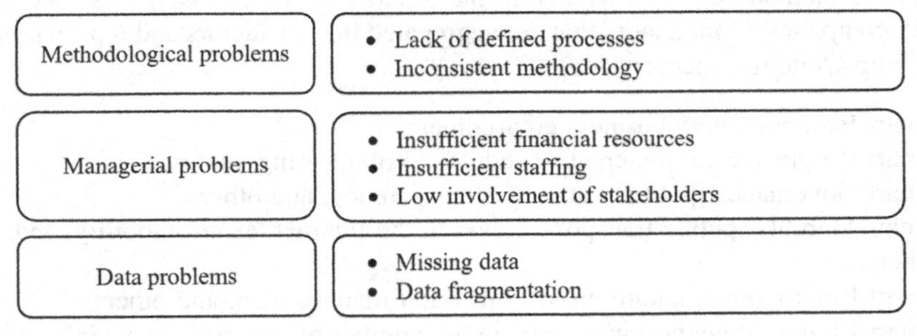

FIG. 4

Problems of implementing sustainable tourism development.

Source: Authors.

interview was carried out among 60 managers of destination management organizations (DMOs) in Slovakia, the Czech Republic, Hungary, and Poland. The results obtained indicate a discrepancy between the managers' opinions and their actual behavior in ensuring sustainable tourism. To investigate the reasons for this discrepancy, we conducted a content analysis of the statements made by the managers of the tourism destinations. Based on their statements, it can be concluded that the barriers to implementing sustainable development can be classified into methodological, managerial, and data problems (as shown in Fig. 4).

A methodological problem refers to the lack of knowledge necessary to successfully implement and monitor sustainable development in the management practices of organizations. The problem mainly arises from undefined monitoring procedures and a lack of uniform methodology. Research shows that, while there is awareness of the need for sustainable development, it has not yet been fully integrated into the development of tourism destinations. Managerial problems include a lack of personnel and financial resources, and inadequate involvement of relevant stakeholders.

The results suggest that destination management organizations should collaborate more effectively with political authorities and interested parties to create partnerships for data acquisition to better ensure sustainable development.

The problems related to data acquisition are the most pressing. Without available data and information, it is challenging and, in some cases, almost impossible to monitor and ensure sustainable tourism development. Destination managers view insufficient and fragmented data as one of the primary obstacles to achieving sustainable development, creating the opportunity to employ the smart approach.

Following the results of this research, the potential of the smart approach to sustainable development of tourism destinations was examined in detail. As the primary goal of tourism development is to ensure sustainable competitiveness of a destination, the emphasis was put on the link between the smart approach and destination life cycle and/or competitive position of a destination.

In order to measure the level of implemented smart solutions in tourism destinations, the methodology and data from the Smart City Index (SCI) was used. The index comprises 74 indicators that are aggregated into 31 factors and 6 primary pillars (https://inteligentnemesta.sk/):

- Smart Economy, e.g., business environment;
- Smart People, education, creative industry, volunteering, and others;
- Smart Governance, political activity, transparency, and others;
- Smart Mobility, public transport, access to the Internet, electromobility, and others;
- Smart Environment, nature protection, water management, and others;
- Smart Living, life expectancy, crime rate, number of overnight stays in accommodation facilities, and more.

The SCI data were aggregated and processed for municipalities that are members of 39 destination management organizations in Slovakia (Fig. 5).

As sustainable tourism development in tourism destinations depends on their life cycle, the models from Butler [28] and Buhalis [29] were used to classify the destinations into individual phases of the life cycle with the help of cluster analysis. To determine their relative competitiveness, the models from Ritchie, Crouch [30] and Dwyer and Kim [31] were used and factor analysis was conducted [32]. The correlation coefficient between these variables was calculated (see Table 1).

Based on the Pearson correlation coefficient, it is possible to observe that there is a moderately strong direct relationship between smart development and the life cycle phase of a destination. This relationship indicates that, while in the initial phases the focus is mainly on building infrastructure and forming the destination's product, later on the emphasis is on planning, sustainability, and quality. Subsequently, it is possible to focus on knowledge, technology, and innovation in the destination, leading to a smart development of a tourism destination. Moreover, a moderately strong direct relationship indicates that smart development has a positive impact on the competitiveness of a tourism destination.

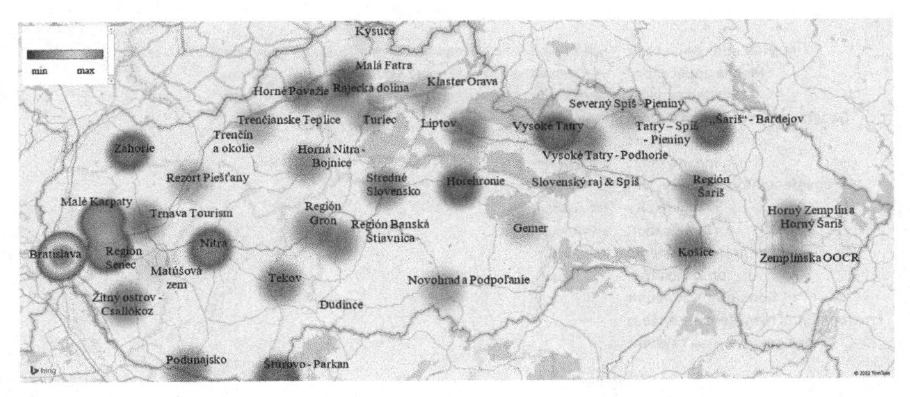

FIG. 5

The level of implemented smart solutions in tourism destinations in Slovakia.
Source: Authors based on the Smart City Index (www.inteligentnemesta.sk, 2022).

Table 1 Pearson correlation results.

Smart City Index	Tourism area life cycle	Tourism destination competitiveness
Pearson correlation	0.551[a]	0.559[a]
Sig. (2-tailed)	0.001	0.000
N	35	35

[a]*Correlation is significant at the 0.01 level (2-tailed).*
Source: Authors.

Furthermore, to determine how information technology contributes to sustainable tourism development, we conducted a survey of 14 destination management organizations in Slovakia in 2021. The most commonly applied activities in line with the principles of sustainable development in destinations include:

- promoting the use of eco-friendly transportation modes,
- involving local residents in tourism development,
- educating tourists and local residents,
- efficient resource use,
- recycling,
- focus on renewable energy sources.

Given the potential for sustainable development in tourism destinations, we find that destination management organizations actually perform only the minimum activities aimed at ensuring sustainable development. Also, sustainable development and its principles should be part of daily activities of DMOs, given the current situation.

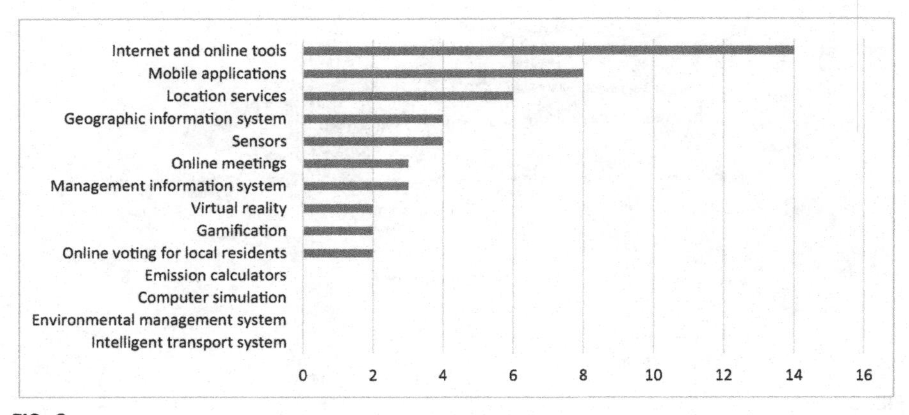

FIG. 6

The use of smart technologies by destination management organizations.

Source: Authors.

The results of the survey (Fig. 6) indicate that destination management organizations have not fully utilized the potential of information technologies that could simplify and streamline the implementation of the principles of sustainable development.

DMO managers stated that they mainly use Internet, mobile applications, location-based services, and geographic information systems. Given the findings on the use of information technology to ensure sustainable development, we recommend that DMOs learn from positive examples in the industry and from international organizations. Destination management organizations that have positive experience supporting sustainable tourism practices are in a good position to share information and give advice to other destinations. This can be further simplified by creating tools to support this process, such as platforms to share ideas, practical guides for application development, and sets of tools to assess the specific sustainable tourism needs of a given destination.

3.1 Göteborg (Sweden)

Gothenburg is currently considered a leading destination for sustainable and smart tourism. The DMO Göteborg & Co′s objective is to ensure that tourism growth is sustainable and considers the well-being of local residents. To achieve this, a city analysis was conducted in 2019 to identify potential strategies for smart and sustainable development. This resulted in the creation of dynamic methodologies to achieve environmental objectives, collaborative resources, strategic partnerships, citizen participation, and improved public-private partnerships focused on sustainability. Göteborg & Co aims to advance its sustainability ambitions while improving the visitor experience and changing the perception of the word "smart" from a technical term to one that values inclusivity and human values. To achieve this, Göteborg

& Co has adopted and supported several smart initiatives that promote sustainable tourism practices. These include the following:

- Digital tourist advisor is available through text or voice messages, improving accessibility for all, including visually impaired. Tourists can also communicate with visitor center staff through social media and live chat features;
- The mobile app "To Go" provides discounted prices on public transportation tickets and a trip planner to encourage tourists to use public transportation more frequently;
- App that simplifies control helps local residents easily provide interpretation or electronic signatures;
- App for reporting barriers allows users to alert city authorities to obstacles such as high curbs and potholes. The app's open-source code makes it easy to use in other locations;
- The event impact calculator is a tool for organizers, destination management organizations, and sponsors to predict and calculate the economic, social, and environmental impacts of planned events, and to test various variables to optimize event sustainability;
- Travelandclimate.org is the online calculator that allows tourists to choose transportation and accommodations with less impact. It provides information on the estimated emissions during their journey and the amount of melting Arctic ice (in square meters) due to their travels;
- "Meeting the Locals" connects tourists with residents who offer services such as guided tours or car sharing;
- Smarta Karten is an interactive tool that promotes sustainable living and responsible consumption among residents and tourists. It provides information on opportunities to give, receive, share, exchange, or rent any items or services, from food to office space;
- Green Gothenburg offers virtual and study trips, advice, contacts, and seminars to delegates from other destinations interested in implementing smart and sustainable offers;
- "101 Sustainable Ideas" is a platform that collects and shares inspiring examples of sustainable tourism from around the world.

Göteborg & Co is now collaborating with Mastercard on developing a joint data platform to share anonymized data with stakeholders in the visitor economy. This will help identify, monitor, and develop key tourist segments by creating a picture of their activities before, during, and after their stay in the city; and facilitate decision-making for sustainable recovery after the COVID-19 pandemic.

3.2 Dubrovnik (Croatia)

Dubrovnik, a city of 42,000 inhabitants, has a historic old town where only 2000 people reside. The increase in tourist numbers (which reached almost 1.3 million tourists in 2018) and the excursion ship industry (with 9500 visitors per day during the peak season) have caused significant problems. These include traffic congestion, overcrowding, pollution, waste management issues, and damage to important sites. The dissatisfaction of local residents is also a concern, with some even leaving the area.

In 2015, UNESCO concluded that the city should not exceed a capacity of 8000 visitors per day to avoid environmental and social damage. They recommended reviewing measures to ensure appropriate visitor management. The city of Dubrovnik aims to identify the impacts of tourism and the associated risks from the increasing number of tourists. To address these issues and sustain tourism growth, the city's tourism management recognizes the importance of collaborating with stakeholders to address the root causes and effects of excessive tourism. The following initiatives were launched in the city to manage visitors more effectively:

- Creation of better conditions for collaboration between the private and public sectors, with a focus on communication and exchange of experiences;
- Launch of the "Respect the City" campaign to protect cultural heritage, preserve the quality of life of locals, promote responsible resource consumption, and provide tourists with the best possible experience;
- Introduction of an online system for public voting, referendums, and feedback on tourism projects, allowing local residents to participate in decision-making;
- Creation and monitoring of a satisfaction index for local residents;
- Implementation of a system for monitoring the number of tourists using six cameras and counting devices at six entrances to the old town. Data, combined with information on overnight stays, arrivals from excursion boats, and daily weather reports, are used to predict the number of people in the historic center using an algorithm. This information is available to residents and tourists on the "Dubrovnik Visitor" platform, helping potential tourists decide when to visit the city to avoid crowded areas;
- Recognizing that addressing the impacts of visitors from excursion boats is crucial to sustainable management of Dubrovnik as a destination, the city has established a working relationship with the Cruise Lines International Association (CLIA). This culminated in a Memorandum of Understanding in 2019, in which CLIA agreed to fund an evaluation of the city using the Global Sustainable Tourism Criteria to create a sustainable tourism plan for the city.

Installing infrastructure (ticketing systems and applications that monitor the number of tourists) provides tourists and businesses with the assurance that tourism is being monitored and managed in a more fair and responsible way.

3.3 Slovenia and the Black Sea region

The tourism industry in the Slovenia and Black Sea region is experiencing growth that has brought about both positive and negative effects on popular destinations. One of the shared challenges facing developing destinations in the region is a lack of data that could provide a comprehensive overview of tourist trends and enable them to predict and address related effects. To address this, the Tourism 4.0 project was established to strengthen collaboration between the tourism sector and a technology company and ensure that tourism decision-making is based on data that are responsive to the needs of the destination. The objectives of the project are to increase the visibility of coastal destinations, reduce the negative impacts of tourism, improve the economic and social outcomes of tourism in destinations, and implement effective visitor management.

The Tourism 4.0 project is built on the Tourism Impact Model (TIM), which is a comprehensive modeling tool used to measure, analyze, and forecast tourist flows, as well as the environmental, economic, and social impacts of tourism activities unique to a specific destination. The model uses multiple sources of data to provide a comprehensive understanding of the effects of tourism. The TIM model focuses on:

- Identification of tourist areas within the specific destination;
- Mapping and collection of data through an advanced questionnaire using more than 300 indicators and up to 100,000 input data relating to topics such as protection of natural and cultural heritage, tourism revenues, visitor-to-resident ratio, inclusion of vulnerable groups, health and safety, tourism employment, CO_2 emissions, quality of life, waste management, collaboration with local suppliers, property costs, water supply and electricity consumption;
- Use of a comprehensive automated evaluation tool to create a clear picture of the impact of tourism in the specific area;
- Creation of a visual "destination character table" illustrating the positive impacts that tourism brings to a specific area, as well as the extent to which different resources are consumed. This helps to signal where tourism becomes exploitative and where it has a relatively small negative impact;
- Development of recommendations for measures aligned with sustainable development goals.

The TIM process enables strategic tourism planning by utilizing detailed open data that are readily available and aligned with sustainable development goals. Additionally, the data is made accessible through result visualization. Online workshops were held in each of the participating countries (Slovenia, Ukraine, Romania, and Georgia) with stakeholders to share information and results.

Using a diverse set of measures to assess the negative effects of tourism, such as resource consumption, tourism businesses can pinpoint their contribution to mitigating these harmful impacts. However, the process of gathering resource usage data can be challenging. Therefore the national tourism organization should encourage local and regional destination management organizations to collect, process, and provide the necessary data for strategic planning purposes.

The analysis conducted offers a comprehensive overview of sustainable and smart tourism development at the level of tourism destinations. The most significant findings are highlighted in the following list:

- Regarding the implementation and monitoring of sustainable development, methodological problems (partial knowledge of implementing sustainable development principles), managerial problems (personnel, financial, inadequate cooperation with involved parties), and problems related to the challenging acquisition of data for measuring and effectively managing sustainable development in tourist destinations are observed;
- There is a positive correlation between the level of smartness and destination life cycle, indicating that mature destinations have been implementing the smart approach more. Moreover, the smart approach positively influences the competitive position of a tourism destination;
- Destination management organizations have not fully utilized the potential of smart technologies, which could help simplify and streamline the provision of sustainable development. They mainly use the Internet and online tools, mobile applications, and location-based services to ensure sustainable development;
- Destination managers see the potential of information technologies in the ability to store and process information and subsequently monitor and regulate tourist behavior.

Based on the findings, it is possible to identify requirements for how the smart approach can help ensure sustainable tourism development. This includes identifying new sources of information and applying smart technologies (such as reservation systems, websites, social media, and mobile applications) to support sustainable tourist behavior. Sensing and networking technologies, such as the Internet of Things, sensors, NFC, and BLE, as well as experience-supporting technologies, such as visitor cards, wearable devices, augmented and virtual reality, and gamification, can be leveraged as tools for sharing data and facilitating exchange. Furthermore, the smart approach should aim to bring the available information and knowledge about sustainable development closer to businesses and organizations, helping to monitor and learn. These requirements should be further developed and incorporated into development planning and tourism policies.

4 Using smart principles for sustainable tourism destination development

Development of information technologies, new data sources, and new possibilities for their processing and knowledge creation can optimize the impacts of tourism on local ecosystems and ensure efficient, transparent planning and development according to the principles of sustainable development. The integration of the principles of smart development should be applied in tourism planning and policy, thus deliberately influencing the sustainable tourism development, shaping appropriate structures, and promoting interaction between stakeholders in tourism development. The goal of smart and sustainable tourism development is to bring added value to tourists, residents, and businesses.

So far, tourism planning and policy have focused mainly on economic development with the goal of increasing the number of tourists, and tourism businesses have preferred quantity over quality. However, the smart approach offers opportunities to expand this perspective and include more sustainability in tourism development. To formulate options, we build on previous findings and develop the idea of Perles Ribes and Ivars Baidal [5] that sustainability relies on data that can be collected and analyzed, using technologies to support decision-making. The use of new data sources allows for a better understanding of tourist behavior and cooperation between tourism businesses. By processing data in real time, relevant information can be obtained, and proactive action can be taken. Intelligent information systems and interactive game elements can be used to support knowledge sharing and improve cooperation between actors. A smart approach also helps to monitor the status quo, which is a starting point for improvement. Improvement can only be achieved through continuous learning, which can be aided by external sources from international organizations.

This new perspective on strategic planning and tourism policy enables faster and more efficient decision-making, positively influencing the tourism industry's transition toward sustainability and resilience.

4.1 From traditional to new data sources in tourism

The smart approach can transform the tourism sector by supporting data-driven decision-making. Currently, the need for sufficient data in the tourism industry has not been fully met. Tourists, businesses, and destinations have relied on data that are retrospective and includes only a small sample set. Such data only allows limited possibilities in strategic planning and management. However, the use of information technologies by tourists and their positive reception by businesses and tourism destinations provides opportunities to change the usual approaches to data collection. These new sources provide a large amount and variety of data that can be used to better understand tourists, optimize processes, and contribute to the sustainable tourism development. The exponential growth of information technologies offers opportunities to use new sources of big data in the tourism industry (see Fig. 7).

Although these new data sources belong to big data, their size does not always guarantee greater accuracy. Not all data collected in new ways are of sufficient quality, and there is often a large amount of "noise" among them. The size of the data and the need to clean it require significant storage and processing costs. Additionally, some data are limited due to tourists' reluctance to share their data. Therefore there is a need to shift the focus to smart data. Smart data is a subset of data (big or small) that is valuable to an organization. While big data focuses mainly on volume (size), velocity, and variety, smart data focuses on the dimensions of veracity and value. Smart data filters out the noise and only keeps the data that are valuable for solving problems. Consequently, smart data are not a new type of data, but are formed by a combination of traditional and big data, providing accurate information and increasing value for tourists, businesses, and tourism destinations.

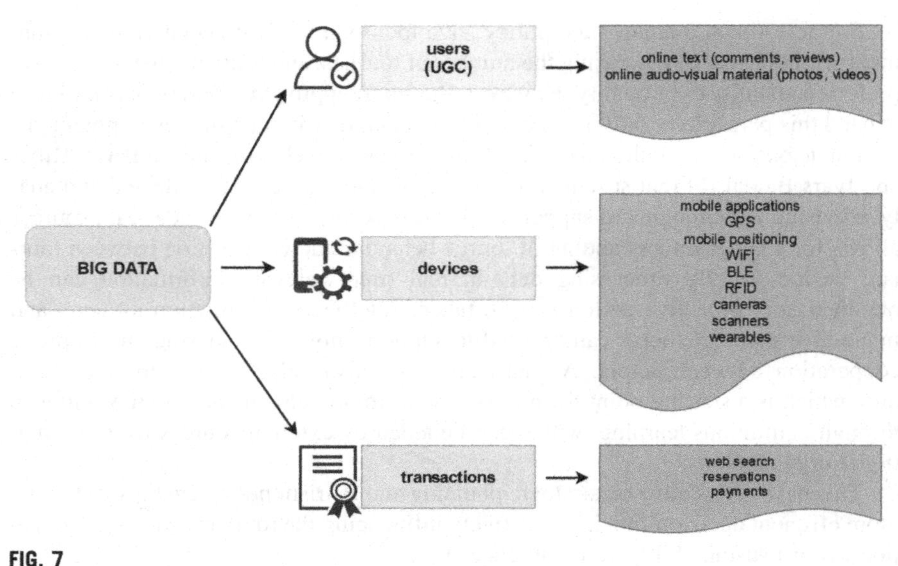

FIG. 7

New sources of data in tourism.

Source: Authors.

4.2 Knowledge transfer and collaboration in an interactive way

Although smart data sources support the smart approach to achieve sustainable tourism, it is necessary that the information and knowledge gained from these data be accessible to all stakeholders. Due to the diversity of data sources, their owners also differ. Therefore it is necessary for the information to reach the stakeholders who can act on it. It is appropriate to use an intelligent information system that will process the data collected, enable sharing, and serve as the basis for the creation of knowledge. At the same time, it is necessary to support cooperation and knowledge sharing, for which interactive game elements can be used. Despite the use of information systems in several tourism destinations, no uniform recommendations have yet been created for the architecture of these systems. The creation of a pilot information system for destinations based on the principle of business intelligence [33] has made a significant contribution. Based on these findings and subsequent investigations [34], the following conceptual model of an intelligent information system consisting of three layers focused on data collection, processing, and exchange is proposed. Destinations interested in using an information system for the creation and sharing of knowledge can take inspiration from this model when specifying requirements for this system (see Fig. 8).

The proposed intelligent information system is based on three key technologies: the Internet of Things, cloud computing, and artificial intelligence. The Internet of Things enhances the efficiency of data transmission by allowing data to be conveyed in real time. However, since not all data can be collected in real time, the system

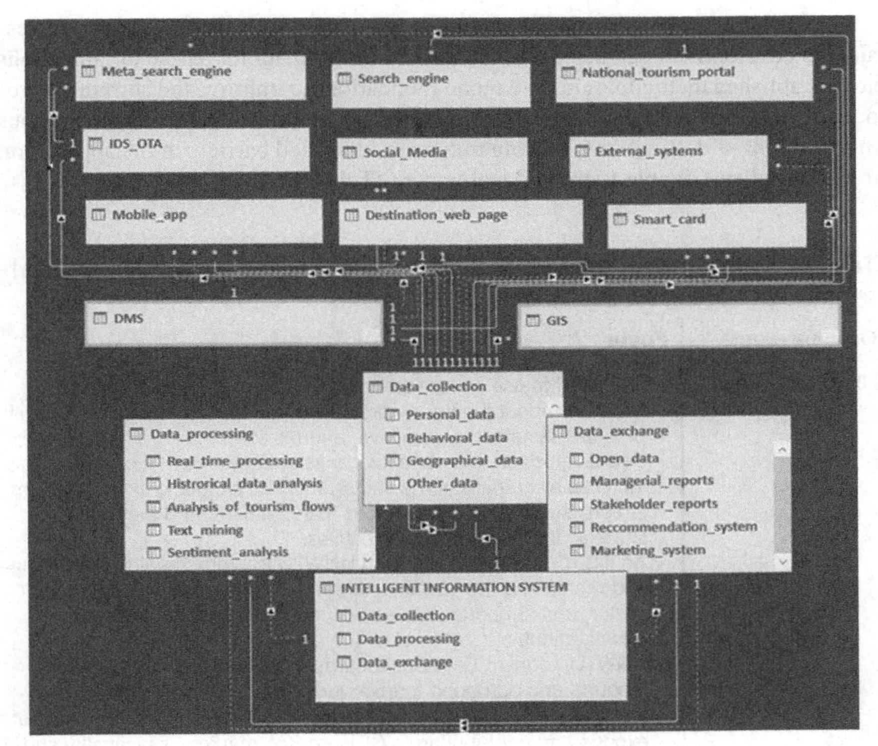

FIG. 8

Architecture of the intelligent destination information system.

Source: Authors.

utilizes both new and traditional data sources. The system's architecture, built on the principles of cloud computing, enables data storage and access, as well as seamless integration of components and layers. Artificial intelligence is also used to process data and provide recommendations.

To make a destination sustainable and competitive, it is necessary that the interested parties cooperate with each other. Applying smart principles in a destination helps create a knowledge-based environment where knowledge transfer is crucial for competitiveness [35].

Although stakeholders may have different interests and may even act as competitors, it is essential that they work together to achieve a synergistic effect. However, significant barriers such as lack of trust, fear of data misuse, and reluctance to communicate can impede cooperation and knowledge sharing between parties. To overcome these obstacles, the use of interactive game elements is recommended. This approach uses play principles to achieve goals, promote learning, and encourage collaboration in a safe and simplified environment. Such an approach can enhance regular stakeholder meetings, discussions, or workshops.

Cooperating with international organizations in the application of smart and sustainable development can be beneficial to tourism destinations. These organizations have established methodologies and recommendations to improve the current state of tourism development. External knowledge can serve as a tool for the continuous improvement of destinations, helping minimize identified barriers in the application of smart and sustainable tourism development (Table 2).

Table 2 Approaches of international organizations to the application of smart technologies to ensure sustainable development.

Organization	Focus
UNWTO	- The need to use smart technologies as a resource for making decisions about the future direction of the tourism sector, which will be more sustainable, inclusive, and resilient; - Data acquisition is essential for measuring tourism in order to increase awareness of the sector, monitor progress, assess tourism effects, support results-oriented management, and highlight strategic issues for policy objectives; - An INSTO initiative is based on UNWTO's commitment to measuring and monitoring the sustainable and resilient growth of the tourism sector, and supports data and experience-based tourism management; - UNWTO Tourism Data Dashboard, which provides statistics on inbound and outbound tourism indicators at global, regional, and national levels. The data includes tourist arrivals, tourism's share of exports and contribution to GDP, source markets, seasonality and accommodation statistics.
OECD	- There is a need to shift the perception of the tourism sector from a focus on performance to a focus on ensuring sustainable development; - It is recommended to use smart solutions, especially for researching the social, cultural, economic, and environmental effects of tourism. This will help to better understand the impact of tourism on destinations and local communities, determine carrying capacities, and evaluate the provision of sustainable development; - The implementation of the smart approach in the tourism industry is considered a key political role of governments. They should create the right framework and conditions for the transformation of the tourism sector.
UN	- Is based on the assumption that data is the lifeblood and basic material for responsibility; - The rapid growth of new technologies has led to an exponential increase in the volume and variety of available data, creating unprecedented opportunities for information and societal transformation while also promoting environmental protection; - Failure to utilize available data and technology for data processing can lead to environmental degradation and the violation of basic human rights;

Table 2 Approaches of international organizations to the application of smart technologies to ensure sustainable development—cont'd

Organization	Focus
	- More and more open data can help ensure knowledge sharing, creating a world of informed and competent citizens and decision-makers who are able to take responsibility for their actions and thus contribute to sustainable development.
European Commission	- Draws attention to the need to create measures and tools necessary to accelerate the sustainable and smart transformation and resilience of the tourism ecosystem, which should be based on comprehensive tourism strategies; - The exchange of data between the public and private sector can help create innovative tourism services that support sustainability (e.g., tourism mobility and transport), manage tourist flows based on real-time data, stimulate tourism demand more effectively, and match supply (e.g., excessive tourism) or services that generate data and statistics for use in policy and administrative decision-making; - To make effective use of the data exchanged, research and innovation are needed to develop and test data-driven destination management tools, practices, and technologies to promote sustainable development and eliminate overtourism.

Source: Authors.

5 Conclusion

Sustainable development has been a concept that has resonated in science and society for nearly four decades. It has been the subject of countless scientific studies, and a popular term in strategic and development documents and political speeches. However, the problem remains that sustainable development is not sufficiently secured in practice. The solutions applied to address negative situations are short-term and one-time solutions. Although numerous approaches, studies, instructions, manuals, and indicators have been developed, ensuring the development of tourism in accordance with the principles of sustainable development remains a challenge. The majority of scientific studies agree that the sustainable development of tourism has so far been applied only as a band-aid to the problems caused by the development of uncontrolled or excessive tourism.

The turbulent changes in the tourism market and increasing uncertainty exert pressure on new approaches to tourism development. In response to these challenges, a smart approach has begun to be applied in various industries and sectors. Although the original idea of a smart approach was based on the goals of sustainable development, its contribution to the sustainable development of tourism has not yet been thoroughly examined and confirmed in theory or practice.

Smart tourism development changes the current situation in the tourism market and revolutionizes the use of technology, innovation, and cooperation to provide tourists with a better experience, increase the quality of life of residents, support the competitiveness of businesses, and promote sustainable tourism development.

As the challenges and pressures on the tourism industry continue to grow, the adoption of a smart tourism development approach is likely to gain momentum. Thus there is a possibility of integrating sustainable development principles more effectively into tourism strategies and practices. This could lead to more balanced and responsible growth of the tourism industry. The success of sustainable tourism development, especially with the integration of a smart approach, will depend on the cooperation and alignment of interests among governments, businesses, local communities, and tourists. There may also be an increased focus on educating and raising awareness among all stakeholders. The development of new technologies may further enhance the potential of smart tourism, making it easier and more efficient to implement sustainable practices. Constant monitoring, learning, and adjustments will be necessary to keep up with changing trends and challenges.

Research shows that a smart approach can positively influence sustainable tourism development at the tourist and business level, as well as at the tourism destination level. However, it should be noted that a smart approach is not a cure for all sustainable development problems. Other principles should be taken into account, and the philosophy of sustainable development should be reflected in the mindset of all interested parties.

Acknowledgments

The research was supported by the research project VEGA 1/0136/23 from resilience to sustainability. The impact of data on sustainable and competitive development of tourism.

References

[1] OECD, Tourism Trends and Policies 2020, OECD, Paris, 2020.

[2] A. Galvani, A.A. Lew, M.S. Perez, COVID-19 is expanding global consciousness and the sustainability of travel and tourism, Tour. Geogr. 22 (3) (2020) 567–576, https://doi.org/10.1080/14616688.2020.1760924.

[3] S. Gössling, Technology, ICT and tourism: from big data to the big picture, J. Sustain. Tour. 29 (5) (2020) 849–858, https://doi.org/10.1080/09669582. 2020.1865387.

[4] U. Gretzel, Smart tourism development, in: P. Dieke, B. King, R. Sharpley (Eds.), Tourism in Development, CABI, 2021.

[5] J.F. Perles Ribes, J. Ivars Baidal, Smart sustainability: a new perspective in the sustainable tourism debate, Investig. Reg. 42 (1) (2018) 151–170.

[6] J. Gelter, et al., A meta-narrative analysis of smart tourism destinations: implications for tourism destination management, Curr. Issue Tour. 24 (20) (2020) 2860–2874, https://doi.org/10.1080/13683500.2020.1849048.

[7] K. Boes, D. Buhalis, A. Inversini, Smart tourism destinations: ecosystems for tourism destination competitiveness, Int. J. Tour. Cities 2 (2) (2016) 108–124, https://doi.org/10.1108/IJTC-12-2015-0032.
[8] U. Gretzel, L. Zhong, C. Koo, Application of smart tourism to cities (Editorial), Int. J. Tour. Cities 2 (2) (2016) 104–107.
[9] S. Joss, et al., The smart city as global discourse: storylines and critical junctures across 27 cities, J. Urban Technol. 26 (1) (2019) 3–34, https://doi.org/10.1080/10630732.2018.1558387.
[10] UNWTO, Tourism and the Sustainable Development Goals—Journey to 2030. Highlights, UNWTO, Madrid, 2017.
[11] R. Sharpley, Tourism, sustainable development and the theoretical divide: 20 years on, J. Sustain. Tour. 28 (11) (2020) 1932–1946, https://doi.org/10.1080/09669582.2020.1779732.
[12] R. Kates, T. Parris, A. Leiserowitz, What is sustainable development? Goals, indicators, values, and practice, Environ. Sci. Policy Sustain. Dev. 47 (3) (2005) 8–21, https://doi.org/10.1080/00139157.2005.10524444.
[13] J. Kučerová, Trvalo udržateľný rozvoj cestovného ruchu, Univerzita Mateja Bela, 1999.
[14] Z. Xiang, I. Tussyadi, D. Buhalis, Smart destinations: foundations, analytics, and applications, J. Destin. Mark. Manag. 4 (3) (2015) 143–144, https://doi.org/10.1016/j.jdmm.2015.07.001.
[15] L. Errichiello, R. Micera, A process-based perspective of smart tourism destination governance, Eur. J. Tour. Res. 29 (2021), https://doi.org/10.54055/ejtr.v29i.2436.
[16] S.H. Lee, Smart industries based on smart technologies in convergence, J. Adv. Inf. Technol. Converg. 3 (1) (2013) 13–20, https://doi.org/10.14801/jaitc.2013.3.1.13.
[17] T. Gajdošík, A. Orelová, Smart technologies for smart tourism development, in: R. Silhavy (Ed.), Advances in Intelligent Systems and Computing, Springer, Cham, 2020, pp. 333–343, https://doi.org/10.1007/978-3-030-51971-1_27.
[18] D. Schmücker, E. Horster, E. Kreilkamp, The Impact of Digitisation and Big Data Analysis on the Sustainable Development of Tourism and Its Environmental Impact, Umweltbundesamt, Dessau-Rosslau, 2019.
[19] S. Gössling, Tourism, information technologies and sustainability: an exploratory review, J. Sustain. Tour. 25 (7) (2017) 1021–1041, https://doi.org/10.1080/09669582.2015.1122017.
[20] P. Benckendorff, Z. Xiang, P. Sheldon, Tourism Information Technology, third ed., CABI, 2019.
[21] A. Ali, A. Frew, Information and Communication Technologies for Sustainable Tourism, Routledge, 2012.
[22] European Commission, Mastering Data: A Toolkit for Tourism Destinations, European Commission, 2023. 19 p. Available at https://smarttourismdestinations.eu/wp-content/uploads/2023/03/Smart-Tourism-Destinations_Toolkit.pdf.
[23] T. Gajdošík, M. Valeri, Complexity of tourism destination governance: a smart network approach, in: M. Valeri (Ed.), New Governance and Management in Touristic Destinations, IGI Global, Hershey, 2022.
[24] T. Gajdošík, Smart tourism: concepts and insights from Central Europe, Czech J. Tour. 7 (1) (2018) 25–44, https://doi.org/10.1515/cjot-2018-0002.
[25] T. Gajdošík, Z. Gajdošíková, DMOs as data mining organizations? Reflection over the role of DMOs in smart tourism destinations, in: Informatics and Cybernetics in Intelligent Systems: Proceeding of 10th Computer Science On-line Conference: Lecture Notes in Networks and Systems, Springer Nature Switzerland AG, 2021, pp. 290–299.

[26] T. Gajdošík, Smart Tourism Destination Governance: Technology and Design-Based Approach, Routledge, 2022, https://doi.org/10.4324/9781003269342.
[27] D. Fan, D. Buhalis, B. Lin, A tourist typology of online and face-to-face social contact: destination immersion and tourism encapsulation/decapsulation, Ann. Tour. Res. 78 (2019), https://doi.org/10.1016/j.annals.2019.102757.
[28] R.W. Butler, The concept of a tourist area cycle of evolution: implications for management of resources, Can. Geogr. 24 (1980) 5–12, https://doi.org/10.1111/j.1541-0064.1980.tb00970.x.
[29] D. Buhalis, Marketing the competitive destination of the future, Tour. Manag. 21 (2) (2000) 97–116, https://doi.org/10.1016/S0261-5177(99)00095-3.
[30] J.R.B. Ritchie, G.I. Crouch, The Competitive Destination: A Sustainability Perspective, CABI International, 2003.
[31] L. Dwyer, C. Kim, Destination competitiveness: a model and determinants, Curr. Issue Tour. 6 (2003) 369–414.
[32] D. Kvasnová, T. Gajdošík, V. Maráková, Are partnerships enhancing tourism destination competitiveness? Acta Univ. Agric. Silvic. Mendel. Brun. 67 (3) (2019) 811–821, https://doi.org/10.11118/actaun201967030811.
[33] Ľ. Štrba, B. Kršák, C. Sidor, P. Blišťan, Destinations business information systems for smart destinations: the case study of Kosice County, Int. J. Bus. Manag. Stud. 5 (1) (2016) 177–180.
[34] T. Gajdošík, Towards a conceptual model of intelligent information system for smart tourism destinations, in: R. Silhavy (Ed.), Software Engineering and Algorithms in Intelligent Systems, Springer, Cham, 2019, pp. 66–74, https://doi.org/10.1007/978-3-319-91186-1_8.
[35] R. Baggio, C. Cooper, Knowledge transfer in a tourism destination: the effects of a network structure, Serv. Ind. J. 30 (10) (2010) 1757–1771, https://doi.org/10.1080/02642060903580649.

CHAPTER

Design of a smart parking space allocation system for higher energy efficiency

16

Fernando Terroso-Saenz[a], Navjot Sidhu[a], Andres Muñoz[b], and Francisco Arcas[a]
[a]*Catholic University of Murcia (UCAM), Polytechnic School, Murcia, Spain,* [b]*University of Cadiz, Higher Polytechnic School, Cádiz, Spain*

1 Introduction

Over the last decade, the transportation sector has been consuming 25% of the world's energy, and a 44% of this share corresponds to personal vehicles [1]. In that sense, several transportation surveys have remarked that the private vehicle is the most prominent commuting means of transport in many parts of the globe. As a matter of fact, 91% of adults commute to work by car in the United States [2] and 70% and 68% of workers do the same in France and Germany, respectively [3, 4]. In addition, these vehicles are estimated to generate 15% of the total European Union (EU) CO_2 emissions [5]. It is therefore a priority in this area to work toward the most efficient optimization in reducing the energy consumption of personal vehicles, especially for cars.

Regarding the energy consumption of the different components of a vehicle, various studies estimate that the heating and cooling system usually represents 2.8% of the vehicle's overall energy consumption [6]. This percentage might increase up to 10% when the refrigeration system is turned on. Note that this percentage is much higher than for other critical elements of a vehicle, such as the brakes (2.5%) or the aerodynamic effect (5.3%).

The efficient use of vehicles' air-conditioning systems is a key factor to achieve more sustainable fuel consumption for cars and, in turn, reduce their greenhouse gas emissions. Indeed, the cabin temperature of a vehicle can exceed 60°C when it is exposed to direct sunlight for a long period of time [7]. However, this temperature can be reduced by more than 15°C if the vehicle is covered by some type of external sunshade (e.g., trees, shades, and so forth). Hence, it is reasonable to assume that the lower the cabin temperature when the driver gets into the vehicle, the lower the energy required by its air-conditioning system to reduce such temperature to a comfortable level. This, in turn, will reduce the overall energy consumption of the car.

Related to this problem, modern societies have promoted the construction of large ecosystems of parking facilities. For instance, there are 21.8 million off-street

Smart Spaces. https://doi.org/10.1016/B978-0-443-13462-3.00016-9

FIG. 1

Examples of surface parking lots. Aerial view of two parking facilities with some covered spaces. The borders of the covered spaces are highlighted.

Source: Google Maps.

parking spaces in municipalities with over 20,000 inhabitants in Europe [8]. Most of these facilities are usually surface parking lots (rather than underground ones) that usually have a certain number of covered spaces and the rest are uncovered, as Fig. 1 shows. In this type of parking lot, a vehicle in a certain space can receive direct solar radiation with different intensity and duration depending on the time of day, day of the year, and location of the parking space itself. This radiation will have an impact on the cabin temperature as discussed previously.

Bearing all of this in mind, this chapter introduces TICKET, an intelligent parking space allocation system for greater energy efficiency. TICKET is designed to operate in a surface parking facility. By means of this system, drivers are informed of the space within the facility that they must use to park their vehicles when they arrive at the entrance. The primary goal is to assign each driver a space to ensure that, when the drivers go back to their vehicle to leave the facility, the temperature in the cabin of the vehicle is as low as possible. The intention is to avoid an intensive usage of the air-conditioning system, therefore reducing the car's energy consumption. In that sense, our work relies on the hypothesis that the temperature will strongly depend on the number of hours that the vehicle has been exposed, or not, to direct sunlight before the driver gets back into the car.

To do so, TICKET follows a module-based architecture that allows multiple environmental factors to be considered that can affect the temperature of a vehicle's cabin. This way, TICKET considers the movement of the roofs' shadows installed in the parking lot and how they cover the different parking spaces. This information is combined with the ambient temperature to perform a final estimation of the cabin temperature of the vehicles for different time horizons. Furthermore, it also predicts the future demand of spaces in the parking lot by means of a hidden Markov model (HMM) that helps to perform a better assignment of places.

The rest of the chapter is structured as follows. Section 2 provides an overview of the current studies focusing on the development of smart solutions for parking facilities. Section 3 is devoted to the description of the TICKET framework, including all its key functional modules. Section 4 shows the main results and evaluation of the framework in a simulation based on a real-world parking lot in the city of Murcia (Spain). Lastly, a summary of the conclusions and future work are included in Section 5.

2 Related work

The TICKET framework relies on three lines of research: (1) the development of mobility predictors, (2) the processing and simulation of temperature in the vehicle cabin, and (3) intelligent parking management mechanisms. The background of each of these fields is given in the following sections.

2.1 Detection and prediction of vehicular mobility

According to recent findings on human mobility prediction, people tend to follow regular commuting routines as they restrict their movements to a few habitual locations [9]. Given these routines, estimating vehicle movements in a particular city is highly predictable [10, 11]. Current solutions for urban mobility prediction focus on comparing the route or trajectory that a user is following at a given moment with a set of movement patterns extracted from observed routes of the same user. From the patterns that match the current route, the prediction mechanism infers the most likely destination.

Numerous studies have focused on probabilistic models that do not require explicit generation of motion patterns. Methods like Bayesian networks [12], Markov models [13], HMMs [14], and the Markov decision process [15] have been proposed for the prediction of individual mobility.

In this field, the most important challenges are related to the development of forecasting mechanisms able to preserve the privacy of the users [16] and the provisioning of prediction outcomes with time horizons large enough to allow applications to better plan their resources. In that sense, both challenges have been considered within the TICKET platform.

2.2 Prediction of the vehicle cabin temperature

Several research studies in recent years have focused on the study of the environmental conditions inside a vehicle in different contexts. In [17], the authors developed an internal temperature predictor from measurements such as outdoor temperature, solar radiation, and cloudiness percentage, taking into account different meteorological conditions but on a specific car model.

A more comprehensive study, covering a longer period of time and different car models, was described in [18]. In this work, authors developed a predictor of cabin temperature using a linear regression model, taking into account the orientation with respect to solar radiation, the vehicle's color, and the level of window opening. However, they did not consider other weather conditions such as wind speed. A similar approach was followed in [19] where six different vehicles were used to compose a dataset to train a linear regression model.

Moreover, some proposals have recently made use of deep learning models. Thus the work described in [20] proposes the use of a graph neural network to predict the temperature inside the vehicle by composing a graph that allows encoding the temperature correlations in different zones of the vehicle cabin. Again, the model was validated using temperature data from a single vehicle.

Finally, some studies have examined the different mechanisms and materials that can be used to reduce the cabin temperature. One interesting work in this line is described in [21] where different passive mechanisms are studied to reduce the temperature of a parked vehicle. Experiments showed that the use of sunshades proved to be the most efficient mechanism for most of the interior areas of a vehicle, with reductions of up to 21% in cabin temperature. Other studies such as the one proposed in [22] also emphasize the impact that tinted windows can have on the interior temperature of the vehicle, with a reduction of up to 10°C in some cases.

As we can see, there are several exploratory studies examining how the cabin temperature evolves under different conditions and environments. However, proposals are lacking that actually leverage such insights to develop utility applications. In that sense, our work defines a cabin temperature predictor used as the primary metric to assign spaces to vehicles in a parking lot.

2.3 Intelligent parking management mechanisms

Several studies have focused on the development of systems that allow the allocation of parking spaces among drivers circulating within an urban area. In that sense, most of the proposals focus on providing mechanisms that indicate to drivers in which parking lots, within an area of interest, they should park given a series of spatial or temporal limitations. The main goal of these approaches is to minimize the monetary and time cost of the drivers and, at the same time, optimize the available parking resources in a city [23]. Thus these types of solutions usually include mechanisms for conflict resolution [24, 25], and different types of algorithms for an optimal driver-parking assignment, such as genetic [26] or game-based ones [27,28].

Another important line for parking management has focused on predicting the level of occupancy that a parking lot will have in the next few minutes or hours. In this setting, the work described in [29] explains a mechanism capable of predicting the occupancy of three public and private car parks in the next 60 min in Geneva (Switzerland). In [30], a mechanism is proposed to perform this prediction in the next 12 min for a set of 14 public parking lots in San Francisco, the United States. In turn, the work presented in [31] develops a prediction mechanism for a parking lot in the city of Shenzhen (China) with a time horizon up to 30 min. Finally, a physical sensor infrastructure capable of making occupancy predictions in 30 different garages in the city of Birmingham (the United Kingdom) has been proposed in [32].

In this field, one of the most important challenges is the definition of parking-lot management systems that are aware of the environmental conditions in their premises and the integration of these conditions as one of the factors that guide some of their management tasks. This will allow, for example, the definition of allocation mechanisms that go beyond a demand-and-offer policy by considering other *exogenous* elements that might be interesting from, for example, a *sustainable* point of view.

All in all, despite this variety of research in the three aforementioned courses of action, there is a scarcity of proposals that provide a holistic solution for parking-lot management based on cabin-temperature and energy-efficiency criteria. In that sense, our work combines different predictors related to varied fields such as human mobility, cabin temperature, and parking lot occupancy to provide users with a fine-grained parking space allocation system. This heterogeneity of mechanisms to achieve more energy-efficient parking lots has not been fully explored yet.

3 TICKET framework

This section describes the TICKET framework in detail. Fig. 2 shows an overview of the architecture of TICKET. As observed, the system has been designed to operate in a particular parking facility, the so-called target parking lot (TPL). Its key goal is to assign cars entering the TPL to the most suitable place where the final cabin temperature is as low as possible by the time the car leaves the facility. Observe that our work assumes that the TPL includes an entry barrier to control the access of the vehicles to the parking lot and a monitor to display information to the incoming drivers.

Fig. 2 also depicts that TICKET is composed of four inner components, namely the itinerary detection module (IDM), the contextual module (CM), the cabin temperature simulator (CTS), and the allocation module (AM). We describe each of these elements next.

3.1 Itinerary detection module

In order for TICKET to allocate a space to a driver in the most efficient manner, it is necessary to know what time the drivers will arrive at the parking premises and when they will return to collect their car. For example, the allocation of parking spaces should be different for a user who is going to pick up the vehicle at 15:00 in August than for another user who will pick it up at 22:00 in the same month. In the former case, it would be necessary to place the vehicle in one of the covered parking spaces, but in the latter the vehicle could be assigned a space without sunshades. This is because, in this case, when the driver returns to the vehicle, it will have already spent several hours without intense sunlight. Hence, the cabin will have had time to cool down naturally.

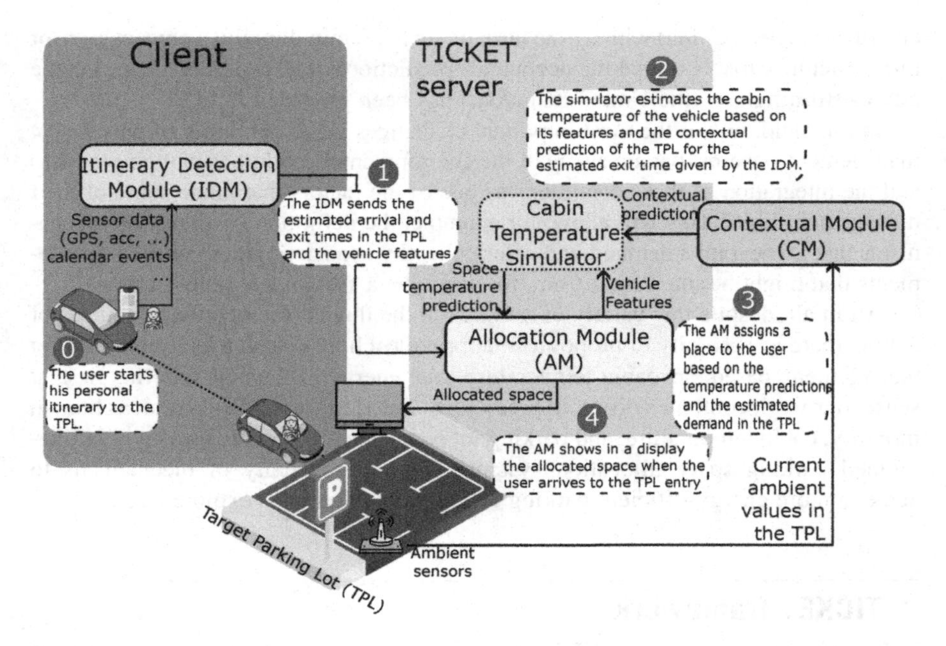

FIG. 2

General overview of the TICKET framework. The system follows a client-server architecture. The client comprises the itinerary detection module (IDM), which allows detecting and predicting the itineraries of each user. The server comprises the allocation module (AM) and contextual module (CM) along with the cabin temperature simulator (CTS) to assign the most efficient place to each user when they arrive at the target parking lot (TPL).

In order to make TICKET as proactive as possible, the IDM is in charge of learning the parking usage routines of each individual user of the TPL. This module is implemented as an Android application to be installed in the personal devices of such users (e.g., smartphone). The pipeline of this mobile application can be summarized in the following steps:

1. First, the IDM mobile app detects whether the target user *u* has initiated a journey in their vehicle. To do so, it relies on the built-in activity recognition API of Android that allows to accurately detect when the smartphone holder has started an automotive activity [33].
2. Once a new vehicular itinerary has started, the IDM app launches a mobility predictor to estimate whether the final destination of the current trajectory will be the TPL. To this end, the IDM relies on the human-mobility predictor proposed in [34]. Briefly, this mechanism is able to predict the final destination of a user based on the sequence of *turning points* of his ongoing trajectory. To detect these turning points, a variant of the DBSCAN clustering algorithm is used. Then, a

Markov process is launched to provide the final destination prediction. These two algorithms are lightweight enough in terms of processing overload to run in energy-constrained devices such as mobile phones.

3. If the predicted destination is the TPL, then the IDM infers the estimated time of arrival (ETA). To compute this time, the module relies on the *directions* API, a built-in module of Android to plan and schedule trips.[a] Given the dth date, the ETA to the TPL of u is represented as $h_{arr}^{(\hat{u},d)}$.
4. Given such arrival time, the IDM forecasts the estimated hour when the user will leave the TPL. In the current version of the module, this is done by means of a HMM [35]. Basically, HMM is a statistical model that allows inferring hidden information from sequences of observations. As Fig. 3 shows, this module infers the estimated exit time of the user u ($h_{ex}^{(\hat{u},d)}$) given the sequence of arrival times of u during the last D days $\langle h_{arr}^{(u,d-D)}, h_{arr}^{(u,d-D+1)}, \ldots, h_{arr}^{(u,d-1)}, h_{arr}^{(u,d)} \rangle$.
5. Finally, the IDL sends $h_{arr}^{(u,d)}$ and $h_{ex}^{(\hat{u},d)}$ to the AM for the space allocation process, as explained in Section 3.4.

3.2 Contextual module

Another key requirement of TICKET is the perception of the weather conditions in the TPL along with the movement of the shadows cast by the sunshades based on the daily movement of the sun. This task is performed by the CM.

More in detail, the CM captures the current and future weather conditions of the TPL's geographical region by means of the *World Weather API*.[b] This third-party

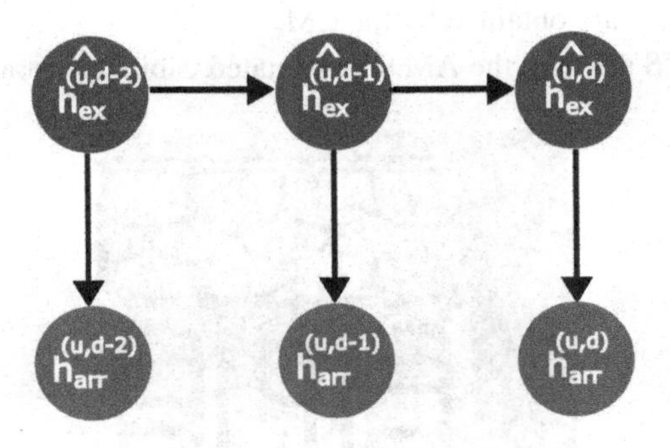

FIG. 3

Graphical model of the hidden Markov model used to predict the exit hour of a user. The hidden states (*upper nodes*) are the exit hours of the user u during a set of consecutive dates ($h_{arr}^{(\hat{u},d)}$), whereas the observations (*bottom nodes*) are the arrival hours of the user to the TPL, $h_{arr}^{(u,d)}$.

[a] See https://developers.google.com/maps/documentation/directions/overview.
[b] See https://www.worldweatheronline.com/weather-api/.

service allows retrieving, among other variables, the current and forecast temperature per hour up to 14 days, $t^h_{forecasted}$.

Besides, this module also returns, for each space s in the TPL, its *coverage* level for each hour h, c^s_h. This parameter is computed as the rate of the space's area that is covered by the shadow of the sunshades as shown in Fig. 4.

Finally, this module returns to the CTS the $t_h^{forecasted}$ and c^s_h values for each required hour h and space s.

3.3 Cabin temperature simulator

The environmental data captured by the CM are used by the CTS. This component estimates the vehicle's cabin temperature in each parking space of the TPL for a particular time interval. Given a user u and a parking space s in the TPL, the estimation is performed using a linear regression model ($\mathcal{P}$) that takes the form

$$\mathcal{P}(h_{arr}^{(u,d)}, \Delta_u, t_{\Delta_u}^{forecasted}, c^s_{\Delta_u}) \rightarrow \hat{t}^s_{(u,\Delta_u)}$$

where the independent variables of the model are the arrival time of u at the TPL ($h_{arr}^{(u,d)}$), the estimated number of hours that the car will remain in the parking lot Δ_u (computed as the difference between the exit and arrival hours, $\hat{h}_{ex}^{(u,d)} - h_{arr}^{(u,d)}$), the estimated average temperature in TPL during such number of hours ($t_{\Delta_u}^{forecasted}$) along with the average sun coverage of s during the same time period ($c^s_{\Delta_u}$). As a result, the predictor generates the estimated cabin temperature of a vehicle in s after Δ_u hours starting at $h_{arr}^{(u,d)}$, $\hat{t}^s_{(u,\Delta_u)}$. It is worth noticing that $h_{arr}^{(u,d)}$ and Δ_u are computed by the IDM and $t_{\Delta_u}^{forecasted}$ and $c^s_{\Delta_u}$ are obtained by the CM.

Finally, the CTS sends to the AM the estimated cabin temperatures as requested.

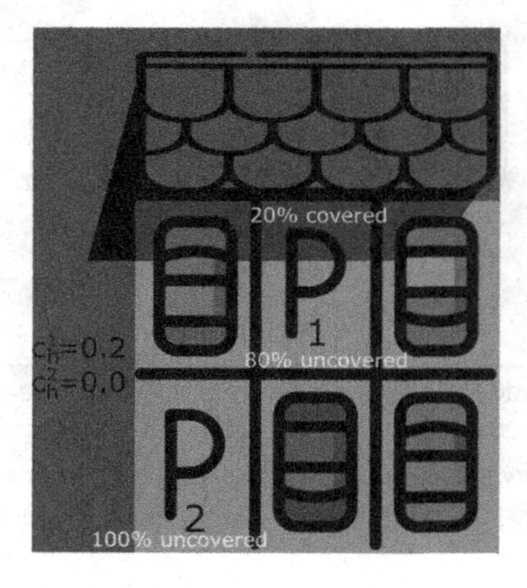

FIG. 4

Example of different coverage levels in two parking spaces given an hour of the day h. In the parking space P1, its coverage level, c^1_h, is 0.2, as only 20% of its area is covered by a roof shadow. On the contrary, place 2 has a coverage level c^2_h of 0 as its entire area is uncovered.

3.4 Allocation module

The AM is responsible for assigning each new incoming driver a corresponding parking space. The main goal of the AM is to reduce as much as possible the number of hours of sun exposure that occurs just before a user leaves the parking lot. The pseudo code for this goal is shown in Algorithm 1. Given a particular date d whose day of the week is d_w, the key steps of this policy are summed up as follows.

1. First of all, the AM computes the average amount of time that users have their cars parked in the TPL. This is done by considering the previous 3 days of the week of type d_w before d. For example, if d is Monday, we use the last three previous Mondays to compute the average. This value is represented as Δ_d^{avg} (line 1 of Algorithm 1).
2. Next, the AM estimates the number of arrivals to the TPL for each hour h of the day. This is computed as the average number of arrivals per hour in the last 3 days of the week d_w before d. Let us call this prediction $\hat{n}_{arr}^h$, $h \in [0, 23]$ (lines 2–3 of Algorithm 1).
3. Then, the parking spaces of the TPL are ranked based on the expected cabin temperature per hour assuming a stay duration of Δ_d^{avg}. To do so, the AM uses the forecast temperature per hour delivered by the CM, $t_h^{forecasted}$ (line 5). This generates a sorted list of the spaces in the TPL for each hour h, $\mathcal{S}_{order}^{(d,h)} = \langle s_1, s_2, \ldots, s_n \rangle$, where the cabin temperature associated with a space s_i will be lower than another space s_j provided that $i < j$ (lines 4–6 of Algorithm 1).
4. Based on the expected number of arrivals, computed in Step 2, the AM books the required number of spaces for each hour by using each ranked list of spaces (lines 7–10). Thus the set $\mathcal{S}_{book}$ enables controlling the spaces that have been already booked to avoid duplicate reservations (line 10). Moreover, the model reserves the spaces from the hour with the highest forecast temperature at date d to the lowest one by using a sorted list of hours of the day $\mathcal{H}_{ordered}$ (line 8). As a result, a list of booked parking spaces per hour, $\mathcal{S}_{book}^{(d,h)}$, is generated (line 9). For instance, $\mathcal{S}_{book}^{(5,8)} = 26$ indicates that the system books 26 spaces to be used by the users arriving at the TPL at 8:00 in the morning.
5. For each incoming driver u, the AM first receives u arrival and exit times from u's IDM (lines 15 and 16). Then, it assigns u a space from the list of booked spaces associated with u exit hour, $\mathcal{S}_{book}^{(d,h_{ex}^{(u,d)})}$. More in detail, the assigned space, s_u, is the one whose estimated cabin temperature is the lowest among the available ones based on the stay duration of the target user (lines 17–22). The selected space is eventually marked as unavailable after the assignment (line 23).

By means of this procedure, the AM books spaces for the hours of the day with high temperature. This ensures that there will be enough spaces in the TPL to cover the demand of users at such hot hours. Hence, the system will be able to allocate an incoming user to a covered space if it is necessary. In this manner, we ensure that there will be enough spaces in the TPL to allocate the expected demand of drivers in suitable spaces during the most critical hours of the day.

Algorithm 1 Pseudocode of the policy followed by the AM.

Input: A date d with day of the week d_w, Set of TPL's spaces $\mathcal{S}$, Historical dataset of entry and exit records in the TPL $\mathcal{R}$, coverage rate per hour of the parking spaces C_d

```
/* 1.  Compute average time duration of the TPL stays.   */
```
1 $\Delta_d^{avg} \leftarrow compute\_avg\_stay\_duration(\mathcal{R}, 3, d_w)$
```
/* 2.  Predict the number of arrivals per hour.           */
```
2 **for each** $h \in [0, 23]$ **do**
3 $\quad \hat{n}_{arr}^{h} \leftarrow predict\_arrivals(\mathcal{R}, h, d_w)$
```
/* 3.  Order spaces based on cabin temperature.           */
```
4 **for each** $h \in [0, 23]$ **do**
5 $\quad t_h^{forecasted} \leftarrow CM.predict\_temperature(d, h)$
6 $\quad S_{order}^{(d,h)} \leftarrow sort\_spaces(\mathcal{S}, h, t_h^{forecasted}, \Delta_d^{avg}, C_{(d,h)})$
```
/* 4.  Book spaces based on the hourly demand.            */
```
7 $\mathcal{S}_{book} \leftarrow \emptyset$
8 **for each** $h \in \mathcal{H}_{ordered}$ **do**
9 $\quad S_{book}^{(d,h)} \leftarrow book\_spaces(S_{order}^{(d,h)} - \mathcal{S}_{book}, \hat{n}_{arr}^{h})$
10 $\quad \mathcal{S}_{book} \leftarrow \mathcal{S}_{book} \cup S_{book}^{(d,h)}$
11 $\mathcal{S}_{unavailable} \leftarrow \emptyset$
12 **for each** *incoming* u **do**
13 $\quad s_u \leftarrow \emptyset$
14 $\quad t_{min} \leftarrow \infty$
```
    /* Get arrival and estimated exit times from IDM.     */
```
15 $\quad h_{arr}^{(u,d)}, h_{ex}^{(\hat{u},d)} \leftarrow IDM_u.get\_info(u)$
```
    /* Compute duration of the stay of u in the TPL.      */
```
16 $\quad \Delta_u \leftarrow \hat{h}_{ex}^{(u,d)} - h_{arr}^{(u,d)}$
17 $\quad$ **for each** $s \in \{S_{book}^{(d,h_{ex}^{(u,d)})} - \mathcal{S}_{unavailable}\}$ **do**
```
        /* The CM computes the coverage rate of each space s
           for the estimated stay duration and the average
           temperature for the estimated stay duration.   */
```
18 $\quad\quad c_{\Delta_u}^{s} \leftarrow CM.get\_coverage(s, C_d, h_{arr}^{(u,d)}, \Delta_u)$
19 $\quad\quad t_{\Delta_u}^{forecasted} \leftarrow CM.predict\_temperature(d, h_{arr}^{(u,d)}, \Delta_u)$
```
        /* The CTS estimates the final cabin temperature for
           the given space.                               */
```
20 $\quad\quad \hat{t}_{(u,\Delta_u)}^{s} \leftarrow CTS.predict(h_{arr}^{(u,d)}, \Delta_u, t_{\Delta_u}^{forecasted}, c_{\Delta_u}^{s}, s)$
21 $\quad\quad$ **if** $\hat{t}_{(u,\Delta_u)}^{s} < t_{min}$ **then**
22 $\quad\quad\quad s_u \leftarrow s$
23 $\quad\quad\quad t_{min} \leftarrow \hat{t}_{(u,\Delta_u)}^{s}$
24 $\quad \mathcal{S}_{unavailable} \leftarrow \mathcal{S}_{unavailable} \cup \{s_u\}$
25 $\quad$ **return** s_u
26 **return** $\mathcal{G}$

4 Evaluation of the framework

In this section, we evaluate TICKET in a real-world setting and compare it with two different baselines.

4.1 Use-case setting

As an evaluation setting, we have used as TPL one of the parking lots that the Catholic University of Murcia (Spain)[c] includes among its premises. This parking lot has a barrier-controlled access and it covers an area of approximately 5000 m^2 with a total of 250 spaces, as shown in Fig. 5. This car park is used by different teaching and research academic staff along with members of administration and service staff of the institution. Therefore its hours of use are highly variable due to the heterogeneity of its users, which makes it an interesting TPL.

Furthermore, the University is located in the city of Murcia (southeastern Spain), where the average maximum temperature in summer is above 32°C in many cases.[d] More in detail, Table 1 shows the average temperature per month since 2015. This made this TPL particularly challenging for TICKET.

As Fig. 5 shows, this TPL does not include any covered spaces. In order to provide a reliable study of TICKET, the evaluation was based on the hypothesis that 50 out of 250 of the TPL spaces were covered by some type of sunshade. Fig. 6 shows the spatial distribution of these simulated covered spaces.

To evaluate our approach in this TPL, we made use of a historical dataset with 32,053 check-in/check-out records of 367 different users spanning a 6-year period

FIG. 5

TPL for the evaluation of the framework. The *right image* shows the access to the parking lot controlled by barriers. The *left figure* shows the aerial view of the parking lot whose spatial boundaries are highlighted with latitude-longitude coordinates (37.992, −1.186).

Source: Google Maps.

[c]See https://ucam.edu.

[d]See https://weatherspark.com/y/40195/Average-Weather-in-Murcia-Spain-Year-Round.

Table 1 Average maximum, minimum, and mean temperatures in Celsius (°C) in the city of Murcia (Spain) from 2015 to 2022.

Values	Jan.	Feb.	Mar.	Apr.	May	Jun.	Jul.	Aug.	Sep.	Oct.	Nov.	Dec.
High	17	18	20	22.8	26.1	30	32.8	32.8	30	25	20	17.2
Mean	10	11.1	13.9	16.1	20	24.4	27.2	27.2	23.9	19.4	13.9	10.5
Low	4.4	5.6	7.2	10	13.9	18.3	20.6	21.1	17.8	13.3	8.3	5.6

FIG. 6

Spatial distribution of the spaces in the TPL. The *simulated* covered spaces heading east are shown in *red*, the ones heading west are shown in *yellow*, whereas the TPL boundaries area are also remarked.

Source: Google Maps.

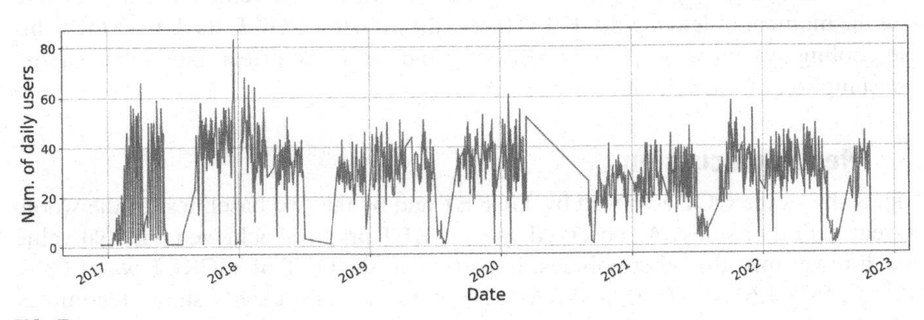

FIG. 7

Number of daily users in the TPL. The gaps in the time series are due to the summer holidays in August, when the academic activity is very limited, and the lockdown from March to June 2020 due to the COVID-19 pandemic.

from September 20, 2016 to October 27, 2022. Fig. 7 shows the daily number of users in the TPL. As observed, there are no records in August, since the university would remain partially closed without any academic activity in that month. Furthermore, it can be also observed that there are no records during the first months of 2020 due to the COVID-19 pandemic.

Finally, we fit the CTS linear regression model (see Section 3.3) with the data of a detailed analysis of the evolution of the cabin temperature of a vehicle under a variety of environmental conditions [36]. As a result, the predictive model $\mathcal{P}$ to infer $\hat{t}^{s}_{(u,\Delta_s)}$ followed this form:

$$\mathcal{P}(h^{(u,d)}_{arr}, \Delta_u, t^{forecasted}_{\Delta_u}, c^{s}_{\Delta_u}) = (0.221 \times \Delta_u) - (7.208 \times c^{s}_{\Delta_s}) + (0.337 \times h^{(u,d)}_{arr}) + t^{forecasted}_{\Delta_s}$$

4.2 Baselines

To assess the reliability of TICKET, we have compared its results to two different allocation policies acting as baselines, namely

1. Random assignment policy (RAP). This policy just assigns each driver a random space among the available ones without taking account of the current or forecast environmental conditions or whether the spaces are covered or not.
2. First covered policy (FCP). This policy first assigns the incoming drivers a covered space, if there are any available. If all the covered spaces are occupied, then it starts to assign any of the uncovered spaces available.

As target metric for the evaluation, we have defined the overall cabin temperature (OCT) as follows:

$$OCT = \sum_{d \in \mathcal{D}} \sum_{u \in \mathcal{U}_d} t^{cabin}_{(u,d)}$$

where $\mathcal{D}$ is the set of dates under consideration, $\mathcal{U}_d$ is the set of drivers who used the TPL at date d, and $t^{cabin}_{(u,d)}$ is the actual cabin temperature of the vehicle of user u at date d at the moment of leaving the TPL. Hence, the higher the OCT, the longer the vehicle cooling system will operate intensively and, as a side effect, the higher energy consumption of the vehicle.

4.3 Result discussion

Fig. 8 shows the OCT achieved by TICKET and by the two baselines for the whole 4-year period of study. As observed, the TICKET proposal achieved a global value much lower than the other policies. In particular, the OCT of TICKET was 1.08×10^6 °C, FCP 1.61×10^6 °C, and RAP 1.53×10^6 °C. This clearly shows the impact of our proposal to reduce the cabin temperature in the vehicles of the TPL.

Furthermore, Fig. 9 shows the OCT per month of the three approaches for the whole study period. Note that these time series exhibited a higher OCT in the summer months and a lower one during the autumn and winter periods. This is consistent with the fact that the cabin temperature is strongly related to the environmental temperature. Besides, this figure also shows that TICKET consistently achieved the lower OCT in all the months under consideration.

Fig. 10 shows the average OCT aggregated by month. Observe that the TICKET framework clearly achieved the lowest OCT in all the months. This confirms that our approach provides consistent and reliable results across time.

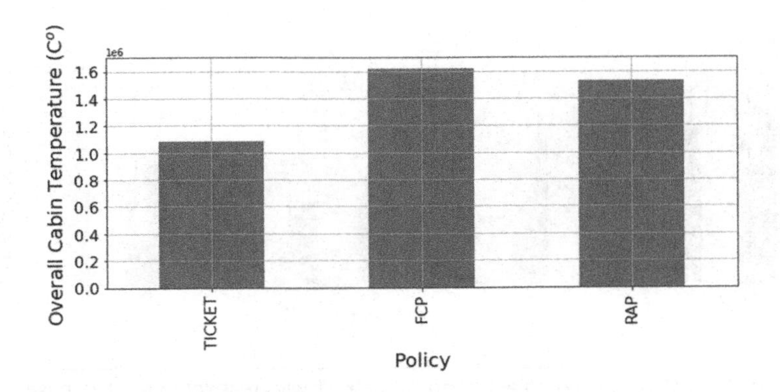

FIG. 8

OCT for the TICKET and baseline policies. The *y*-axis is defined in the range of 10^6 for a better visualization.

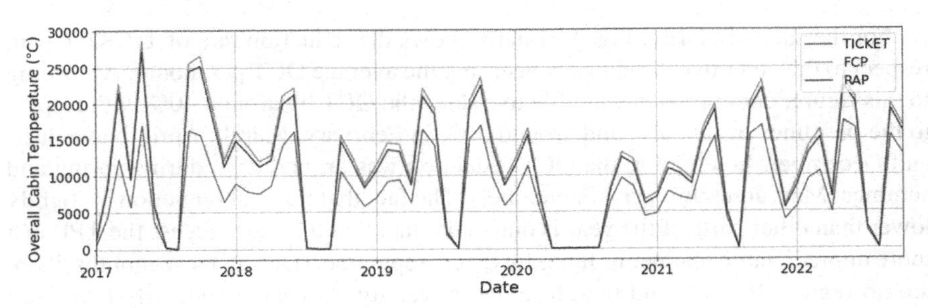

FIG. 9

OCT evolution per month. OCT time series of the three approaches reporting the OCT per month during the whole period of study.

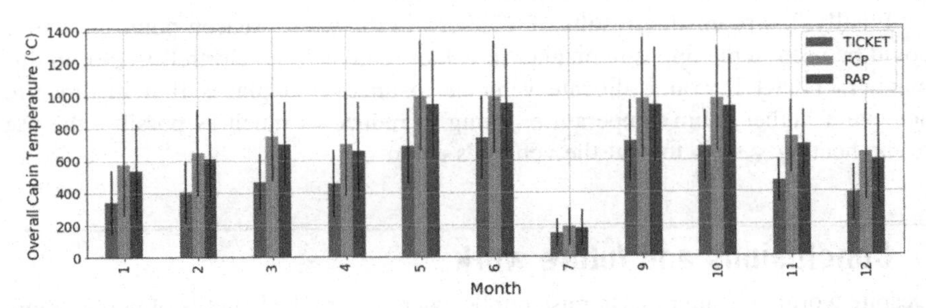

FIG. 10

Average OCT per month. *Each bar* indicates the mean of the OCT along with its standard deviation depicted as error bars. Month 8 (August) is not included in the plot as there is no activity in the TPL due to the summer holidays. Moreover, the low values of the three policies in July are because, during that month, many members of the university tend to move to other universities due to different internships and conferences. As a result, the number of regular users of the TPL decreases in this month.

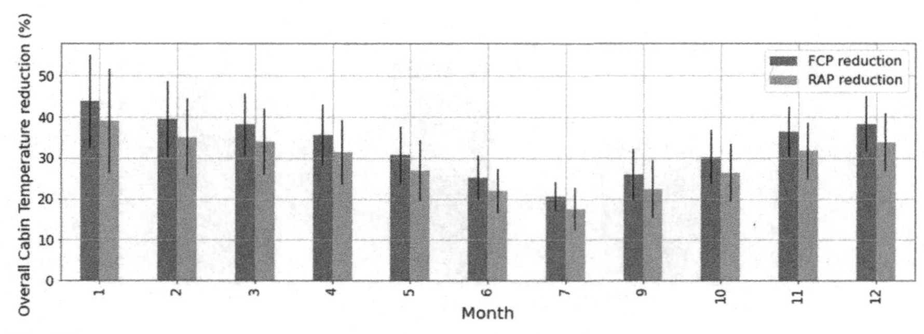

FIG. 11

Reduction of the OCT per month achieved by TICKET with respect to the two baselines. The *bars* with legend "FCP reduction" depict the percentage of reduction of TICKET with respect to the FCP baseline, and the ones with legend "RAP reduction" represent the same for the RAP policy.

For the sake of clarity, Fig. 11 finally shows the reduction rate of TICKET with respect to the other two baselines concerning the average OCT per month. According to this figure, our approach was able to reduce the OCT by around 40% with respect to the baselines in January, and around 35% in February, March, April, November, and December. In addition, the OCT reduction was around 25% during spring and summer (May, June, July, and September). The fact that this last reduction is slightly lower than other parts of the year is due to the fact that the users access the TPL in a more unpredictable manner in June, July, and September. During these months, there are no regular lessons, and therefore the university staff have a slightly less fixed schedule of activities and events. This makes their check-in/check-out hours more varied with respect to the rest of the year. As a result, the IDM is not able, in many cases, to provide an accurate estimation of the exit hours of the users. Eventually, this leads to a slightly less efficient allocation of spaces.

Finally, the promising results of TICKET in the winter season point out that it could be used in the opposite manner in colder regions (e.g., North Europe). In this manner, TICKET would allocate vehicles in uncovered places first in order to achieve a higher cabin temperature, trying to reduce as much as possible the use of the heating system to heat the vehicle's cabin.

5 Conclusions and future work

Despite worldwide attempts to raise public awareness of the benefits of public transport, the use of private vehicles continues to be a "necessary evil" in all countries. Thus efforts should continue to focus on optimizing the energy consumption, and therefore the related pollution, of these vehicles. One of the factors that most affect this consumption is the car's cooling system, especially in countries with high temperatures during most of the year.

This work introduces TICKET, an intelligent system for the optimal distribution of vehicles in surface parking lots with a proportion of shaded parking spaces to minimize the use of the car's air conditioning system once the car leaves the parking lot. From the fusion of parking occupancy pattern data, weather data, evolution of the shaded areas of the parking lot during the day, and data on drivers' parking usage habits, the goal of TICKET is to assign the most convenient parking space to incoming cars so that their cabin temperature is as low as possible when leaving the parking lot. To do this, TICKET simulates the temperature that will be in the cabin through a linear regression model using the described data and then applies an algorithm to implement the space allocation policy based on the results of that simulation.

The TICKET system has been evaluated in a real parking lot belonging to the Catholic University of Murcia (Spain) with data obtained during 6 years, being a convenient scenario due to its different usage patterns by more than 300 users. The results obtained have been compared with two baseline models, namely a model that assigns parking spaces randomly and another model whose policy is to assign the shaded spaces first as long as they are available regardless of the rest of the available data. An improvement of between 25% and 40% was obtained in the reduction of the temperature in the cabin for our proposal compared to the baseline models, depending on the season of the year. These results allow us to confirm that the TICKET system can help to reduce the energy consumption of the car in a simple manner and without an extra cost in the parking infrastructure. Regarding the limitations of this work, it must be taken into account that the cabin temperature has been obtained through a simulation, which, although it has been considered valid, can be improved with the installation of temperature sensors in the cars for more accurate results. It should also be noted that this proposal is aimed at parking lots in locations with high temperature most of the year. For parking lots located in areas with mostly cold temperature, a different policy to be applied in TICKET should be studied. Finally, it should be noted that the parking lot used for this research did not have shaded parking areas, so they had to be simulated.

As future work, the TICKET system will be implemented in different parking lots with different usage patterns to continue analyzing the improvements obtained. We will also acquire several temperature sensors to install in pilot cars and improve the accuracy of the temperature reduction obtained. Finally, we will explore the integration of a blockchain mechanism in TICKET to fully automate the system in the reservation of parking spaces.

Acknowledgments

Financial support for this research has been provided under grant PID2020-112827GB-I00 funded by MCIN/AEI/10.13039/501100011033.

References

[1] J. Conti, P. Holtberg, J. Diefenderfer, A. LaRose, J.T. Turnure, L. Westfall, International Energy Outlook 2016 with projections to 2040, USDOE Energy Information Administration (EIA), Washington, DC (United States), 2016. Technical Report.

[2] Household, Individual, and Vehicle Characteristics, Bureau of Transportation Statistics, 2017. Technical Report.

[3] C. Brutel, J. Pages, The Car Remains the Main Mode of Transport to Go to Work, Even for Short Distances, Institut national de la estatistique et des études économiques, 2021. Technical Report.

[4] Zahl der Woche Nr. 42 vom 18. Oktober 2022, Statistischen Bundesamtes, 2022. Technical Report.

[5] D. Komnos, S. Tsiakmakis, J. Pavlovic, L. Ntziachristos, G. Fontaras, Analysing the real-world fuel and energy consumption of conventional and electric cars in Europe, Energy Convers. Manag. 270 (2022) 116161, https://doi.org/10.1016/j.enconman.2022.116161.

[6] V.H. Johnson, Fuel used for vehicle air conditioning: a state-by-state thermal comfort-based approach, Fuel 1 (2002) 1957.

[7] K.I. Scott, J.R. Simpson, E.G. McPherson, Effects of tree cover on parking lot microclimate and vehicle emissions, J. Arboric. 25 (3) (1999) 129–142.

[8] B. Pierik, I. Königsberg-Brouns, J. Klinge, Car Parks in Towns and Cities – Future-Proof Investment or Soon a Thing of the Past?, Bouwfonds Investment Management, 2017. Technical Report.

[9] L. Pappalardo, F. Simini, S. Rinzivillo, D. Pedreschi, F. Giannotti, A.-L. Barabási, Returners and explorers dichotomy in human mobility, Nat. Commun. 6 (1) (2015) 8166.

[10] C. Song, Z. Qu, N. Blumm, A.-L. Barabási, Limits of predictability in human mobility, Science 327 (5968) (2010) 1018–1021.

[11] M. Lin, W.-J. Hsu, Z.Q. Lee, Predictability of individuals' mobility with high-resolution positioning data, in: Proceedings of the 2012 ACM Conference on Ubiquitous Computing, 2012, pp. 381–390.

[12] J. Krumm, R. Gruen, D. Delling, From destination prediction to route prediction, J. Locat. Based Serv. 7 (2) (2013) 98–120.

[13] A.Y. Xue, R. Zhang, Y. Zheng, X. Xie, J. Huang, Z. Xu, Destination prediction by sub-trajectory synthesis and privacy protection against such prediction, in: 2013 IEEE 29th International Conference on Data Engineering (ICDE), IEEE, 2013, pp. 254–265.

[14] J. Zhou, A.K. Tung, W. Wu, W.S. Ng, A "semi-lazy" approach to probabilistic path prediction in dynamic environments, in: Proceedings of the 19th ACM SIGKDD International Conference on Knowledge Discovery and Data Mining, 2013, pp. 748–756.

[15] B.D. Ziebart, A.L. Maas, A.K. Dey, J.A. Bagnell, Navigate like a Cabbie: probabilistic reasoning from observed context-aware behavior, in: Proceedings of the 10th International Conference on Ubiquitous Computing, 2008, pp. 322–331.

[16] R. Pellungrini, L. Pappalardo, F. Pratesi, A. Monreale, A data mining approach to assess privacy risk in human mobility data, ACM Trans. Intell. Syst. Technol. 9 (3) (2017) 1–27.

[17] A. Grundstein, V. Meentemeyer, J. Dowd, Maximum vehicle cabin temperatures under different meteorological conditions, Int. J. Biometeorol. 53 (2009) 255–261.

[18] I.R. Dadour, I. Almanjahie, N.D. Fowkes, G. Keady, K. Vijayan, Temperature variations in a parked vehicle, Forensic Sci. Int. 207 (1–3) (2011) 205–211.

[19] J. Horak, I. Schmerold, K. Wimmer, G. Schauberger, Cabin air temperature of parked vehicles in summer conditions: life-threatening environment for children and pets calculated by a dynamic model, Theor. Appl. Climatol. 130 (2017) 107–118.

[20] X. He, Y. Wang, F. Guo, X. Zhang, X. Duan, J. Pei, Modeling for vehicle cabin temperature prediction based on graph spatial-temporal neural network in air conditioning system, Energy Build. 272 (2022) 112229.

[21] M.A. Jasni, F.M. Nasir, Experimental comparison study of the passive methods in reducing car cabin interior temperature, in: Malaysia: International Conference on Mechanical, Automobile and Robotics Engineering (ICMAR), Penang, 2012, pp. 229–233.

[22] C.-Y. Tseng, Y.-A. Yan, J.C. Leong, Thermal accumulation in a general car cabin model, J. Fluid Flow Heat Mass Transfer 1 (2014) 48–56.

[23] A.O. Kotb, Y.-C. Shen, X. Zhu, Y. Huang, iParker—a new smart car-parking system based on dynamic resource allocation and pricing, IEEE Trans. Intell. Transp. Syst. 17 (9) (2016) 2637–2647.

[24] A. Muñoz, J.A. Botía, Developing an intelligent parking management application based on multi-agent systems and semantic web technologies, in: Hybrid Artificial Intelligence Systems: 5th International Conference, HAIS 2010, San Sebastián, Spain, June 23–25, 2010. Proceedings, Part I, Springer, 2010, pp. 64–72.

[25] H. Errousso, J. El Ouadi, S. Benhadou, H. Medromi, et al., A hybrid modeling approach for parking assignment in urban areas, J. King Saud Univ. Comput. Inf. Sci. 34 (6) (2020) 2405–2418.

[26] S. Abidi, S. Krichen, E. Alba, J.M. Molina, A new heuristic for solving the parking assignment problem, Procedia Comput. Sci. 60 (2015) 312–321.

[27] N. Mejri, M. Ayari, R. Langar, F. Kamoun, G. Pujolle, L. Saidane, Cooperation versus competition towards an efficient parking assignment solution, in: 2014 IEEE International Conference on Communications (ICC), 2014, pp. 2915–2920. https://doi.org/10.1109/ICC.2014.6883767.

[28] D. Ayala, O. Wolfson, B. Xu, B. Dasgupta, J. Lin, Parking slot assignment games, in: Proceedings of the 19th ACM SIGSPATIAL International Conference on Advances in Geographic Information Systems, 2011, pp. 299–308.

[29] S.L. Tilahun, G. Di Marzo Serugendo, Cooperative multiagent system for parking availability prediction based on time varying dynamic Markov chains, J. Adv. Transp. 2017 (2017).

[30] J. Xiao, Y. Lou, J. Frisby, How likely am I to find parking?—A practical model-based framework for predicting parking availability, Transp. Res. B Methodol. 112 (2018) 19–39.

[31] F. Bock, S. Di Martino, A. Origlia, Smart parking: using a crowd of taxis to sense on-street parking space availability, IEEE Trans. Intell. Transp. Syst. 21 (2) (2019) 496–508.

[32] G. Ali, T. Ali, M. Irfan, U. Draz, M. Sohail, A. Glowacz, M. Sulowicz, R. Mielnik, Z.B. Faheem, C. Martis, IoT based smart parking system using deep long short memory network, Electronics 9 (10) (2020) 1696.

[33] A.A. Ismael, M. Jayabalan, D. Al-Jumeily, A study on human activity recognition using smartphone, J. Adv. Res. Dyn. Control Syst. 12 (05 Special Issue) (2020), https://doi.org/10.5373/JARDCS/V12SP5/20201818.

[34] F. Terroso-Saenz, M. Valdes-Vela, A.F. Skarmeta-Gomez, Online route prediction based on clustering of meaningful velocity-change areas, Data Min. Knowl. Disc. 30 (6) (2016) 1480–1519.
[35] L.E. Baum, T. Petrie, Statistical inference for probabilistic functions of finite state Markov chains, Ann. Math. Stat. 37 (6) (1966) 1554–1563.
[36] Efecto de la radiación solar en la temperatura interior del vehículo, RACC-Real Automovil Club de Cataluña, 2015. Technical Report.

CHAPTER

17 Smart cities as spaces of flows and the digital turn in architecture and urban planning: Big data vis-à-vis environmental and social equity

Marianna Charitonidou

Faculty of Art Theory and History, Athens School of Fine Arts, Athens, Greece

1 Introduction

The chapter intends to examine two issues: first, the mutation of the status of the architectural artifact due to the form being generated through the use of digital tools; second, the implications of the possibility of real-time data visualization for the reconceptualization of the notion of spatiality. The chapter's main objective is to render explicit how digital design tools and hybrid use of software and hardware provide the conditions for more mutable and open-ended generative processes than those provided by conventional methods of architectural design.

The distinction between the digital and computation is pivotal for grasping the epistemological mutations that are pinpointed here. The digital refers to a kind of state of being, or a condition, while the computation concerns active processes [1]. Another issue that is examined in this chapter is the interaction between physical, virtual, and augmented reality and the real-virtual relationship in the case of augmented reality. All the case studies that are analyzed in this chapter are based on experimentation with geometry. The reasons for which they have been chosen to be examined are mainly two: first, they exemplify an ontological shift of the design process; second, they illustrate a reinvention of the established hierarchies of the design process. A common parameter of the case studies examined in this chapter is their ambition to invert the role between the architectural profession and architectural academia. The chapter also places particular emphasis on how urban scale digital twins help develop new data-driven scenarios, promote sustainable development goals, and shape new participatory design methods.

Smart Spaces. https://doi.org/10.1016/B978-0-443-13462-3.00003-0

To better grasp the implications of perceiving space as a flux, we could bring to mind Manuel Castells's approach and the following three concepts on which his theory is based: "space of flows," "space of places," and "timeless time" [2 (p. 1674),3]. In *The Rise of the Network Society*, which is devoted to the spaces of flows, Castells analyzes "[t]he relationships between the space of flows and the space of places, between simultaneous globalization and localization" [3,p. 458]. He argues that "function and power in our societies are organized in the space of flows" [3,p. 458]. According to Castells, the network society is organized around these three concepts. Castells, through these concepts, intends to render explicit how the "incorporation of the impact of advanced forms of networked communication" [2,pp. 1673–1674] calls for a new understanding of societies. He places particular emphasis on the fact that in network society there are no boundaries, and suggested that contemporary urbanization and networking dynamics should be studied conjointly. Additionally, he argues that transport and digital communication infrastructures should also be examined in relation to each other. The main objective of Castells's approach is to render explicit how urban dynamics work. In contrast with Saskia Sassen's global city [4], Manuel Castells's informational city emphasizes the significance of the "incessant flows of information, goods, and people" [5,p. 163]. A turning point for his work is the theory he develops in *The Informational City: Information Technology, Economic Restructuring and the Urban-Regional Process* [6]. As Felix Stalder has highlighted, according to Castells's theory, cities should be understood as processes and not as places [5].

The chapter analyzes the emergence of the "paperless studios" at Columbia University's Graduate School of Architecture, Planning and Preservation (GSAPP), Greg Lynn's Embryological House, the New York Stock Exchange project by Asymptote Architecture, and Aegis Hyposurface by dECOi architects. Moreover, the chapter examines the implications of the first digital turn in architecture for the ontological status of architectural artifacts, the second digital turn, and the role of urban scale digital twins in shaping sustainable urban policies.

2 The emergence of the "paperless studios" at Columbia University's GSAPP

"Paperless studios" refer to a new pedagogical agenda concerning the teaching of design studios at the Graduate School of Architecture, Planning and Preservation (GSAPP) at Columbia University during the mid-1990s. They were founded in 1992 by Bernard Tschumi, who was then dean at that institution [7–13]. The "paperless studios" established a new set of terms for the ongoing conversation on the role of digital tools in architecture. They are related to the emergence of new concepts of spatiality thanks to an ensemble of experimentations with geometry and virtual reality. These experimentations should be understood beyond their formalistic characteristics since their very force lies in their capacity to transform architectural artifacts' ontological status. As becomes evident in Tschumi's article "The School's New Computing Facilities," published in *Newsline* in 1994, the main aspiration of the so-called "paperless studios" was to create the circumstances that would permit schools of architecture to acquire a more protagonist role concerning their relationship with the existing conditions in the profession. According to Tschumi, this would become possible through the creation of new conditions of architectural production regarding the design process and the generation of forms, but also their construction and the relationship between the design process and the construction. Tschumi was convinced that these new conditions would have an important impact on the way architectural design practices function. Similarly, Mark Goulthorpe, Mark Burry, and Grant Dunlop remark, in "The Bordering of University and Practice," that thanks to the use of digital design tools, "practice becomes reliant on the universities to solve design methodology problems" [7–12,p. 345, 14].

Elizabeth Diller and Ricardo Scofidio's installation called *Para-site*, which was displayed in 1989 at the Museum of Modern Art in New York, represents a significant moment regarding the fusion between digital culture and architecture. This installation contributed significantly to the viewer's experience of space through the transmission of fragmented images. Princeton's architecture journal *Fetish* edited by Greg Lynn, Edward Mitchell, Sarah Whiting, and Lois Nesbitt played an important role in this reorientation of architecture's epistemological interest from Jacques Derrida's approaches and the attraction to the notion of disjunction to Gilles Deleuze's concepts and the concern about the notion of connectivity [15]. Around the same period, Greg Lynn edited the seminal issue of *Architectural Digest* titled *AD: Folding in Architecture* [16].

An aspect of the "paperless studios" and especially of Hani Rashid's studio is the intensification of the role of augmented reality and its contribution to the transformation of the experience of spatiality. The distinction between augmented reality and augmented virtuality is of great significance for comprehending what is at stake in the projects under study in this chapter. In the case of augmented reality, users interact with physical objects, while in the case of augmented virtuality users interact with the virtual environment within a context characterized by the fusion of the virtual and physical objects. In other words, in the case of augmented reality, virtual and physical objects are displayed seamlessly. As Xiangyu Wang notes in "Augmented Reality in Architecture and Design: Potentials and Challenges for Application," "[a]ugmented [r]eality [...] is a technology or an environment where the additional information generated by a computer is inserted into the user's view of real world scene" [17 (p. 311),18]. Both virtual and augmented reality extend the sensorial environment of an individual by mediating reality through technology [19].

The central aspiration of the "paperless studios" was to provide a terrain of experimentation rendering explicit that the conception of architectural design processes according to its conventional phases and categories should be challenged. Tschumi's agenda as dean of Columbia University's GSAPP was based on his conviction that schools of architecture should establish new strategies concerning the transmission of knowledge and skills. He believed that these new strategies should aim to render architects "instrumental in the construction of the new computerized technologies that are already transforming building and design processes" [20,p. 8]. The very force of his vision lies in the way he related the emergence of "a new social conscience" after the 1968 protests to the necessity to take distance from "a laissez-faire acceptance of today's design conditions" [20,p. 8]. He was convinced that architects should "design new conditions" and go "[b]eyond the construction of technology." He related the creation of these new conditions for architecture to the emergence of "new attitudes toward the activities that take place in architectural spaces" [20,p. 8].

Bernard Tschumi's aspirations regarding the "paperless studios" went far beyond a technophile vision. They were based on the intention to embrace "a new attitude toward programs and the production of events, so as to reconfigure and to provide a rich texture of experience start will redefine architecture and urban life" [20,p. 8]. Even though at the beginning the "paperless studios" were only 2 out of 12 or 13 design studios at Columbia University's GSAPP, their impact on architecture's epistemological reorientation was significant. They marked a turning point concerning the dissemination of digital tools in architectural education and the profession. Apart from Stan Allen and Greg Lynn, who taught the first two "paperless studios," other educators involved in them were Jesse Reiser, Hani Rashid, Keller Easterling, Scott Marble, Richard Plunz, and Laurie Hawkinson.

The computer studios used for the first "paperless studios" were located on a mezzanine in McKim, Mead, and White's 1912 Avery Hall in a space that was designed by Stan Allen in collaboration with Lyn Rice, Kathy Kim, and Anna Mueller. The name of this space was "Avery 700-Level Computer Studios." The GSAPP, thanks to a $1.4 million combined grant and loan from Columbia University, constructed three physically separate but electronically linked environments for learning and research: the "Paperless Design Studio," the "Multimedia/CAD Lab," and the "Digital Design Lab" (DDL). The "Paperless Design Studio" was completely new, while the other two expanded existing facilities. The "Paperless Design Studio" denoted a significant reversal of the standard notion of the student's working space. Columbia University's GSAPP was the first architecture school to provide students with their own SGI machines with state-of-the-art visualization software such as Softimage.

As Eden Muir and Rory O'Neill remark, in their article titled "The Paperless Studio: A Digital Design Environment," published in *Newsline* in 1994, after Tschumi's invitation to integrate digital technologies in teaching architecture, they "proposed a seamless electronic infrastructure with a complete suite of state-of-the-art digital design and presentation tools" [21,p. 10]. They also mention that "[t]he electronic configuration of the Paperless Studio was derived from experiments conducted by the Digital Design Lab (DDL), a GSAP research group concerned with electronic environments, both real and virtual" [21,p. 10]. As we can read in the issue of *Columbia Daily Spectator* of October 11, 1995, the "paperless studios" aimed to respond "to changes in the way architects work since the introduction of computers" [22,p. 1]. Tschumi was convinced that new tools are intrinsically linked to changes regarding the ways of understanding the very processes of architectural design. He believed that "[t]he tools that they have today mean that people don't think in the same way." He claimed that this shift was related to the fact that "[a]rchitects can [...] think three-dimensionally in both time and space" [22,p. 1].

In the pages of *Newsline*, we can see some of the projects of the students of Hani Rashid, Scott Marble, and Greg Lynn's "paperless studios" [23] (Figs. 1 and 2). Stan Allen analyzes the reorientations of architectural pedagogy that occurred in conjunction with the emergence of the "paperless studios" in "The Paperless Studios in Context," where he distinguishes the term digital and the term computation, claiming that the former refers to a kind of state of being, or a condition, while the latter concerns active processes [1]. According to Allen, a book that triggered the interaction between architecture and digital culture was *Cyberspace: First Steps* edited by Michael Benedikt, in which cyberspace is defined as "an infinite artificial world where humans navigate in information-based space" and as "the ultimate computer-human interface" [24].

Bernard Tschumi invited the Japanese architect Shoei Yoh, who designed the Galaxy Toyama Gymnasium (1990–96), to teach at Columbia University's GSAPP

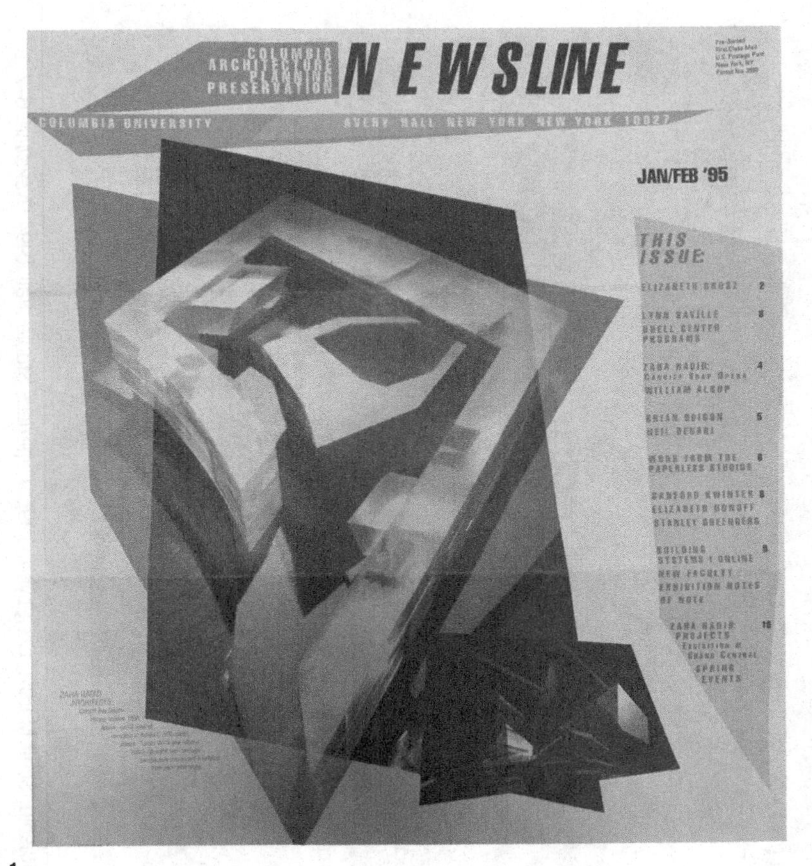

FIG. 1

Cover of *Newsline* (January/February 1995).

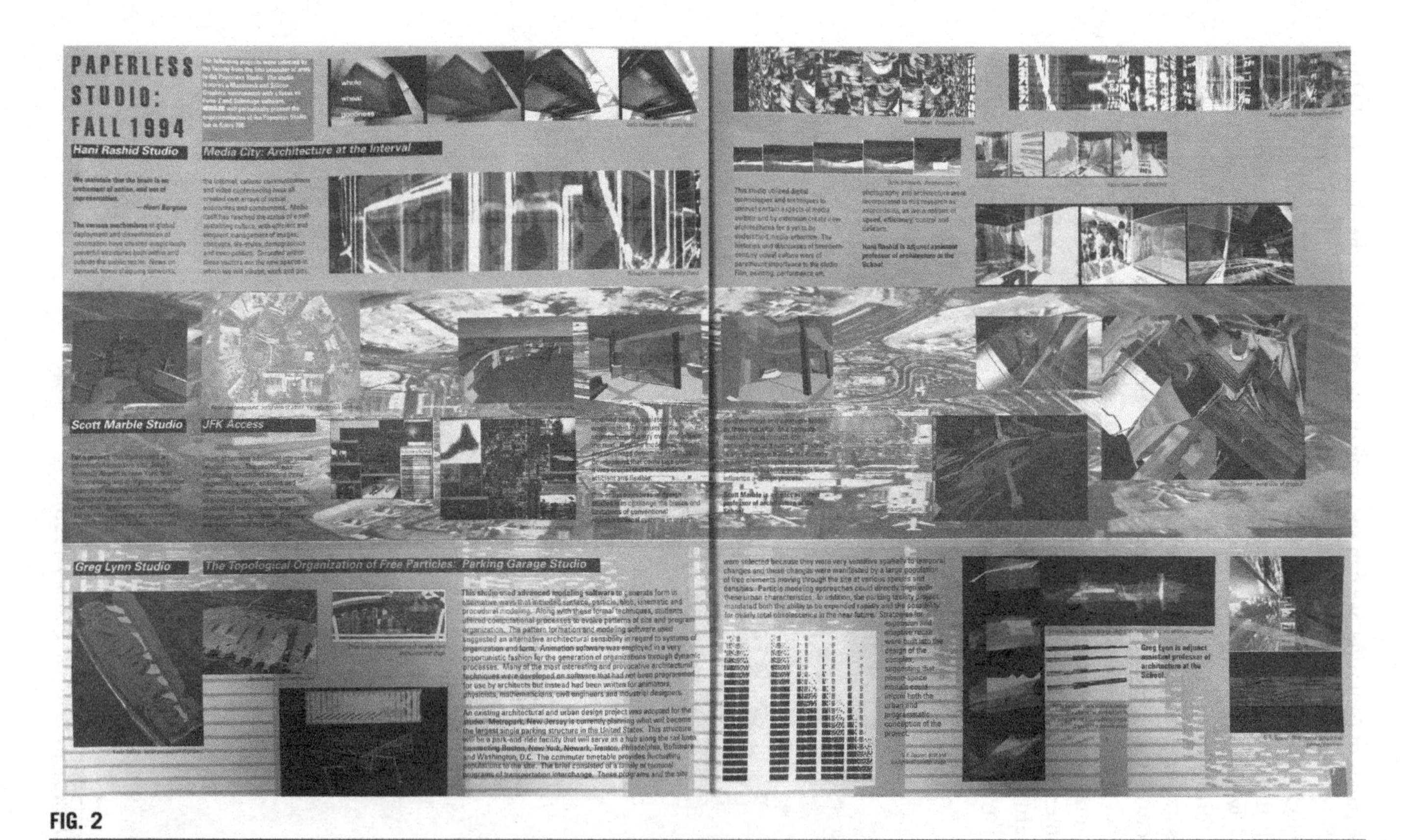

FIG. 2

Hani Rashid, Scott Marble, Greg Lynn. 1995. Paperless Studio: Fall 1994. *Newsline*: 6–7.

in the early 1990s. The Galaxy Toyama Gymnasium in Imizu in Japan was developed as a theme pavilion for the first Japan Expo Toyama in 1992. In the case of this project, "[d]igital modelling and simulation done in collaboration with structural engineers and contractors resulted in complex geometries of space frame structures" [25]. Shoei Yoh conceived this project as a "3D topology." Its roof was characterized by a textile-like quality. It was designed with the purpose of responding to the problems caused because of the weight of snowfall and the predominant winds that characterize its site. For this purpose, its dimensioning was based on calculations and verifications conducted by the consulting engineering company Taiyo Kogyo.

The main objective of the "paperless studios" was to incorporate emerging digital technologies. The first "paperless studios" were based on the use of Macintosh computers and FormZ, a Boolean-driven solids software, and three or four Silicon Graphics machines, which were running Softimage. Maya was introduced to the "paperless studios" later than FormZ. The "paperless studios" were addressed to third-year students and incorporated the use of Alias and Softimage, two software packages that were also used in Hollywood at the time. The software, instead of being only a rendering tool, informed and transformed the design process [26]. Marco Frascari in his essay "A reflection on paper and its virtues within the material and invisible factures of architecture," which is included in the book *From Models to Drawings: Imagination and Representation in Architecture*, refers to the following description of the "paperless studios" published on the website of Columbia University's GSAPP:

> *The 'digital imperative' to switch from analog to digital mode will manifest itself this year at the architecture school in the form of the Paperless Studio. Projecting ahead, we envision the inevitable and ubiquitous presence of advanced digital design and communication technologies. Architecture students will routinely use the best of new technologies within an information-rich and fully networked, multimedia environment [27,p. 27].*

Among the first "paperless studios" taught at Columbia University's GSAPP was that of Hani Rashid, which was called "Media City: Architecture at the Interval" [23]. This studio was focused on the production of video installation pieces. Rashid introduced interactivity to the virtual reality experiments in the framework of his "paperless studio." For this purpose, he used virtual reality markup language (VRML). Moreover, he placed particular emphasis on the use of animation software that was created to be used by animators, physicists, mathematicians, civil engineers, and industrial designers to generate organizations through dynamic processes. Greg Lynn's "paperless studio," taught during the fall semester of 1994, was called "The Topological Organization of Free Particles: Parking Garage Studio" [23] and focused on an existing architectural and urban design project—Metropark in New Jersey—that had been conceived as one of the largest single parking structures in the United States along the rail lines connecting Boston, New York, Newark, Trenton, Philadelphia, Baltimore, and Washington, DC. Among the students were

Kevin Collins, Jason Payne, and Oliver Lang [23]. According to what Bernard Tschumi claims in "The Making of a Generation: How the Paperless Studios Came About," Scott Marble used fly-through, Stan Allen used datascapes, Hani Rashid did a lot of collage, and Greg Lynn introduced experimentation with fluid mechanics [28,p. 415]. Tschumi, in the same text, addressed the following questions: "to what extent did computer enable architecture to develop new concepts?" [28,p. 415]; "To what extent [...] the digital tool we use is more than a formal device that has a fantastic ability to be translated in construction terms?" [28,p. 415].

An issue that is of great interest is the impact of the introduction of film and animation techniques on the pedagogies of architecture. During the spring semester of 1994, in 206 Fayerweather, the DDL assembled an early prototype of the "paperless studio," an electronic design environment that resembled a special-effects film studio. Advanced equipment was granted from Silicon Graphics Inc.; Columbia University's Center for Telecommunications Research provided additional SGI hardware and video capability. The use of a variety of software was granted by leading vendors, including Alias and Softimage, best known for modeling and animating the dinosaurs in the movie *Jurassic Park*. Each student in the Paperless Studio had access to a machine dedicated to his or her sole use and files could be transferred over the network to the machines as required for special shared or computation-intensive tasks. As Eden Muir and Rory O'Neill underscored in "The Paperless Studio: A Digital Design Environment," "learning the protocols of telecommunications and multi-user situations [was] [...] a valuable part of the experience of the Paperless Studio, where electronic library resources, online databases and global internet access will be delivered digitally to each student's desk" [21].

3 Greg Lynn's Embryological House and space as a flux and the different variations

Greg Lynn's Embryological House project (1997–2001) is worth analyzing to better grasp what was at stake in his efforts to establish strategies aiming to generate geometric surfaces (Figs. 3–6). It is worth mentioning that the design process of this project was initiated with Microsoft Excel. Lynn designed the Embryological House some years after the emergence of the "paperless studios" using animation software. More specifically, he first established the parameters for the Embryological House geometry, that is to say its primitive curves, using MicroStation software. At a later stage, he imported these files concerning the geometrical parameters to Maya. Maya permitted him to produce smoothly rendered surfaces. Lynn was a pioneer in using geometric modeling software, such as MicroStation software, to establish the geometrical parameters—the primitive curves—of an architectural artifact. The form of the Embryological House consisted of vector-based surfaces that emerged through experimentation according to certain prescribed limits with 12 control points attached to an established geometry. Lynn manipulated the 12 control points to experiment with the curves, which were articulated as splines. According to Greg

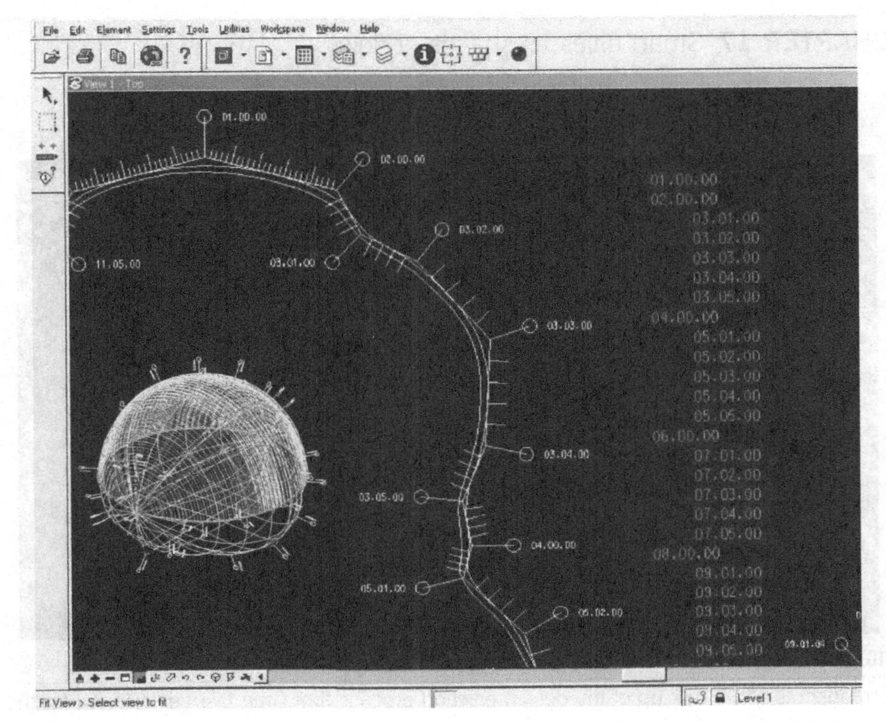

FIG. 3

Microstation was used to animate the House's transformation through its key iterations, creating a potentially infinite number of intermediate forms.

© *Greg Lynn.*

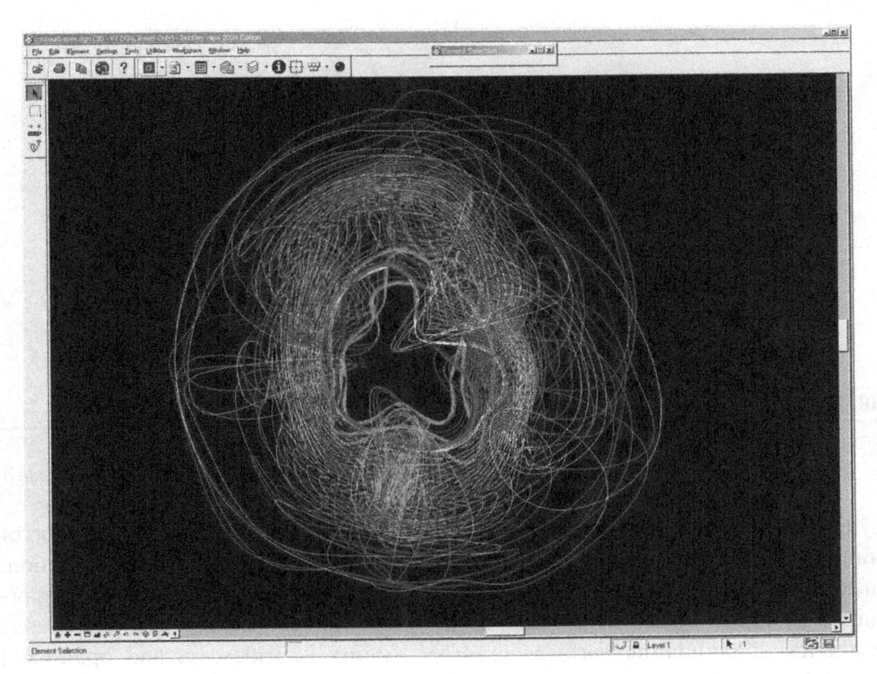

FIG. 4

The Embryological House visualized as a set of overlaid contours.

© *Greg Lynn.*

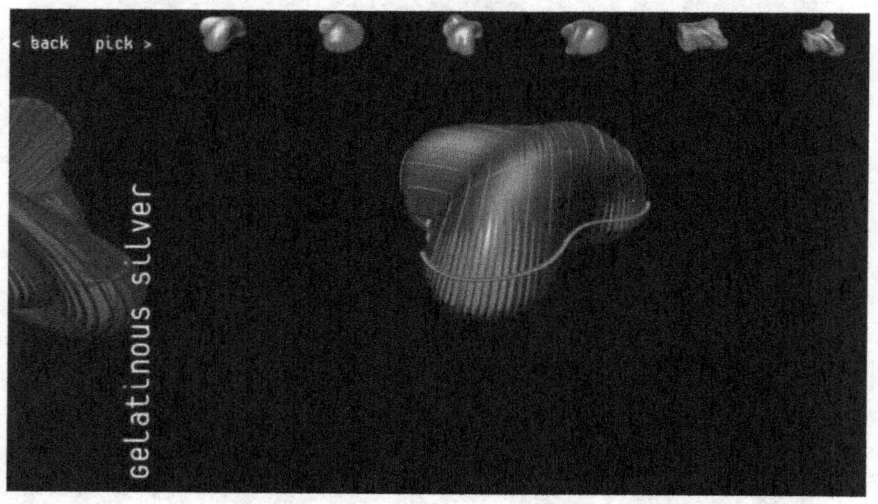

FIG. 5

Screenshots of a mock-up of the design election process that Greg Lynn envisioned for the Embryological House project.

© Greg Lynn.

FIG. 6

Greg Lynn, Embryological House: Size "A" eggs, ca. 1999.

© Greg Lynn. Credits: CCA Collection.

Lynn "[t]he Embryologic Houses can be described as a strategy for the invention of domestic space that engages contemporary issues of brand identity and variation, customisation and continuity, flexible manufacturing and assembly and, most importantly, an unapologetic investment in the contemporary beauty and voluptuous aesthetics of undulating surfaces rendered vividly in iridescent and opalescent colours" [29,p. 31].

The combination of different software was pivotal for the design process of the Embryological House. This hypothesis can be easily verified thanks to its digital files, which are conserved at the Canadian Centre for Architecture in Montreal. As Lawrence Bird and Guillaume Labelle remark in "Re-Animating Greg Lynn's Embryological House: A Case Study in Digital Design Preservation," "[t]he CCA developed [...] a speculative timeline of the evolution of the Embryological House" [30, p. 248]. It is worth noting that the experimentation with the House's landscaping involved the application of principles of animate form to the Villa Corner by Andrea Palladio (1552–53). The experimental status of Lynn's Embryological House lies in the exploration of potentially unlimited iterations derived from a basic form. It is also related to the intention to introduce difference into the system without creating a disjunctive break.

The significance of this project for the epistemological mutations of architecture is related to the fact that it is one of the first cases that showed how architectural form can be conceived as a snapshot of a process of metamorphosis that can be described with precision geometrically using the aforementioned software. This ontological shift of the conception of the design process should be understood in relation to the reorientation of the architects' interest from the approach of Jacques Derrida and his disjunctive theory to the approach of Gilles Deleuze and his appeal to connectivity. The space was perceived as a flux and the different variations of the Embryological House were conceived as snapshots of a continuous transformative process.

4 The New York Stock Exchange project by Asymptote Architecture and the production of new types of spatial flux

Asymptote Architecture, consisting of Hani Rashid and Lise Anne Couture, was invited in 1997 by the New York Stock Exchange to design, in collaboration with the Securities Industry Automation Corporation (SIAC), the Virtual New York Stock Exchange (3DTF), which was a virtual reality environment (Figs. 7 and 8). Asymptote Architecture's New York Stock Exchange "Virtual Trading Floor" (1997–99) visualized real-time numerical and statistical data to detect suspicious trading activity and tracked the impact of global news events on the market. The very force of the "Virtual Trading Floor" lies in its aspiration to display "live feed" data. According to what Hani Rashid claims in "Learning from the Virtual," "the 3DTF Virtual Trading Floor (VTF) is the architectural manifestation of a data ecosystem" [31] and it was conceived as a virtual environment designed to visualize real-time numerical and statistical data in the New York Stock Exchange.

For this project, which represents a turning point in digital culture's impact on architecture, Asymptote Architecture used Silicon Graphics Incorporated (SGI) RealityEngine hardware for the demanding needs of visualization; Softimage and Alias Software for the modeling, rendering, and animation; and Virtual Reality Modeling Language (VRML) and Macromedia Flash for the integration of data in

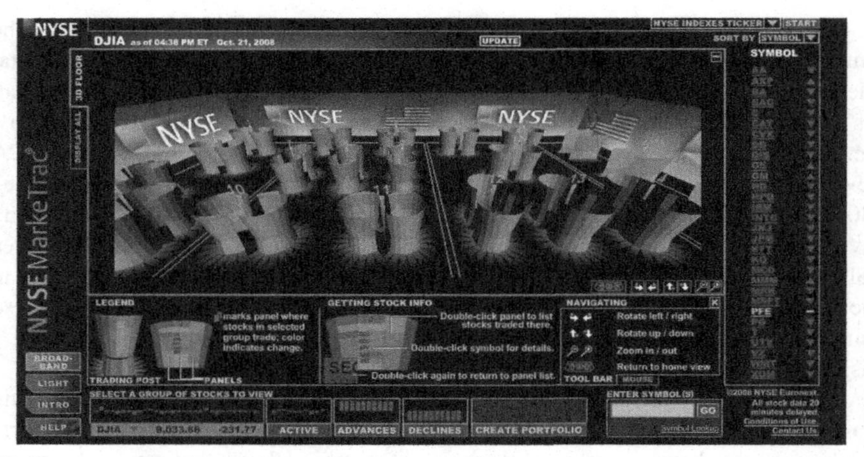

FIG. 7

Hani Rashid and Lise Anne Couture, an electrostatic print from New York Stock Exchange "Virtual Trading Floor" project, 1997–99.

© *Asymptote Architecture.*

FIG. 8

Hani Rashid and Lise Anne Couture, New York Stock Exchange "Virtual Trading Floor" project, 1997–99.

© *Asymptote Architecture.*

real time into a three-dimensional environment. Alias and Softimage were also employed in the first "paperless studios." It is worth mentioning that in the mid-1990s Rashid, in the framework of his "paperless studio" at Columbia University's GSAPP, had produced in collaboration with his students a plethora of video installations aiming at addressing spatiality in innovative ways and at augmenting space. The main objective of Rashid's studio was the production of new types of spatial flux, using Softimage, Macromedia, and QuickTime 1.0. As he remarks, the production of three-dimensional navigable cinematic architecture based on the digitization of films such as Alfred Hitchcock's film *Rope* and an ensemble of "early experiments in digitally manufactured spatiality allowed [them] [...] to look at space and architecture in a new way and to see things that we had previously only been able to theorize" (Rashid in Ref. [32]).

The usable surface area of the "Virtual Trading Floor" was 111.480 m^2. The animation code of the project was RT-SET, its operating system and hardware were Silicon Graphics, and its information display solution was PixelVision. In September 1998, Steve Silberman announced in *Wired* that a "revolutionary data-display hub designed by Asymptote [...] [would] begin operation on the floor of the exchange" [33]. Asymptote Architecture used for this project high-resolution plasma monitors, Silicon Graphics Onyx rendering engines, and VRM to visualize trading information in real time and in three dimensions, providing the possibility of simultaneous tracking of activity on several stock exchanges. An array of 6-ft diagonal PixelVision plasma screens displayed a real-time image of trading activity called 3DTF, while the "systemscape" tracked server activity and allowed brokers and NYSE computer operators to tweak remote operations without leaving the trading floor. It was possible to zoom in on particular stocks. Characteristically, Hani Rashid underscored in an interview he gave to Greg Lynn for the book *NYSE Virtual Trading Floor, Asymptote Architecture: Archaeology of the Digital 07*:

> *All of our experiments were in anticipation of the next two or three generations of these technologies: we were working in anticipation of Virtual Reality, interactive web, spatial and temporal augmentation, digital imaging in terms of creating new spatiality and realities, and the possibility of augmented reality in the later stages (Rashid in Ref. [32]).*

At the center of Asymptote Architecture's vision is the fusion between virtual reality environments and real spatiality. The 3DTF Virtual Trading Floor was conceived as a large-scale real-time virtual-reality environment, which exhibited flows and data for use by the operations team at the NYSE. It brought together information flows, data, and correlation models into a single seamless three-dimensional architectural environment. In parallel, it permitted users to monitor and correlate the stock exchange's daily trading activity and present the information within an interactive environment (Fig. 9).

FIG. 9

Hani Rashid and Lise Anne Couture, New York Stock Exchange "Virtual Trading Floor" project, 1997–99.

5 Aegis Hyposurface by dECOi architects and real-time response

The Aegis Hyposurface project by dECOi architects was an interactive metal surface that was able to respond with shape changes to multiple stimuli from its environment. The Aegis Hyposurface is a product of the collaboration of experts from various disciplines (Fig. 10). Mark Goulthorpe, Mark Burry, and Grant Dunlop have underscored the importance of the "collaborative participation of a very large, diverse and globally dispersed group of specialists" [14,p. 345]. The multidisciplinary group that created this project consisted of architects, mathematicians, computer programmers, rubber research engineers, pneumatic engineers, adhesive research engineers, electronic engineers, mechatronics engineers, and structural engineers, among other specialists [34]. Dr. Alex Scott, Peter Wood, and Xavier Robitaille worked on the mathematics and programming for the project, while among the collaborators for the project were Mark Burry and Grant Dunlop. Ove Arup & Partners and Sean Billings were the technical consultants for the project.

The basic concept of the Aegis Hyposurface was a dynamic system and an interactive faceted metallic skin. Thanks to a structure of 896 pneumatic pistons, the Aegis Hyposurface was capable of generating dynamic terrains as a result of real-time calculations. The Aegis Hyposurface was conceived as a faceted metallic surface having the capacity to be reshaped according to real-time response to

FIG. 10

dECOi Architects, Hyposurface, 2004. Gift of Mark Goulthorpe.

© Mark Goulthorpe. Credits: Mark Goulthorpe fonds, CCA Collection.

electronic stimuli from the environment, such as movement, sound, light, and so on. The sensors of the Aegis Hyposurface could be influenced by the movement of people next to it, generating visual effects on it. The Aegis Hyposurface was displayed at the exhibition "Non Standard Architectures," held at the Centre Georges Pompidou in Paris from December 10, 2003 through March 1, 2004 [35].

The project included the creation of a virtual simulacrum of the physical structure on the screen. The first working prototype of the Aegis Hyposurface was unveiled in March 2001. The main objective of this project was to radicalize architecture by announcing the possibility of dynamic form. Thanks to its use of nanotechnology, the fusion of information and matter became possible. Both dECOi architects' Aegis Hyposurface and Asymptote Architecture's "Virtual Trading Floor" were based on the intention to take advantage of the possibilities of real-time response. The surface of Aegis Hyposurface, thanks to its use of 576 pistons, was able to expand up to 200% of its planar area. Mathematica surfaces and formulae were used to experiment on the elasticity of its surface. The first prototype—HypoSurface 1.0—was displayed in the foyer of McCormick Place in Chicago in the framework of the International Manufacturers Technology Show (IMTS). Thanks to its 560 actuators, it was capable of responding to the sound and movement of the visiting public and was able to be physically reshaped in response to its environment's stimuli such as movement, sound, and light. A centralized mathematical function was used to control its surface. The significance of this project lies in the "environmental capabilities of the polyvalent wall [...] [and its] fluidity" [36].

6 The first digital turn and the ontological shift of architectural artifacts

Mario Carpo's *The Alphabet and the Algorithm* is useful for better grasping the epistemological and ontological shifts related to the questions addressed in this chapter. Carpo remarks, referring to the digital turn of the 1990s, that "[t]he new organicist and morphogenetic theories that crossed paths with the mathematical ones [...] would eventually become staples of digital design theory" [37,p. 40]. Carpo considers as very central for this first digital turn Greg Lynn, especially his *Folds, Bodies & Blobs: Collected Essays* [38] and *Animate Form* [39], and Bernard Cache with his book *Earth Moves: The Furnishing of Territories*, where he develops his reflections around the concept of the "objectile" [40,41]. Cache's book had an important impact on introducing an ensemble of concepts of Gilles Deleuze's work, and especially of certain of his reflections in *The Fold: Leibniz and the Baroque* in architectural discourse [42,43].

The reinvention of the relationship between architectural pedagogy and architectural practice due to the establishment of "paperless studios" and the incorporation of new software and advanced digital technologies in the teaching procedure and design process was accompanied by a transformation of the ontological status of architectural artifacts [8–12]. Thérèse Tierney analyzes this mutation of the ontological status of architectural artifacts in her book *Abstract Space: Beneath the Media Surface*. She explains in which sense the architectural drawings that are produced through the use of digital tools differ ontologically from those that are produced analogically [44]. More specifically, she claims that "[e]volutionary form-generating software had transformed the architecture image into event and performance, either by understanding architecture as the epigenesis of spatial conditions, or by the object being conceptualized as the actualization of built-up potentials" [44,p. 1]. To better grasp the ontological shift of architectural artifacts due to their production through the use of computation tools, one should bear in mind that "[d]igital architecture objects cannot be said to represent architecture in the same sense that the drawings or models that make up conventional architectural collections do" [30,p. 245].

7 The second digital turn and the role of urban scale digital twins in shaping sustainable urban policies

In *The Second Digital Turn: Design Beyond Intelligence*, Mario Carpo draws a distinction between the first and the second digital turn, claiming that the first digital turn was characterized by the invention and interpretation of "a new cultural and technical paradigm [...] [and the creation of] a visual style that defined an epoch and shaped technological change" [45,p. 8]. One could claim that the second digital turn, which concerns the present state of architecture, has to do mostly with the ongoing debates around the role of big data, especially in relation to questions concerning the notions of sovereignty, democracy, and the public realm, and less with the formal and visual experimentations that were at the center of the first digital turn. The current state of digital turn in architecture is more oriented toward social aspects, placing particular emphasis on questions concerning the democratization of data and the issues related to the role of "digital commons" [46,p. 139]. Within such a context, most of the efforts are concentrated on sharpening the visualization techniques and on using them to transform top-down design strategies into more bottom-up ones.

An ensemble of tendencies that try to incorporate the advantages of big data in the very design processes, in both architecture and urban planning, "share an optimism towards the flexibility offered by technology [...] [and] are based on the conviction that big data offers citizens the possibility to 'make connections [...] in a more visible way and acquire more insights about the ubiquitous presence of digital and data technologies in the city'" [47,48]. A risk that is present in our data-driven societies, which are based on the myths of "digital universalism," is to neglect that all data have complex attachments to place [49,50]. To avoid this, one should take into account the following six principles that are examined by Yanni Alexander Loukissas in *All Data Are Local: Thinking Critically in a Data-Driven Society*: all data are local; data have complex attachments to place; data are collected from heterogeneous sources; data and algorithms are inextricably entangled; interfaces recontextualize data; and data are indexes to local knowledge [50]. The second digital turn in architecture to which Mario Carpo refers [45] is related to the generalized use of "digital twins," which refers to digital representations enabling comprehensive data exchange and can contain models, simulations, and algorithms describing their counterpart in the real world [51]. The term "digital twin" was coined in the early 2000s by Michael Grieves and refers to digital simulation models that run alongside real-time processes [52].

Digital twins are conceptualized as digital replicas of physical entities. Despite their aspirations to enhance the participation of citizens in the decision-making processes and to incorporate their input into urban planning strategies, the fact that digital twins are based on a limited set of variables and processes makes them problematic. The creation of urban scale digital twins is based on the use of advanced technological applications, such as sensing, processing, and data transmission. Urban scale digital twins are used in the field of urban analytics, as well as in the field of computational social sciences. They enhance evidence-based operational decisions and experimentation on urban policies. Recently, within the field of smart cities, the notion of urban scale digital twin has acquired a central place [51,53,54]. Yanni Loukissas and Anita Say Chan criticize "digital universalism" [49,50,55]. Their critiques are useful for understanding that "digital twins," despite their potentials, entail the risks of neglecting the social aspects of urban fabric since they are based on the abstraction of sets of variables and processes. The current state of research concerning the role of digital twins in shaping urban policies is characterized by a dichotomy between scholars who focus on the technological and sustainable benefits of the use of urban scale digital twins and researchers who criticize "digital universalism" [51,56]. Noteworthy are the following characteristics of urban scale digital twins: their "scalability"; their "predictability," which becomes possible thanks to the use of simulation algorithms; their capacity to integrate new elements thanks to the use of IoT sensors, and in situ real-time data; and their capacity to enhance cooperation because they can be broadly accessible [57].

A recent program that is focused on the role of urban scale digital twins in promoting sustainability in urban planning is the "National Digital Twin programme" (NDTpNot). This program belongs to the programs of the Centre for Digital Built Britain, and is based on a collaboration between the University of Cambridge and the Department for Business, Energy and Industrial Strategy. The so-called "Gemini Principles" aim "to build consensus on foundational definitions and guiding [...] and to begin enabling alignment on the approach to information management across the built environment" [58]. The aforementioned program was launched in July 2018. At the core of this program is the intention to contribute to society, economy, business, and sustainability. The NDTp runs the Digital Twin Hub, a collaborative and supportive web-enabled community for those who own, or who are developing, digital twins within the built environment. According to the Institution of Engineering and Technology (IET), the environmental benefits of the use of urban scale digital twins should focus on the endeavors to minimize waste, achieve resource efficiency, and enhance the circular economy in the built environment. The IET places particular emphasis on the advantages of the use of urban scale digital twins that focus on the effort to minimize waste for the economy [11].

8 Conclusions: On how to combine environmental and social equity in smart cities

At the core of the debates around smart cities is the concept of the self-organizing city, which refers to the role of bottom-up mechanisms in the development of cities. The potential of smart cities and urban scale digital twins should be understood within the context of the endeavors of responding to the global ecological crisis. Two programs that play a major role in shaping sustainable urban planning methods are the European New Green Deal and the Agenda for Sustainable Development. The latter included the so-called Sustainable Development Goals (SDGs). The European Union has set the following goals regarding sustainable urban planning strategies: firstly, the empowerment of "urban actors towards common goals; secondly, the development of people-oriented urban planning strategies that aim to contribute to the social equity of communities; thirdly, the development of digital platforms and other digital tools that intend to enhance interactive and proactive approaches in urban planning decision-making," and "the creation of integrated, open, and functional technological infrastructures for the development of programmes and the provision of services (data-driven planning)" [57,p. 6263].

Among existing urban scale digital twins that are either in operation or under development are the twins of the following cities or districts: Athens in Greece, Plzeň in the Czech Republic, Dublin Docklands in Ireland, Herrenberg in Germany, Vienna in Austria, Zurich in Switzerland, New York City in the United States, London in the United Kingdom, and Helsinki in Finland. Other noteworthy urban scale digital twins are those of Cambridge, Gothenburg, Munich, Newcastle, Paris, Rennes, and Rotterdam [57]. The current debates around the role of urban scale digital twins in shaping sustainable urban policies intend to take into consideration the goals of the aforementioned programs. However, a risk that is not easy to challenge is that of celebrating the technological benefits of advanced technologies without placing particular emphasis on the fact that different social groups have uneven access to data. A remark of Manuel Castells that can help us better understand the risks of uneven access to data is the following: "the network of decision-making and generation of initiatives, ideas and innovation is a micro network operated by face-to-face communication concentrated in certain places" [59,p. 2742].

A challenge to which both smart cities and urban scale digital twins are invited to respond within the current context is the necessity to share design performance-based systems that are informed and tested through scenarios based on performance simulations. Adaptive ecologies and homeostatic urbanism are based on the analysis of urban development through adaptive models of ecology. Another aspect concerning generative approaches to architectural design and self-organizing computation that we should not neglect is the role of pattern-generation tools. Another issue to which we are invited to respond is that concerning the uneven access to the technologies and data that make possible smart cities and urban scale digital twins. To go beyond the myth of "data universalism," we should explore ways that would offer possibilities to combine environmental and social equity, on the one hand, and to understand the debates around the smart cities beyond their technological aspects, on the other hand. Within such a framework, it is of pivotal importance to shape methodological tools that would offer the possibility to develop new forms of social advocacy around big data, and would create a shared terrain of reflections between the debates on smart cities and the debates around urban commons [60–62]. By incorporating urban scale digital twins in the decision-making processes concerning urban planning, urban planners can shape new participatory design methods. This could become possible through the interaction of citizens with the visualization of data concerning the existing situation of urban environments.

Acknowledgments

The research project was supported by the Hellenic Foundation for Research and Innovation (H.F.R.I.) under the "3rd Call for H.F.R.I. Research Projects to support Post-Doctoral Researchers" (Project Number: 7833). I am grateful to the Hellenic Foundation for Research and Innovation (H.F.R.I.) for the support and to Greg Lynn, Asymptote Architecture, dECOi Architects, Mark Goulthorpe, the Canadian Centre for Architecture (CCA), and the Institution of Engineering and Technology (IET) for the figures accompanying the chapter.

References

[1] S. Allen, The paperless studios in context, in: A. Goodhouse (Ed.), When Is the Digital in Architecture? Canadian Centre for Architecture/ Sternberg Press, Montreal; Berlin, 2017, pp. 383–404.

[2] A. White, Manuel Castells's trilogy the information age: economy, society, and culture, Inf. Commun. Soc. 19 (12) (2016) 1673–1678, https://doi.org/10.1080/1369118X.2016.1151066.

[3] M. Castells, The Rise of the Network Society, Second edition with a new preface, Wiley-Blackwell, Chichester, West Sussex, 2010.

[4] S. Sassen, The Global City: New York, London, Tokyo, Princeton University Press, Princeton, NJ, 2013.

[5] F. Stalder, Manuel Castells. The Theory of the Network Society, Polity Press, Cambridge, 2008.

[6] M. Castells, The Informational City: Information Technology, Economic Restructuring and the Urban-Regional Process, Basil Blackwell, Oxford, 1989.

[7] M. Charitonidou, Paperless studios and the articulation between the analogue and the digital: geometry as transformation of architecture's ontology, in: Paper Presented at the International Symposium "Scaffolds: Open Encounters with Society, Art and Architecture", Brussels, Belgium, 22-23 November 2018, 2018, https://doi.org/10.3929/ethz-b-000442820.

[8] M. Charitonidou, Architectural Drawings as Investigating Devices: Architecture's Changing Scope in the 20th Century, Routledge, London; New York, 2023, https://doi.org/10.4324/9781003372080.

[9] M. Charitonidou, The role of virtual worlds (VWs) in online architectural design studio teaching: from paperless studios to collaborative computer-aided strategies of distance learning, in: M. McVicar (Ed.), Productive Disruptive: Spaces of Exploration In-Between Architectural Pedagogy and Practice. Association of Architectural Educators 7th International Conference, the Welsh School of Architecture, Cardiff University, UK, 12-15 July 2023, Cardiff University, Cardiff, 2023, pp. 102–112, https://doi.org/10.17613/21hw-d949.

[10] M. Charitonidou, Cosmopolitics and virtual environments in architectural design studio teaching: collaborative computer-aided strategies and social and environmental equity, in: Paper Presented at the 2023 ACSA/EAAE Teachers Conference: Educating the Cosmopolitan Architect, 22-24 June 2023. The Association of Collegiate Schools of Architecture (ACSA) and the European Association for Architectural Education (EAAE), Biennial Joint Teachers Conference, Iceland University of the Arts (IUA), Reykjavik, 2023, https://doi.org/10.17613/ys1p-b985.

[11] M. Charitonidou, Urban scale digital twins vis-à-vis complex phenomena: datafication and social and environmental equity, in: W. Dokonal, U. Hirschberg, G. Wurzer (Eds.), Digital Design Reconsidered, Proceedings of the 41st eCAADe Conference, 20-22

September 2023, Graz University of Technology, Graz, Austria, vol. 2, Education and Research in Computer Aided Architectural Design in Europe and Graz University of Technology Faculty of Architecture, Brussels; Graz, 2023, pp. 821–830. https://doi.org/10.52842/conf.ecaade.2023.2.821.
[12] M. Charitonidou, The digital turn and the transformation of architecture's ontology: experimenting with geometry, virtual reality and big data, in: Paper Presented at the FILARCH 2023 Conference, 25-27 May 2023, University of Patras, 2023, https://doi.org/10.17613/3t8d-d218.
[13] M. Charitonidou, Drawing and Experiencing Architecture: The Evolving Significance of Cit's Inhabitants in the 20th Century, Transcript Verlag, Bielefeld, 2022. https://doi.org/10.1515/9783839464885.
[14] M. Goulthorpe, M. Burry, G. Dunlop, Aegis Hyposurface©: the bordering of university and practice, in: Proceedings of ACADIA 2001, 2001, pp. 334–349.
[15] M. Wigley, Theoretical slippage, Fetish, The Princeton Architectural Journal 4 (1992) 88–129.
[16] G. Lynn, AD: Folding in Architecture, Wiley-Academy, Chichester, 1993.
[17] X. Wang, Augmented reality in architecture and design: potentials and challenges for application, Int. J. Archit. Comput. 7 (2) (2009) 309–326.
[18] X. Wang, M.A. Schnabel, Mixed Reality in Architecture, Design, And Construction, Springer, 2008.
[19] M. Charitonidou, Interactive art as reflective experience: imagineers and ultra-technologists as interaction designers, Vis. Resour.: An International Journal on Images and Their Uses 36 (4) (2020) 382–396, https://doi.org/10.1080/01973762.2022.2041218.
[20] B. Tschumi, The school's new computing facilities, Newsline (1994) 8–9.
[21] E. Muir, R. O'Neill, The paperless studio: a digital design environment, Newsline (1994) 10–11.
[22] C. Sisk, Computers change ways architects learn, teach, Columbia Daily Spectator 119 (105) (1995) 1.
[23] H. Rashid, S. Marble, G. Lynn, Paperless Studio: Fall 1994, Newsline (1995) 6–7.
[24] M. Benedikt (Ed.), Cyberspace: First Steps, The MIT Press, Cambridge, MA, 1991.
[25] G. Lynn, Sports Complex, Galaxy Toyama/Odawara Gymnasium: Archaeology of the Digital 03, Canadian Centre for Architecture, Montreal, 2014.
[26] A. Andia, Integrating digital design and architecture during the past three decades, in: Proceedings of the Seventh International Conference on Virtual Systems and Multimedia (VSMM'01), IEEE Computer Society, Berkeley, CA, 2001, pp. 677–686.
[27] M. Frascari, A reflection on paper and its virtues within the material and invisible factures of architecture, in: M. Frascari, J. Hale, B. Starkey (Eds.), From Models to Drawings: Imagination and Representation in Architecture, Routledge, London; New York, 2008, pp. 23–33.
[28] B. Tschumi, The making of a generation: how the paperless studios came about, in: A. Goodhouse (Ed.), When Is the Digital in Architecture? Canadian Centre for Architecture/Sternberg Press, Montreal; Berlin, 2017, pp. 405–420.
[29] G. Lynn, Greg Lynn: Embryological Houses, Archit. Des. 70 (3) (2000) 26–35.
[30] L. Bird, G. LaBelle, Re-animating Greg Lynn's Embryological House: a case study in digital design preservation, Leonardo 43 (3) (2010) 243–249, https://doi.org/10.1162/leon.2010.43.3.243.

[31] H. Rashid, Learning From the Virtual, E-flux, 2017. 25 July 2017 https://www.e-flux.com/architecture/post-internet-cities/140714/learning-from-the-virtual/. (Accessed 14 January 2021).
[32] G. Lynn, NYSE Virtual Trading Floor, Asymptote Architecture: Archaeology of the of the Digital 07, Canadian Centre for Architecture, Montreal, 2015.
[33] S. Silberman, NYSE Gets 'Ramped' Up, Wired, 1998. 17 September 1998 https://www.wired.com/1998/09/nyse-gets-ramped-up/. (Accessed 14 January 2021).
[34] G. Lynn, Mark Goulthorpe, HypoSurface: Archaeology of the Digital 06, Canadian Centre for Architecture, Montreal, 2014.
[35] F. Migayrou (Ed.), Architecture non standard, Centre Georges Pompidou, Paris, 2003.
[36] O. Ataman, J. Rogers, A. Ilesanmi, Redefining the wall: architecture, materials and macroelectronics, Int. J. Archit. Comput. 4 (4) (2006) 125–136.
[37] M. Carpo, The Alphabet and the Algorithm, The MIT Press, Cambridge, MA, 2011.
[38] G. Lynn, Folds, Bodies & Blobs: Collected Essays, La lettre volée, Brussels, 1998.
[39] G. Lynn, Animate Form, Princeton Architectural Press, New York, 1999.
[40] B. Cache, in: M. Speaks (Ed.), Earth Moves: The Furnishing of Territories, The MIT Press, Cambridge, MA, 1995 (A. Boyman Trans.).
[41] G. Lynn, Bernard Cache, Objectile: Archaeology of the Digital 09, Canadian Centre for Architecture, Montreal, 2015.
[42] G. Deleuze, Le Pli: Leibniz et le Baroque, Les Editions de Minuit, Paris, 1988.
[43] G. Deleuze, The Fold: Leibniz and the Baroque (T. Conley Trans.), University of Minnesota Press, Minneapolis, 1992.
[44] T. Tierney, Abstract Space: Beneath the Media Surface, Taylor & Francis, London; New York, 2007.
[45] M. Carpo, The Second Digital Turn: Design Beyond Intelligence, The MIT Press, Cambridge, MA; London, 2017.
[46] E. Isin, E. Ruppert, Being Digital Citizens, Rowman & Littlefield International, London, 2020.
[47] M. Charitonidou, Takis Zenetos's electronic urbanism and tele-activities: minimizing transportation as social aspiration, Urban Sci. 5 (1) (2021) 31, https://doi.org/10.3390/urbansci5010031.
[48] A. Ersoy, K.C. Alberto, Understanding urban infrastructure via big data: the case of Belo Horizonte, Reg. Stud. Reg. Sci. 6 (2019) 374–379.
[49] A.S. Chan, Networking Peripheries Technological Futures and the Myth of Digital Universalism, The MIT Press, Cambridge, MA, 2014.
[50] Y.A. Loukissas, All Data Are Local: Thinking Critically in a Data-Driven Society, The MIT Press, Cambridge, MA, 2019.
[51] M. Charitonidou, Urban scale digital twins in data-driven society: challenging digital universalism in urban planning decision-making, Int. J. Archit. Comput. 20 (2) (2022) 238–253, https://doi.org/10.1177/14780771211070005.
[52] M. Grieves, Digital Twin: Manufacturing Excellence through Virtual Factory Replication, White paper, 1, 2014, pp. 1–7.
[53] M. Charitonidou, Mobility and migration as constituting elements of urban society: migration as a gendered process and how to challenge digital universalism, in: J.A. Scelsa, J.J. Tandberg (Eds.), 2021 ACSA Teachers Conference, Curriculum for Climate Agency: Design in Action, Association of Collegiate Schools of Architecture (ACSA), Washington DC, ACSA, 2021, pp. 194–199. https://doi.org/10.35483/ACSA.Teach.2021.27.

[54] M. Charitonidou, Commoning practices and mobility justice in data-driven societies: urban scale digital twins and their challenges for architecture and urban planning, in: M. Blanco Lage, O. Atalay Franck, N. Marine, R. de la O Cabrera (Eds.), Towards a New European Bauhaus: Challenges in Design Education: EAAE Annual Conference Madrid 2022, Springer Nature, Cham, 2024.

[55] C.L. Borgman, Big Data, Little Data, No Data: Scholarship in the Networked World, The MIT Press, Cambridge, MA, 2015.

[56] M. Charitonidou, Public spaces in our data-driven society: the myths of digital universalism, in: Paper Presented at the International Latsis Symposium "Deep City—Climate Crisis, Democracy and the Digital", 24-27 March 2021, EPFL, Lausanne, Switzerland, 2021, https://doi.org/10.3929/ethz-b-000465249.

[57] G. Caprari, G. Castelli, M. Montuori, M. Camardelli, R. Malvezzi, Digital twin for urban planning in the green deal era: a state of the art and future perspectives, Sustainability 14 (1) (2022) 6263, https://doi.org/10.3390/su14106263.

[58] A. Bolton, L. Butler, I. Dabson, M. Enzer, M. Evans, T. Fenemore, F. Harradence, et al., Gemini Principles (CDBB_REP_006), 2018, https://doi.org/10.17863/CAM.32260.

[59] M. Castells, Globalisation, networking, urbanisation: reflections on the spatial dynamics of the information age, Urban Stud. 47 (13) (2010) 2737–2745.

[60] M. Charitonidou, Urban scale digital twins and commoning practices: mobility justice and sharing ground resources, in: Paper Presented at the 2022 AHRA Conference "Building Ground on Climate Collectivism: Architecture after the Anthropocene", 17-19 November 2022, 2022, https://doi.org/10.17613/t3gq-hd32.

[61] M. Charitonidou, Housing programs for the poor in Addis Ababa: urban commons as a bridge between spatial and social, J. Urban Hist. 48 (6) (2022) 1345–1364, https://doi.org/10.1177/0096144221989975.

[62] M. Charitonidou, Urban commons as a bridge between the spatial and the social, in: R.C. Brears (Ed.), The Palgrave Encyclopedia of Urban and Regional Futures, Palgrave Macmillan, Cham, 2022, https://doi.org/10.1007/978-3-030-51812-7_290-2.

CHAPTER

18

Generative adversarial network (GAN) assisted IoT search engine for disaster damage assessment

Hengshuo Liang[a], Cheng Qian[a], Chao Lu[a], Guobin Xu[b], and Wei Yu[b]

[a]*Department of Computer and Information Sciences, Towson University, Towson, MD, United States,* [b]*Department of Computer Science, Morgan State University, Baltimore, MD, United States*

1 Introduction

Machine learning (ML) and especially deep learning (DL)-based applications have greatly improved various aspects of life. ML and DL techniques have garnered significant attention and have been widely adopted across a variety of fields, such as image/voice/language processing, computer networks, security, and cyber-physical systems, among others [1–6]. They have proven to be particularly effective in numerous areas, leading to widespread use and applications [7–9]. For example, DL technology has significantly enhanced the capabilities of cameras, enabling them to perform real-time object detection on-site [10], which allows for more focused and accurate surveillance [11]. It has also expanded the possibility of searching for similar items based on images [12]. Moreover, DL technology can significantly improve network management and security, such as providing a more efficient method for network routing [13], enhancing the security for more effectively identifying suspicious behaviors [14–16], and maintaining a greater throughput for network capacity even under heavy loads [17], among others. In addition, DL technology can be used to assist in creating a real-time rendering image filter for TikTok [18], predicting energy use forecasting [19] and weather [20], as well as seamlessly clipping and merging multiple videos into a new one [21].

As a leading-edge technology, DL has a significant impact on disaster response systems and damage assessment during or after natural disasters, as well as weather forecasting [22,23]. Its ability to perform assertive and effective data mining and analysis has made it a crucial tool in disaster risk management, including disaster early warning, risk identification, disaster rescue, and recovery efforts [24]. Besides that, DL technology has the potential to improve efficiency of damage assessment

Smart Spaces. https://doi.org/10.1016/B978-0-443-13462-3.00012-1

following a disaster [23] by automating the search and detection processes. This can save human resources and reduce the workload of disaster response teams. However, the learning effectiveness in this context is still dependent on the quality and quantity of the available data, as well as the performance of the network used to deliver that data.

A high-quality, extensive dataset is necessary for building an accurate learning model. The Internet of Things search engine (IoT-SE) can greatly enhance disaster damage assessment efforts by providing real-time data on affected areas. IoT devices, such as sensors and cameras, can be deployed in areas prone to natural disasters to gather information on environmental conditions, structural integrity, and other factors that may impact the extent of damage [23,25,26]. The use of the IoT-SE in disaster damage assessment allows for a more comprehensive and accurate understanding of disaster damage, enabling more effective and efficient response efforts.

Insufficient data can lead to underfitting, which can compromise the accuracy of the model. When the available dataset is relatively small or of poor quality, a learning model cannot be well trained [27]. One way to overcome the issue of insufficient data is to use generative adversarial networks (GANs) that can generate additional data samples based on the distribution of the available data to assist in the training of learning models. GANs have proven to be an effective method for augmenting small or low-quality datasets, making the training of learning models more accurate and robust.

In this chapter, a new disaster damage assessment framework is proposed that combines the use of GAN with the IoT-SE to enhance performance in situations in which the available dataset is small. The GAN is used to generate additional data samples based on historical data, which are then merged with the collected data to form a new, larger dataset. This enhanced dataset is then used to train DL models for the purpose of assessing disaster damage. The goal of this approach is to improve the accuracy and efficiency of damage assessment efforts when there is insufficient data, which is a common practice. We conduct extensive evaluations and comparisons of the performance of several different learning models.

To do so, we first preprocess the dataset into two groups for GAN and DL models to assess the damage. For the GAN, the dataset has been reclassified based on damage targets (buildings, vehicles, people, and others). For DL models to assess the damage, we randomly crop, resize, and rotate them to make them uniform. The baseline case is when DL models are completely trained with real-world data. To evaluate the efficacy of the images generated by GAN, we keep the size of the training set unchanged (60%), but gradually increase the generated data from 10% of the 60% training set to 90% of the 60% training set (in an interval of 10%). The performances between the baseline and different sizes of the GAN-generated dataset for each model are compared. Our experiments confirm that the use of GAN can improve the performance of DL models with complex structures, but it can reduce the performance of DL models with a lightweight structure.

The remainder of this chapter is organized as follows: In Section 2, we introduce damage assessment issues, deep learning, GAN, and IoT-SE. In Section 3, we provide the framework design, scenarios, and the reason why these scenarios are essential for validation. In Section 4, we provide an overview of the methodology and analyze the performance in the scenarios with different settings. In Section 5, we discuss the future research direction. Finally, we conclude the paper in Section 6.

2 Preliminaries

In this section, the challenges related to damage assessment are discussed. A brief background study of deep learning, as well as GAN and IoT-SE, is presented.

2.1 Damage assessment challenges

In general, damage assessment is concerned with two aspects. One is efficient high-volume data collection, and the other is fast and highly accurate data analysis for damage assessment. Human evaluation techniques [28], social media-based assessment techniques [29], and deep learning-based assessment approaches [30] have all been examined as viable ways of assessing disaster damage. In particular, human evaluation techniques could be involved by sending individuals into an affected region so that questionnaires can be conducted or photographs taken. After that, the resulting data can be further analyzed for assessment purposes. However, it is obvious that such an approach necessitates high labor costs and additional time and effort to complete data collection, which could further slow down the damage recovery plan. It is also worth noting that the social media-based approach involves gathering data from social media platforms such as Twitter, Facebook, and Instagram, which could expedite the data collection process. However, the performance of social media-based assessment approaches largely depends on the network status of affected areas and other factors. The DL-based strategies require collecting data from social media and completing the evaluation by feeding the collected data to DL models to obtain predictions, which can improve the accuracy of assessment scenes. Nonetheless, the accuracy of DL-based methods is greatly dependent on the size of the collected datasets. Generally speaking, greater precision is achieved with larger datasets, which obviously incur additional computing cost.

2.2 Deep learning (DL)

DL is the process of constructing a neural network for analytic learning that permits the interpretation of data such as images, audio, and text [1,2,7]. Extensive research efforts have been devoted to DL regarding network security, multitarget classification, and data analytics. For example, Liang et al. proposed a DL-based scheme to detect flooding attacks in the named data network (NDN)-based IoT-SE [31]. Cheng et al. proposed a new neutral network structure with a feature map over the dataset so that better performance in image classification can be obtained [32]. Likewise, Liang et al. designed a convolutional neural network (CNN)-based scheme to automatically recognize industrial components in industry IoT systems [33]. For our research, we leverage the most prevalent DL models (VGG-19 [34], ResNet50 [35], GoogleNet [36], and MobileNetV2 [37]) and well-defined training sets, which are discussed in detail in Section 4.1. These datasets are prepossessed and fine-tuned for training simulation, as discussed in Section 3.1.

2.3 Generative adversarial network (GAN)

Generally speaking, the GAN is composed of two entities. One is *Generator* that attempts to generate fake and new sample data, while the other is *Discriminator* that is responsible for determining whether the data being examined is real data taken from a training dataset or a sample data generated by the generator [38]. As a relatively new technique, GAN has been applied to a variety of uses, including creating Anime characters [39], posing guided person image generation [40], creating super-resolution images [41], and others. To this end, some research efforts have been conducted on GAN. For example, Qian et al. proposed GAN as an effective way of increasing dataset volume and improving DL model performance [27]. GAN was offered as a new way of generating standard network traffic, which is indistinguishable from packets generated by network flow [42]. In this study, we propose to make use of the GAN generator to create more related images as part of the training dataset for the DL model to improve the disaster damage assessment. Recall that if less data is collected from social media due to constrained network performance, GAN can enhance the size of the dataset in this case.

A number of GAN variants have been proposed to accommodate various scenarios. For example, convolution-based GANs [43] were frequently employed for image generation based on supervised machine learning. Generally speaking, condition-based GAN [43] utilizes a conditional variable as a label (video, audio, etc.) to tune the generated samples. Since our objective is to generate image samples precisely and efficiently, we leverage convolution-based GAN (i.e., FastGAN) to generate new image samples based on the original data. In addition, FastGAN [44] utilizes the skip-layer channel-wise excitation (SLE) module to enhance robust model weight adjustment and accelerate the data training process. Meanwhile, feature maps are utilized to collect additional details from image samples to further train the generator. Moreover, both the SLE module and feature maps enable FastGAN to efficiently generate image samples with a minimum computational cost.

2.4 Internet of Things search engine (IoT-SE)

A vast number of IoT devices are continuously being deployed around the world. The heterogeneous network of devices makes it exceedingly difficult to share data due to the diverse formats used by the numerous devices from a variety of different companies. IoT-SE intends to develop a way of efficiently transmitting data and building intelligence within IoT systems by establishing a universal format for users. This would speed up the process of data sharing among a wide variety of devices [25].

There are some existing research efforts on IoT-SE. For example, Hatcher et al. proposed a long short-term memory (LSTM)-based learning scheme in IoT-SE so that incoming query volume can be predicted, which leads to query efficiency [25]. Likewise, Cheng et al. designed an IoT-SE platform based on constrained application protocol (COAP) and evaluated the efficacy of the proposed query optimization algorithms [45]. In this chapter, we consider IoT-SE as the data collector to capture all relevant data and formalize it into the same format for subsequent actions.

3 Our approach

In this section, we propose an approach for assisting IoT-SE in disaster damage assessment by leveraging GAN.

The framework of leveraging GAN to support the disaster damage assessment is presented as an application of the IoT-SE-based architecture. As shown in Fig. 1, there are four components in GAN assisting IoT-SE architecture: (i) *GAN data generator*, (ii) *Source of data*, (iii) *IoT-SE*, and (iv) *DL application*. Each GAN data generator focuses on a certain data type. For example, in the disaster damage assessment scenario, it represents a specific damage level, such as severe damage, mild damage, little or no damage, etc. These GAN data generators are fed with prelabeled historical data to train on each damage level. The generated data will be transmitted to IoT-SE for further processing. When a disaster occurs, all newly collected data is sent to IoT-SE for formalization and storage. Then, IoT-SE will combine the generated data samples with collected data and send them to the DL application to perform assessments.

GAN-based disaster damage assessments rely on the following three essential aspects: (i) *Dataset Volume:* The volume of datasets will dramatically affect the performance of DL models. In general, the larger the dataset, the superior the performance of DL models. (ii) *IoT-SE Management:* When huge amounts of data

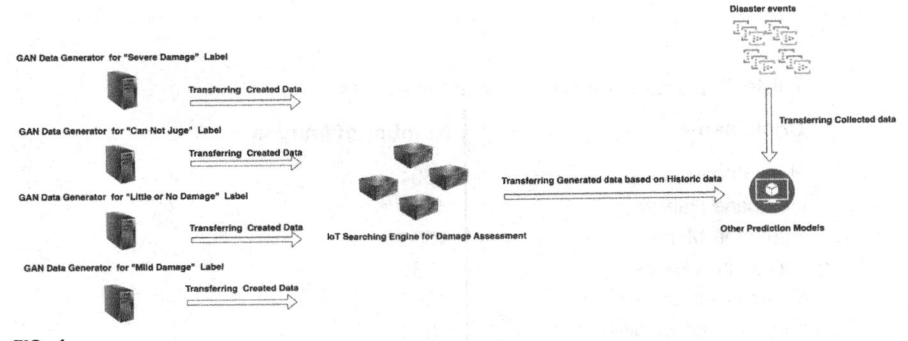

FIG. 1

GAN assisting DL-based IoT-SE architecture for damage assessment.

reach the IoT-SE, it will be crucial to determine how to store the data quickly and efficiently. When the IoT-SE's upper layer demands a vast volume of metadata, the IoT-SE's ability to efficiently manage data access will be crucial. (iii) *DL Model Performance:* The size, training time, and cost of the DL model will affect its ultimate performance in specific tasks.

3.1 Data collection

In this section, we utilize the labeled data of natural disasters from the Artificial Intelligence for Disaster Response (AIDR) [46] platform to demonstrate the efficacy of our proposed approach. AIDR is a dataset website of natural events created by crawling related data from social media and manually labeling the data. We chose the datasets from the following events: Hurricane Irma, Hurricane Harvey, Hurricane Maria, California Wildfires, Mexico Earthquake, Iraq-Iran Earthquake, and Sri Lanka Floods [47]. Table 1 lists all the datasets with the total number of images initially collected.

All of these natural disasters cause physical destruction, including shattered buildings, damaged vehicles, etc. The image data may indicate one of five levels of damage: (i) *Severe Damage:* massive buildings are destroyed; (ii) *Mild Damage:* a portion of the building has been damaged, although it is still functional; (iii) *Little or No Damage:* the buildings are damage-free; (iv) *Cannot Judge:* The image offers limited data concerning building damage; (v) *NaN (Not Related):* The image contains no visible damage to the building. Some examples of these damage levels are depicted in Fig. 2.

With the levels of damage defined here, the number of each level of damage contained in each dataset is given in Table 2. As seen in the table, most of the image data is classified as NaN (not related). More likely, the data is typically collected via social media platforms such as Twitter. NaN and CanNotJudge labeled data is meaningless for damage assessment. Even with the excellent network performance, the size of the dataset is still insufficient for DL models to perform adequately throughout the training and testing phases. Thus we focus on generating data for these three

Table 1 Dataset for seven disaster events.

Crisis name	Number of images
Hurricane Irma	4504
Hurricane Harvey	4434
Hurricane Maria	4556
California wildfires	1589
Mexico earthquake	1380
Iraq-Iran earthquake	597
Sri Lanka floods	1022

FIG. 2

Examples of labeled images.

levels of damage: Severe Damage, Mild Damage, and Little or No Damage. These three categories of data are essential for damage assessment and must be identified immediately.

To accomplish this, GAN was leveraged to generate extra data to fill the dataset volume in constrained networks. Six datasets from Table 1 (except the Hurricane Harvey dataset) were selected and combined as a new dataset for the GAN data generator. We reclassified the new dataset based on damaged targets into several categories: buildings, people, vehicles, and others. Table 3 shows the summary of labeled image data for the GAN dataset. As we can see from the table, the majority of the image data shows three levels of building damage during the disaster, and only a small number of images reveal the serious damage to people and vehicles. Fig. 3 shows the GAN-generated image samples.

3.2 Damage assessment scenarios

Recall in Section 2.1, two factors affect the effectiveness of disaster damage assessment: (i) efficient high-volume data collection, and (ii) fast and highly accurate data analysis for damage assessment. In the training phase, the performance of DL models highly depends on the volume of the dataset. Low dataset volume may mislead the DL classifier and result in an erroneous prediction. For DL models, the amount of useful feature information that can be extracted during the training phase will depend on the number of neural network layers and the number of features. This will affect how well the model works when making predictions. The incorporation of additional

Table 2 Number of labeled images for each dataset.

Classes	California wildfires	Sri Lanka floods	Hurricane Harvey	Hurricane Irma	Mexico earthquake	Iraq-Iran earthquake	Hurricane Maria
Severe damage	465	60	556	316	148	158	509
Mild damage	51	30	220	229	25	11	273
Little or no damage	15	5	116	250	5	4	80
Cannot judge	14	1	22	19	4	0	41
NaN (not related)	1044	926	3520	3690	1198	424	3653
Total	1589	1022	4434	4504	1380	597	4556

Table 3 Number of labeled images for GAN dataset.

Damage target	Damage level		
	Severe damage	**Mild damage**	**Little or no damage**
Buildings	872	578	525
People	177	0	0
Vehicles	99	0	0
Others	889	0	0
Total	1703	578	525

a) Fire b) Mild Flood c) Severe Flood

FIG. 3

Examples of GAN-generated images.

neural network layers and features in the model can facilitate the extraction of additional information during the learning phase, which, in turn, improves the accuracy of the prediction.

Based on those two indicators, we present a new framework that combines GAN with DL-based IoT-SE in order to improve the efficiency of damage assessment, while working within constrained network environments. We assume that under constrained network environments, an insufficient amount of data can be collected for disaster damage assessment purposes. For the purposes of our case study, we have assumed that only 60% of a disaster event dataset can be publicly collected. Using a reclassifying dataset from Table 3, the GAN model can be well trained from scratch, and more data can be generated to fill the volume of the dataset. Further, a portion of the dataset generated by GAN and a portion of the real-world dataset collected from IoT-SE are combined and fed into DL models for damage assessment. We use the two types of datasets for model training: (i) *Baseline Dataset:* we use 60% real samples from the Hurricane Maria dataset to evaluate how well the DL model performs. (ii) *GAN-assisted Dataset:* we use GAN to generate data to evaluate the performance of DL models to simulate the scenarios where there are insufficient data samples. The size of the training datasets remain constant. Initially, there is no data generated by

GAN. We then gradually increase the generated dataset from 10% of the training set to 90% of the training set (in an interval of 10%). In this way, we can validate the impact of the amounts of generated data in the training dataset on the learning model.

4 Performance evaluation

We conduct extensive performance evaluations to validate the performance of the GAN-assisted IoT-SE damage assessment approach. The performances of the baseline and GAN-filled datasets on four DL models (i.e., VGG-19, ResNet50, GoogleNet, and MobileNetV2) are compared. In the following, we first introduce our evaluation methodology, and then present the evaluation results.

4.1 Methodology

The GAN-assisted IoT-SE uses PyTorch [48] as the main platform to be leveraged for carrying out experiments. The following subsection introduces dataset preprocessing, learning schemes, learning settings, and key performance indicators.

(1) *Data Preprocessing:* We preprocess the image data by resizing the images into 256×256 size and then randomly rotating the images 30 degrees. After that, the image is randomly central cropped into 224×224. All images are normalized with the suggested mean value and standard deviation by PyTorch [49]. Additionally, we perform multiple resizing of images to fit the input of different CNN models.

(2) *Learning Schemes:* Fast-GAN [44] is employed as the GAN model and the dataset from Table 3 is fed into the model and trained from scratch. In addition, four pretrained CNN based models—VGG-19 [34], ResNet50 [35], GoogleNet [36], and MobileNetV2 [37]—are utilized in the learning phase. Those models are well trained with ImageNet [50] that contains 10 million images over a thousand categories. The last layer of those models is retrained with our data to improve their applicability for our case.

(3) *Learning Settings:* We conduct multiple rounds of training by using different sized datasets (from 60,000 to 100,000), increasing by 10,000 each round. The image output size is 512×512. An NVIDIA Geforce RTX 2080 Ti is used as our computing resource for training. The environment is set with a batch size of 64, momentum of 0.9, and a learning rate of 10^{-3}. Only 25 epochs are provided to better simulate disaster response scenarios. Then, each dataset is partitioned into two subsets: training set (60%) and test set (40%).

(4) *Performance Metrics:* For the GAN Data Generation, we adopt Fréchet inception distance (FID), which can be used to compare the similarity between generated data and real data. The smaller the FID, the more efficient the GAN. To evaluate DL models for damage assessment, we evaluate the performance of the defined scenarios and consider the following four metrics: (i) *Best Training*

Accuracy: Each training session obtains a recording of the best training accuracy (i.e., classification accuracy), and updates the current learning weight. (ii) *Average Testing Accuracy:* The test session obtains the average results based on the whole testing results to measure the overall accuracy. (iii) *Average Precision and Average Recall in Testing:* The test session keeps the test with precision and recall each time, obtaining an average value of all testing results. It is worth noting that the precision represents the false positive rate of the DL model, while the recall represents the true positive rate of the DL model. (iv) *Macro F1-Score:* Since this is a multiclassifying problem, the macro F1 score would be a fitting metric to determine performance, based on the previous two metrics in testing. (v) *Confusion Matrix:* It computes the precision and recall for each class, as well as the true positive rate and false positive rate. The matrix consists of rows and columns that represent the true labels and predicted labels, respectively.

4.2 Results

We now provide the experimental results and evaluate the efficacy of the proposed approach in damage assessment.

(1) *GAN Data Generation:* As seen in Table 4, with five distinct iterations, each GAN model exhibits distinct performance. Those highlighted numbers in the table indicate the minimum score in each GAN model. We found that 100,000 iterations for the severe damage GAN model, 100,000 iterations for the mild damage GAN model, and 90,000 iterations for the little or no damage GAN model are the best iteration setting. Thus these three models are employed in our evaluation to generate more new data for further damage assessment.

(2) *DL Models for Damage Assessment:* For the VGG-19 model result shown in Fig. 4, the x-axis and y-axis represent the ratio of the GAN dataset in the entire training dataset and the accuracy in the test, respectively. It shows that no GAN dataset setting can outperform the baseline for VGG-19, and it keeps closing to the baseline. The best performance can be achieved when the GAN dataset takes a 90% proportion of the entire dataset. Table 5 demonstrates that as the proportion of GAN-generated data in the dataset increases, the F1 score is

Table 4 FID score in GAN models with multiple iterations.

GAN-generated samples	Iteration for training				
	60,000	70,000	80,000	90,000	100,000
Severe damage	84.1576	83.097	83.6316	81.3254	**80.9208**
Mild damage	98.8591	99.6222	96.1093	97.3648	**95.8899**
Little or no damage	107.9931	109.1285	108.1482	**105.56**	107.6415

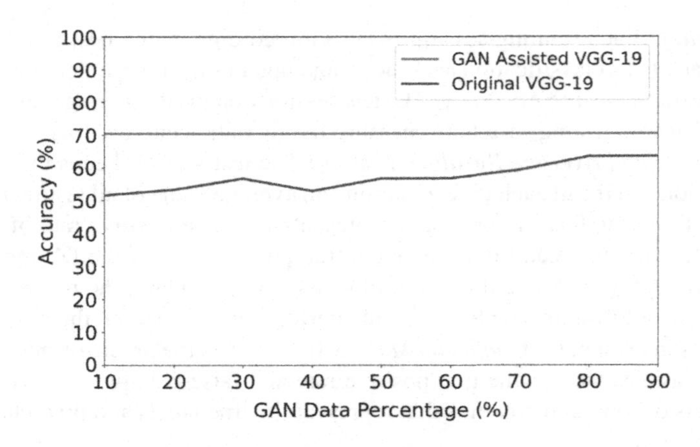

FIG. 4

The accuracy in testing phase for VGG-19.

relatively stable, confirming that VGG-19 maintains the same learning performance throughout the session training period. We found out that the Best Training Accuracy is higher than the Accuracy in Test over all cases. Due to the fact that the content of created data is less rich than that of actual data, less information can be extracted from generated data than from real-world data.

The result of the ResNet50 model is shown in Fig. 5, which has a similar trend to the VGG-19 model result. As a heavy structure of models, ResNet50 learns more features than VGG-19, and the best performance can be reached when the GAN dataset takes a 70% proportion of the entire dataset. Table 6 shows that the F1 score remains consistent with the increment of the ratio of the GAN dataset in the entire dataset, and it validates that RestNet50 has a steady learning performance. The best training accuracy for the ResNet50 model is averagely higher than that of the VGG-19 model in all cases, which indicates that, as a DL model with a more complex structure than VGG-19, ResNet50 has superior learning capacity.

The result of the GoogleNet model is shown in Fig. 6. We can see from the figure that by increasing the volume of GAN-generated data in the training dataset, the Accuracy in Test dramatically changes. In most cases, it is significantly below the baseline. The F1 score in Table 7 unpredictably changes with an increment of GAN-generated data in the training dataset. The best training accuracy fluctuates at random. As a lightweight DL model, GoogleNet has a weaker learning performance when its dataset contains limited information.

Fig. 7 demonstrates that in every case, the MobileNet model Accuracy in Test is significantly lower than the baseline. Even if it reaches the best performance, in which the entire dataset contains 50% GAN-generated data, the performance of the baseline still outperforms its performance by roughly 20%. The performance of the MobileNet model is significantly worse than that of GoogleNet due to its

Table 5 Result for VGG-19.

Metrics	Percentage of GAN dataset in entire dataset								
	10%	20%	30%	40%	50%	60%	70%	80%	90%
Best training accuracy	0.765222	0.71534	0.748766	0.752057	0.729567	0.661547	0.697202	0.763028	0.72463
Precision in test	28.09881	28.45522	29.49433	27.5358	28.65822	27.17934	26.70398	26.23061	25.5408
Recall in test	28.37855	28.57348	29.75313	27.78163	29.57069	28.65095	28.05857	27.81486	27.3116
Accuracy in test	51.61817	53.3349	56.7525	52.9478	56.7213	56.81926	59.31312	63.43473	63.49554
F1 score in test	27.7699	27.9816	29.3639	26.6482	28.34158	27.05622	26.18505	26.04888	25.06815

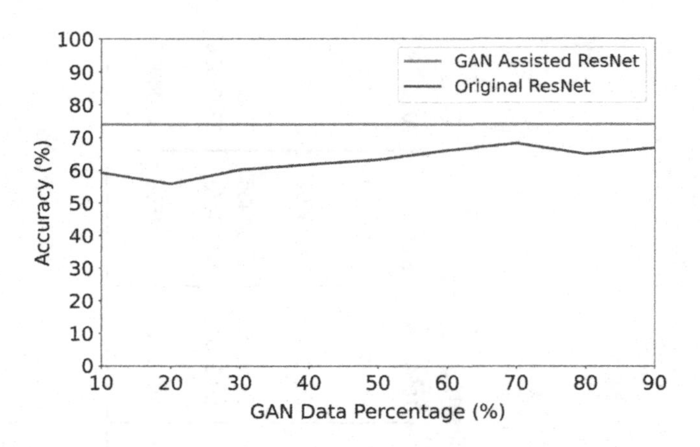

FIG. 5

The accuracy in testing phase for ResNet50.

lighter structure, which results in fewer learning outcomes. The MobileNet model F1 scores shown in Table 8 decline as the size of the GAN dataset grows, indicating that its lightweight model cannot perform well with less information.

Based on the preceding discussion, we now investigate the average worst performance model (MobileNet) and average best performance model (ResNet50) for each label learning performance in depth. Upon that, we chose three data points from each model (when the GAN dataset takes the proportion of 10%, 50%, and 90% of the entire training dataset) to compare the performance of those two models.

Fig. 8 indicates the confusion matrix of the average worst performance model (MobileNet). When the increased GAN dataset is 10%, the NaN labeled data only achieves 14% accuracy. Even with more GAN datasets fed into training (taking 90% of the training dataset), the accuracy is only 34%. Since the majority of the data in the dataset is labeled as NaN, the model learning performance has been significantly hindered, resulting in poor learning performance overall. As for the severe damage label performance, it remains consistent at approximately 30% average. Overall, we believe that this lightest-weight model cannot effectively process the heavy portion of specific labels well.

The average best performance model (ResNet50) confusion matrix is shown in Fig. 9. The severe damage labeled data learning maintains a similar score as that of MobileNet, which is around 30% accuracy. For the NaN labeled data learning performance, the performance improves dramatically as additional GAN datasets are added, increasing from 17% to 34%. This is because the complex structure of the ResNet50 model makes itself more robust to handle the large amount of NaN labeled data well.

Table 6 Result for ResNet50.

Metrics	Ratio of GAN dataset in entire dataset								
	10%	**20%**	**30%**	**40%**	**50%**	**60%**	**70%**	**80%**	**90%**
Best training accuracy	0.81898	0.733406	0.778387	0.7674	0.779484	0.80362	0.80372	0.73944	0.74769
Precision in test	31.5761	28.0694	30.2065	30.64019	31.14311	32.28136	31.76283	26.91232	27.41319
Recall in test	30.64473	27.95755	30.301	31.11666	31.32401	31.16874	30.63207	27.4778	28.551
Accuracy in test	59.24181	55.8053	60.1428	61.73494	63.15096	66.02513	68.20316	64.8827	66.7194
F1 score in test	30.82966	27.4685	29.9565	30.56072	30.91494	31.399	30.82663	25.78885	26.10381

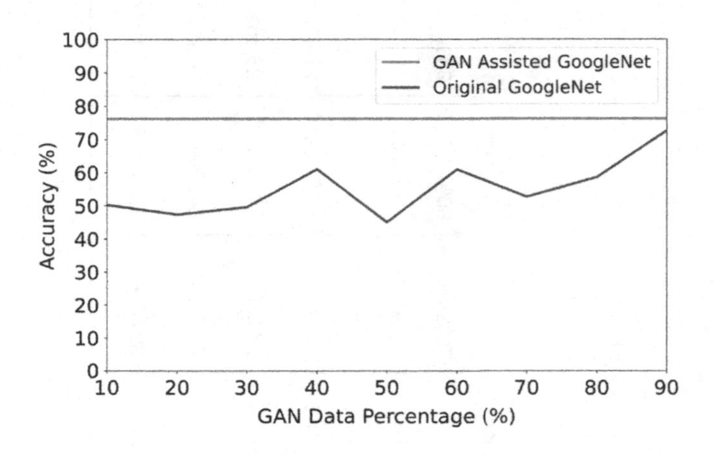

FIG. 6

The accuracy in testing phase for GoogleNet.

5 Discussion

Performance and security issues are two important issues for GAN-assisted DL-based IoT-SE for damage assessment.

Performance Issue: Through the performance evaluation, we can see that the majority of the models are close to the upper bound, while others are far behind. However, their training performance is always superior to the test performance. There are two possible strategies available for addressing the issue: (i) improving dataset size and quality from GAN training, and/or (ii) refining the four deep learning models, and including some new designs and features to be able to learn more effectively from datasets having limited information.

Security Issue: We discovered, based on the declining performance of GoogleNet and MobileNet, that GAN-generated datasets drag down their learning performance, which can be used as an attack technique to poison model performance. To account for this, these lightweight models should introduce new features, such as the addition of a small batch of feeding datasets and the preservation of a specific ratio of sources as its dataset. In this situation, if an attack occurs, the system can quickly perform a traceback, identify the attack source, and block the attack source by adding it to a blacklist.

Table 7 Result for GoogleNet.

	Percentage of GAN dataset in whole dataset								
Metrics	**10%**	**20%**	**30%**	**40%**	**50%**	**60%**	**70%**	**80%**	**90%**
Best training accuracy	0.570488	0.394953	0.518925	0.73944	0.291278	0.6029	0.50576	0.659901	0.800329
Precision in test	26.27479	24.69629	24.64	27.72165	23.1001	25.62666	23.95132	22.68536	27.88339
Recall in test	26.32483	25.19542	25.1737	28.11809	24.34647	26.627	25.58244	24.95939	26.88051
Accuracy in test	50.25359	47.34444	49.5893	61.02792	45.02398	60.9729	52.66306	58.59506	72.61544
F1 score in test	25.50461	23.74356	24.0181	27.29889	22.2371	25.03	23.34942	22.09126	26.93732

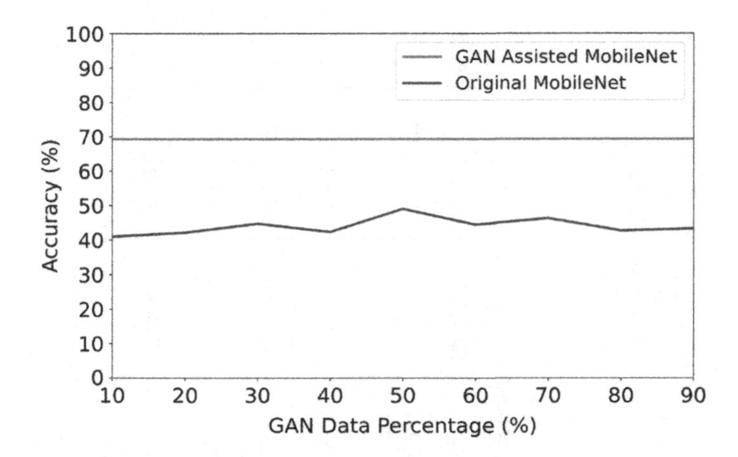

FIG. 7

The accuracy in testing phase for MobileNet.

6 Conclusion

This chapter focuses on addressing the issue of inadequate data volume due to constrained networks in DL-based IoT-SE damage assessment. The chapter considers insufficient data volume collection in disaster damage assessment scenarios. The performances of the baseline and different sizes of GAN-filled datasets on four DL models (VGG-19, ResNet50, GoogleNet, and MobileNetV2) are evaluated. The evaluation results have shown that complex architecture models like ResNet50 benefited from GAN-generated datasets, while lightweight models such as GoogleNet and MobileNetV2 faced performance degradation.

The observed gap between training and testing performance suggests the need for advanced GAN model training methods to improve dataset quality. It also calls for revisions in DL model architectures to adapt to datasets with limited information. Additionally, the vulnerability of lightweight models to potential adversarial attacks highlights the need for robust security measures. These measures could include diverse data sourcing, batched data ingestion, and rapid threat detection mechanisms. The GAN model can potentially handle data-limited scenarios in IoT-SE damage assessment. Nonetheless, their integration requires careful optimization for performance while safeguarding against security threats.

Table 8 Result for MobileNet.

Metrics	Ratio of GAN dataset in whole dataset								
	10%	**20%**	**30%**	**40%**	**50%**	**60%**	**70%**	**80%**	**90%**
Best training accuracy	0.403182	0.568843	0.611081	0.244652	0.710916	0.402633	0.665935	0.312671	0.282501
Precision in test	23.14808	24.8879	23.286	24.59494	23.74079	23.03533	23.15062	22.74431	20.68623
Recall in test	22.8703	23.73205	24.51	24.94266	24.72727	23.9363	23.61364	21.56391	20.30712
Accuracy in test	40.9555	42.11925	44.6974	42.3096	48.96453	44.3944	46.2534	42.6617	43.25864
F1 score in test	21.5753	22.72029	22.6167	22.98535	23.02785	21.51379	21.05	19.36927	17.67289

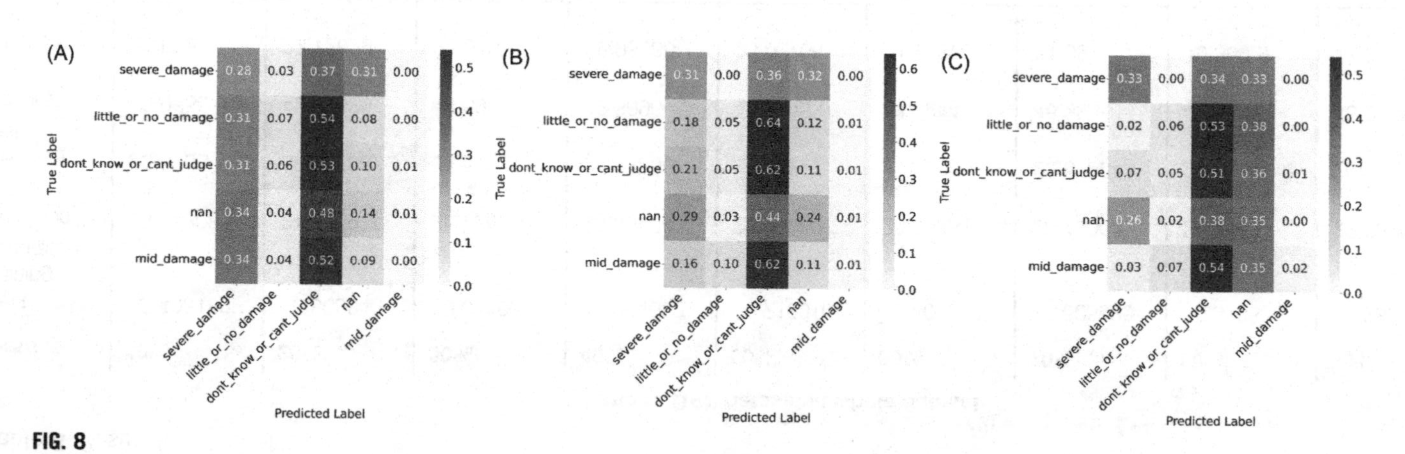

FIG. 8

MobileNet confusion matrix. (A) 10% added GAN dataset. (B) 50% added GAN dataset. (C) 90% added GAN dataset.

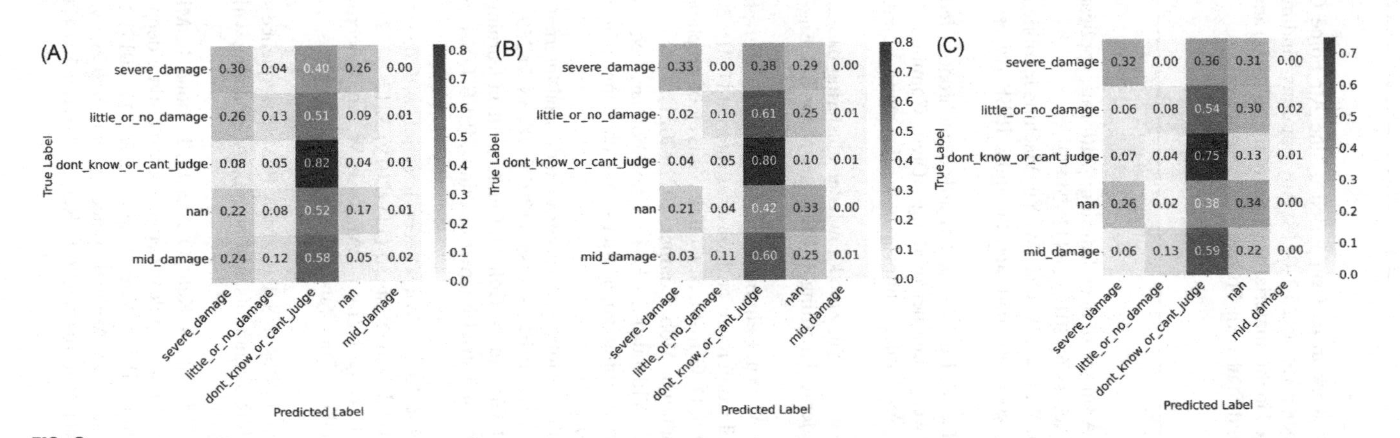

FIG. 9

ResNet50 confusion matrix. (A) 10% added GAN dataset. (B) 50% added GAN dataset. (C) 90% added GAN dataset.

Acknowledgment

This material is based upon work in part supported by the Air Force Office of Scientific Research under award number FA9550-20-1-0418. Any opinions, findings, and conclusions or recommendations expressed in this material are those of the author(s) and do not necessarily reflect the views of the United States Air Force.

References

[1] W.G. Hatcher, W. Yu, A survey of deep learning: platforms, applications and emerging research trends, IEEE Access 6 (2018) 24411–24432.

[2] F. Liang, W.G. Hatcher, W. Liao, W. Gao, W. Yu, Machine learning for security and the Internet of Things: the good, the bad, and the ugly, IEEE Access 7 (2019) 158126–158147.

[3] Z. Chen, W. Liao, K. Hua, C. Lu, W. Yu, Towards asynchronous federated learning for heterogeneous edge-powered internet of things, Digit. Commun. Netw. 7 (3) (2021) 317–326. [Online]. Available: https://www.sciencedirect.com/science/article/pii/S2352864821000195.

[4] Z. Chen, P. Tian, W. Liao, W. Yu, Zero knowledge clustering based adversarial mitigation in heterogeneous federated learning, IEEE Trans. Netw. Sci. Eng. 8 (2) (2021) 1070–1083.

[5] X. Liu, H. Xu, W. Liao, W. Yu, Reinforcement learning for cyber-physical systems, in: 2019 IEEE International Conference on Industrial Internet (ICII), 2019, pp. 318–327.

[6] F. Liang, C. Qian, W. Yu, D. Griffith, N. Golmie, Survey of graph neural networks and applications, Wirel. Commun. Mob. Comput. 2022 (2021).

[7] M.M. Najafabadi, F. Villanustre, T.M. Khoshgoftaar, N. Seliya, R. Wald, E. Muharemagic, Deep learning applications and challenges in big data analytics, J. Big Data 2 (1) (2015) 1–21.

[8] H. Xu, W. Yu, D. Griffith, N. Golmie, A survey on industrial Internet of Things: a cyber-physical systems perspective, IEEE Access 6 (2018) 78238–78259.

[9] H. Xu, X. Liu, W. Yu, D. Griffith, N. Golmie, Reinforcement learning-based control and networking co-design for industrial internet of things, IEEE J. Sel. Areas Commun. 38 (5) (2020) 885–898.

[10] Z.-Q. Zhao, P. Zheng, S.-t. Xu, X. Wu, Object detection with deep learning: a review, IEEE Trans. Neural Netw. Learn. Syst. 30 (11) (2019) 3212–3232.

[11] Q. Fang, H. Li, X. Luo, L. Ding, H. Luo, T.M. Rose, W. An, Detecting non-hardhat-use by a deep learning method from far-field surveillance videos, Autom. Constr. 85 (2018) 1–9.

[12] E. Bramucci, A. Paiardini, F. Bossa, S. Pascarella, PyMod: sequence similarity searches, multiple sequence-structure alignments, and homology modeling within PyMOL, BMC Bioinform. 13 (4) (2012) 1–6.

[13] F. Tang, B. Mao, Z.M. Fadlullah, N. Kato, O. Akashi, T. Inoue, K. Mizutani, On removing routing protocol from future wireless networks: a real-time deep learning approach for intelligent traffic control, IEEE Wirel. Commun. 25 (1) (2017) 154–160.

[14] H. Zhao, H. Liu, W. Hu, X. Yan, Anomaly detection and fault analysis of wind turbine components based on deep learning network, Renew. Energy 127 (2018) 825–834.

[15] J. Booz, J. McGiff, W.G. Hatcher, W. Yu, J. Nguyen, C. Lu, Tuning deep learning performance for android malware detection, in: 2018 19th IEEE/ACIS International Conference on Software Engineering, Artificial Intelligence, Networking and Parallel/Distributed Computing (SNPD), 2018, pp. 140–145.
[16] P. Tian, Z. Chen, W. Yu, W. Liao, Towards asynchronous federated learning based threat detection: a DC-Adam approach, Comput. Secur. 108 (2021) 102344. [Online]. Available: https://www.sciencedirect.com/science/article/pii/S0167404821001681.
[17] Q. Mao, F. Hu, Q. Hao, Deep learning for intelligent wireless networks: a comprehensive survey, IEEE Commun. Surv. Tutor. 20 (4) (2018) 2595–2621.
[18] Z. Yan, H. Zhang, B. Wang, S. Paris, Y. Yu, Automatic photo adjustment using deep neural networks, ACM Trans. Graph. 35 (2) (2016) 1–15.
[19] F. Liang, W.G. Hatcher, G. Xu, J. Nguyen, W. Liao, W. Yu, Towards online deep learning-based energy forecasting, in: 2019 28th International Conference on Computer Communication and Networks (ICCCN), 2019, pp. 1–9.
[20] J. Booz, W. Yu, G. Xu, D. Griffith, N. Golmie, A deep learning-based weather forecast system for data volume and recency analysis, in: 2019 International Conference on Computing, Networking and Communications (ICNC), 2019, pp. 697–701.
[21] A. Garcia-Garcia, S. Orts-Escolano, S. Oprea, V. Villena-Martinez, P. Martinez-Gonzalez, J. Garcia-Rodriguez, A survey on deep learning techniques for image and video semantic segmentation, Appl. Soft Comput. 70 (2018) 41–65.
[22] W. Yu, H. Xu, J. Nguyen, E. Blasch, A. Hematian, W. Gao, Survey of public safety communications: user-side and network-side solutions and future directions, IEEE Access 6 (2018) 70397–70425.
[23] H. Liang, L. Burgess, W. Liao, E. Blasch, W. Yu, Deep learning assist IoT search engine for disaster damage assessment, Cyber-Phys. Syst. (2022) 1–25.
[24] X.-W. Chen, X. Lin, Big data deep learning: challenges and perspectives, IEEE Access 2 (2014) 514–525.
[25] W.G. Hatcher, C. Qian, W. Gao, F. Liang, K. Hua, W. Yu, Towards efficient and intelligent Internet of Things search engine, IEEE Access 9 (2021) 15778–15795.
[26] F. Liang, C. Qian, W.G. Hatcher, W. Yu, Search engine for the internet of things: lessons from web search, vision, and opportunities, IEEE Access 7 (2019) 104673–104691.
[27] C. Qian, W. Yu, C. Lu, D. Griffith, N. Golmie, Toward generative adversarial networks for the industrial Internet of Things, IEEE Internet Things J. 9 (19) (2022) 19147–19159.
[28] M. Imran, F. Alam, U. Qazi, S. Peterson, F. Ofli, Rapid Damage Assessment Using Social Media Images by Combining Human and Machine Intelligence, arXiv preprint arXiv:2004.06675, 2020.
[29] D.T. Nguyen, F. Ofli, M. Imran, P. Mitra, Damage assessment from social media imagery data during disasters, in: Proceedings of the 2017 IEEE/ACM International Conference on Advances in Social Networks Analysis and Mining 2017, 2017, pp. 569–576.
[30] Z.M. Hamdi, M. Brandmeier, C. Straub, Forest damage assessment using deep learning on high resolution remote sensing data, Remote Sens. 11 (17) (2019) 1976.
[31] H. Liang, L. Burgess, W. Liao, C. Lu, W. Yu, On detecting interest flooding attacks in named data networking (NDN) based IoT search, AI, Machine Learning and Deep Learning: A Security Perspective, CRC Press, 2022, p. 16. In press.
[32] G. Cheng, Z. Li, J. Han, X. Yao, L. Guo, Exploring hierarchical convolutional features for hyperspectral image classification, IEEE Trans. Geosci. Remote Sens. 56 (11) (2018) 6712–6722.

[33] F. Liang, W. Yu, X. Liu, D. Griffith, N. Golmie, Toward edge-based deep learning in industrial internet of things, IEEE Internet Things J. 7 (5) (2020) 4329–4341.

[34] K. Simonyan, A. Zisserman, Very Deep Convolutional Networks for Large-Scale Image Recognition, arXiv preprint arXiv:1409.1556, 2014.

[35] K. He, X. Zhang, S. Ren, J. Sun, Deep residual learning for image recognition, in: Proceedings of the IEEE Conference on Computer Vision and Pattern Recognition, 2016, pp. 770–778.

[36] C. Szegedy, W. Liu, Y. Jia, P. Sermanet, S. Reed, D. Anguelov, D. Erhan, V. Vanhoucke, A. Rabinovich, Going deeper with convolutions, in: Proceedings of the IEEE Conference on Computer Vision and Pattern Recognition, 2015, pp. 1–9.

[37] A.G. Howard, M. Zhu, B. Chen, D. Kalenichenko, W. Wang, T. Weyand, M. Andreetto, H. Adam, MobileNets: Efficient Convolutional Neural Networks for Mobile Vision Applications, arXiv preprint arXiv:1704.04861, 2017.

[38] I. Goodfellow, J. Pouget-Abadie, M. Mirza, B. Xu, D. Warde-Farley, S. Ozair, A. Courville, Y. Bengio, Generative adversarial nets, Adv. Neural Inf. Process. Syst. 27 (2014).

[39] Y. Jin, J. Zhang, M. Li, Y. Tian, H. Zhu, Z. Fang, Towards the Automatic Anime Characters Creation With Generative Adversarial Networks, arXiv preprint arXiv:1708.05509, 2017.

[40] L. Ma, X. Jia, Q. Sun, B. Schiele, T. Tuytelaars, L. Van Gool, Pose guided person image generation, Adv. Neural Inf. Process. Syst. 30 (2017).

[41] A. Bulat, J. Yang, G. Tzimiropoulos, To learn image super-resolution, use a GAN to learn how to do image degradation first, in: Proceedings of the European Conference on Computer Vision (ECCV), 2018, pp. 185–200.

[42] A. Cheng, PAC-GAN: packet generation of network traffic using generative adversarial networks, in: 2019 IEEE 10th Annual Information Technology, Electronics and Mobile Communication Conference (IEMCON), IEEE, 2019, pp. 0728–0734.

[43] W. Xia, Y. Zhang, Y. Yang, J.-H. Xue, B. Zhou, M.-H. Yang, Gan inversion: a survey, IEEE Trans. Pattern Anal. Mach. Intell. 45 (2022) 3121–3138.

[44] B. Liu, Y. Zhu, K. Song, A. Elgammal, Towards faster and stabilized GAN training for high-fidelity few-shot image synthesis, in: International Conference on Learning Representations, 2020.

[45] C. Qian, W. Gao, W.G. Hatcher, W. Liao, C. Lu, W. Yu, Search engine for heterogeneous internet of things systems and optimization, in: 2020 IEEE Intl Conf on Dependable, Autonomic and Secure Computing, Intl Conf on Pervasive Intelligence and Computing, Intl Conf on Cloud and Big Data Computing, Intl Conf on Cyber Science and Technology Congress (DASC/PiCom/CBDCom/CyberSciTech), IEEE, 2020, pp. 475–482.

[46] Artificial Intelligence for Disaster Response (AIDR). [Online]. Available: http://aidr.qcri.org/.

[47] F. Alam, F. Ofli, M. Imran, CrisisMMD: multimodal twitter datasets from natural disasters, in: Proceedings of the 12th International AAAI Conference on Web and Social Media (ICWSM), June 2018.

[48] A. Paszke, S. Gross, F. Massa, A. Lerer, J. Bradbury, G. Chanan, T. Killeen, Z. Lin, N. Gimelshein, L. Antiga, A. Desmaison, A. Kopf, E. Yang, Z. DeVito, M. Raison, A. Tejani, S. Chilamkurthy, B. Steiner, L. Fang, J. Bai, S. Chintala, PyTorch: an imperative style, high-performance deep learning library, in: H. Wallach, H. Larochelle, A. Beygelzimer, F. d'Alché-Buc, E. Fox, R. Garnett (Eds.), Advances in Neural

Information Processing Systems 32, Curran Associates, Inc., 2019, pp. 8024–8035. [Online]. Available: http://papers.neurips.cc/paper/9015-pytorch-an-imperative-style-high-performance-deep-learning-library.pdf.
[49] Welcome to PyTorch Tutorials—PyTorch Tutorials 1.8.1+cu102 Documentation. [Online]. Available: https://pytorch.org/tutorials/.
[50] J. Deng, W. Dong, R. Socher, L.-J. Li, K. Li, L. Fei-Fei, ImageNet: a large-scale hierarchical image database, in: 2009 IEEE Conference on Computer Vision and Pattern Recognition, IEEE, 2009, pp. 248–255.

CHAPTER

19

Design and development of enhanced algorithm for the demarcation of ocean wakes from SAR imagery of rough sea conditions

P. Subashini[a], P.V. Hareesh Kumar[b], S. Lekshmi[b], M. Krishnaveni[a], and T.T. Dhivyaprabha[c]

[a]*Department of Computer Science, Centre for Machine Learning and Intelligence, Avinashilingam Institute for Home Science and Higher Education for Women, Coimbatore, India,* [b]*Naval Physical & Oceanographic Laboratory, Defence Research and Development Organization, Kochi, India,* [c]*Centre for Machine Learning and Intelligence, Avinashilingam Institute for Home Science and Higher Education for Women, Coimbatore, India*

1 Introduction

Synthetic aperture radar (SAR) is a type of sensor used to observe and characterize the surface of the earth. SAR sensors have a number of benefits, including the capacity to capture images with a high spatial resolution, the ability to observe in both daylight and darkness, and all-weather capability. One of the most crucial applications for environmental monitoring and surveillance is the recognition of ship wakes in SAR images. Because SAR has the ability to provide ship wake detection at high resolution over wide bands in choppy sea conditions, it is used to overcome the limitations of other surveillance systems, such as coastal zone monitoring, agricultural monitoring, climate monitoring, tropical forest monitoring, and so forth. Long streaks are frequently visible in SAR photos of ship wakes taken by an aerial radar system; these linear traces appear in high-resolution SAR photos when there is a ship wake. The turbulent wake's distinctive central black line, which is parallel to the ship's longitudinal axis, may be seen. Within a half-angle spanning from 1.5 to 4 degrees with regard to the turbulent dark line, two bright linear features (narrow-V wakes) can be seen. The Kelvin arms, which form at an angle with an aperture of 19.50 with regard to the turbulent dark line, are represented by the most distant couple of wakes. The various linear characteristics in SAR images are shown

Smart Spaces. https://doi.org/10.1016/B978-0-443-13462-3.00007-8

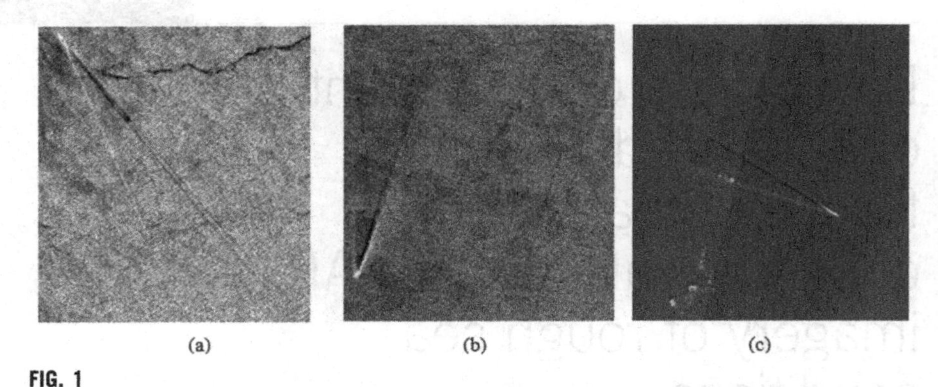

FIG. 1

Real-time ship-wake SAR images. (A) turbulent wake, (B) V-wake, (C) Kelvin wake.

in Fig. 1. Internal wave wakes, an alternating dark and light wave pattern roughly parallel to the ship track, are a rare sight at best. However, the persistence of the longer L-band Bragg waves in the wake detection process is more pronounced in L-band than in X-band or C-band. With regard to ship tracks, these streaks are attributed to turbulent wake, narrow V-wake, and Kelvin wake. In the rough sea surface of the actual world maritime environment, wind-generated waves and ship wakes combine. The presence of oil spills, marine currents, undersea terrain, icebergs, ambiguities, oceanic waves, and oceanic internal waves also makes it more difficult to discern ship wakes beneath a rough sea surface in SAR images. With spaceborne radar, the ocean surface may be discernible, but these linear structures are incorrectly recognized as wakes. The propagation of radar energy and the consequent radar coverage are greatly impacted by the numerous atmospheric effects, such as strong winds, weather, oceanic currents, and waves, that create a rough sea surface. It is believed to be a very difficult task and a never-ending challenge to develop an algorithm to detect ship wakes in SAR images in rough sea conditions because of the coherent emission of backscattering waves and the presence of multiplicative speckle noise in the SAR image.

Studies of the ship wakes are very helpful in predicting dangerous maritime storms and in weather forecasting. As a result, the detection of ship wakes and estimates of their parameters have become highly active study areas. In this study, a stated path has been proposed to create a very effective algorithm to find ship wakes in unknown sea patterns. This section explains SAR images in shifting sea conditions and their parameters.

1.1 Synthetic aperture radar (SAR)

SAR, or coherent airborne or spaceborne radar, is a technology that produces high-resolution images for remote sensing by using the flight path of the platform to imitate an incredibly large antenna or aperture. The radar antenna successively collects backscattered echoes after transmitting electromagnetic waves across the target area. Images of these echoes are saved for later use. The SAR image is often shown as intensity values, where each pixel represents the reflectance of the associated spot on the ground. It is well known that SAR's travel distance over the target region during the period required for the radar pulses to return to the antenna results in high-resolution synthetic images [1]. Range resolution and azimuth resolution are two different subcategories of SAR picture resolution. The bandwidth of the transmitted pulse is the primary constraint on the range resolution of a pulsed radar system. A pulse with a brief duration might have a broad bandwidth. Unfortunately, the transmitted energy and radiometric resolution decrease with pulse length. SAR systems generate a lengthy pulse with a linear frequency modulation to maintain the radiometric resolution. The range resolution is enhanced once the received signal has been reduced without sacrificing radiometric resolution. By using the same pulse compression approach used for range direction, SAR artificially improves the antenna's size to improve azimuth resolution in comparison to range resolution. Synthetic aperture processing is a challenging data-processing method using a tiny antenna to gather signals and phases from moving targets. Synthetic aperture length is the phrase used to describe the conversion of the effects of signals to the effects of a big antenna. It is possible to project the range of beam width in the azimuth direction for a radiometric resolution of the same length. Tables 1 and 2 list the features of SAR images taken with other earth observational devices.

1.2 Characteristics of SAR images

Synthetic aperture radar (SAR) sends microwave signals initially, after which it receives the signals that are reflected or backscattered from the earth. It calculates the separation between the sensor and the location on the surface of the earth from

Table 1 Description of SAR imagery.

Characteristics	Lidar	Optical multispectral	SAR
Platform	Spaceborne	Airborne	Airborne/ spaceborne
Radiation	Own radiation	Reflected sunlight	Own radiation
Spectrum	Infrared	Visible/infrared	Microwave
Frequency	Single frequency	Multifrequency	Multifrequency
Polarimetry	Not applicable (NA)	NA	Polarimetric phase
Interferometry	NA	NA	Interferometric phase
Acquisition time	Day/night	Day time	Day/night
Weather	Blocked by clouds	Blocked by clouds	Visible through clouds

Table 2 SAR vs other earth observation instruments.

Wavelength	Instrument	Satellite
P-band = ~65 cm	Airborne	AIRSAR
L-band = ~23 cm	Spaceborne	JERS-1 SAR, ALOS PALSAR
S-band = ~10 cm	Spaceborne	Almaz-1
C-band = ~5 cm	Spaceborne	ERS-1/2 SAR, RADARSAT-1/2, ENVISAT ASAR
X-band = ~3 cm	Spaceborne	TerraSAR-X, COSMO-SkyMed
K-band = ~1.2 cm	Airborne	Military domain

which the signal is backscattered. This distance can be projected onto the ground, which is represented as slant range. The direction of the flight is also known as the along-track or azimuth direction, while the across-track or range direction is the direction that is perpendicular to the flight path. The incidence angle is the angle formed by the center of the radar beam and the normal to the topography of the area. Range resolution and azimuth resolution are two different subcategories of SAR picture resolution. The bandwidth of the transmitted pulse is the primary constraint on the range resolution of a pulsed radar system. SAR systems produce a lengthy pulse with a linear frequency modulation to maintain the radiometric resolution. The range resolution is enhanced once the received signal has been reduced without sacrificing radiometric resolution. By using the same pulse compression approach used for range direction, SAR artificially increases the antenna's size to improve azimuth resolution in comparison to range resolution. Synthetic aperture processing is a challenging data-processing method using a tiny antenna to gather signals and phases from moving targets. Synthetic aperture length refers to the process by which signals are transformed into the effect of a big antenna. It is possible to project the range of beam width in the azimuth direction for a radiometric resolution of the same length. The most specific SAR image properties, such as wavelength, polarization, and incidence angle, are also examined. A brief description is provided in the subsections that follow [2].

1.2.1 Wavelength

Radio waves used in radar transmission have a wavelength that is significantly longer than that of visible light. The most important factor in choosing a wavelength is penetration, and the relationship between wavelength and frequency is inverse, with longer wavelengths corresponding to lower frequencies. The wavelengths that are typically employed in SAR images are shown in Table 2.

1.2.2 Polarization

In order to detect and explore the polarimetric features of artificial and natural scatters, SAR polarimetry is a generally used technology for the computation of qualitative and quantitative physical information such as land, snow and ice, ocean, and urban applications [2]. Polarimetry, which employs a scattering matrix to identify an object's behavior using electromagnetic waves, is used in radio wave measurement. It is represented by the intersection of transmitted and received signals' horizontal and vertical polarization states.

$$S = \begin{bmatrix} S_{HH} & S_{HV} \\ S_{VH} & S_{VV} \end{bmatrix} \tag{1}$$

where *HH* is the horizontal convey and horizontal accept; *HV* is the horizontal convey and vertical accept; *VH* is the vertical convey and horizontal accept; *VV* is the vertical convey and vertical accept.

Both horizontal (*H*) and vertical (*V*) electric-field vectors can be transmitted by radar signals, which can also receive both horizontal (*H*) and vertical (*V*) signals. Similar-polarized (*HH* or *VV*) signals are produced by fundamental physical processes called quasispecular surface reflection. For instance, water is quiet (i.e., waveless) that appears black. Because of factors like surface roughness, the cross-polarized (*HV* or *VH*) signals are often weak and frequently coupled with various reflections.

1.2.3 Incidence angle

The angle created by the radar beam and a line perpendicular to the surface is referred to as the incidence angle. When the incidence angle is large, the backscattered waves are powerful, and when it is small, the backscattered waves are faint. Tables 3 and 4 provide information on the polarization of space-based SAR sensors and the frequency bands that SAR systems employ the most frequently.

The incident angle of the radar pulse on the surface determines how microwaves interact with the surface. The incident angle for ERS SAR is 23 degrees at the scene's center. The first spaceborne SAR with several beam modes, RADARSAT is capable of microwave imaging at various incident angles and resolutions. For the ERS SAR, a 23-degree incident angle is ideal for detecting ocean waves and other surface features.

2 Literature survey

Images produced by synthetic aperture radar (SAR) suffer greatly from speckle, a form of multiplicative noise. Due to the surface's roughness, speckle is created by the random interference of backscattered electromagnetic waves. In the majority of identification systems when speckle is present, reduction of the speckle noise is crucial since speckle generally has the property of concealing image details. It typically occurs in RADAR and results from arbitrary fluctuations in the power of backscattered waves from objects. By making it more difficult to identify spatial patterns and limiting the capacity to detect ground targets, speckle may reduce the usefulness of SAR imaging. As a result, it makes automatic picture classification challenging and complicates the interpretation of visual images [3].

Table 3 Overview of spaceborne SAR sensors and their main characteristics.

Sensor	Operation	Frequency band (polarization)	Institution, country
Seasat	1978	L (HH)	NASA/JPL, United States
ERS-1/2	1991–2000/ 1995–2011	C (VV)	ESA, Europe
J-ERS-1	1992–98	L (HH)	JAXA, Japan
SIR-C/X-SAR	April and October 1994	L & C (quad) X (VV)	NASA/JPL, United States DLR, Germany ASI, Italy
Radarsat-1	1995–today	C (HH)	CSA, Canada
SRTM	February 2000	C (HH+VV) and X (VV)	NASA/JPL, United States DLR, Germany ASI, Italy
ENVISAT/ASAR	2002–12	C (dual)	ESA, Europe
ALOS/palSAR	2006–11	L (quad)	JAXA, Japan
TerraSAR-X/ TanDEM-X	2007/2010	X (quad)	DLR/Astrium, Germany
Radarsat-2	2007–today	C (quad)	CSA, Canada
COSMO-SkyMed-1/4	2007–10	X (dual)	ASI/MiD, Italy
RISAT-1	2012	C (quad)	ISRO, India
HJ-1C	2012	S (VV)	CRESDA/CAST/ NRSCC, China
Kompsat-5	Launch scheduled in 2013	X (quad)	KARI, Korea
PAZ	Launch scheduled in 2013	X (quad)	CDTI, Spain
ALOS-2	Launch scheduled in 2013	L (quad)	JAXA, Japan
Sentinel-1a/1b	Launch scheduled in 2013/2015	C (dual)	ESA, Europe
Radarsat Constellation-1/2/3	Launch scheduled in 2017	C (quad)	CSA, Canada
SAOCOM-1/2	Launch scheduled in 2014/2015	L (quad)	CONAE, Argentina

Table 4 Commonly used frequency bands for SAR systems and the corresponding frequency and wavelength ranges.

Frequency band	Ka	Ku	X	C	S	L	P
Frequency (GHz)	40–25	17.6–12	12–7.5	7.5–3.75	3.75–2	2–1	0.5–0.25
Wavelength (cm)	0.75–1.2	1.7–2.5	2.5–4	4–8	8–15	15–30	60–120

During the SAR picture creation process, speckle noise is produced due to coherent radiation. This noise, which is typically characterized as multiplicative noise, has an undesirable impact that lowers the quality of photographs. Moreover, SAR images feature statistical characteristics that mostly result from the multiplicative noise model. Eq. (2) multiplicative noise model can be used to create this image:

$$I(t) = R(t) \cdot v(t) \tag{2}$$

where $I(t)$ is the noisy image, $R(t)$ is the original image or the radar backscatter property of the ground targets without noise, and $v(t)$ is the speckle noise and it is independent of $R(t)$. The mean value of $v(t)$ for SAR speckle, which is caused by a zero-mean random phase of echo signals, is 1, and its variance is relevant with respect to the same number of SAR images. The technique of wake detection and classification is complicated by the presence of speckle noise in SAR images. As a result, during the preprocessing of SAR images, speckle noise must be removed. This section provides a thorough analysis of the research on SAR images and denoising algorithms, including the traditional Lee, Frost, Kuan, top-hat, Wiener, mean, median, adaptive, and gamma algorithms.

The impressive literature culled from a thorough investigation of several noise filtering methods is presented in the following text.

A nonlinear filtering method was created by Mark Schulze and Qing Wu in 1995 to reduce speckle noise in SAR images. This approach outperforms existing filtering methods on both simulated and real-time SAR images, according to experimental data. It was tested with simulated speckle noise introduced in original SAR images collected by the JERS-1 satellite [4].

In order to despeckle SAR images, Nelson Mascarenhas [5] conducted a survey on noise filters. Above Freiburg, Germany, an airborne SAR-580 L-Band, 1-1.11 polarization, one-look, linear detected image was collected for analysis and contrasted with several filters, including Lee, Kuan, Frost, adaptive, geometrical, and MAP filters. The findings showed that enhanced Lee filters performed better for visual interpretation because edges, linear features, point targets, and texture data are preserved. During the segmentation process, geometrical and adaptive filters produced good smoothing capabilities [5].

By adjusting the parameters based on geographical or textural variations in the images, Masayoshi Tsuchida et al. [6] introduced the Kalman filter for the reduction of speckle noise in SAR images. Real-time SAR images acquired from a lake surrounded by mountains for a quantitative assessment of noise reduction and information preservation were used to validate the suggested method. The SNR values of the tested results signified that the introduced method successfully kept the image's details while bringing the noise level down to a manageable level [6].

In order to improve the speckle noise in SAR, RADARSAT, and JERS-1 image data, Fang Qiu et al. [3] developed a local adaptive median filter. A numerical analysis of the results has shown that a local adaptive median filter outperforms other filters for both the best speckle noise suppression and preservation of the information contained in it [7].

For the purpose of cutting down on speckle noise and maintaining the information in SAR images, Josaphat Tetuko et al. [8] presented a speckle filter. The proposed speckle filter was found to perform marginally better than frequently employed filters including Kuan, gamma, enhanced Lee, and enhanced Frost filters when its performance was assessed using PALSAR (new Japanese sensor) and JERS-1 images that contain the northern forest region of Iran [8].

For the purpose of removing speckle noise from SAR ship photos, Shi-qi Huang et al. [9] suggested a unique method that combines the coherence reduction speckle noise (CRSN) algorithm with a coherence constant false-alarm ratio (CCFAR) detection algorithm. When used as a first stage in the ship detection of SAR imaging, the results indicate that it is both practical and efficient [9].

In real SAR imageries with desert and grass regions, Ahmed Mashaly et al. [10] presented an adaptive mathematical morphological filter to remove speckle noise. A number of despeckling filters, including the Lee and enhanced Lee filters, the Frost and enhanced Frost filters, the Kuan filter, and the gamma MAP filter, are compared to the proposed filter. According to the results, the adaptive mathematical morphological filter outperformed all other filtering strategies in terms of target to clutter ratio, normalized mean, standard deviation to mean, and edge index [10].

Level set filter, a type of curvature flow propagation, was created by Hongga Li et al. [11] for the purpose of suppressing speckle noise in ocean and land SAR data. The results show that the suggested approach produces better results when compared to other existing filters, such as the Lee filter, extended Lee filter, Kuan filter, Frost filter, enhanced Frost filter, and gamma filter [11]. The proposed method was tested using simulation images and ERS-2 SAR data.

A study of the effectiveness of speckle noise reduction filters in RADAR images was carried out by Meenakshi and Punitham [12]. Conventional filters including mean, median, Lee-sigma, Local-region, Lee, gamma MAP, Frost, and speckle reducing anisotropic (SRAD) filters were used to validate the effectiveness using simulated and real-time images. According to the empirical findings, the Lee-sigma filter outperformed other statistical filters in terms of preserving image details and reducing speckle noise. In image repositories, the Frost and gamma MAP filters produced the lowest MSE and maximum SNR values, respectively [12].

In order to reduce speckle noise in SAR images, Milindkumar Sarode and Prashant Deshmukh [13] created a revolutionary wavelet-based noise thresholding algorithm. It was evaluated using ultrasound and satellite photos, and PSNR and MSE were used to confirm the filter's effectiveness. The results demonstrated that the suggested filter performs better than the Kuan, Frost, Wiener, VisuShrink, and BayesShrink filters [13].

Three despeckling techniques—recursive filter-based despeckling, model-based despeckling, and anisotropic diffusion—were introduced by Rajeshwari et al. [14] for the purpose of reducing speckle noise in both simulated and real-time SAR images. According to the results of the analysis, anisotropic diffusion was computationally effective for actual SAR images, and the recursive filter produced superior results for both artificial and genuine SAR images [14].

Ji Yuan et al. [15] used a variety of filters, including mean, median, statistics Lee, Kuan, Frost, gamma MAP, and wavelet transform based filters, to reduce the amount of speckle noise in SAR images that interferes with ship detection. Studies show that, compared to other common filters, wavelet analysis-based filters are more effective in achieving the goal of speckle noise filtering [15].

The curvelet transform was proposed by Wenbo Wang et al. [16] to reduce speckle noise in SAR images, and the results of the analysis indicate that the method enhances the visual quality of the images [16].

For the purpose of removing multiplicative noises present in SAR, photographic, ultrasound, PET, CT scan, and MRI imagery, Gopinathan and Poornima [17] constructed a number of noise filters, including Lee, Frost, Kuan, Weiner, median, and SRAD (speckle reducing anisotropic diffusion) filters. The statistical analysis and visual evaluation of the outcomes show that the Lee, Kuan, and SRAD filters offer promising outcomes for CT and photographic pictures. SRAD and median filters produced incredibly high-quality findings for PET, SAR, and ultrasound images. Last but not least, the Wiener filter improves MRI image quality [17].

A hybrid mean-median filter with speckle-reducing anisotropic diffusion (SRAD) was proposed by Masume Rahimi and Mehran Yazdi [18] for despeckling Ku-band (15 GHz) synthetic aperture radar images from the Sandia Twin Otter aircraft, and the results show that the proposed filter achieved significantly improved outcomes over wavelet thresholding methods, anisotropic diffusion, and speckle reducing anisotropic diffusion [18].

Pranali Hatwar and Heena Kher [19] applied the Lee filter, the Frost filter, the Kuan filter, the gamma MAP filter, the enhanced Lee filter, the enhanced Frost filter, and the enhanced Kuan filter for the reduction of speckle noise in SAR texture and land surface pictures. When the effectiveness of the aforementioned noise filters was measured, it became clear that the improved Kuan filter and the Gamma map filter produced better results for the reduction of speckle in SAR images [19].

Kupidura [20] carried out a comparative examination of various filtering techniques for SAR picture speckle noise reduction. The tests were conducted on photos from the RadarSat-2 satellite, and trials were conducted with morphological operations-based filters, such as the simple alternate filter, multiple structuring elements, and multiple structuring element by reconstruction. The findings of the experiment were compiled to show that morphological operations-based filters give outcomes that are significantly more successful than those of other adaptive filters [20].

2.1 Following are the remarkable works in the literature pertaining to wavelet filtering techniques

Standard filters were used by Yonghong and Van Genderen [21] to remove speckle noise from SAR images, including the moving average filter, median filter, Lee filter, enhanced Lee filter, Frost filter, enhanced Frost filter, Kuan filter, and gamma Map filter. The ERS-1 and ERS-2 data were used in the investigation, and the findings showed that standard filters produce better results—especially the gamma Map filter, which outperforms other filters in terms of visual quality [21].

For the purpose of removing the speckle noise present in SAR images, Gagnon and Jouan [7] validated the sophisticated wavelet coefficient shrinkage (WCS) filter as well as common filters such as the Lee filter, Kuan filter, Frost filter, geometric filter, Kalman filter, and Gamma filter. Real-time SAR photos and generated speckle noise were used to evaluate it. The WCS filter performed equally well with standard filters for low-level noise suppression and outperformed other traditional filters for high-level noise degradation, according to the qualitative and quantitative results [7].

For the purpose of removing speckle noise from SAR data, Hua Xie et al. [22] suggested a unique algorithm by fusing the wavelet Bayesian denoising technique with Markov random field modeling. It was examined using test photos, and the experimental data demonstrate that the suggested technique performs better than the enhanced Lee filter [22].

For despeckling SAR pictures, Lav Varshney [23] introduced the contourlet transform, which used a multiscale transform to have the input image's edges identified before employing a local directional transform to identify contours. The results showed that the contourlet transform with cycle spinning approach eliminates artifacts and enhances the visual features in SAR ship wake images [23]. This was tested using synthetic and real-time RADARSAT SAR images.

Using various image-processing approaches, Ajay Kumar Boyat and Brijendra Kumar Joshi [24] carried out an evaluation on noise models. Gaussian noise, white noise, Brownian noise, impulse valued noise, periodic noise, quantization noise, speckle noise, photon noise, Poisson-Gaussian noise, gamma noise, Rayleigh noise, and structured noise are some of the image noise models that are discussed. This work examines the numerous types of noise found in digital photographs and discusses the quantitative measure of several noise models [24].

Noise filters for despeckling SAR images were studied by Jyoti Jaybhay and Rajveer Shastri in 2015. They included filters such as the scalar, median, mean, adaptive, Frost, Lee, Kuan, gamma MAP, and Wiener filters, as well as performance metrics like the signal-to-noise ratio (SNR), mean absolute error (MAE), peak signal-to-noise ratio (PSNR), average peak signal-to-noise ratio (APSNR), Pratt's figure of merit (FoM), and contrast-to-noise ratio (CNR) [25].

For the purpose of reducing speckle noise in SAR images, Murali Mohan Babu et al. [26] presented a hybrid BM3D approach by merging the spatial domain and transform domain. The first phase consists of constructing a denoised image for calculating statistics using wavelet filter-based hard thresholding in the transform domain. The second method applies an image denoising technique employing a Wiener filter in the spatial domain. It was evaluated using RISAT-1 and TerraSAR-X images of various noise levels (variations of 0.1, 0.25, and 0.5) and standard sizes (256 × 256 and 512 × 512). In terms of assessing quality variables like equivalent number looks (ENL), speckle suppression index (SSI), correlation coefficient (CC), and edge saving or preserving, the results reveal that the suggested despeckling methodology outperforms existing methods [26].

The potential literature references for the application of optimal wavelet filtering techniques are listed in the following text.

A hybrid denoising technique was created by Somnath Mukhopadhyaya and Mandal [27] that combines the discrete wavelet transform (DWT) and genetic algorithm (GA). The inverse wavelet transform was used to filter noisy pixels, and GA was used to select the best Bayesian threshold wavelet coefficients. The suggested approach was tested on an ultrasound medical image that had been tainted by Gaussian noise, and the results demonstrated that the suggested filter outperformed current filters by 8.18 dB PSNR values [27].

A novel denoising method based on blending linear and nonlinear filters with an optimization algorithm was proposed by Memoona Malik et al. in 2014. The cuckoo search method was utilized as the optimization technique, and it was used to find the best order of filters with each type of noise. With the suggested method, noises including Gaussian, speckle, and salt-and-pepper noises were removed from photos. Analysis and comparisons of the denoising effectiveness of nonlinear filters and wavelet shrinkage threshold approaches were also conducted. This showed that the proposed filter's resilience gives better results on the basis of peak signal-to-noise ratio and image quality index when compared to state-of-the-art methods [28].

Medical picture denoising using a wavelet filter based on a genetic algorithm was proposed by Yanxia Liu et al. [29]. The best threshold was determined using a genetic algorithm in conjunction with the clustering histogram of the picture information. This computation was done without using a priori knowledge, such as noise variance, and it increased search effectiveness. According to experimental findings, the conventional wavelet threshold denoising approach can enhance the peak signal-to-noise ratio and the visual impact of a medical image after denoising [29].

The discrete wavelet transform for image denoising based on the evolutionary algorithm was introduced by Sonali Singh and Sulochana Wadhwani [30]. The Bayesian method's optimal threshold wavelet coefficients were chosen using GA. The findings of the investigation, which used X-ray images tainted by Gaussian noise, showed that the improved filter performs better than previous approaches [30].

The important academic references illustrating the implementation of the Radon transform for wake identification in SAR images are listed as follows.

For the purpose of identifying ship wakes in SAR data, Anthony Copeland et al. [31] proposed the feature space line detector (FSLD) algorithm. It attempted to extract linear feature space (line segments) and suppress the false alarm rate using a Radon transform-based technique. It was put to the test using a variety of synthetic and real-time SAR images, and the analysis of the data revealed that the FSLD algorithm is capable of accurately detecting the linear signatures of maritime wake patterns [31].

The wavelet-based Radon transform was developed by Krishnaveni et al. [32] to identify wakes and calculate vessel velocity. The best threshold for despeckling image datasets was selected using the SURE-LET shrinkage approach, and ship wakes were identified using the Radon transform. The suggested method provided better wake detection and significantly lowered the false alarm rate, according to experimental results [32].

A fast discrete Radon transform was used by Gregory Zilman et al. [33] to identify the edges of ship wake boundaries in the rough sea surface. For this exploratory work, the Michell theory was used to model Kelvin wakes. The impact of ship wake, wind, and SAR parameters on the probability of wake missed detection (PMD) and false alarm rate (FAR) was discussed. It showed that the sea state [33] is a major determinant of the likelihood of false alarm rate and missed detection.

A Radon transform was suggested by Maria Daniela Graziano et al. [34] for the wake tracking on X-band SAR data from COSMO/SkyMed and TerraSAR-X. To confirm the algorithm's performance in relation to the frequency and resolution of space-borne radar, 28 pictures taken by the Sentinel-1 mission with varying incidence angles and polarization were used for testing. Results analysis showed that the majority of wakes were accurately detected and provided false confirmation rates of 15.8% and 18.5% on the X-band and C-band pictures, respectively [34].

For the ship wakes in SAR imaging, Jin and Wang [35] used the Radon transform and morphological image-processing techniques. An image that had been converted to grayscale or binary was subjected to the Radon transform to increase the likelihood of spotting ship wakes. It was tested using SEASAT SAR images, and the results demonstrate that the algorithm obtained precise ship wake detection [35].

To comprehend the SAR image properties and the difficulties in wake detection in SAR images, a brief literature review was conducted. According to the literature, the coherent nature of backscattered radar signals inserted speckle noise into the SAR images, giving them a noisy look. This had an impact on the visual quality of ship wake SAR images for wake identification. It can be concluded that the most crucial preprocessing step for wake detection in the SAR image is noise removal.

In order to learn more about the best filtering methods for despeckling SAR images, a variety of literature articles were reviewed. It was shown that mean, median, Kuan, adaptive, gamma, and wavelet methods had been widely used to reduce speckle or multiplicative noises found in different kinds of SAR images, and the encouraging findings showed their efficacy. In SAR photos, the effect of a rough sea causes a significant quantity of multiplicative speckle noise. Hence, an optimization strategy is suggested to boost the denoising algorithm's effectiveness. The results of the earlier wake detection algorithms show that the Radon transform is thought to be the best technique for locating linear features in SAR images of ship wakes. It becomes difficult to eliminate artifacts in SAR photos without losing the information they convey. Hence, an optimization approach is suggested to boost the Radon transform's ability to find weak linear signals in real-time SAR images.

3 Methodology

Fig. 2 shows the suggested methodology's basic structure. It entails a number of stages, including importing an original SAR image, putting noise-reduction techniques to use to reduce speckle noise in the SAR image, and using wake detection algorithms to find wakes in the SAR imagery. Following is a description of the methodology's general steps.

Step 1: *Load an original SAR image*—The real-time SAR image is given as input.
Step 2: *Noise removal*—An optimized discrete wavelet transform (DWT) method is developed based on the synergistic fibroblast optimization (SFO) algorithm for despeckling SAR images.
Step 3: *Wake detection*—An optimized Radon transform (RT) method is developed based on synergistic fibroblast optimization (SFO) for the detection of linear signatures in SAR images under rough sea conditions.
Step 4: *Target SAR image*—The linear features identified by the optimized wake detection algorithm are indicated in the resultant image for visual assessment.

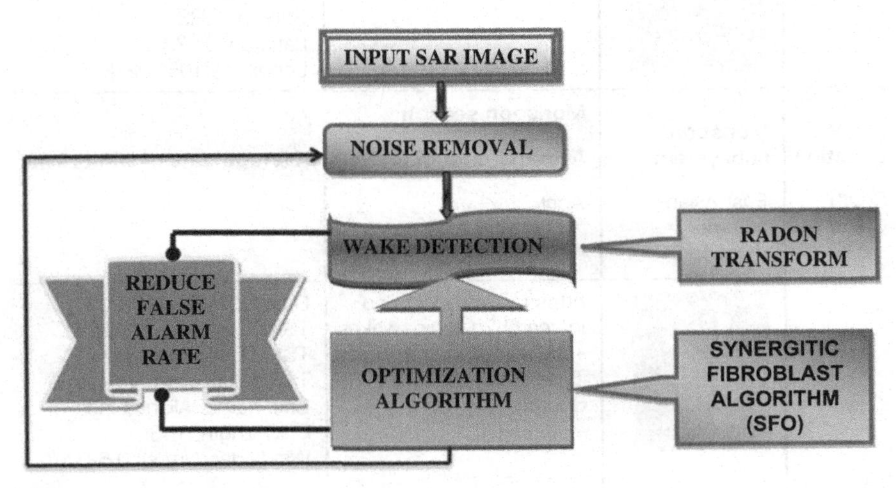

FIG. 2

Block diagram for proposed wake detection algorithm under rough sea conditions.

3.1 Step 1: Load an original SAR image

For the experimental study, two different forms of data collecting were conducted.

1. SAR images of choppy seas were gathered from the ESA Sentinel Copernicus archive and are shown in Table 5.
2. Table 6 lists SAR images taken in hazardous water conditions.

Table 5 Ship wake SAR images with rough sea condition collected from web sources.

S.no.	Image	Source details	Description
1.		Type: Ship wakes Place: South China Sea Data Source: [36]	Date: 13-Apr-1996 Time: 03:14 Orbit: 05125 Frame: 3483 Satellite: ERS-2 Latitude: 6° 07′ N Longitude:107° 15′ E
2.		Type: Ship wakes Place: South China Sea Data Source: [36]	Date: 09-Apr-1996 Time: 03:08 Orbit: 24755 Frame: 3447 Satellite: ERS-1 Latitude: 7° 54′ N Longitude: 109° 04′ E
3.		Type: Ship wakes Place: South China Sea Data Source: [36]	Date: 15-Apr-1996 Time: 03:20 Orbit: 24841 Frame: 3555 Satellite: ERS-1 Latitude: 2° 32′ N Longitude:105° 02′ E
		Monsoon season	
Location	**Monsoon/ subsystem**	**Average date of arrival**	**Average date of withdrawal**
South China	East Asian summer monsoon	April	July
4.		https://www.google.co.in/search?q=ship+wakes+SAR+images Type: Ship wakes Source: [37]	Date: 23-May-2008 TSX image time: 23:22:02 TSX mode: Stripmap TSX polarization: VV Ship name: Alianca Inca Ship length: 148 Wind speed (m/s): 0.5/0.5/0.4
		Monsoon season	
Location	**Monsoon/ subsystem**	**Average date of arrival**	**Average date of withdrawal**
Straits of Florida	Humid subtropical	May	October

Table 6 Test images.

S.no.	Image	Source details	Description
1.		Contributed by NPOL scientists	Not revealed
2.		Contributed by NPOL scientists	Not revealed
3.		Contributed by NPOL scientists	Not revealed

The properties of the gathered SAR information were inferred, and the results show that the sea state determines the sea surface condition, which is correlated with wave height and wind speed. Rough sea conditions are defined as those that have waves that reach a height of 2.0m or more [38]. According to evidence and discussions among scientists, the rough seas are especially high during monsoon season (submitted by NPOL research scientists on February 6, 2018, at the DRDO NPOL, Kochi, Kerala, India, during the project review meeting). Based on the geographic characteristics of the obtained images, it was determined that ship wake SAR images that fulfill the rough sea surface condition would be selected as the training dataset for the real-time ship wake SAR image database from the ESA Sentinel Copernicus database. In order to confirm the rough sea surface, SAR ship wake imagery was acquired based on inferential theories. The source information and description of these images include the orbit, frame, satellite, type, date, time, latitude, and longitude as well as topography [36,37].

3.2 Step 2: An optimized discrete wavelet transform based on synergistic fibroblast optimization for despeckling SAR images

3.2.1 Conventional discrete wavelet transform (DWT) for denoising SAR images

A hierarchical subband system with subbands that are logarithmically separated in frequency and reflect octave-band decomposition is exactly what a discrete wavelet transform (DWT) has produced. Fig. 3 illustrates the block schematic of the

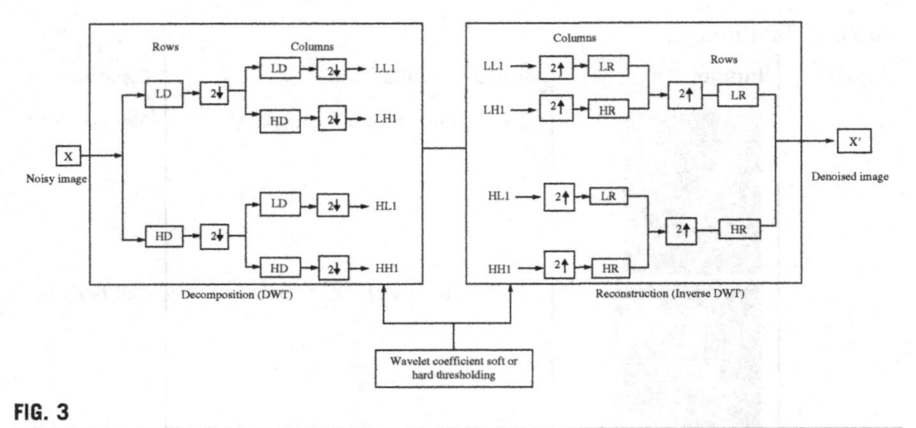

FIG. 3

Wavelet transform for one-level decomposition and one-level reconstruction.

traditional DWT, put into use to get rid of the speckle noise that appeared in the SAR images. It divided the raw image into four subbands, which were created by combining the rows and columns with the low-pass filter L and the high-pass filter H, as well as downsampling it by two. It performed picture reconstruction using the inverse wavelet transform after decomposing the image into approximation coefficients for the reduction of speckle noise at three orientations: horizontal, vertical, and diagonal. In this work, wavelet families that were taken into consideration for the noise removal of SAR images were chosen, and they are listed in Table 7. This part provides the results of a thorough evaluation of the wavelet filter's performance using the wavelet families Daubechies, Coiflets, Symlets, discrete Meyer, biorthogonal, and reverse biorthogonal.

Table 7 Wavelet families in MATLAB.

Wavelet families	Wavelets (MATLAB notation)
Daubechies	"db1," "db2," ..., "db10,"..., "db45"
Coiflets	"coif1," ..., "coif5"
Symlets	"sym2," ..., "sym8," ..., "sym45"
Discrete Meyer	"dmey"
Biorthogonal	"bior1.1," "bior1.3," "bior1.5" "bior2.2," "bior2.4," "bior2.6," "bior2.8" "bior3.1," "bior3.3," "bior3.5," "bior3.7" "bior3.9," "bior4.4," "bior5.5," "bior6.8"
Reverse Biorthogonal	"rbio1.1," "rbio1.3," "rbio1.5" "rbio2.2," "rbio2.4," "rbio2.6," "rbio2.8" "rbio3.1," "rbio3.3," "rbio3.5," "rbio3.7" "rbio3.9," "rbio4.4," "rbio5.5," "rbio6.8"

3.2.2 Optimized discrete wavelet transform based on synergistic fibroblast optimization for denoising SAR images

Fig. 4 illustrates the workflow of the suggested improved denoised wavelet filter utilizing SFO. Here, the ideal threshold is selected using SFO, which enhances the performance of the traditional discrete wavelet transform. In order to improve the effectiveness of traditional DWT for the elimination of speckle noise found in SAR images, the synergistic fibroblast optimization (SFO) technique is utilized in this section.

3.3 Step 2—Procedure 1

Step 1: *Provide the SAR image input*: The real-time SAR image taken under choppy sea conditions is provided.

Step 2: *Preprocessing—Noise removal*: Speckle noise found in image data is suppressed using the discrete wavelet transform (DWT) approach. It consists of three processes, which are thoroughly explained in the following sections: image decomposition, picture reconstruction, and thresholding of the DWT coefficients.

Step 2.1: Image decomposition is performed using a wavelet filter. A hierarchical subband structure, with its subbands spread logarithmically apart in frequency, represents octave-band decomposition. As demonstrated in Fig. 5, when DWT is applied to a noisy image, it divides the input image into four subbands (LL1, LH1, HL1, and HH1), where the subbands result from the convolution of the rows and columns with the low-pass filter (L) and high-pass filter (H). Whereas LL1 denotes the coarse level coefficients, or approximation of an image, LH1, HL1, and HH1 represent the finest scale coefficients, or picture details. In order to acquire the subsequent coarse level of wavelet coefficients, the subband LL1 is further separated into a two-level decomposition and significantly sampled, as shown in Fig. 6.

Step 2.2: *Image reconstruction*—The approximate input image is referred to as the subband LL1. It was created using low-pass filtering that was applied both horizontally and vertically. Detail subbands are used to characterize the remaining bands. The one-dimensional low-pass filter (LPF) and high-pass filter (HPF) for image decomposition are represented by the filters L and H. As it retains

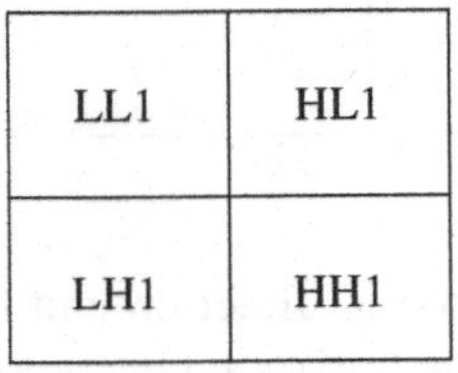

LL1	HL1
LH1	HH1

FIG. 4

Life cycle of optimized denoised wavelet filter using SFO.

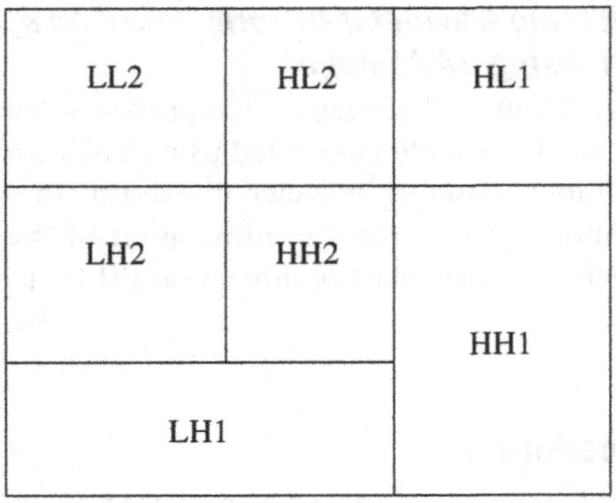

<table>
<tr><td>LL2</td><td>HL2</td><td rowspan="2">HL1</td></tr>
<tr><td>LH2</td><td>HH2</td></tr>
<tr><td colspan="2">LH1</td><td>HH1</td></tr>
</table>

FIG. 5

One-level decomposition.

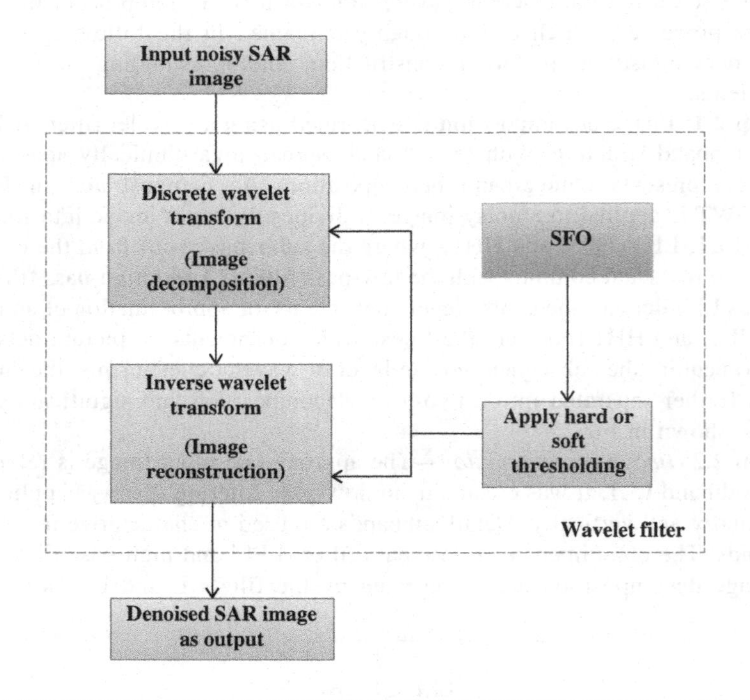

FIG. 6

Two-level decomposition.

high-pass filtering in the horizontal direction and low-pass filtering in the vertical direction, HL1 stands for the horizontal decomposition. LH1 is known as vertical decomposition, since it was created by applying low-pass and high-pass filters in opposite directions. Given that it was produced by high-pass filtering in both the horizontal and vertical axes, HH1 is also known as diagonal decomposition. LL1 is broken down into four subbands, including LL2, LH2, HL2, and HH2. Until a certain ultimate scale is reached, the process is repeated. After L decompositions, D (L) = 3 *L +1 subbands are obtained. To reconstruct the image from the decomposition of image subbands, low-pass (L) and high-pass (H) reconstruction filters are used.

Step 2.3: *Thresholding*—The essential element in the wavelet domain for picture denoising is the concept of hard or soft thresholding. It decreases the quantity of wavelet coefficients with low signal-to-noise ratios (SNRs) throughout the wavelet transform's breakdown and reconstruction phases. Each wavelet coefficient will be given a value of zero in hard thresholding if it is below a predetermined threshold. Otherwise, the coefficient's value stays the same. Soft thresholding is the process of reducing the wavelet coefficient to its absolute value by a small amount below the threshold. Therefore, for the denoising process, soft thresholding is desirable. Wavelet domain denoising is focused on the discrete wavelet transform's (DWT) sparsity property, which states that although noise is represented by many small coefficients, information is stored in a few big coefficients. The shift-variant nature of the critically sampled DWT and its inability to comprehend directional imagery data are also problems. The number of coefficients pertaining to the signal grows as a function of the amount of decomposition, so an increase in the number of coefficients preserved is necessary. Hence, the level dependent threshold is computed using the Birge-Massart technique and then soft thresholding is utilized to apply it.

The Birge-Massart strategy is represented mathematically as follows:

$$K_j = \frac{m}{(J_0 + 1 - J)^\wedge \alpha} \tag{3}$$

where J_0 is the represents decomposition level; m is the length of the coarsest approximation over 2; α is the sparsity (ALPHA) parameter.

When using the Birge-Massart strategy of wavelet decomposition and reconstruction structure, ALPHA (α), a sparsity parameter, is crucial for the determination of the penalized threshold [39]. The range of the ALPHA parameter is between 2.5 and ALPHA 10, 1.5 and ALPHA 2, and 1 and ALPHA 2, respectively, for high, medium, and low penalty thresholds. It is implied that choosing the fittest sparsity parameter would speed up the wavelet filter's denoising procedure. Hence, for a new optimization method to select the ideal sparsity value that can be used in wavelet coefficients in both image decomposition and picture reconstruction based on the Birge-Massart thresholding method, the synergistic fibroblast optimization (SFO) algorithm is used [40]. Procedure 2 explains the various algorithmic methods used to select the sparsity parameter.

Step 3: *Filtered image*—The wavelet threshold approach, which carries out image decomposition and image reconstruction, makes use of the ideal sparsity value selected by SFO to produce the highly qualitative filtered image. The original SAR picture and denoised SAR image based on the optimum wavelet filter [41] and Daubechies for visual analysis are explained in the sample in Figs. 7 and 8.

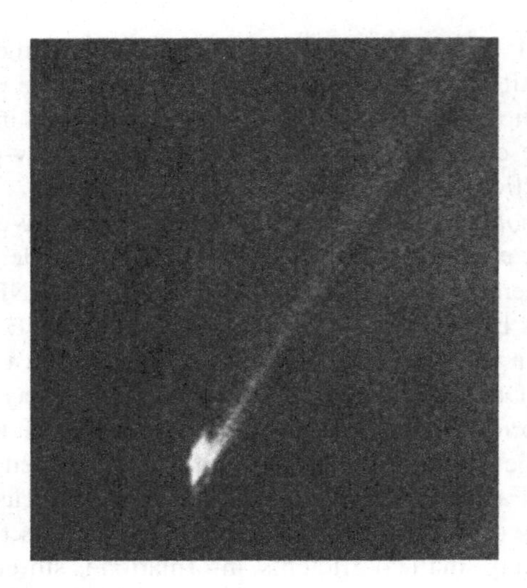

FIG. 7

Original image—1.

FIG. 8

Denoised image based on wavelet filter under Daubechies family.

3.4 Step 2—Procedure 2

Fig. 9 shows the life cycle of the SFO method used for optimization, including the algorithmic steps for choosing the best parameter.

3.4.1 Algorithm: SFO applied to choose optimal sparsity parameter

Step 1: The first step is to stochastically generate position (xn) and velocity (vn) values for the population of discrete fibroblasts (n size) in the two-dimensional problem space. Collagen deposition in extracellular matrix is the range of the ALPHA variable for the three levels (high, medium, and low) of penalized techniques (ecm). Collagen best (Cbest), diffusion coefficient ()=0.8, cell speed ($s = s/(k$ ro $L)$), and number of cells ($L = n$) are some of the established parameters.

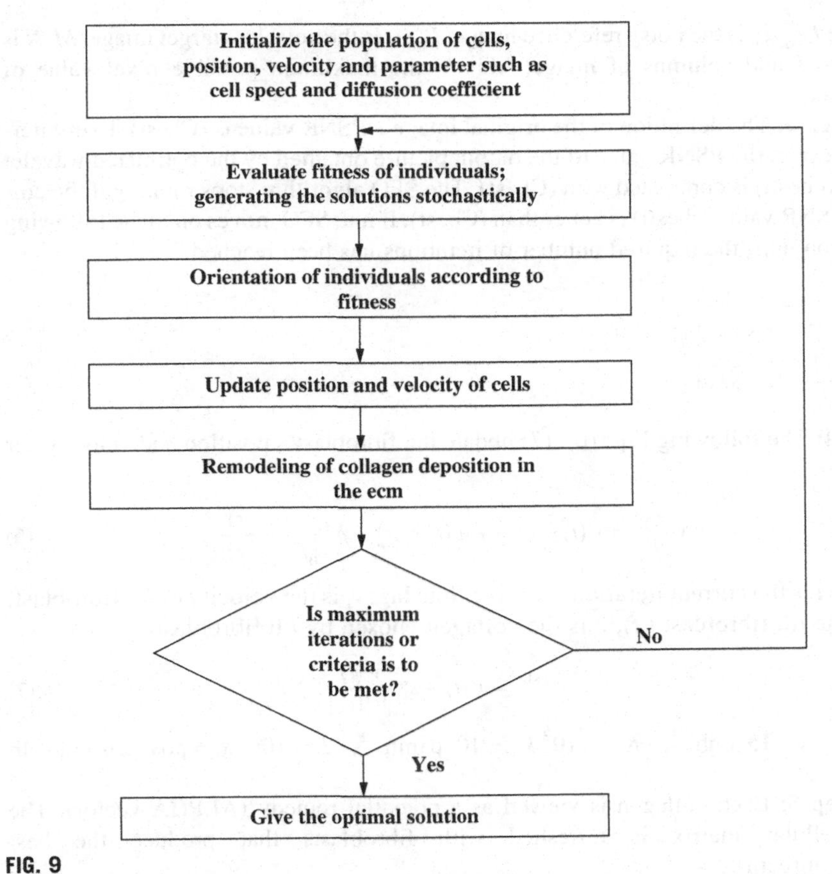

FIG. 9

Life cycle of synergistic fibroblast optimization (SFO) algorithm.

Step 2: The Griewank benchmark function is used to assess the fibroblast's fitness. Peak signal-to-noise ratio (PSNR), used as the objective function, was compared with an ideal ALPHA value that was applied to a wavelet filter.

Griewank function (continuous, differentiable, nonseparable, scalable, multimodal) [42]:

$$f(x) = \sum_{i=1}^{n} \frac{x_i^2}{4000} - \Pi \cos\left(\frac{x_i}{\sqrt{i}}\right) + 1 \tag{4}$$

where $i = i$th iteration; $x =$ ALPHA parameter.

$$\text{MSE} = \frac{1}{MN} \sum_{i=1}^{M} \sum_{i=1}^{N} \left[I_{\text{original}}(M,N) - I_{\text{target}}(M,N)\right]^2 \tag{5}$$

$$\text{PSNR} = 10 \log_{10}\left(\frac{\text{Max}^2}{\text{MSE}}\right)$$

where I_{original} is the noisy reference image; I_{target} is the noiseless target image; M, N is the rows and columns of image; Max is the maximum possible pixel value of the image.

Step 3: The definition of the original image's PSNR value is (Cbest). Every iteration cycle, the PSNR value of the output picture obtained by the optimized wavelet filter (cbesti) is contrasted with (Cbest). The SFO algorithm stops running if the current PSNR value (cbesti) is better than (Cbest). If not, SFO moves on to the following iteration until the required number of iterations has been reached.

```
if (cbest_i>Cbest)
Cbest = cbest_i;
else
Cbest = Cbest;
```

Step 4: The following Eqs. (6), (7) update the fibroblast's position and velocity per cycle.

$$v^{i(t+1)} = v^i(t) + (1-\rho)c\left(f^i(t), t\right) + \rho * \frac{f^i(t-\tau)}{\left\|f^i(t-\tau)\right\|} \tag{6}$$

where t is the current iteration t; τ is the time lag; v_i is the velocity of ith fibroblast; f_i is the ith fibroblast (f); c is the collagen chosen by ith fibroblast.

$$x^{i(t+1)} = x^i(t) + s * \frac{v^i(t)}{\left\|v^i(t)\right\|} \tag{7}$$

where $s = 15\,\mu\text{mh}^{-1}$; $K_{ro} = 10^3\ k_{ro} = 10^3\,\mu\,\text{min}^{-1}$; $L = 10$; $x_i =$ position of ith fibroblast.

Step 5: Each collagen is viewed as a potential remedy (ALPHA vector). The extracellular matrix is refreshed with fibroblasts that produce the best collagen (ecm).

Step 6: The steps from steps 2 to 5 are continued until the program has run through 1000 generations or an ideal solution (sparsity value) has been found. Steps 2.1 and Step 2.2 of Procedure 1 apply the ideal sparsity value selected by SFO.

3.4.2 Experimental results for optimal denoising method

A singular SAR image is used as the training set for an experiment that tests the best wavelet filter among its relatives. The visual assessment findings are listed from Figs. 10 through 22.

Performance evaluation

To improve the inquiry of this study, the performance effectiveness of the suggested wavelet filtering technique using the SFO is assessed using the standard performance metrics provided in the following text and the obtained results are demonstrated in Figs. 23 through 26.

FIG. 10

Original image—1.

FIG. 11

Denoised image based on wavelet filter under Daubechies family.

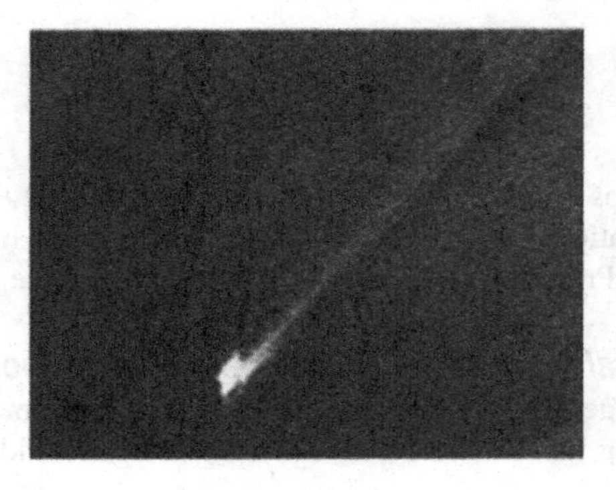

FIG. 12

Denoised image based on optimal wavelet filter under Daubechies family.

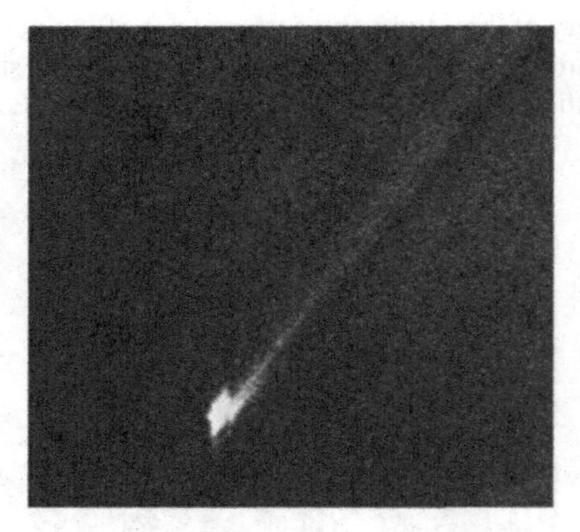

FIG. 13

Denoised image based on wavelet filter under Coiflet family.

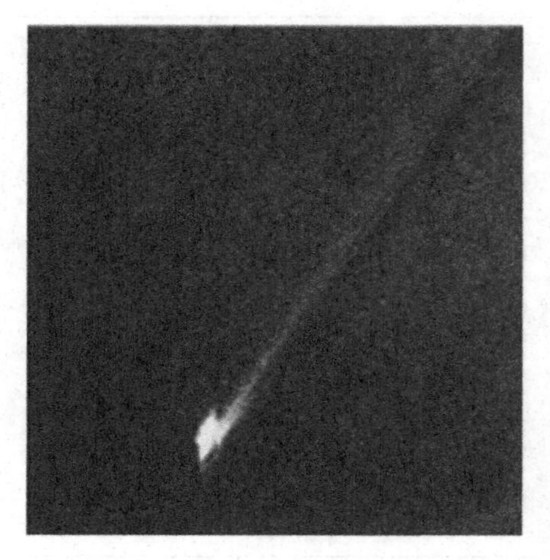

FIG. 14

Denoised image based on optimal wavelet filter under Coiflet family.

FIG. 15

Denoised image based on wavelet filter under Symlet family.

FIG. 16

Denoised image based on optimal wavelet filter under Symlet family.

Experimental conditions and metrics: The experiment was run in MATLAB (R2017a) using an Intel (R) Core (TM) i7-4790 processor clocked at 3.60 GHz and 4 GB of memory. A Windows 7 Professional 64-bit computer was the operating system platform. Peak signal-to-noise ratio (PSNR), mean squared error (MSE), mean absolute error (MAE), and correlation coefficient (r) were used to evaluate the denoising performance. The mathematical expressions for these performance metrics are provided as follows [43,44].

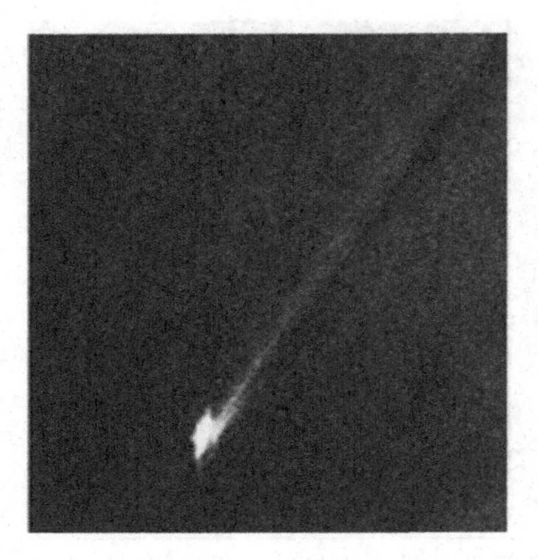

FIG. 17

Denoised image based on wavelet filter under discrete Meyer family.

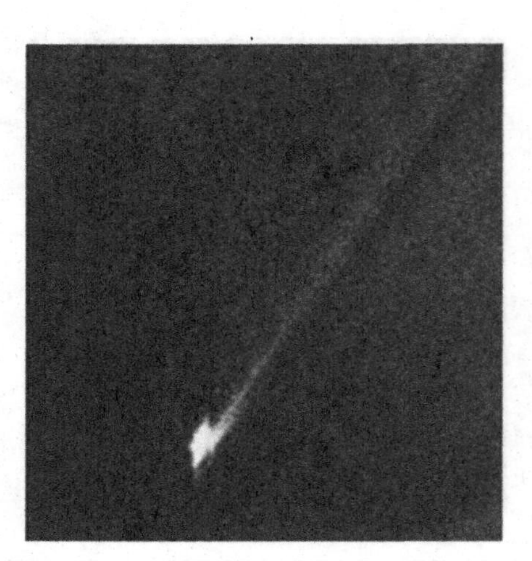

FIG. 18

Denoised image based on optimal wavelet filter under discrete Meyer family.

Mean square error (MSE): The difference between the original picture (I) and the restored image (I') is known as the mean square error (MSE). It is determined by employing the formula:

$$MSE = \frac{1}{MN}\Sigma_{i=1}^{M}\Sigma_{j=1}^{N}\left[I_{i,j} - I'_{i,j}\right]^2 \tag{8}$$

where M and N represent the number of rows and columns in the original image ($I_{i,j}$) and restored image ($I'_{i,j}$), respectively.

FIG. 19

Denoised image based on wavelet filter under biorthogonal family.

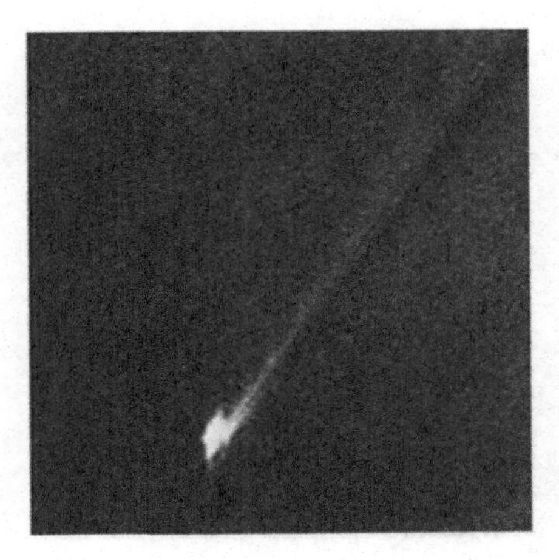

FIG. 20

Denoised image based on optimal wavelet filter under biorthogonal family.

Peak signal-to-noise ratio (PSNR): Determined using the following equation, it is the ratio between the image's original and restored noisy pixel values.

$$PSNR = 10\log_{10}\left(\frac{255^2}{MSE}\right) \tag{9}$$

Mean absolute error (MAE): The average difference in absolute error between the original image (I) and the restored image (I') is known as MAE.

FIG. 21

Denoised image based on wavelet filter under reverse biorthogonal family.

FIG. 22

Denoised image based on optimal wavelet filter under reverse biorthogonal family.

$$MAE = \frac{1}{MN}\Sigma_{i=1}^{M}\Sigma_{j=1}^{N}\left[I_{i,j} - I'_{i,j}\right] \tag{10}$$

where M and N represent the number of rows and columns in the original image ($I_{i,j}$) and restored image ($I'_{i,j}$), respectively.

Pearson correlation coefficient (*r*): The chance that the original image (I) and the filtered output image (I') have a linear connection is known as the Pearson correlation coefficient (r). The value of r is 1, 0, or −1 if the two images are

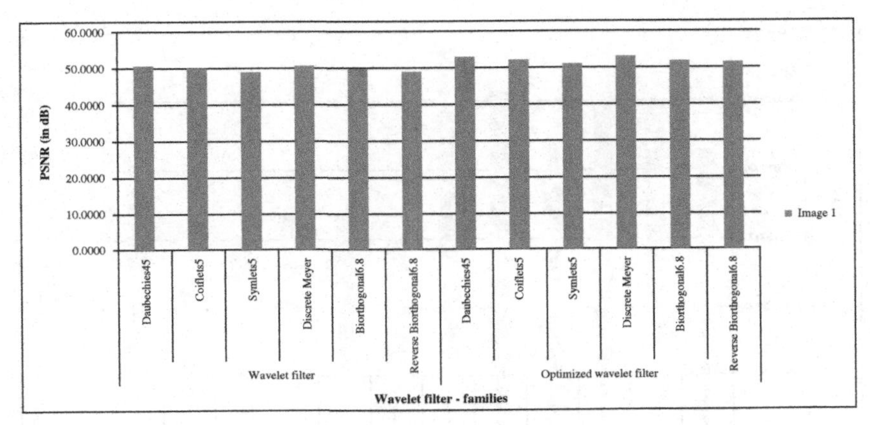

FIG. 23

PSNR values of wavelet families.

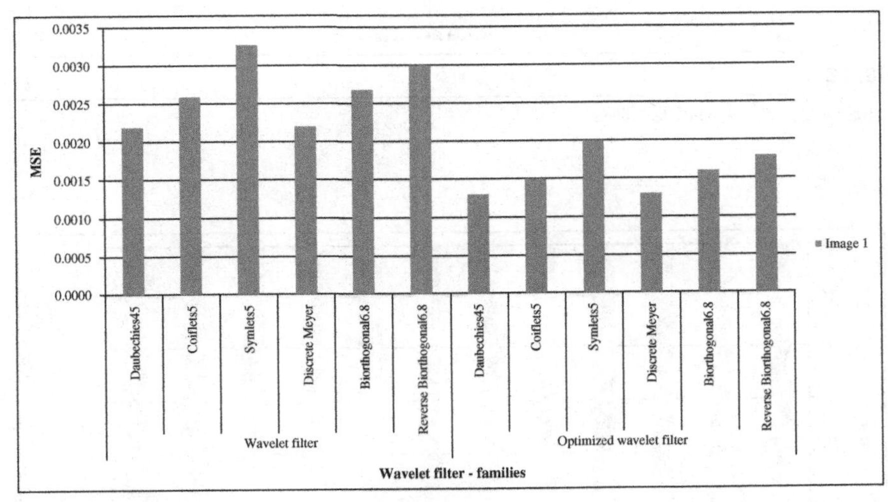

FIG. 24

MSE values of wavelet families.

totally correlated, completely uncorrelated, or completely anticorrelated. It is determined by:

$$r = \frac{\Sigma_i (I_i - I_m)(I'_i - I'_m)}{\sqrt{\Sigma_i (I_i - I_m)^2} \sqrt{\Sigma_i (I'_i - I'_m)^2}} \tag{11}$$

where I_i is the intensity values of ith pixel in original image; I'_i is the intensity values of ith pixel in filtered image; I_m is the mean intensity values of original image; I'_m is the mean intensity values of filtered image.

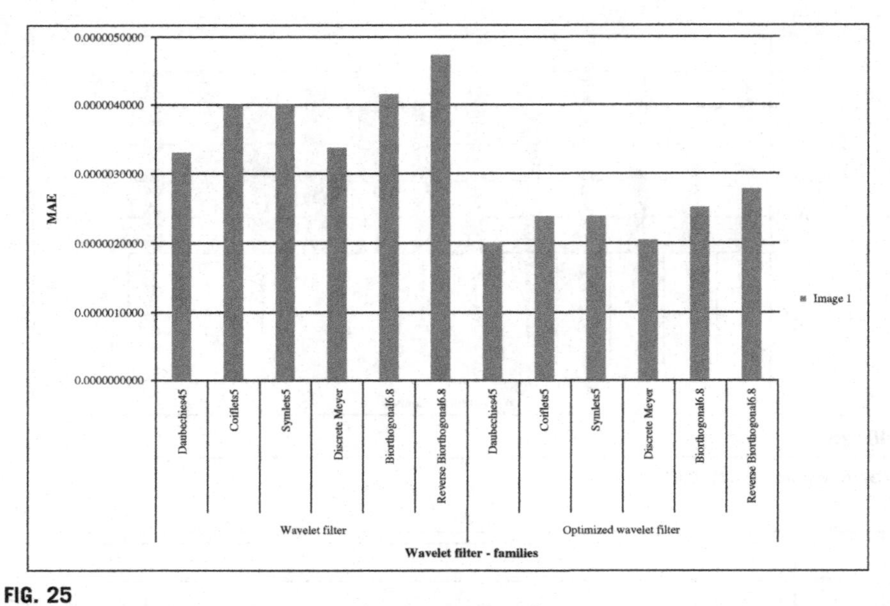

FIG. 25

MAE values of wavelet families.

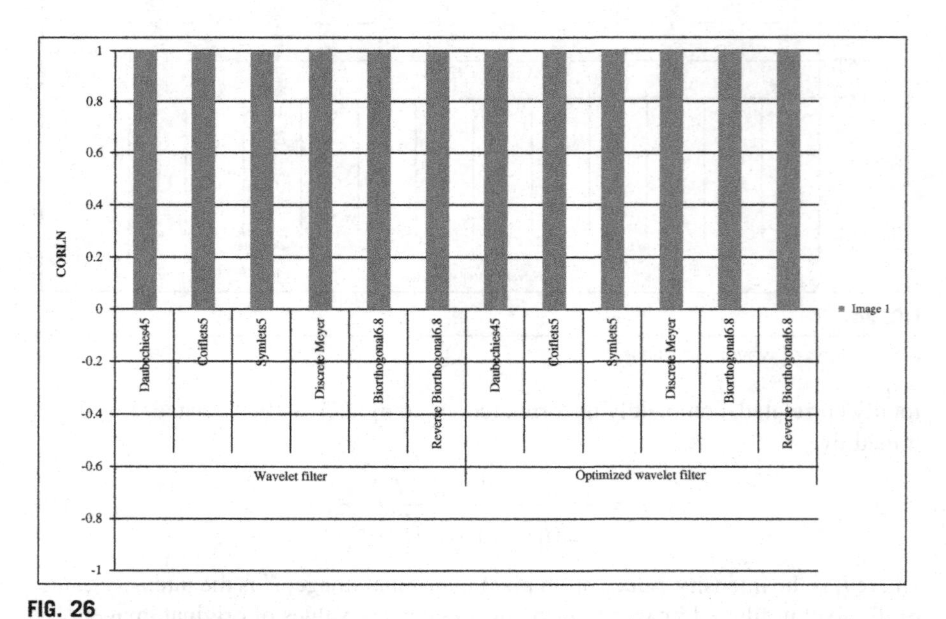

FIG. 26

CORLN values of wavelet families.

Inference from experimental results: The subjective analysis of the denoised images produced by both conventional DWT and optimized DWT is depicted in Figs. 11 through 22 and from objective assessment illustrated in Figs. 23 through 26 it is inferred that the SFO-based wavelet filter under the Daubechies family produces better results compared to other wavelet families. Henceforth, denoised images obtained by optimal DWT under Daubechies have been taken as the input image for wake detection in the next phase.

Step 3: *Wake detection—An optimized Radon transform (RT) method is developed based on synergistic fibroblast optimization (SFO) for the detection of linear signatures in SAR images*

Two factors make it crucial to identify wake patterns in SAR images: (1) Wake patterns are larger and more pronounced than ship signatures, and they can provide a more accurate estimate of the ship's true location. (2) The direction and speed of the ship can be inferred from the ship wake pattern. However, due to factors including varied SAR sensor orientations, variations in ship size and speed, and also varying ocean and wind conditions, it is possible to discern internal waves on the surface in SAR images. However, not all wake elements are visible in every image [31]. Consequently, the project's main difficulty is to create a method for locating and identifying ship wakes in important SAR image characteristics, particularly in choppy seas.

Applying conventional Radon transformation for wake detection

The wake detection block takes the advantage of applying the Radon transform to detect the ship wake [33]. The Radon transform calculates the angle that a straight line perpendicular to the track makes with the x-axis in the center of the image [38]. Knowing this, simply add 90 degrees to the value obtained to find the angle of the wake arm. If an image is considered as I, with dimensions $M \times M$, Eq. (12) contains the Radon transformation:

$$\hat{I}(x_\theta, \theta) = \sum_{y\theta=-M/2}^{M/2} I(x_\theta \cos\theta - y_\theta \sin\theta, x_\theta \sin\theta + y_\theta \cos\theta) \tag{12}$$

where $(x_\theta, y_\theta) \in Z$ and $\theta \in [0; \pi]$.

A two-dimensional function $f(x, y)$ called the discrete Radon transform (DRT) is used to represent data points that are projected at angle of, 0. Lines can be effectively detected in the full image domain using the DRT mathematical transform. The recognition of ship wakes in SAR imagery is carried out in this case using the DRT Eq. (13).

The equation of the line is $y = sx + t$. ($|s| <= 1$)

$$\text{Radon}(\{y = sx + t\}, I) = \sum_{u=-n/2}^{\frac{n}{2}-1} \widetilde{I^1}(u, su + t) \tag{13}$$

where

$$\tilde{I}(u,y) = \sum_{v=-n/2}^{\frac{n}{2}-1} I(u,v)D_m(y-v)$$

$D_m(t) = \dfrac{\sin(\pi t)}{m\sin\left(\frac{\pi t}{m}\right)}$, the Dirichlet kernel with $m=2n+1$.

Where n is the mean value of pixels in the image ($-n < t < n$).

The subjective evaluation of wake detection using the conventional Radon transform for a single SAR image is presented in Figs. 27 and 28.

There are a few limitations associated with the traditional Radon transform for linear features. It can be challenging to detect line segments that are much shorter than the image dimensions, since the intensity integration is carried throughout the entire length of the image [45]. The transform is not able to provide data on line length or the locations of the ends of these shorter line segments. The linear features might not produce appropriate peaks or troughs in the domain of the traditional Radon transform, which causes linear fingerprints in SAR images to be incorrectly identified.

When the transform is implemented on a picture with a large amount of noise, these issues become even more serious. These shortcomings may negatively affect efforts to examine SAR imagery for the purpose of identifying ship wakes in rough

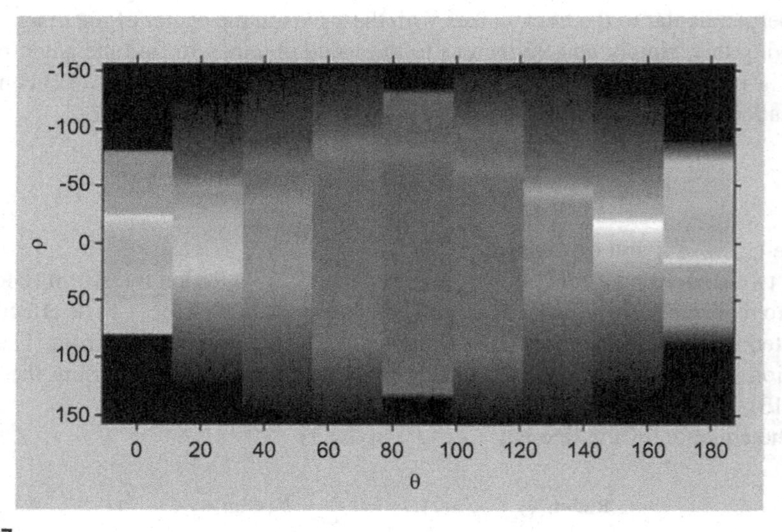

FIG. 27

Linear features angle projection based on conventional Radon transform image 1.

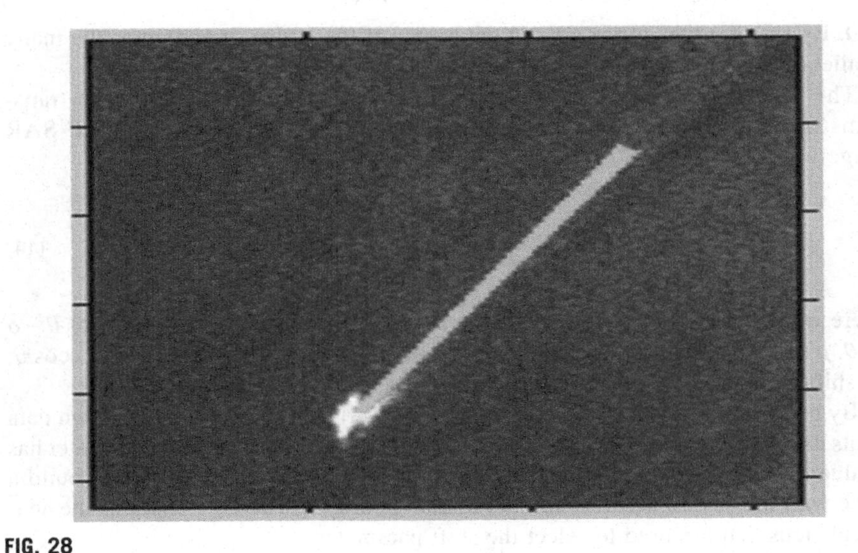

FIG. 28
Resultant image 1 based on conventional Radon transform for wake detection.

seas, despite the fact that over time, owing to variations in ship headings, water depths, or ocean conditions, ship wake components typically appear straight. The generation of linear curvature that is not related to peaks or troughs in the SAR data was the main issue found during the experiment utilizing the Radon transform, due to the enormous impact that oil leaks, tide, wind speed, and wind direction have on ship wakes in choppy seas. Due to the high multiplicative speckle noise, attempts to detect ship wake signals is negatively impacted.

Optimized Radon transform for wake detection using SFO

The optimization method used in the Radon transform for locating linear signatures existing in the denoised SAR pictures is described in the following section. Procedure 1 introduces the SFO algorithm to select the ideal shift parameter value that will produce accurate line detection in the image. The algorithmic method of the improved Radon transform used for wake detection in SAR imaging is described in the following procedure.

3.5 Step 3—Procedure 1

3.5.1 Procedure 1: An optimized Radon transform based on SFO

Step 1: *Denoised image*—The input is the filtered SAR image.

Step 2: *Wake detection*—The linear features in the denoised SAR images are found using an improved discrete Radon transform (DRT) approach based on

SFO. By rotating the source around the center of the image, it computes the many parallel-beam projections of the image at various angles.

The line equation $0 = x \cdot \cos 0 + y \cdot \sin 0$ is used to represent the Radon transform (0, 0), which is comparable to the integral of the input denoised SAR image.

$$RI(\theta, \rho, \sigma) = R_{Loc}\{RI\} = f \int_{x\min}^{x\max} \int_{y\min}^{y\max} f(x,y)\delta(\rho - x\cos\theta - y\sin\theta)dydx \quad (14)$$

where $x\min = \min(\rho \cos \theta - \sigma \sin \theta, \rho \sin \theta - (\sigma + \lambda) \sin \theta)$; $x\max = \max(\rho \cos \theta - \sigma \sin \theta, \rho \sin \theta - (\sigma + \lambda) \sin \theta)$; $y\min = \rho \sin \theta + \sigma \cos \theta$; $y\max = \rho \sin \theta + (\sigma + \lambda) \cos \theta$; σ = shift parameter; λ = length of the line segment of integration (LSOI).

By taking the angle and offset values from a series of discrete, well-known data points, the shift parameter is used to create sinusoidal waves. The shift parameter has a value that falls between 0.0 and 10.0 within a 0.25 interval. It is used to build a curve with the best fit across a number of data points. Procedure 2 explains the analytical steps that are used to select the shift parameter.

Step 3: *Wake detection in SAR images—Ship wakes found in SAR imagery are used as the acquired target image.*

3.6 Step 3—Procedure 2

3.6.1 Algorithm: SFO implemented to choose σ value to be utilized in Radon transform

Step 1: Initialize the population of fibroblast cells f_i, $i = 1, 2, \ldots, 10$; with randomly generated position (x_i), velocity (v_i), and the range of σ ($0 \leq 0.25 \leq 10$) values are defined as collagen deposition (c_i) in extracellular matrix (ecm) in the n-dimensional problem space. A set of parameters involve collagen best (C_{best}), diffusion coefficient (ρ) = 0.8, cell speed $\left(s = \frac{s}{k_{ro}L}\right)$, and number of cells ($L = n$) are predefined.

Repeat

Step 2: Evaluate the fitness of fibroblast using the objective function F (f_i).
Alpine 2 function (continuous, differentiable, separable, scalable, multimodal):

$$f(x) = \prod_{i=1}^{D} \sqrt{x_i} \sin(x_i) \quad (15)$$

Step 3: At each cycle, the reorientation of the cell is performed to find the optimal candidate solution (cbest_i) in the evolutionary space.

Step 4: Update the velocity (v^i) and position (x^i) of a cell using the Eqs. (16) and (17):

$$v^{i(t+1)} = v^i(t) + (1-\rho)c\left(f^i(t), t\right) + \rho^* \frac{f^i(t-\tau)}{\left\|f^i(t-\tau)\right\|} \tag{16}$$

where t is the current time; τ is the time lag; v^i is the velocity of ith cell; $\rho = 0.5$.

$$x^{i(t+1)} = x^i(t) + s^* \frac{v^i(t)}{\left\|v^i(t)\right\|} \tag{17}$$

where $s = \frac{S}{k_{ro}L}$, $k_{ro=10^3}$ μmin^{-1}, L is the cell length.

Step 5: Remodeling of collagen deposition (c_i) is synthesized in the extracellular matrix (ecm), **until** the predetermined condition(s)/maximum iterations are met.

3.6.2 The optimal shift parameter value chosen by SFO is implemented in Procedure 1

Outcome of the algorithm: *If the intensity of pixels at point (ρ_0, θ_0) over a line is relatively bright or dark, then it is considered as a ship wake. Otherwise, it is not. The optimal parameter value of the Radon transform enables the user to detect at a particular point and provide the analytical representation of the line. The obtained results are portrayed in Figs. 29 and 30.*

Step 4: *Experimental results for proposed optimal wake detection method are tabulated from* Tables 8–11.

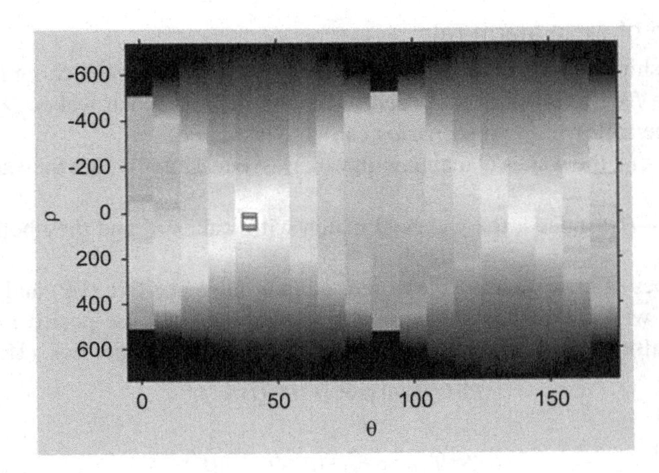

FIG. 29

Linear features angle projection based on optimized Radon transform image 1.

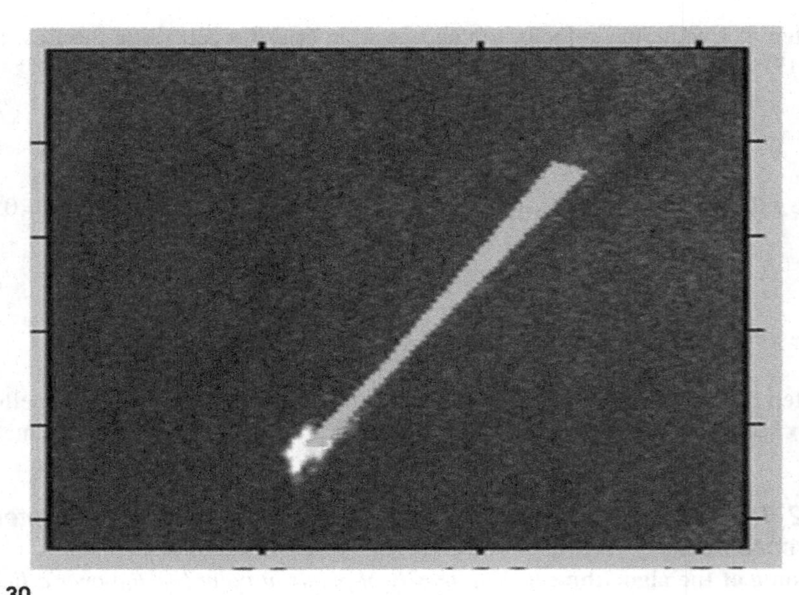

FIG. 30

Resultant image 1 based on optimized Radon transform for wake detection.

Evaluation of false alarm rate

Consider a ship wake B_0 in the reference object set $\{B\}$, with a matching function F, in an image I with N wakes that need to be tested [35]. When wakes $Q \in I$ match wakes B, the following two scenarios can be considered.

Type I—H_0: the wakes Q match with wakes B' occasionally, but they are actually different.

Type II—H_1: indeed, the wakes Q match with wakes B' and they both describe ship wakes.

Two types of errors can be measured, namely false positive (FP) and false negative (FN), with a threshold value δ. The probability of false-positive detections (PFP) and false negative detections (PFN) can be calculated using Eqs. (18) and (19):

$$PFP(B', \delta) \equiv P_r(F(B', Q) < \delta| \tag{18}$$

$$PFN(B', \delta) \equiv P_r(F(B', Q \geq \delta|H_1)) \tag{19}$$

where $P_r(\cdot)$ is a probability function defined on the detection wakes Q.

The validation of algorithms for the detection of both single target and multiple targets are illustrated in Tables 12 through 15.

Table 8 Visual assessment of SAR images (training dataset: single target)—Comparative analysis.

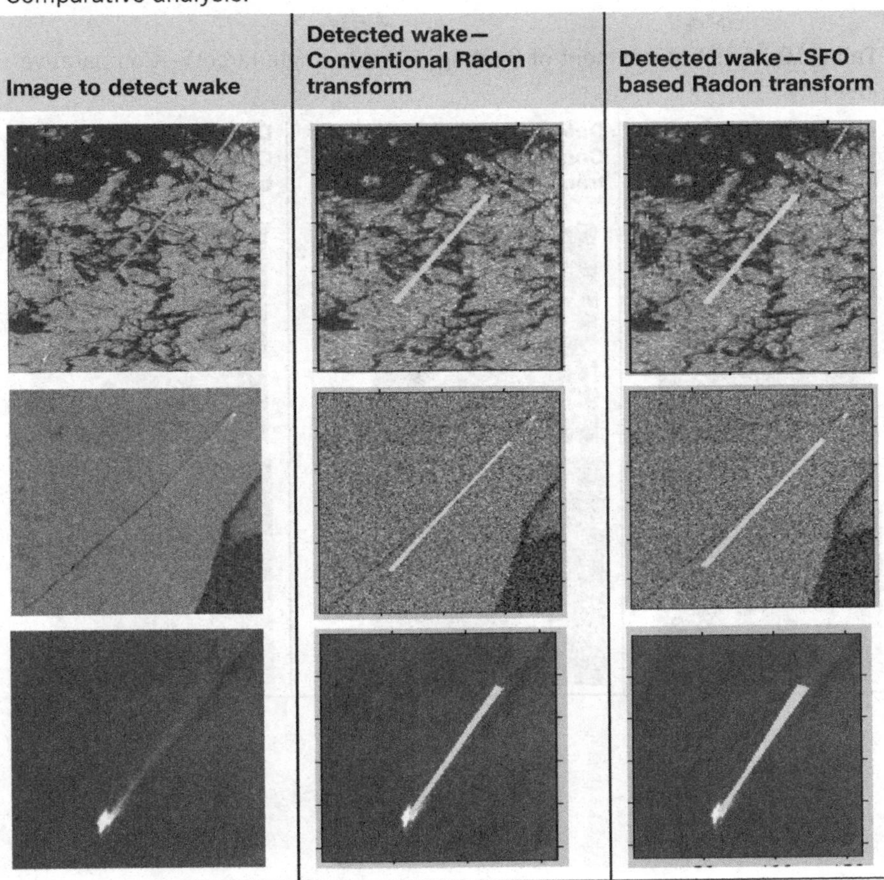

Image to detect wake	Detected wake—Conventional Radon transform	Detected wake—SFO based Radon transform

Table 9 Visual assessment of (training dataset: multiple targets)—Comparative analysis.

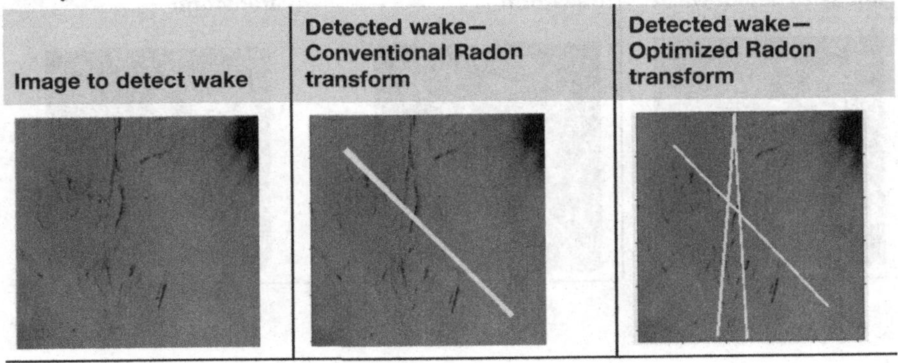

Image to detect wake	Detected wake—Conventional Radon transform	Detected wake—Optimized Radon transform

Table 10 Visual assessment of (testing dataset: single target)—Comparative analysis.

Image to detect wake	Detected wake—Conventional Radon transform	Detected wake—Optimized Radon transform

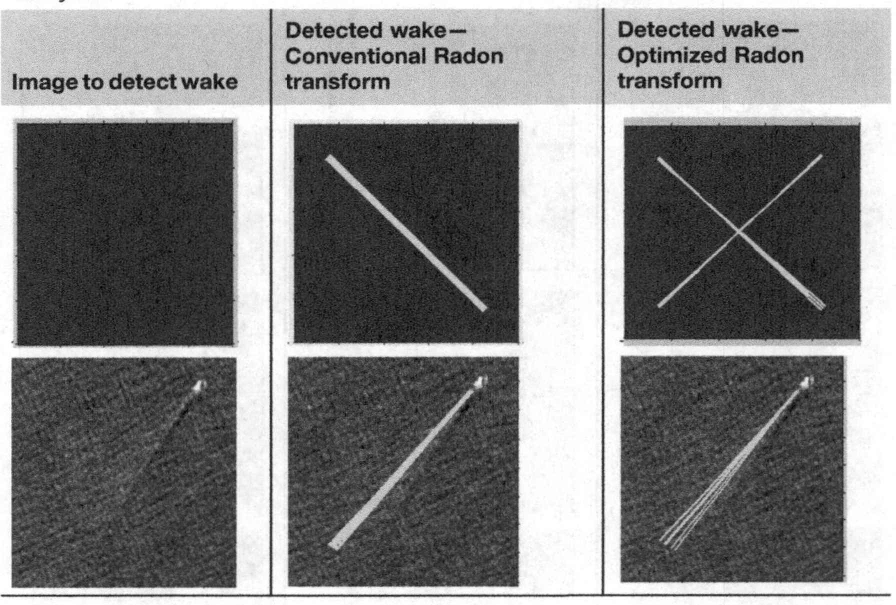

Table 11 Visual assessment of (testing dataset: multiple targets)—Comparative analysis.

Image to detect wake	Detected wake—Conventional Radon transform	Detected wake—Optimized Radon transform

Table 12 Assessment of wake detection algorithms (0^0–179^0)—Training dataset (single target).

Input image	Visual observation	Linear feature identified		Misidentified linear feature		False alarm rate	
		Conventional Radon transform (RT)	Optimized Radon transform (ORT)	Conventional Radon transform (RT)	Optimized Radon transform (ORT)	Conventional Radon transform (RT)	Optimized Radon transform (ORT)
	1	1	1	0	0	0	0
	1	1	1	0	0	0	0
	1	1	1	0	0	0	0

Table 13 Assessment of wake detection algorithms (0^0–179^0)—Training dataset (multiple targets).

Input image	Visual observation	Linear feature identified		Misidentified linear feature		False alarm rate	
		Conventional Radon transform (RT)	Optimized Radon transform (ORT)	Conventional Radon transform (RT)	Optimized Radon transform (ORT)	Conventional Radon transform (RT)	Optimized Radon transform (ORT)
	14	1	5	13	9	0	0

Table 14 Assessment of wake detection algorithms (0^0–179^0)—Testing dataset (single target).

Input image	Visual observation	Linear feature identified		Misidentified linear feature		False alarm rate	
		Conventional Radon transform (RT)	Optimized Radon transform (ORT)	Conventional Radon transform (RT)	Optimized Radon transform (ORT)	Conventional Radon transform (RT)	Optimized Radon transform (ORT)
	2 (1 wake + 1 oil slick)	1	2	1	0	0	0
	1	1	1	0	0	0	0

Table 15 Assessment of wake detection algorithms (0^0–179^0)—Testing dataset (multiple targets).

Input image	Visual observation	Linear feature identified		Misidentified linear feature		False alarm rate	
		Conventional Radon transform (RT)	Optimized Radon transform (ORT)	Conventional Radon transform (RT)	Optimized Radon transform (ORT)	Conventional Radon transform (RT)	Optimized Radon transform (ORT)
	6 (3 wakes + 3 oil clutter)	1	3	5	3	0	0

4 Discussion

Three training datasets and one testing dataset are used to validate the proposed optimization algorithm. Four photos from the ESA Sentinel Copernicus database were taken from this dataset and are used for training, while the remaining three images are used for testing. Wavelet filter and Radon transform optimization are the first and second steps in the application of optimization techniques to enhance the performance of the proposed algorithm. In order to conduct the wavelet filter, the Birge-Massart strategy's optimal sparsity (ALPHA) parameter is first selected using the SFO algorithm. Using high-pass and low-pass filters, the wavelet coefficients make use of the ideal sparsity parameter to reduce noise along the image's horizontal, vertical, and diagonal axes. Significant speckle noise content is reduced, which preserves the information it contains and enhances the quality of the SAR image for wake detection. High penalized threshold (2.5 ALPHA 10), medium penalized threshold (1.5 ALPHA 2.5), and low penalized threshold (1 ALPHA 2) are the three categories that make up the sparsity parameter range. Based on a trial-and-error methodology, it can be deduced that selecting the highest sparsity parameter has a significant impact on the wavelet decomposition structure, which negatively reduces pixel values in SAR images. Hence, Griewank's fitness is assessed in order to select the least sparsity parameter as the objective value. In order to remove the speckle noise present in SAR images, the optimum sparsity (ALPHA) parameter selected by SFO is used in the wavelet decomposition and wavelet reconstruction of DWT.

Second, the use of Radon transform and inverse Radon transform techniques for wake detection in the denoised SAR images verifies the efficacy of the speckle noise reduction process carried out by the improved wavelet filter. The previously mentioned statistics, which depict subjective evaluations of the Radon transform's findings, show that the conventional approach is unable to yield highly qualitative results for some SAR images. When ship wakes and wind-generated waves are combined in SAR photographs of choppy seas, the difficulty of ship wake recognition in SAR images has increased. It was discovered that the best fit to a set of data points is the main focus of the sinusoidal wave created by the Radon transform. Instead of concentrating on the range of a discrete set of known data points, linear interpolation is a method of curve fitting that builds sinusoidal waves that roughly fit data points. The range of a discrete collection of known data points utilizing an offset value and an angle is between 0 and 10 across a 0.25 interval. Here, a data point is optimized using an optimization approach used in the curve-fitting technique that maximizes the coverage of the Radon-produced data points. It is accomplished by using the Alpine benchmark function to select precisely calibrated interpolated functional values for the curve fitting method's linear interpolation. The Alpine test suite is used to evaluate the SFO's fitness and select the best data point as the benchmark function. The application of the optimization technique has significantly improved the performance of the Radon transform, and the acceptable result produced proves its effectiveness, according to the visual inspection of the results. Conclusion: The suggested optimization algorithm performs better than the traditional approach, and the results are generally acceptable.

5 Conclusion

An optimization approach is suggested in this study to detect wakes in rough sea conditions, a topic that has received the least attention in the literature. The effectiveness of the suggested method is confirmed using ground truth SAR imagery from the ESA Sentinel Copernicus database, and the algorithm's advancement is assessed using conventional performance indicators. The analysis of the experimental data shows that the optimization strategy is the most effective technique for despeckling SAR images and that it is capable of detecting wakes in SAR photos in rough sea conditions. Given how dependable and effective the offered procedures are, it is expected that they will produce positive results in an uncontrolled setting (rough sea condition). The suggested method's dependability, simplicity, and cheap computational cost are key characteristics. When combined with new algorithms to get the best results, the findings of this detection algorithm are found to be reliable and adequate.

Acknowledgment

This work is supported by the project titled "An algorithm for the demarcation of ocean wakes from SAR imagery of rough sea condition" (No.NPOL/18CR002/PR-240) funded by Defence Research & Development Organisation (DRDO)—Naval Physical & Oceanographic Laboratory (NPOL), Kochi, Kerala, India.

References

[1] K. Tomiyasu, Tutorial review of synthetic-aperture radar (SAR) with applications to imaging of the ocean surface, Proc. IEEE 66 (1978) 563–583.
[2] A. Moreira, P. Prats-Iraola, M. Younis, G. Krieger, P. Irena Hajnsek, K. Papathanassiou, A tutorial on synthetic aperture radar, IEEE Geosci. Remote Sens. Mag. (2013) 6–43.
[3] F. Qiu, R. Judith Berglund, J. Jensen, Speckle noise reduction in SAR imagery using a local adaptive median filter, GISci. Remote Sens. 41 (2004) 244–266.
[4] A. Mark Schulze, X. Qing Wu, Nonlinear edge-preserving smoothing of synthetic aperture images, in: Proceedings of the New Zealand Image and Vision Computing '95 Workshop, Christchurch, New Zealand, 1995, pp. 65–70.
[5] D.A. Nelson Mascarenhas, An overview of speckle noise filtering SAR images, in: Proceedings of the First Latino-American Seminar on Radar Remote Sensing - Image Processing Techniques, Buenos Aires, Argentina, 1997, pp. 71–79.
[6] M. Tsuchida, M. Haseyama, H. Kitajima, A kalman filter using texture for noise reduction in SAR images, Electron. Commun. Jpn 86 (2003) 266–277.
[7] L. Gagnon, A. Jouan, Speckle filtering of SAR images—a comparative study between complex-wavelet-based and standard filters, in: SPIE Proc. Conference Wavelet Applications in Signal and Image Processing V, San Diego, 1997, pp. 1–12.

[8] A. Josaphat Tetuko, S. Sumantyo, J. Amini, A model for removal of speckle noise in SAR images (alos palsar), Can. J. Remote. Sens. 34 (2008) 503–515.
[9] S.-q. Huang, D.-z. Liu, G.-q. Gao, X.-j. Guo, A novel method for speckle noise reduction and ship target detection in SAR images, Pattern Recogn. 42 (2009) 1533–1542.
[10] S. Ahmed Mashaly, F.E.E. AbdElkawy, A.T. Mahmoud, Speckle noise reduction in SAR images using adaptive morphological filter, in: IEEE International Conference on Intelligent Systems Design and Applications, Cairo, Egypt, 2010, pp. 260–265, https://doi.org/10.1109/ISDA.2010.5687254.
[11] H. Li, B. Huang, X. Huang, A level set filter for speckle reduction in SAR images, EURASIP J. Adv. Signal Process. (2010) 1–14.
[12] A.V. Meenakshi, V. Punitham, Performance of speckle noise reduction filters on active radar and SAR images, Int. J. Technol. Eng. Syst. 2 (2011) 111–114.
[13] V. Milindkumar Sarode, R. Prashant Deshmukh, Reduction of speckle noise and image enhancement of images using filtering technique, Int. J. Adv. Technol. 2 (2011) 30–38.
[14] G.S. Rajeshwari, R. Simon, R. Sulochana, J.B. Bhattacharjee, Despeckling of SAR images using recursive filter based model based and anisotropic diffusion based methods, Anale Seria Informatica 10 (2012) 15–26.
[15] W. Ji Yuan, Y.Y. Bin, H. Qingqing, C. Jingbo, R. Lin, Speckle noise reduction in SAR images ship detection, Proc. SPIE 8532 (2012) 1–8, https://doi.org/10.1117/12.974375.
[16] W. Wang, X. Zhang, X. Wang, Speckle suppression method in SAR image based on curvelet domain Bivashrink model, J. Softw. 8 (2013) 947–954.
[17] S. Gopinathan, S. Poornima, Enhancement of images with speckle noise reduction using different filters, Int. J. Appl. Sci. Eng. Res. 4 (2015) 333–352.
[18] M. Rahimi, M. Yazdi, A new hybrid algorithm for speckle noise reduction of SAR images based on mean-median filter and SRAD method, in: IEEE International Conference on Pattern Recognition and Image Analysis, 2015, pp. 1–6, https://doi.org/10.1109/PRIA.2015.7161623.
[19] A. Pranali Hatwar, R. Heena Kher, Analysis of speckle noise reduction in synthetic aperture radar images, Int. J. Eng. Res. Technol. 4 (2015) 508–512.
[20] P. Kupidura, Comparison of filters dedicated to speckle suppression in SAR images, Int. Arch. Photogramm. Remote Sens. Spatial Inf. Sci. XLI-B7 (2016) 269–276, https://doi.org/10.5194/isprsarchives-XLI-B7-269-2016.
[21] Y. Huang, J.L. Van Genderen, Evaluation of several speckle filtering techniques for ERS-1&2 imagery, Int. Arch. Photogramm. Remote Sens. 31 (1996) 164–169.
[22] E. Hua Xie, T. Leland Pierce, F. Ulaby, SAR speckle reduction using wavelet denoising and Markov random field modeling, IEEE Trans. Geosci. Remote Sens. 40 (2002) 2196–2212.
[23] R. Lav Varshney, Despeckling synthetic aperture radar imagery using the contourlet transform, Appl. Signal Process. 4 (2004) 1–6.
[24] A.K. Boyat, B.K. Joshi, A review paper: noise models in digital image processing, Sig. Img. Process. Int. J. 6 (2015) 63–75.
[25] J. Jaybhay, R. Shastri, A study of speckle noise reduction filters, Signal Image Process. Int. J. 6 (2015) 71–80, https://doi.org/10.5121/sipij.2015.6306.
[26] Y.M.M. Babu, M.V. Subramanyam, M.N.G. Prasad, A new approach for SAR image denoising, Int. J. Electr. Comput. Eng. 5 (2015) 984–991.
[27] S. Mukhopadhyaya, J.K. Mandalb, Wavelet based denoising of medical images using subband adaptive thresholding through genetic algorithm, Procedia Technol. 10 (2013) 680–689, https://doi.org/10.1016/j.protcy.2013.12.410.

[28] M. Malik, F. Ahsan, S. Mohsin, Adaptive image denoising using cuckoo algorithm, Soft. Comput. (2014) 1–14.
[29] Y. Liu, Y. Ma, F. Liu, X. Zhang, Y. Yang, The research based on the genetic algorithm of wavelet image denoising threshold of medicine, J. Chem. Pharm. Res. 6 (2014) 2458–2462.
[30] S. Singh, S. Wadhwani, Genetic algorithm based medical image denoising through sub band adaptive thresholding, Int. J. Sci. Eng. Technol. Res. 4 (2015) 1481–1485.
[31] C. Anthony Copeland, G. Ravichandran, M.M. Trivedi, Localized transform-based detection of ship wakes in SAR images, IEEE Trans. Geosci. Remote Sens. 33 (1995) 1–11.
[32] M. Krishnaveni, S.K. Thakur, P. Subashini, An optimal method for wake detection in SAR images using radon transformation combined with wavelet filters, Int. J. Comput. Sci. Inf. Secur. 6 (2009) 66–69.
[33] G. Zilman, A. Zapolski, M. Marom, On detectability of a ship's kelvin wake in simulated SAR images of rough sea surface, IEEE Trans. Geosci. Remote Sens. 53 (2015) 609–618.
[34] M.D. Graziano, M. Grasso, M. D'Errico, Performance analysis of ship wake detection on sentinel-1 SAR images, Remote Sens. 9 (2017) 1–11, https://doi.org/10.3390/rs9111107.
[35] Y.-Q. Jin, S.-Q. Wang, An algorithm for ship wake detection from the synthetic aperture radar images using the radon transform and morphological image processing, Imaging Sci. J. 48 (2000) 159–163, https://doi.org/10.1080/13682199.2000.11784357.
[36] https://earth.esa.int/web/guest/missions/esa-operational-eo-missions/ers/instruments/sar/applications/tropical/-/asset_publisher/tZ7pAG6SCnM8/content/ship-wakes-south-china-sea.
[37] A. Soloviev, M. Gilman, K. Young, S. Brusch, S. Lehner, Sonar measurements in ship wakes simultaneous with TerraSAR-X overpasses, IEEE Trans. Geosci. Remote Sens. 48 (2) (2010) 841–851.
[38] A. Scherbakov, R. Hanssen, G. Vosselman, R. Feron, Ship wake detection using radon transforms of filtered SAR imagery, in: Proceedings of SPIE Microwave Sensing and Synthetic Aperture Radar, 1996, pp. 1–11, https://doi.org/10.1117/12.262684.
[39] S. Sidhi, Comparative study of Birge–Massart strategy and unimodal thresholding for image compression using wavelet transform, Optik 126 (2015) 5952–5955.
[40] P. Subashini, T.T. Dhivyaprabha, M. Krishnaveni, Synergistic fibroblast optimization, in: S. Dash, K. Vijayakumar, B. Panigrahi, S. Das (Eds.), Artificial Intelligence and Evolutionary Computations in Engineering Systems, Advances in Intelligent Systems and Computing, 517, Springer, Singapore, 2017, pp. 285–294, https://doi.org/10.1007/978-981-10-3174-8_25.
[41] P. Subashini, P.V. Hareesh Kumar, S. Lekshmi, M. Krishnaveni, T.T. Dhivyaprabha, Improved noise filtering technique for wake detection in SAR image under rough sea condition, in: 2021 IEEE World AIIoT Congress (AIIoT), IEEE, Seattle, 2021, pp. 0174–0180.
[42] M. Jamil, X.-S. Yang, A literature survey of benchmark functions for global optimization problems, Int. J. Math. Model. Numer. Optim. 4 (2013) 150–194, https://doi.org/10.1504/IJMMNO.2013.055204.
[43] S. Jansi, P. Subashini, Particle swarm optimization based total variation filter for image denoising, J. Theor. Appl. Inf. Technol. 57 (2013) 169–173.
[44] T. Saba, A. Rehman, G. Sulong, An intelligent approach to image denoising, J. Theor. Appl. Inf. Technol. 6 (2005) 32–36.
[45] L.M. Murphy, Linear feature detection and enhancement in images via the radon transform, Pattern Recogn. Lett. 4 (1986) 279–284.

CHAPTER

Smart spaces and ensuring their information security

20

Vladimir N. Shvedenko[a], Oleg Shchekochikhin[b], and Dmitriy Alexeev[c]

[a]*FSBUN All-Russian Institute of Scientific and Technical Information of the Russian Academy of Sciences (VINITI RAS), Moscow, Russia,* [b]*Department of Analytics at PJSC "Softline", Moscow, Russia,* [c]*Department of Information Security, Kostroma State University, Kostroma, Russia*

1 Introduction

At present, smart spaces have a pronounced national and regional character, and they have developed in different locations based on the peculiarities of the region. Thus in Asia, Africa, the Americas, and Europe the focus is ecology and health [1–3]. In India it is energy, smart homes (due to growing urban population and a great deal of uncomfortable housing), and water issues [4–7]. China is most characterized by smart homes, smart cities, transportation, and public administration [8–10].

Obviously, the functioning and efficiency of a smart space and its use will depend on the flow of data that comes to it, both from the outside, as well as the data that is formed within the space itself. The quality and objectivity of the data will determine the effective functioning of the elements of the smart space and its state. We should not rule out the fact that smart space infrastructures may become the target of hacker attacks.

It is clear that one cannot discount the desire to use information resources as a means of waging war or destroying competitors, or simply carrying out malicious intent on the part of blackmailers, dictators, or criminals. Therefore ensuring information security must be systematic and comprehensive, understanding that the manipulation of data can become an element of armed struggle.

Threats to the smart space are very diverse. Studies on the security of smart spaces have identified threats ranging from hardware and network attacks of various types, to threats arising from the human factor [11]. In all cases, a solution is needed to ensure information security. This should be envisaged at the design stage of the smart space, taking into account the requirements of the standard architecture. A detailed description of the architecture for the Internet of Things (IoT), which meets the requirements of ISO/IEC/IEEE 42010:2011, is given in the international standard IEEE 2413-2019. The standard provides a four-tiered smart space system architecture (Fig. 1): the device level, the communication network level, the Internet

Smart Spaces. https://doi.org/10.1016/B978-0-443-13462-3.00021-2

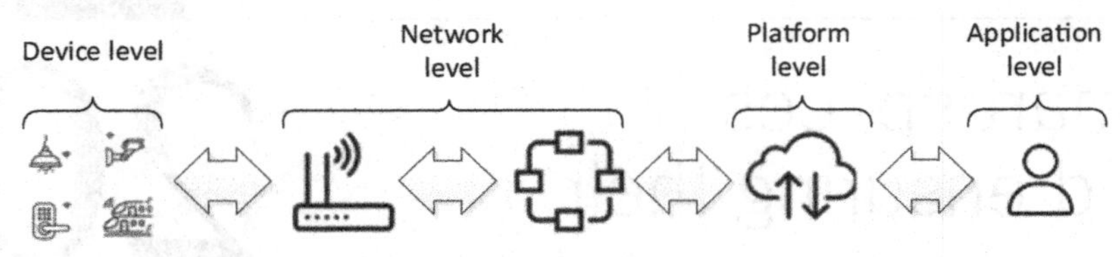

FIG. 1

Standard four-tier smart space architecture.

of Things platform level, and the application level. The standard also defines interconnections and attributes specific to the cloud computing center, advanced computing technologies, and big data analytics associated with the IoT.

The typical structure of the existing architecture does not take into account the information security requirements of a smart space. Therefore each smart space that is considered as a project involves original design, taking into account the elements used to receive and transport data. However, it is of practical interest to develop a reference model for information security of a smart space that summarizes the diversity of threats.

The function of a smart space is accompanied by the transportation and processing of a large amount of data. Therefore in this chapter we propose to consider a universal system for identifying threats to ensure information security of the smart space, as well as diagnosing and transmitting signals to eliminate abnormal situations that arise and occur in it. We also propose a reference model and architecture of the information security system for the smart space.

2 Definitions and designations

In what follows, these concepts and definitions will be used:

Information security: preservation of confidentiality, integrity, and availability of information.

Information system: a system that organizes the processing of information about the subject area and its storage.

Computer system: a set of hardware controlled by software (operating system) as a single module.

Anomaly: a deviation from the norm, from a general pattern; an irregularity.

ISS (information security system): information system for detecting threats and finding data anomalies.

DSS (decision support system): a computer system whose purpose is to help decision-makers in complex environments to fully and objectively analyze subject activities.

Reference model: a general or basic model that represents the general characteristics of a given class of systems.

3 Literature review

The current state of smart space information security issues is reflected not only in scientific articles by individual groups of researchers, but also in analytical reviews by companies. These works also focus on suggestions for improving the architecture of smart spaces to improve their qualitative characteristics.

For example, Alcatel-Lucent Enterprise [12] presented a guide on the design, implementation, and maintenance of smart city networks with a set of typical architectures and typical solutions for information security of IoT devices. The main purpose of this guide is to present smart space developers with requirements and recommendations for using typical architectures, as well as design options, best practices, and system configuration recommendations.

Some of the authors of the various articles offer different options for smart space architectures, with projects focused on exploring interactions between different layers and other tasks not directly related to data protection issues.

Diaconita et al. in Ref. [13] presented a smart city architecture with IoT support, which consists of three main levels: the internal level (data storage and processing), the level of peripheral IoT nodes (sensors, actuators and other embedded systems), and the middle level (gateways).

According to a study by Syed [14], two types are distinguished from the great variety of existing smart city architectures: a centralized operating platform with a pyramidal architecture and a three-layer architecture consisting of an instrumental layer, an interconnected layer, and an intelligent layer.

The architecture of smart space in the work of Ouafiq et al. [15] is considered through a decomposition into two vertical layers and one horizontal layer, without solving the problem of data protection.

A number of works by groups of scientists contain private solutions for information security of the smart space. For example, Siddiqui [16] in his work considers, among other things, the security problems of heterogeneous smart city networks. The architecture proposed by the authors is based on three levels: perception level, control level, and application level. For distributed SDN controllers, Kalkan and Zeadally [17] proposed a security architecture consisting of an intrusion controller, a key controller, and a crypto controller. Each SDN controller communicates with a domain controller using access rules.

Yamamoto [18] proposes to analyze the reference architecture of a smart city using ArchiMate. Information security issues are considered as aspects that pass through the architecture hierarchy. It is noted that the smart city reference architectures are based on the enterprise architecture.

In spite of the proposed methods of data protection, the assessment of Rahman et al. [19] noted that the issues of information security of smart space services remain open.

According to Aryavalli and Kumar [20], the main reason for the lack of information security is that the architecture used in the systems based on the Internet of Things is not secure enough or that the controls are not sufficient to protect applications.

Table 1 State of the smart space architecture levels.

Architecture	Device level	Network level (Wi-Fi, LTE, 5G)		Platform level IoT	Application level (IP)
Physical Environment	IoT	Gateway IoT	Network connection	Server (cloud)	Software
Disadvantages	Low processing power, small memory size	Open	Open	Dissimilar data	Lack of software certification
Solutions	Encryption, key generators	Cryptography, authentication	VPN, traffic monitoring	Antiviruses, firewall	Software certification

Considered from the preceding review of sources, the existing shortcomings of the smart space architecture layers are summarized in a table along with their solutions (Table 1). For example, IoT devices have low computing power and low memory capacity, and the network layer is open. Given the characteristic disadvantages of IoT devices, it is difficult to implement an effective data protection system at this level of the architecture space by means of the devices themselves.

The network level protection solutions designed and implemented are effective and do an excellent job. Modern in-line active security tools not only detect threats but also block them in real time. However, when implementing active means, researchers face problems of network fault tolerance [21], further scalability of security means, and the need to reduce packet transmission delays when the volume of traffic in the network increases. It is impossible to consider its individual elements and cyber-physical devices in isolation from the overall state of the smart space. The solution to this problem is very relevant to the smart space.

Existing application layer deficiencies must be addressed by certifying and licensing the software used for user interaction with IoT devices. Only reliable software vendors should enter the market for IoT products.

IoT platform level equipment is protected by its owners with antivirus protection and firewalling. The layer is protected against distributed denial of service attacks by specialized hardware and software tools from distributed denial of service (DDoS) protection or anti-DoS systems.

4 The problem

The standard smart space architecture does not provide at any level an unambiguous place for the information security component. There are private solutions for data protection, but the summation effect of these does not provide a synergistic effect for information security.

In terms of information security, a smart space should be viewed as a system of interacting services. The set of services is not finite and tends to expand its elements during the life cycle of a smart space.

Information threats are a single-type phenomenon with no uniqueness, almost always with the same goal—to damage cyber-physical devices. Therefore it is necessary to identify the basic principles of detection and suppression of information threats. At the same time, ensuring information security should be an integral part of smart space design.

Any fragment of a smart space through which the flow of information takes place should use a reference model of an information security system, invariant to the place of its application. Such an approach can significantly reduce the cost of data processing, as well as reduce the cost of data protection tools. It is necessary to create a reference model of the information security system of a smart space, which by tuning or modification can be applied at any level of the architecture, to as many services as possible.

5 Proposed solution

The concepts of behavioral models embedded in the design methodology of the open information security system of smart spaces implement technical and software tools that perform human intellectual functions. This methodology has the tools to implement behavioral models and frees humans from routine work. It is clear that artificial intelligence will not be able to compete with natural intelligence, but some groups of processes and interactions with cyber-physical objects are entrusted to intelligent information systems.

The simplest and least complicated control tasks in the smart space need to be put into automatic mode, just as, for example, at one time the administration of information systems was transferred to an automated mode by Microsoft Corporation, keeping many things hidden from users and not allowing them to interfere in the process of information security.

An automated smart space information security system should give signals only to the support and decision-making system (decision-making center). This system must have the properties of intelligent behavior.

Indirect signs of nonobvious, unusual behavior from IoT devices are perceived in their operation as anomalous behavior. There must be a system to respond to emerging anomalies in data behavior. We need to understand the extent to which these anomalies affect the functioning of smart space elements and the threats that arise.

The reference model for the information security of smart spaces includes tactical, operational, and strategic types of management, which are inherent in the corresponding loop of the preparation and management decision-making system. Tactical management is reactive, operational management is active, and strategic management is proactive.

The loop with reactive control constantly monitors the state of information flows and issues commands to eliminate the possibility of information threats to the viability of the smart space in automatic mode.

Active control provides for possible response options and automated activation of the relevant functions of the management decision-making system. The system under active control must be open and provide identification and additions to the loop of newly appearing abnormal situations, as well as provide methods for their elimination.

Proactive management is associated with changes in the organization of smart space, its modernization, and equipping it with new devices, by analyzing the latest advances in science and technology relevant to improving the smart space functioning.

There are two notional groups of data flow in the operation of a smart space. The first group is the general impersonal data flow, which is associated with supporting the functioning of the technical and software tools of the smart space. The second group is the data separated from the general flow that comes to the information security system of the smart space.

In addition to identifying threats and classifying them by type, the system's indirect analysis of the behavior of the allocated data stream will identify anomalies in data behavior. The absence of anomalies in the data stream will characterize the normal functioning of the smart space. Analysis of data streams from individual devices can point to specific causes of emerging threats to elements of the smart space. It should be noted immediately that the data-processing algorithms of all data streams will be the same and represent an invariant information system.

Identification of anomalies in the system is carried out by indicators. The role of indicators is played by the criterion indicators of the smart space information security system. The use of indicators ensures the selection of management decisions corresponding to the current state of the system.

It is proposed to use, not direct, but indirect data, which may be, for example, the percentage of CPU load; percentage of physical disk idle time; percentage of bytes of RAM used; number of bytes sent and received through the network adapter; and number of connections via TCP/IP.

Detection of information security threats relies on two key proxy indicators—the number of bytes sent and received through the network adapter. Anomalies in the operation of cyber-physical devices are detected based on other indirect indicators of system operation.

Automatic and automated modes of operation provide for the implementation of certain behaviors in various abnormal situations.

When using the indirect signs of the smart space data flow, the information security system uses three types of models of behavior corresponding to tactical, operational, and strategic types of management in the contours of the system of support and management decision-making.

The first model of behavior, with tactical (stationary) control, should provide an automatic response to deviation of the criteria indicators of the information data flow and priority control over the behavior of deviated indicators. Suppose, for example, in a greenhouse with a given temperature regime there was an excess of temperature. Incorrect temperature control can kill the crop and this will be accompanied by economic damage. The job of a smart space is to regulate the system and keep it functioning properly. Obviously, the nature of the data flow of an IoT device performing the tasks of a temperature regulator will be different from the steady state. The behavior model must issue a control signal to bring the temperature regime to a normal, stationary state (homeostasis).

The triggered behavior model identifies multiple device deviations in the stream of data being analyzed. It does not select a single solution. A behavior model for each abnormal state of a system or device can be represented by frames. The case having multiple decision choices involves automated decision-making and involving a specialist in the control loop for this purpose. For example, consider the task of ensuring the comfortable state of the microclimate of a smart house when using heterogeneous equipment from different manufacturers (air conditioning, ventilation, and heating) together. Disruption will create uncomfortable working conditions for people, reducing their productivity. Management of joint function of various cyber-physical devices is entrusted to a single information center for coordination of their interaction. Registration of anomalies, depending on the behavior model, will entail either disabling the faulty device, or switching to manual control with the overall preservation of system performance.

Proactive management is associated with the anticipation of variants of events, changes in the organization of the smart space, and its updating. Therefore the smart space system of information security should be built as an open system, with the possibility of improving it without disrupting its performance. It is laid down as a direction of prospective development of the system of information security of the smart space.

Frames are one of the methods of knowledge representation and implementation of the behavior model with complex logic, which make it possible to build complex artificial intelligence systems. Frames can refer to other frames or can be elementary. The need to perform actions can be unordered or ordered. In the latter case, if several actions are possible under a certain combination of conditions, the frame specifies their priority. The frame defines the dependence function of a combination of input system states—values of criterion indicators, with the corresponding output data—control actions in the decision-making center.

Let us consider an example of making a control decision for the model of behavior presented in Figs. 2–4. Suppose that the information security system has indicated an excess of receiving data quantity, over the threshold level (Table 2), which leads to the transition to the next level of knowledge at address 1.5 (highlighted in *green, gray in print version*).

Transition to address 1.5 leads to the frame of the value of the correlation coefficient module received bytes. The value of the correlation coefficient of a very weak data link (Table 3), for example, equal to 0.07 provides a transition to address 1.5.2 (highlighted in *blue, dark gray in print version*).

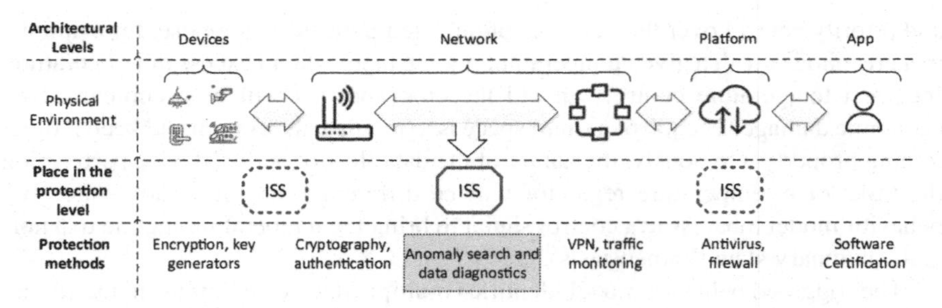

FIG. 2

A variant of smart space architecture with network-level data diagnostics.

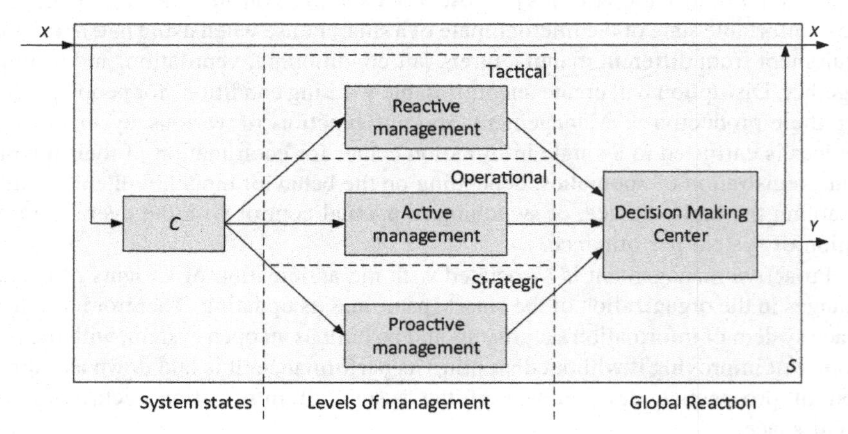

FIG. 3

Reference model of the smart space information security system.

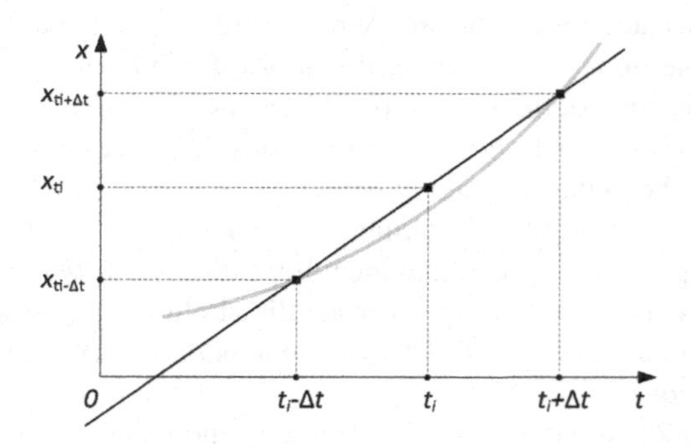

FIG. 4

Graphical interpretation of the derivative.

Table 2 Frame of indirect characteristics of the computer system.

1. Computer system characteristic		
Slot	**Filler**	**Address**
Percentage of CPU load exceeded	Correlation coefficient CPU	1.1
The percentage of active disk (physical disc) is exceeded	Correlation coefficient physical disc	1.2
Percentage of used allocated memory is exceeded	Correlation coefficient memory	1.3
Sent bytes over the allowable threshold	Correlation coefficient sent bytes	1.4
Received bytes per second (received bytes) exceeds the allowable threshold	Correlation coefficient received bytes	1.5

Table 3 Correlation coefficient frame received bytes.

1.5. The value of the modulus of the correlation coefficient received bytes		
Slot	**Filler**	**Address**
$r = 0$	No data linkage	1.5.1
$0 < r < 0{,}2$	Very weak data linkage	1.5.2
$0{,}2 \leq r < 0{,}5$	Weak data linkage	1.5.3
$0{,}5 \leq r < 0{,}7$	Average data connectivity	1.5.4
$0{,}7 \leq r < 0{,}9$	High data connectivity	1.5.5

Going to address 1.5.2 results in a frame of very weak received bytes data communication. For a correlation coefficient of 0.07, the behavior model provides (Fig. 5), for example, activation of the IoT device internal verification procedure, an additional firewall function, and a self-check of the information security system (highlighted in *gray*).

Knowledge retrieval in frames is performed by the knowledge base up to the elementary frame (Table 4). Obtaining knowledge by navigating to the addresses of each frame is shown in Fig. 5.

The fillers of the elementary frame are commands (functions) issued to the decision-making center. Examples of commands are, for example, enabling the self-check of the information security system, enabling the internal verification procedure of an IoT device, or enabling an additional firewall function (Figs. 4 and 6).

The smart space is subject to seasonal cycles. This can be related to natural, economic, social, technological, and other cycles. The basic idea of anomaly monitoring is to cluster time series analysis to find the maximum similar sample in cycles. The

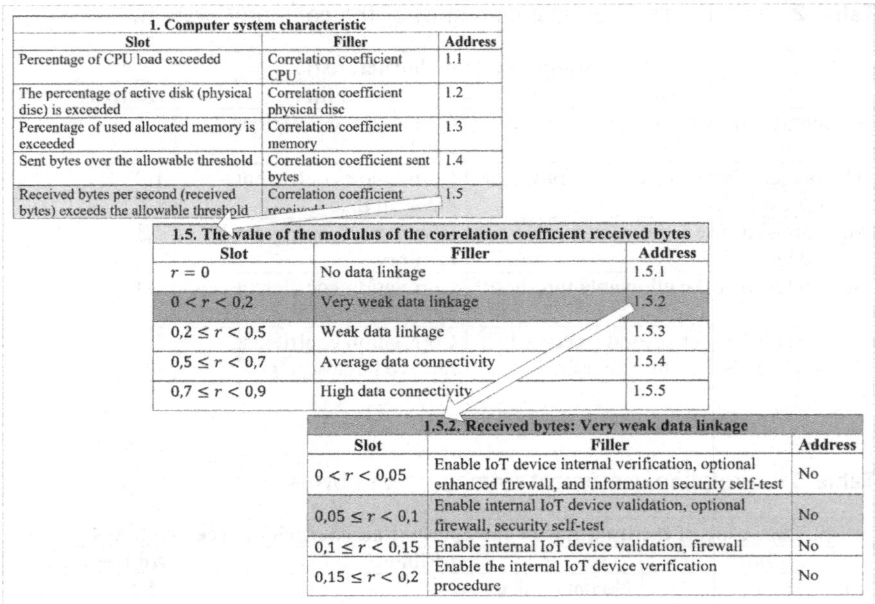

1. Computer system characteristic		
Slot	**Filler**	**Address**
Percentage of CPU load exceeded	Correlation coefficient CPU	1.1
The percentage of active disk (physical disc) is exceeded	Correlation coefficient physical disc	1.2
Percentage of used allocated memory is exceeded	Correlation coefficient memory	1.3
Sent bytes over the allowable threshold	Correlation coefficient sent bytes	1.4
Received bytes per second (received bytes) exceeds the allowable threshold	Correlation coefficient	1.5

1.5. The value of the modulus of the correlation coefficient received bytes		
Slot	**Filler**	**Address**
$r = 0$	No data linkage	1.5.1
$0 < r < 0{,}2$	Very weak data linkage	1.5.2
$0{,}2 \leq r < 0{,}5$	Weak data linkage	1.5.3
$0{,}5 \leq r < 0{,}7$	Average data connectivity	1.5.4
$0{,}7 \leq r < 0{,}9$	High data connectivity	1.5.5

1.5.2. Received bytes: Very weak data linkage		
Slot	**Filler**	**Address**
$0 < r < 0{,}05$	Enable IoT device internal verification, optional enhanced firewall, and information security self-test	No
$0{,}05 \leq r < 0{,}1$	Enable internal IoT device validation, optional firewall, security self-test	No
$0{,}1 \leq r < 0{,}15$	Enable internal IoT device validation, firewall	No
$0{,}15 \leq r < 0{,}2$	Enable the internal IoT device verification procedure	No

FIG. 5

Example of transition to behavioral models using frames.

Table 4 Frame of the correlation coefficient of a very weak relationship of data received bytes.

1.5.2. Received bytes: Very weak data linkage		
Slot	**Filler**	**Address**
$0 < r < 0{,}05$	Enable IoT device internal verification, optional enhanced firewall, and information security self-test	No
$0{,}05 \leq r < 0{,}1$	Enable internal IoT device validation, optional firewall, security self-test	No
$0{,}1 \leq r < 0{,}15$	Enable internal IoT device validation, firewall	No
$0{,}15 \leq r < 0{,}2$	Enable the internal IoT device verification procedure	No

time series analysis in the information security system is performed by a forecasting algorithm based on exponential smoothing with trend and multiplicative seasonality, also known as the Holt-Winters model. The main idea of this model is that an open system is subject to dialectical laws, develops in a spiral, and the stages are repeated, but with changing properties.

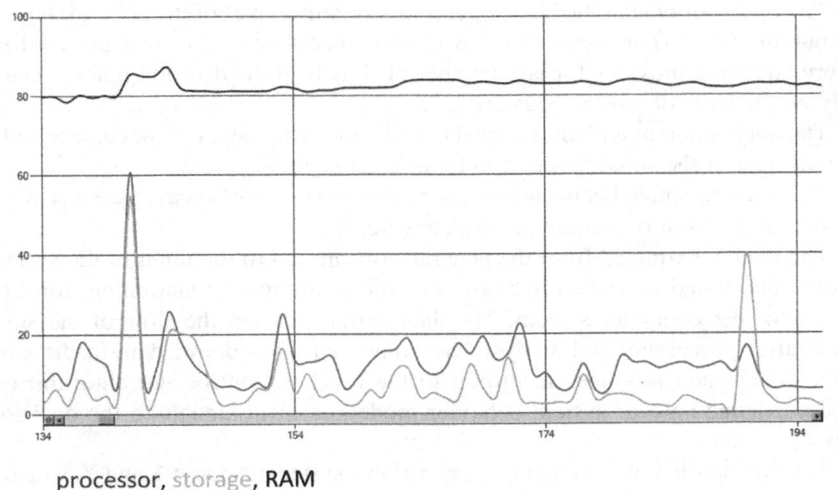

FIG. 6

Sampling data of three devices from the hourly load monitoring of the computer system.

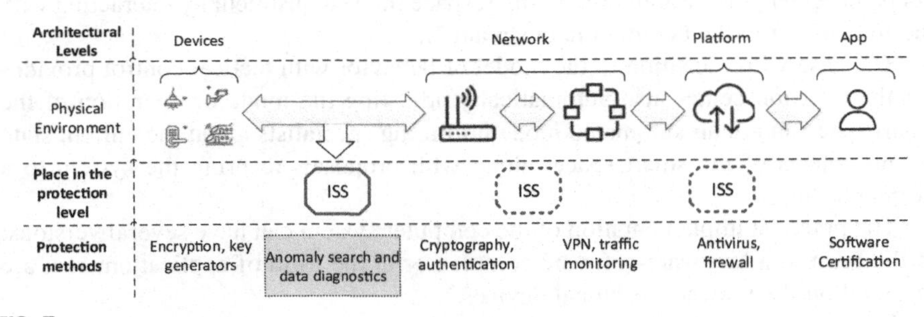

FIG. 7

A variant of smart space architecture with data diagnostics between device and network layers.

If there is seasonality in the time series and a sufficiently large amount of available data, it is likely to find two segments that are maximally similar to each other, which are expressed by the formula.

The models of behavior and algorithms of data processing considered here are included in the system of diagnostics of smart space data. The system approach of the developed information security system provides the possibility of its placement at almost any level of the smart space architecture (Figs. 7 and 2).

Fig. 7 shows a variant of placing the data diagnostics system between the device and network layers. This option provides processing of nonpersonalized information from IoT devices by the system and the possibility of communication of the system directly with them.

The option of placing the data diagnostics system as part of the network layer at the output of the IoT-device gateway performs processing of already depersonalized information and provides for issuing control signals to the decision-making center only on the basis of indirect signs (Fig. 2).

The application of a platform-level data diagnostics system allows assessing the overall state of the smart space, also by indirect signs.

The reference model of the information security system of smart space S is implemented in the form of a computer system (Fig. 3).

The data X extracted from the general flow are fed to the input of the system. These data, based on the indirect signs of the smart space functioning, form the state C of the computer system. The data extracted from the flow of the smart space after processing and analysis are converted into control data in the computer system and produce, according to the level of control and algorithms of the knowledge base of system behavior models, control signals to the decision-making center.

The decision-making center produces a global system response X and Y, which in the analysis of nonpersonalized data are transmitted in a common stream X control commands cyber-physical devices and control signals Y to the specialist.

If the system state defines a transition to a tactical control behavior model, the decision center performs automatic smart space mode adjustment by interacting with the IoT device in the common data stream X.

The case with transition to the model of behavior with tactical control provides for the decision center also automatically adjusting the mode of operation of the smart space and in an automated mode informing specialists about the current state of the objects of the smart space, along with proposals to bring the system to a normal state.

The practical implementation of the computer system can have several versions: in the form of a hardware-software complex or in the form of applications that are installed on the relevant technical devices.

Decisions can be made after analyzing the data in streams from each device on anomalies not only in terms of information security, detecting threats, but also in the case of abnormal situations in the operation of the smart space. Abnormal operation of a smart device will manifest itself in the violation of its seasonality in operation. Signs of seasonality will allow controlling the operation of IoT devices.

The proposed information security system for a smart space, in addition to solving the main task—information security—also solves the problem of diagnostics when any abnormal situation occurs during the IoT device function and sends a control signal to cyber-physical devices or a specialist. Searching not only for differences (anomalies), but also for similarities in the analyzed data provides a systematic view of the state of the smart space. Adjustment of the information security system is provided by adjustments of criterion indicators and adaptation of behavior patterns.

The proposed system can claim to be a reference model of a smart space information security system and can be embedded in its architecture.

6 Analysis

The data security system performs two tasks. First, it assesses the threat level to the smart space by implicit features extracted from the overall data stream and, second, it assesses the threat level by implicit features of the computer information security system's functional characteristics.

Functional characteristics of IoT-device data flow at the gateway output in the network layer of the smart space architecture are depersonalized. It is necessary to establish outlier detection capabilities for the depersonalized data stream. Dynamic changes in the data flow of IoT devices become most easily detected either at the input of the network layer or at the output of the gateway in the network layer of the smart space architecture, when the informative data flow is not yet mixed with the overall network space flow. A synergistic effect of detecting anomalies in the functioning of IoT devices in such a generalized flow is possible, at the expense of indirect signs. The first problem is the actual detection of anomalies in the time series of the data stream. Second are the indicators of dynamic characteristics of the flow itself, expressed through the first and second derivatives, carrying the physical meaning of the rate of change of the data flow and acceleration of the data flow.

Each indirect criterion of the first problem is considered from three points of view: in terms of the data flow, the rate of change of data in the flow, and the acceleration or deceleration of the data flow.

If the dataset at some i-point in time $t_i \in \{t|t \geq 0\}$ is defined as a function of x: $X(t_i) \subset \{t|t \geq 0\} \rightarrow \{t|t \geq 0\}$, then the derivative of the data stream will be the number k and the function in the vicinity of $X(t_i)$ will have the form:

$$x_{t_i+\Delta t} = x_{t_i-\Delta t} + k \cdot 2\Delta t,$$

if k exists.

Calculation of the derivative for a discrete set of data from a time series expressed as a central difference approximation is performed according to the formula:

$$\left.\frac{dx}{dt}\right|_i \approx \frac{x_{t_i+\Delta t} - x_{t_i-\Delta t}}{2\Delta t}.$$

The second derivative in the form of the central difference approximation has the form:

$$\left.\frac{d^2x}{dt^2}\right|_i \approx \frac{x_{t_i+\Delta t} - 2x_{t_i} + x_{t_i-\Delta t}}{(\Delta t)^2}.$$

The application of computed dynamic data flow metrics in the information security system processing algorithms enables detection and control of nonobvious behavior of IoT devices.

The threat level assessment of the second task is performed using indirect indicators of the characteristics of the functioning of the information security of the computer system. As noted earlier, such attributes include quantitative parameters of the processor, data storage medium, RAM, and network devices.

The analysis of the operation of the prototype smart space information security system is performed in an application that performs characteristic diagnostics while simulating the operation of the computer system. The actual data for monitoring the hourly load of the processor, storage, and RAM are shown in Fig. 6.

The graphical representation of the data parameters of the three analyzed devices in Fig. 6 allows us to see that, during the selected time interval, the periodicity of load indicators does not appear and from the results obtained we can judge only concerning the absolute values of the data.

The picture of the displayed data changes significantly when the studied period increases up to 32 hours (Fig. 8).

The seasonality of the indicators begins to be traced in Fig. 8. It becomes possible to visually separate the periods of operation and standby mode of the computer system. If we assume that these indicators refer to the periodic mode of operation of the IoT devices, it is obvious that the nature of the load will repeat with a given periodicity.

The operation of the computer system has its own patterns, which are manifested in the data analysis. Using the method of time series analysis by the prediction algorithm based on exponential smoothing with trend and multiplicative seasonality, the value of the confidence interval of each data indicator is calculated with a given accuracy. The emergence of anomalies in the behavior of the data, manifested in their departure from the boundaries of the confidence interval of the indicators, determines which value of each indicator is stable and which is not. The absence of anomalies in the data behavior is interpreted as a sign of stable operation of the IoT devices.

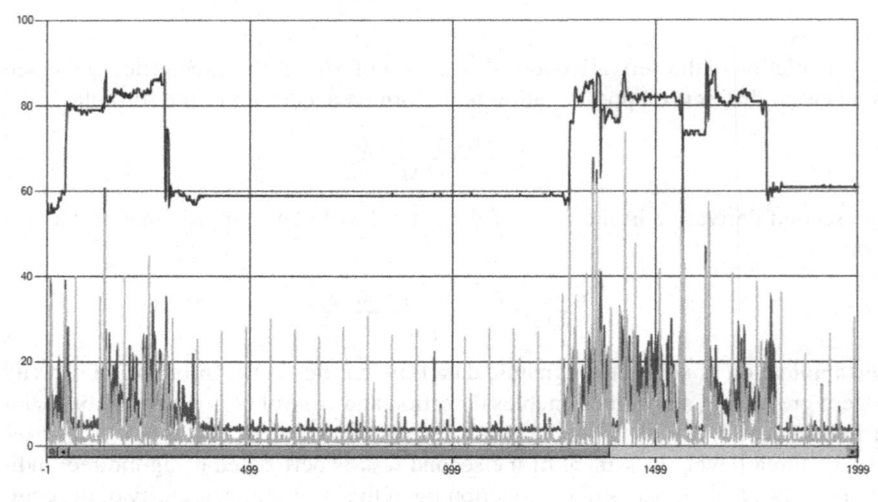

FIG. 8

Monitoring data of the continuous load period.

One possible scenario for the use of this method can be represented as follows: an attacker obtaining selective query data from the general flow of information will be meaningless and will not damage the operation of the smart devices. However, an attacker working with large amounts of data will significantly affect the load of the computer system, which is successfully detected and the right class of control mechanism is included in the behavior model.

Fig. 9 shows a series of load values on the computer system's processor. The model using maximum likelihood sampling assumes that if history repeats, then for every sample A preceding prediction B, there is a similar sample A1 contained in the actual values of the same time series.

The time series of load on the drive for 2 days is shown in Fig. 10.

Fragment A1 in the right part of the graph represents the defined period of the study preceding the prediction. Fragment C in the left part of the graph shows the period of the past monitoring extended to the left and right by the specified interval. The period is chosen based on a certain assumption: for example, if the fragment that will precede the forecast reflects the monitoring during the interval from 9:00 until 10:00 and the search should be performed with the borders expanded by 2 hours, then you should search for such a sample from 7:00 until 12:00 of the previous day. During the extended analysis period, the probability of finding a sample of indicators reflecting the same type of activity as at the current moment, becomes higher. Thus the prediction accuracy increases and the amount of data under study decreases, which provides an overall increase in data processing productivity.

Clustering of the analyzed period is carried out by the brute-force method. If you perform the search for such a sample among the values for the entire monitoring period, it will significantly affect the increase in the analysis time as a whole.

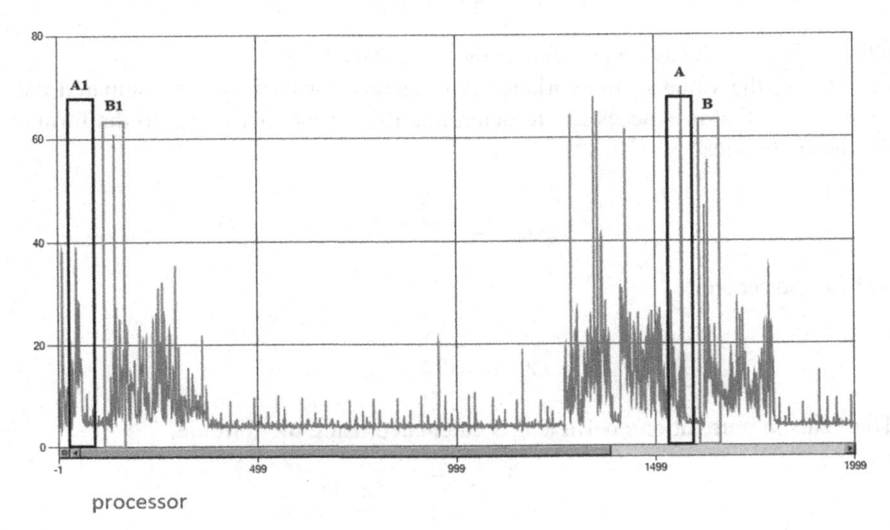

FIG. 9

Two-day monitoring of CPU load indicators.

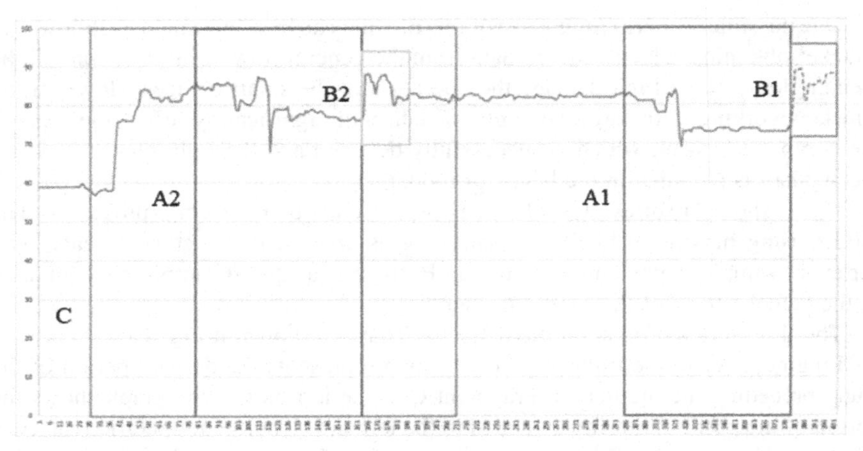

FIG. 10

Load monitoring graph for 2 days.

The brute-force method involves passing through a period equal to the monitoring interval on the day before the forecast, with expanded values of the same time of the previous day. The behavior of the data in the fragment A2 on the left side of Fig. 10 is almost identical to the data in the fragment A1 on the right side. When comparing the two samples, finding the linear correlation coefficient is used. For each data series it is necessary to find the dispersion coefficient by the formula:

$$S^2 = \frac{1}{n-1} * \left(\sum_1^n \left(\frac{1}{n} \sum_1^n x_j - x_j \right)^2 \right),$$

where $\frac{1}{n}\sum_1^n x_j$ is the average value of the indicators.

Finding the value of the variance is necessary for each of the compared data series. After that it is necessary to determine the Z-score according to the formula for the first series

$$Z_x = \frac{x_j - \frac{1}{n}\sum_1^n x_j}{S_x}$$

and for the second:

$$Z_y = \frac{y_j - \frac{1}{n}\sum_1^n y_j}{S_y}.$$

The Pearson correlation coefficient is calculated using the formula:

$$r = \frac{1}{n-1} * \sum_{i=1}^n \left(Z_{x_j} * Z_{y_j} \right).$$

This indicator assesses the similarity of the two data series, where the case of $-1 \leq r < 0$ is a negative relationship, and $0 \leq r \leq 1$ is a positive relationship.

Correlation does not imply a causal relationship between data series. The strength of the relationship between data series is independent of its direction and is determined by the absolute value of the correlation coefficient. If the correlation coefficient is 0, both data series are linearly independent of each other. Relying on the indicator of the absolute value of the linear correlation coefficient, the series are subjected to cluster distribution. The values of the module value of the correlation coefficient are indicators for the information security system. The case of absence of dependence of data series is the basis for formation of a signal to the decision-making center about the presence of anomaly.

Such an approach makes it possible to create an algorithm for detecting anomalies in data behavior and it becomes a constituent element of the smart space reference model.

7 Example of use

The basic functions of the information security reference model are embedded in a computer program that includes a number of computational procedures. Their use makes it possible to detect anomalies in the data, assess the level of threats, and issue appropriate signals for their elimination. As an example, let's consider a typical procedure for calculating prediction error. This procedure evaluates both threats and anomalies in data behavior. The behavior model, depending on the state of the system, selects the RMSE or the MAPE percentage average absolute error.

The calculation formulas for the Holt-Winters model predictions use coefficients that are chosen to minimize the error. To calculate the error, it is necessary to shift the studied samples by several values backward and compare the predictions with the actual indicators (Fig. 11).

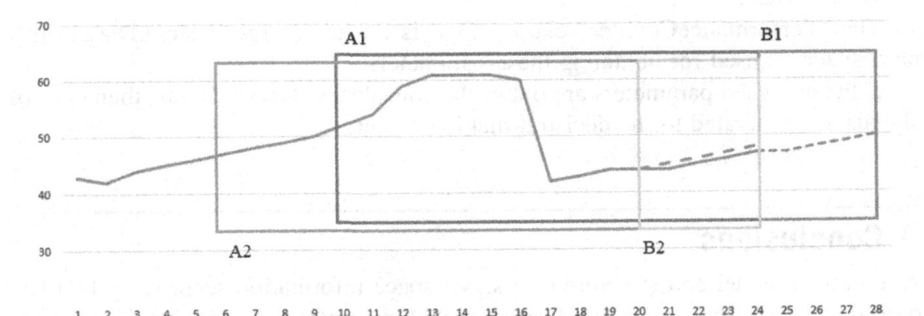

FIG. 11

Demonstration of coefficient selection using sample bias (A1—the sample preceding the prediction; A2—shifted interval, equal in duration to A1; B1—forecast period; B2—overlay of the predicted indicator *(dashed line)* on the current one).

The smallest difference in the indices is a sign that the coefficients are selected successfully. Shift of samples occurs on the value equal to the period of predicted values. This is due to the fact that the shift will affect the linear correlation coefficient between similar series. The minimal possible change of samples allows the small deviation of the coefficient to be neglected.

The coefficients are selected by minimizing the prediction error (RMSE or MAPE percentage average absolute error).

The processing algorithms for security threat assessment use the RMSE mean square error in the calculations, and the MAPE absolute error in the anomaly detection algorithms. The RMSE is the most sensitive to anomaly occurrence. The absolute MAPE percentage error has no dimensionality and is very easy to interpret. It can be expressed both in fractions and percentages, which is objectively used when detecting threats.

The example shows that universal algorithms embedded in the reference model or their combination can solve the corresponding class of problems in the field of information security of the smart space.

8 Application

The application, which performs the activation of the appropriate class of control mechanism when characteristic security threats are detected, is based on periodic monitoring of the hardware characteristics of the computer system. The program uses the Windows operating system performance counters (Fig. 12).

The data output on the graphs in the main window of the program is demonstrative in nature and is presented in order to visualize the analysis and prediction of values. The data is displayed periodically separately for the percentage of CPU load (CPU), percentage of disk activity (physical disk), percentage of allocated memory usage (memory), number of bytes sent (sent bytes), number of bytes received (received bytes).

The "PerformanceCounter" class, which is located in the System.Diagnostics namespace, is used for obtaining these parameters.

If the analyzed parameters are out of the confidence interval limits, then control signals are generated to the decision-making center.

9 Conclusions

A reference model and structure of a smart space information security system has been developed. Based on the reference model, a computer system was developed, which can be built into the architecture of a smart space depending on the needs.

The proposed information security model for smart spaces provides three levels of control. The current state of the smart space in the model is analyzed by indirect signs. These attributes are extracted from the information flows of the space and the

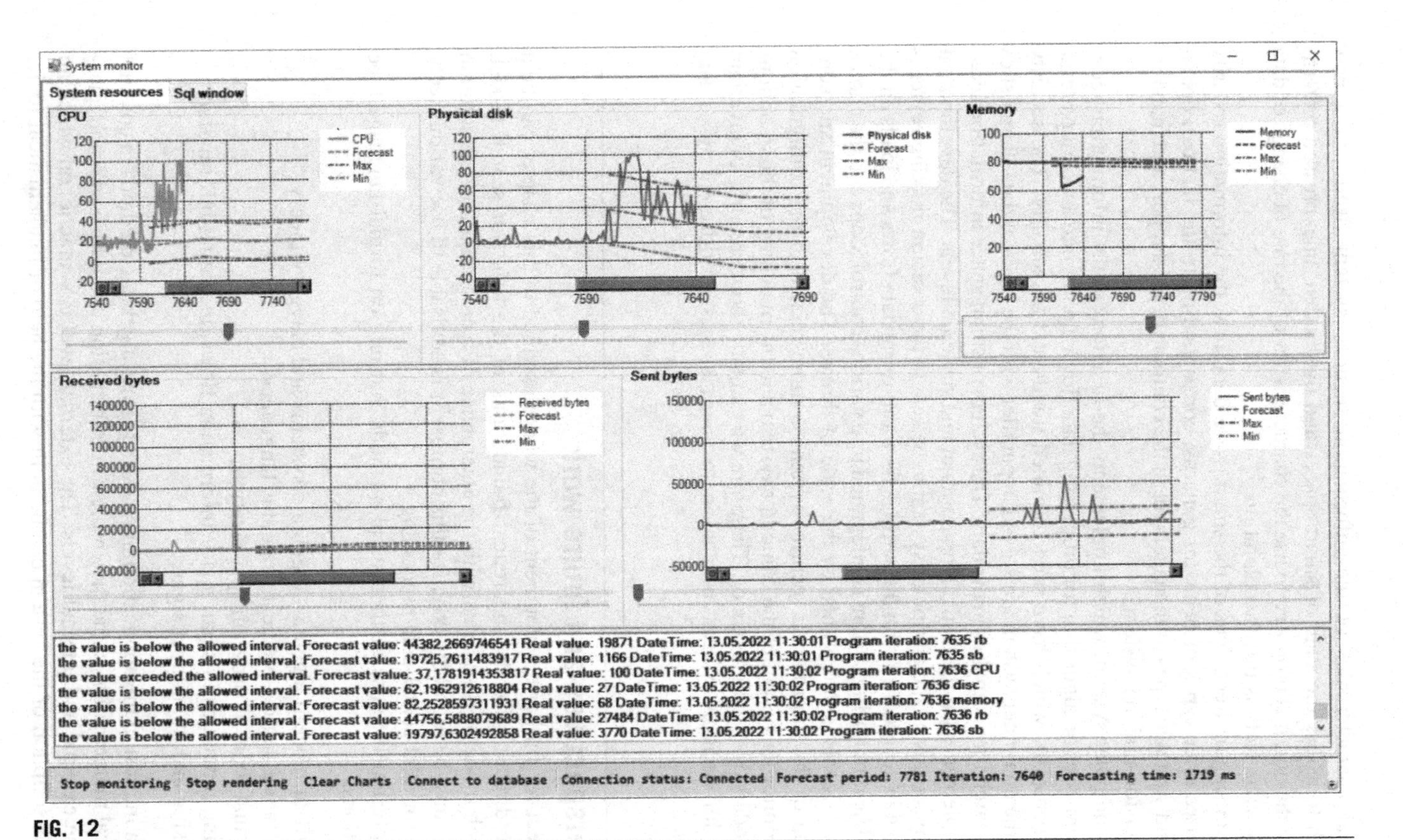

FIG. 12

Example of the application operation.

functional characteristics of the information security system itself. As a new approach to assess the state of smart spaces and the hidden, unexpressed patterns of their behavior, it is proposed to use the characteristics of the rate of change of data in the flow, as well as their acceleration.

When using indirect signs of the smart space data flow, the information security system uses three types of behavior patterns, corresponding to the tactical, operational, and strategic types of management in the contours of the support and management decision-making system.

The computer system, which implements the functions of the information security system, uses an algorithm for processing data flows, taking into account patterns in their behavior, which have predetermined trends and seasonality. The security tasks in the computer system are solved in parallel with the tasks of data diagnostics and the search for anomalies in the data. The system of smart space information security is implemented at the tactical, operational and strategic levels. The development of control actions by the computer system to be sent to the decision-making center is carried out in accordance with the settings for monitoring data flows and loads on the system. The monitoring system is integrated into various control loops and the choice of control signals is determined by the scale of threats. The development and configuration of the information security system is more related to the configuration of the monitoring system of the computer system and the matching of threat options to the models of system behavior. Therefore we can talk about creating an invariant security information system in relation to smart spaces of different types and purposes.

10 Perspectives and future work

A prospect for further development of the work described in this chapter can be a scientific direction for the theoretical foundations of the behavior search model. The search model is associated with the emergence of a new situation, the solution of which must be found quickly; for this purpose it is advisable to use various technologies for intellectual data processing.

It is presumed to use mathematical methods that rely on formal language-type constructions:

- methods of modern algebra—algebraic systems and category theory,
- mathematical logic—predicate calculus languages,
- conceptual tools of program specification,
- modeling semantics in smart space information flows, including information processing and its indirect signs.

The main interest in the application of modern logical-algebraic methods in contrast to classical mathematics is that they represent information about the system in question at the external and internal levels. The external level is syntactic, on which an abstract description of the system can be built with the help of mathematical logic.

The internal level is represented by an algebraic system. This allows definitions of some mapping between the algebraic system as its realization and the abstract syntactic theory of the system, which allows defining a semantic interpretation for the abstract theory of smart space.

One of the components of this model should be a system for ensuring its information security, which provides for the maximum exclusion of humans from the control loop. A future possibility is the transfer of automated functions, previously performed only by humans, to artificial intelligence technologies; these functions should be transferred to automatic mode. The transition to manual control mode should then become an exceptional, unique phenomenon.

Creation of a typical model of a smart space information security system intruder and requirements for information system response to intruder actions will provide significant help in detailed system design and contribute to the pursuit of absolute smart space security.

References

[1] Y.T. Negash, L.S.C. Sarmiento, Smart product-service systems in the healthcare industry: intelligent connected products and stakeholder communication drive digital health service adoption, Heliyon (2023), https://doi.org/10.1016/j.heliyon.2023.e13137.

[2] G. Gaobotse, E. Mbunge, J. Batani, B. Muchemwa, Non-invasive smart implants in healthcare: redefining healthcare services delivery through sensors and emerging digital health technologies, Sens. Int. (2022), https://doi.org/10.1016/j.sintl.2022.100156.

[3] I. Keshta, AI-driven IoT for smart health care: security and privacy issues, Inform. Med. Unlocked (2022), https://doi.org/10.1016/j.imu.2022.100903.

[4] H. Kumar, M.K. Singh, M.P. Gupta, A policy framework for city eligibility analysis: TISM and fuzzy MICMAC-weighted approach to select a city for smart city transformation in India, Land Use Policy (2019), https://doi.org/10.1016/j.landusepol.2018.12.025.

[5] D. Prasad, T. Alizadeh, R. Dowling, Multiscalar smart city governance in India, Geoforum (2021), https://doi.org/10.1016/j.geoforum.2021.03.001.

[6] G.H. Krishnan, L.S. Ganesh, Renewable energy for electricity use in India: evidence from India's smart cities mission, Renew. Energy Focus (2021), https://doi.org/10.1016/j.ref.2021.05.005.

[7] K.M. Shahanas, P.B. Sivakumar, Framework for a smart water management system in the context of smart city initiatives in India, Procedia Comput. Sci. 92 (2016) 142–147.

[8] M. Wanga, T. Zhou, Does smart city implementation improve the subjective quality of life? Evidence from China, Technol. Soc. (2023), https://doi.org/10.1016/j.techsoc.2022.102161.

[9] J. Liu, N. Chen, Z. Chen, L. Xu, W. Du, Y. Zhang, C. Wang, Towards sustainable smart cities: maturity assessment and development pattern recognition in China, J. Clean. Prod. (2022), https://doi.org/10.1016/j.jclepro.2022.133248.

[10] H. Jiang, S. Geertman, P. Witte, The contextualization of smart city technologies: an international comparison, J. Urban Manag. (2022), https://doi.org/10.1016/j.jum.2022.09.001.

[11] A. Abdulmalik, et al., A survey on security and privacy issues in edge-computing-assisted Internet of Things, IEEE Internet Things J. 8 (2020) 4004–4022.

[12] Smart City Network Architecture Guide, Alcatel-Lucent Enterprise, 2019. https://www.al-enterprise.com/-/media/assets/internet/documents/smart-city-network-architecture-guide-en.pdf.
[13] V. Diaconita, A.R. Bologa, R. Bologa, Hadoop oriented smart cities architecture, Sensors 18 (4) (2018) 1181, https://doi.org/10.3390/s18041181.
[14] S.W. Shah, et al., Comprehensive survey on smart cities architectures and protocols, EAI Endorsed Trans. Smart Cities 6 (18) (2022) e5, https://doi.org/10.4108/eetsc.v6i18.2065.
[15] E.M. Ouafiq, et al., Data architecture and big data analytics in smart cities, Procedia Comput. Sci. 207 (2022) 4123–4131, https://doi.org/10.1016/j.procs.2022.09.475.
[16] S. Siddiqui, et al., Smart contract-based security architecture for collaborative services in municipal smart cities, J. Syst. Archit. (2022), https://doi.org/10.1016/j.sysarc.2022.102802.
[17] K. Kalkan, S. Zeadally, Securing Internet of Things with software defined networking, IEEE Commun. Mag. 56 (9) (Sept. 2018) 186–192, https://doi.org/10.1109/MCOM.2017.1700714.
[18] S. Yamamoto, Analysis of smart city reference architecture by ArchiMate, Procedia Comput. Sci. 207 (2022) 514–521, https://doi.org/10.1016/j.procs.2022.09.106.
[19] A. Rahman, M.J. Islam, S.S. Band, G. Muhammad, K. Hasan, P. Tiwari, Towards a blockchain-SDN-based secure architecture for cloud computing in smart industrial IoT, Digit. Commun. Netw. (2022), https://doi.org/10.1016/j.dcan.2022.11.003.
[20] S.N.G. Aryavalli, H. Kumar, Top 12 layer-wise security challenges and a secure architectural solution for Internet of Things, Comput. Electr. Eng. (2022), https://doi.org/10.1016/j.compeleceng.2022.108487.
[21] I. Sadek, J. Codjo, S.U. Rehman, B. Abdulrazak, Security and privacy in the internet of things healthcare systems: toward a robust solution in real-life deployment, Comput. Methods Programs Biomed. Update (2022), https://doi.org/10.1016/j.cmpbup.2022.100071.

Index

Note: Page numbers followed by *f* indicate figures, *t* indicate tables, and *b* indicate boxes.

E

Q

R

S

T

U

Printed in the United States
by Baker & Taylor Publisher Services